The Chemistry of Hyperpolarized Magnetic Resonance Probes

Advances in Magnetic Resonance Technology and Applications Series

Series Editors

In-Young Choi, PhD
Department of Neurology, Department of Radiology, Department of Molecular and Integrative Physiology, Hoglund Biomedical Imaging Center, University of Kansas Medical Center, Kansas City, KS, United States

Peter Jezzard, PhD
Wellcome Centre for Integrative Neuroimaging, Nuffield Department of Clinical Neurosciences, University of Oxford, Oxford, United Kingdom

Brian Hargreaves, PhD
Department of Radiology, Department of Electrical Engineering, Department of Bioengineering, Stanford University, Stanford, CA, United States

Greg Zaharchuk, MD, PhD
Department of Radiology, Stanford University, Stanford, CA, United States

Titles published:

Volume 1 – Quantitative Magnetic Resonance Imaging – Edited by Nicole Seiberlich and Vikas Gulani

Volume 2 – Handbook of Pediatric Brain Imaging – Edited by Hao Huang and Timothy Roberts

Volume 3 – Hyperpolarized Carbon-13 Magnetic Resonance Imaging and Spectroscopy – Edited by Peder Larson

Volume 4 – Advanced Neuro MR Techniques and Applications – Edited by In-Young Choi and Peter Jezzard

Volume 5 – Breast MRI – Edited by Katja Pinker, Ritse Mann, and Savannah Partridge

Volume 6 – Motion Correction in MR – Edited by André Van Der Kouwe and Jalal B. Andre

Volume 7 – Magnetic Resonance Image Reconstruction – Edited by Mehmet Akçakaya, Mariya Doneva, and Claudia Prieto

Volume 8 – MR Linac Radiotherapy – Edited by Enis Ozyar, Cem Onal, and Sara L. Hackett

Volume 9 – Imaging Neuroinflammation – Edited by Cornelia Laule and John D. Port

Volume 10 – Ultra-High Field Neuro MRI – Edited by Karin Markenroth Bloch, Maxime Guye, and Benedikt A. Poser

Volume 11 – Cheng and Strijkers: *Quantitative Perfusion MRI*, ISBN: 9780323952095 August 2023

Visit the Series webpage at https://www.elsevier.com/books/book-series/advances-in-magneticresonance-technology-and-applications

Advances in Magnetic Resonance Technology and Applications

The Chemistry of Hyperpolarized Magnetic Resonance Probes

Volume 12

Edited by

Eul Hyun Suh
Department of Pharmaceutical Sciences, North Texas Eye Research Institute, University of North Texas Health Science Center, Fort Worth, TX, United States
Advanced Imaging Research Center, The University of Texas Southwestern Medical Center, Dallas, TX, United States

Zoltan Kovacs
Advanced Imaging Research Center, The University of Texas Southwestern Medical Center, Dallas, TX, United States

ELSEVIER

ACADEMIC PRESS
An imprint of Elsevier

Academic Press is an imprint of Elsevier
125 London Wall, London EC2Y 5AS, United Kingdom
525 B Street, Suite 1650, San Diego, CA 92101, United States
50 Hampshire Street, 5th Floor, Cambridge, MA 02139, United States

Notices

Knowledge and best practice in this field are constantly changing. As new research and experience broaden our understanding, changes in research methods, professional practices, or medical treatment may become necessary.

Practitioners and researchers must always rely on their own experience and knowledge in evaluating and using any information, methods, compounds, or experiments described herein. In using such information or methods they should be mindful of their own safety and the safety of others, including parties for whom they have a professional responsibility.

To the fullest extent of the law, neither the Publisher nor the authors, contributors, or editors, assume any liability for any injury and/or damage to persons or property as a matter of products liability, negligence or otherwise, or from any use or operation of any methods, products, instructions, or ideas contained in the material herein.

ISBN: 978-0-323-91842-8
ISSN: 2666-9099

For information on all Academic Press publications visit our website at https://www.elsevier.com/books-and-journals

Publisher: Mara Conner
Acquisitions Editor: Tim Pitts
Editorial Project Manager: Emily Thomson
Production Project Manager: Erragounta Saibabu Rao
Cover Designer: Vicky Pearson Esser

Typeset by TNQ Technologies

Contents

5. Technical considerations of MRI methods for validating DNP probes in small animals

Crystal E. Harrison, Jun Chen, Sung-Han Lin, Jae Mo Park

Contributors

Daniel Anable, Department of Physics, University of Texas at Dallas, Richardson, TX, United States

Asiye Asaadzade, Department of Physics, University of Texas at Dallas, Richardson, TX, United States

Erika Cerutti, Molecular Biotechnology and Health Sciences, University of Turin, Turin, Italy

Jun Chen, Advanced Imaging Research Center, The University of Texas Southwestern Medical Center, Dallas, TX, United States

Roozbeh Eskandari, Department of Biochemistry, Albert Einstein College of Medicine, Bronx, NY, United States

Wirya Feizi, Department of Physics, University of Texas at Dallas, Richardson, TX, United States; McLean Imaging Center, McLean Hospital, Harvard Medical School, Belmont, MA, United States

Crystal E. Harrison, Advanced Imaging Research Center, The University of Texas Southwestern Medical Center, Dallas, TX, United States

Mai T. Huynh, Advanced Imaging Research Center, The University of Texas Southwestern Medical Center, Dallas, TX, United States

Fatemeh Khashami, Department of Physics, University of Texas at Dallas, Richardson, TX, United States; Advanced Imaging Research Center, University of Texas Southwestern Medical Center, Dallas, TX, United States

Chalermchai Khemtong, Department of Medicine, Division of Endocrinology, Diabetes and Metabolism, University of Florida, Gainesville, FL, United States; Department of Biochemistry and Molecular Biology, University of Florida, Gainesville, FL, United States

Zoltan Kovacs, Advanced Imaging Research Center, The University of Texas Southwestern Medical Center, Dallas, TX, United States

Cody Larsen, Department of Physics, University of Texas at Dallas, Richardson, TX, United States

Sung-Han Lin, Advanced Imaging Research Center, The University of Texas Southwestern Medical Center, Dallas, TX, United States

Lloyd Lumata, Department of Physics, University of Texas at Dallas, Richardson, TX, United States; Department of Neuroscience, University of Texas at Dallas, Richardson, TX, United States; Department of Bioengineering, University of Texas at Dallas, Richardson, TX, United States

Mor Mishkovsky, Laboratory of Functional and Metabolic Imaging, Ecole Polytechnique Fédérale de Lausanne (EPFL), Lausanne, Switzerland

Olivier Ouari, Institute of Radical Chemistry, Aix-Marseille University, CNRS, ICR, Marseille, France

Jae Mo Park, Advanced Imaging Research Center, The University of Texas Southwestern Medical Center, Dallas, TX, United States; Department of Biomedical Engineering, The University of Texas Southwestern Medical Center, Dallas, TX, United States; Department of Radiology, The University of Texas Southwestern Medical Center, Dallas, TX, United States

Francesca Reineri, Molecular Biotechnology and Health Sciences, University of Turin, Turin, Italy

Gaurav Sharma, Advanced Imaging Research Center, The University of Texas Southwestern Medical Center, Dallas, TX, United States; Department of Cardiovascular and Thoracic Surgery, UT Southwestern Medical Center, Dallas, TX, United States

Eul Hyun Suh, Department of Pharmaceutical Sciences, UNT System College of Pharmacy, University of North Texas Health Science Center, Fort Worth, TX, United States; Advanced Imaging Research Center, The University of Texas Southwestern Medical Center, Dallas, TX, United States

Qing Wang, Department of Physics, University of Texas at Dallas, Richardson, TX, United States; Department of Imaging Physics, University of Texas MD Anderson Cancer Center, Houston, TX, United States

Hikari A.I. Yoshihara, Laboratory of Functional and Metabolic Imaging, Ecole Polytechnique Fédérale de Lausanne (EPFL), Lausanne, Switzerland

Preface

Nuclear magnetic resonance (NMR) is a fascinating and immensely useful spectroscopic method with diverse applications in chemistry, biochemistry, biology, biophysics, and various other fields. Additionally, it is the underlying principle of magnetic resonance imaging (MRI). Despite notable advancements in both hardware and software, NMR has inherently low sensitivity posing limitations on its application to studying biological systems. Hyperpolarization encompasses technologies that enhance the sensitivity of conventional NMR experiments by creating nonequilibrium nuclear spin polarizations.

Admittedly, the adoption of hyperpolarization demands extensive and often costly instrumentation. Undoubtedly, there remains ample room for improvement. Nevertheless, the potential rewards are substantial. The ability to detect and image nuclei beyond ^{1}H opens avenues for real-time visualization and monitoring enzyme-catalyzed biochemical reactions in healthy and diseased tissues offering promising biomedical and clinical applications. There are several excellent recent reviews and books on hyperpolarized NMR and MRI predominantly approaching the subject from physical and/or biological perspectives. However, the pivotal role of chemistry is often overlooked.

Our motivation for writing this book was to give a comprehensive overview on the chemistry aspects of hyperpolarization and to offer insights into how chemistry can be used to achieve high polarization, extend the lifetime of hyperpolarized signal, and/or design more efficient probes. Given the special significance of carbon in both chemistry and life sciences, our primary focus in this book is on the synthesis, hyperpolarization, and in vivo applications of ^{13}C-labeled compounds.

The first chapter introduces conventional and hyperpolarized NMR, with a particular emphasis on the most widely used hyperpolarization technique, dynamic nuclear polarization (DNP). It provides an overview of the diverse nuclei that have been subjected to hyperpolarization. Additionally, the chapter summarizes the synthesis of ^{13}C-labeled compounds employed in ^{13}C DNP experiments. Chapter 2 delves into factors influencing solid-state polarization, while Chapter 3 explores the applications of hyperpolarized probes from a biological perspective. Finally, Chapter 4 outlines the design and execution of preclinical hyperpolarized magnetic resonance spectroscopy/imaging experiments.

This book has been three years in the making largely due to the challenges of compiling and reviewing the wide array of papers associated with this subject. The editors express their sincere gratitude to the contributing authors for dedicating their time and effort, thereby making this book possible.

Chapter 1

Hyperpolarized tracer design, synthesis, and characterization

Eul Hyun Suh[1,3], Roozbeh Eskandari[2], Mai T. Huynh[3], Gaurav Sharma[3,4], Chalermchai Khemtong[5,6], Olivier Ouari[7] and Zoltan Kovacs[3,*]

[1]*Department of Pharmaceutical Sciences, UNT System College of Pharmacy, University of North Texas Health Science Center, Fort Worth, TX, United States;* [2]*Department of Biochemistry, Albert Einstein College of Medicine, Bronx, NY, United States;* [3]*Advanced Imaging Research Center, The University of Texas Southwestern Medical Center, Dallas, TX, United States;* [4]*Department of Cardiovascular and Thoracic Surgery, UT Southwestern Medical Center, Dallas, TX, United States;* [5]*Department of Medicine, Division of Endocrinology, Diabetes and Metabolism, University of Florida, Gainesville, FL, United States;* [6]*Department of Biochemistry and Molecular Biology, University of Florida, Gainesville, FL, United States;* [7]*Institute of Radical Chemistry, Aix-Marseille University, CNRS, ICR, Marseille, France*

**Corresponding author: E-mail: Zoltan.Kovacs@UTSouthwestern.edu*

1.1 Background

Spin is a fundamental property of particles, including the electron and atomic nuclei [1]. Particles with nonzero spin are endowed with intrinsic angular momentum. The angular momentum imparts a magnetic moment (μ) to the particle, which is directly proportional to the angular momentum L. Quantum mechanics tells us that L is quantized (we are only concerned about components in the direction of an applied field, z) (Eq. 1.1):

$$\mu = \gamma L = \gamma \hbar I \tag{1.1}$$

Where γ proportionality constant is the gyromagnetic ratio of the particle, I is the spin quantum number, and $\hbar$ is the reduced Planck's constant ($h/2\pi$). The gyromagnetic ratio is specific for the particle (Table 1.1) and essentially expresses how strongly the particle interacts with an external magnetic field (B_0).

The electron, the neutron, and the proton have a spin quantum number (I) of ½. Atomic nuclei have zero I if they are composed of an even number of neutrons and protons. I is a positive integer (1, 5, 8) for nuclei with an odd number of protons and odd neutrons, while I is a half-integer (1/2, 3/2, 5/2, 7/2, 9/2) for nuclei that have odd/even and even/odd combinations of protons and neutrons. For particles with nonzero I, the angular momentum generates a magnetic moment. In a zero external magnetic field, the spin state (2I + 1)-fold degenerates. However, when nonzero spins are placed in a B_0 field, the degeneracy of the spin states is broken (Zeeman effect), and 2I + 1 spin states are formed. Thus, for a spin ½ nucleus with a positive gyromagnetic ratio, the Zeeman splitting gives rise to two energy levels. Note that the magnetic energy is proportional to the B_0 external magnetic field and given as $E = \mu B_0 = -\gamma\hbar\, IB_0$. Thus, the energy difference between the two states can be expressed as:

$$\Delta E = E_\beta - E_\alpha = \gamma \hbar B_0 \tag{1.2}$$

The two energy levels are traditionally denoted as the alpha state (low energy, spin up, aligned with B_0) and the beta state (high energy, spin down, opposed to B_0) (Fig. 1.1).

Magnetic resonance occurs when the spin system is perturbed by a radiofrequency (rf) field whose frequency matches this energy difference. Under resonance conditions, the spin system absorbs energy ($h\nu$) from the rf field:

$$\Delta E = h\nu = \hbar\omega = \gamma \hbar B_0 \tag{1.3}$$

The Chemistry of Hyperpolarized Magnetic Resonance Probes. https://doi.org/10.1016/B978-0-323-91842-8.00006-9

TABLE 1.1 The gyromagnetic ratio of the free electron and selected nuclei [2,3].

Particle	γ (MHz/T)[a]
Free electron	−28,024.95
^{1}H	42.58
^{19}F	40.08
^{31}P	17.24
^{129}Xe	−11.78
^{13}C	10.71
^{29}Si	8.47
^{2}H	6.54
^{6}Li	6.27
^{133}Cs	5.62
^{15}N	−4.32
^{89}Y	−2.09
^{109}Ag	−1.99
^{107}Ag	−1.73

[a] *Note that the SI unit of the gyromagnetic ratio is rad/sT, however, it is usually given in MHz/T. These values can be interconverted by a factor of 2π: γ (10^7 rad/sT)/$2\pi = \gamma$ (MHz/T).*

Where ν and ω are the Larmor frequencies in Hz and rad/s, respectively. Thus, the basic equation of NMR can be written as:

$$\omega = \gamma B_0 \tag{1.4}$$

As seen from Eq. (1.3), the resonance frequency is linearly proportional to the magnetic field. Note that the Larmor frequency is usually given in MHz, even when denoted as ω. The resonance frequency in MHz at a given B_0 can be calculated with the γ values in Table 1.1. A peculiar property of the resonance frequency is that it is influenced in a predictable way by the molecular environment. The existence of chemical shift has made nuclear magnetic resonance (NMR) an immensely useful tool for studying the chemical structure and molecular dynamics of small inorganic, organometallic, and organic compounds, as well as proteins, polymers, and supramolecular structures. In fact, without the chemical shift, NMR would have remained an obscure nuclear physics phenomenon. NMR is also the underlying principle of magnetic resonance imaging (MRI), which has become one of the most important clinical imaging modalities. However, NMR has a significant disadvantage compared to other analytical techniques: inherently low sensitivity. The low intensity of the NMR signal is the consequence of the low nuclear spin polarization (P) at ambient temperature. P is defined as the

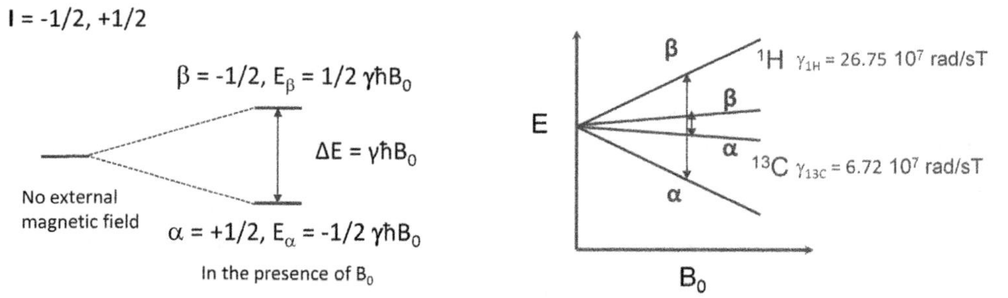

FIGURE 1.1 The magnitude of Zeeman splitting depends on the B_0 external magnetic field and the gyromagnetic ratio.

normalized difference in the nuclear spin populations on the Zeeman levels (Eq. 1.5) and is usually given as a percentage, or ppm. Thermal equilibrium P is easily calculated using classical thermodynamics since the nuclear spins populate the Zeeman levels (alpha and beta states) according to the Boltzmann distribution. Thus, the population ratio of the alpha and beta states can be expressed as (Eq. 1.5), and then the polarization is given as (Eq. 1.6):

$$\frac{N^{\alpha}}{N^{\beta}} = e^{\frac{\left(E^{\beta}-E^{\alpha}\right)}{kT}} \tag{1.5}$$

$$P_1 = \frac{N_{\alpha} - N_{\beta}}{N_{\alpha} - N_{\beta}} = tanh\left(\frac{\gamma_1 \hbar B_0}{2K_B T}\right) \tag{1.6}$$

Since the thermal energy of the spins at ambient temperature (kT) is much higher than the energy of the Zeeman splitting ($\gamma\hbar B_0$) at common magnetic field strengths, the Zeeman levels are almost equally populated at thermal equilibrium, and consequently, the detectable NMR signal at thermal equilibrium is weak.

The above overview of NMR is rather simplified. It was not meant to be exhaustive; its purpose is to aid the subsequent discussions in this chapter. More detailed discussions of magnetic resonance are widely available in classic textbooks [4–10].

1.2 Sensitivity of NMR experiments

In NMR spectroscopy, the sensitivity of an NMR experiment is characterized by the signal-to-noise ratio (SNR), which is defined as the peak height S (not the peak area) divided by the standard deviation (root-mean-square value, σ_N) of the noise ($2\sigma_N$ has also been used) [6,11]. In MRI, the SNR and the contrast-to-noise ratio are defined using signal intensities [12]. Here, we will use the classic definition used in spectroscopy (Eq. 1.7) without factor 2 [6], and we will consider a single-scan experiment.

$$\text{SNR} = \frac{S}{\sigma_N} \tag{1.7}$$

The SNR can be expressed by a rather complex formula involving several parameters such as the field, the gyromagnetic ratio, temperature, the effective transverse relaxation time (T_2^*), the number of spins in the coil volume, the number of scans, the flip angle, and several parameters related to the coil (quality factor, filling factor, etc.). Detailed discussions on how the SNR is dependent on these parameters in conventional NMR experiments have been published [13–15]. As seen from Eq. (1.7), the SNR of conventional NMR experiments can be improved by increasing S and/or reducing the noise [11,16]. A Stronger signal can be obtained by increasing the field and optimizing the receiver coil, in particular, by increasing the filling factor and decreasing the diameter of the coil [17,18]. The thermal (Johnson–Nyquist) noise can be reduced by using cryoprobes, in which the rf and preamplifier electronics are cooled by cold helium gas [19]. From an HP-NMR perspective, however, it will be more relevant to consider how the SNR depends on nuclear spin polarization. It will be assumed that the excited and detected spins are the same. In thermally polarized samples, the signal amplitude is proportional to the Larmor frequency and the net macroscopic magnetization (M_0), which can be expressed as the product of the magnetic moment and the polarization (P), for which both the classical and the quantum mechanical treatment give the same approximation [11,20]. The effect of T_2^* will be ignored. The net z-magnetization parallel to B_0 is given by Eq. (1.8):

$$M_0 = N\mu P \tag{1.8}$$

Here, N is the number of spins per unit volume, γ is the gyromagnetic ratio, ħ is the reduced Planck's constant ($h/2\pi$), T is the absolute temperature, and k is the Boltzmann's constant:

Thus, for an I = ½ nucleus:

$$M_0 = N\left(\frac{\gamma\hbar}{2}\right)P \tag{1.9}$$

The hyperbolic tangent expression of P (Eq. 1.6) can be approximated by $\gamma\hbar B_0/2\,kT$:

$$M_0 = N\left(\frac{\gamma\hbar}{2}\right)\left(\frac{\gamma\hbar B_0}{2kT}\right) = \frac{N\gamma^2\hbar^2 B_0}{4kT} \tag{1.10}$$

Strictly speaking, this is valid only when the polarization is relatively small. The classical approach to the calculation of M_0 based on the summation of the individual magnetic moments on the Zeeman levels according to the Boltzmann distribution gives the following general expression for M_0 [4].

$$M_0 = \frac{N\gamma^2\hbar^2 I(I+1)B_0}{3kT} \tag{1.11}$$

Which affords the same formula for M_0 as Eq. (1.10) when I = ½. From these equations, the proportionality of S to the Larmor frequency (γB_0) and M_0 can be expressed as:

$$S \propto (\gamma B_0)\left(\frac{\gamma^2 h^2 B_0}{4kT}\right) = \frac{\hbar^2\gamma^3 B_0^2}{4kT} \tag{1.12}$$

Thus, at a given temperature, the signal amplitude is proportional to $\gamma^3 B_0^2$ in thermally polarized samples. Since the noise-induced voltage is proportional to the square root of γB_0, the SNR in a conventional single-scan experiment will be proportional to $\gamma^{5/2}B_0^{3/2}$. [4,11,21] An HP-NMR experiment differs from the conventional one in that the P polarization is not at thermal equilibrium at the B_0 field of the spectrometer but is artificially created outside the spectrometer magnet. Thus, its value will be the same in any field, i.e., in any spectrometer magnet. Therefore, P is not dependent on γB_0, and the SNR of HP-NMR will be proportional to ($\gamma^{3/2}B_0^{1/2}$)P. It is also evident that the SNR of an HP experiment will be largely determined by the value of P, which can be several orders of magnitude greater than the thermal polarization.

It must be emphasized that the SNR alone cannot be used to compare the sensitivities of different instruments due to its dependence on various coil parameters. The limit of detection (LOD) is more suitable for this purpose. The LOD is defined, rather arbitrarily, as the amount of sample (molar amount or concentration) that is needed to achieve an SNR of 3 [22,23]. Although systematic studies to determine the detection limits of HP-NMR have not been performed yet, it was estimated that for HP-^{13}C, it is around 1 μM based on the SNR of a 2 mM HP[1-^{13}C]acetate solution at 9.4 T (SNR = 840), which had a ^{13}C-polarization of 17%. This allowed the observation and kinetic analysis of transient intermlediates present at low steady-state concentrations in consecutive enzymatic reactions [24]. One can assume that the detection limit could be further improved by at least one or two orders of magnitude by increasing the polarization (P) and reducing the noise by using cryogenically cooled probes. In HP-NMR, usually low flip angle (θ) pulses are used to acquire an array of HP spectra, which reduces the HP magnetization by $cos^n\theta$ after the acquisition of n spectra. It was shown that the sensitivity of detecting metabolic intermediates by HP-^{13}C NMR could be significantly enhanced by applying carefully timed 90-degree pulses rather than a series o f low-flip-angle excitations [25]. HP-^{15}N NMR has a similar, somewhat lower sensitivity; a detection threshold of 5 to 1 μM was demonstrated with about 6% ^{15}N polarization at 9.4 T [26].

1.3 Conventional NMR technologies to improve the signal intensity

The low sensitivity has motivated much of the development of NMR technology development over the past 60 years. As seen from Eq. (1.6), stronger B_0 improves the polarization, and this motivated the design and manufacture of stronger and stronger magnets, eventually culminating in the development of superconducting NMR magnets that are widely used in modern NMR spectrometers as well as in preclinical and clinical MR imaging systems. Recently, spectroscopy magnets operating above 1 GHz ^{1}H resonance frequency (24 T) have become available [27,28]. As we will discuss later, for a given nucleus, the sensitivity is roughly proportional to the 3/2 power of B_0, and in addition, a stronger magnetic field also offers higher spectral resolution [11,13]. The advent of Fourier transform (FT)/pulsed NMR spectrometers in 1966 offered a significant advantage over continuous wave instruments by greatly facilitating signal averaging [29–31]. Cryoprobes, in which the electronics are cooled to 20 K to reduce thermal noise, provide about a 4-fold sensitivity improvement over conventional probes [19,32]. Although these technological inventions combined with optimized NMR pulse sequences provided about a four-order-of-magnitude sensitivity improvement over the early iron magnet-based instruments, the sensitivity of NMR falls short compared to other spectroscopy techniques.

1.4 Hyperpolarized NMR

Hyperpolarized NMR technologies take an entirely different approach to improve sensitivity. Rather than enhancing thermal equilibrium polarization or reducing noise, HP-NMR aims at increasing the polarization by generating nonthermal equilibrium spin distributions on the Zeeman levels [33]. In hyperpolarized NMR, the nuclear polarization is field-independent, and consequently, a given nuclear polarization will afford higher signal enhancement (defined as the ratio

of the hyperpolarized signal intensity to the thermally polarized signal intensity) at lower fields. Nonequilibrium spin distributions can be generated by several experimental methods, including the "brute force" approach, dynamic nuclear polarization (DNP), spin-exchange optical pumping (SEOP), and parahydrogen-induced polarization (PHIP) [11,34–42]. Among these, the DNP is the most versatile because it can be applied to a wide range of nuclei. DNP is a phenomenon in which the spin polarization of electrons in the presence of a magnetic field is transferred to coupled nuclei by microwave irradiation at or near the electron spin resonance (ESR) frequency [43]. DNP can be performed at low (1K, 100K) or room temperature in low or high magnetic fields. Depending on the conditions and polarizing agents used, the polarization transfer occurs by different mechanisms (thermal mixing (TM), solid effect (SE), cross effect (CE), Overhauser effect (OE)) [44–60] and the experimental setup as well as the achievable NMR signal enhancements vary widely. The phenomenon was first predicted theoretically by Overhauser in 1953 and soon thereafter experimentally demonstrated by Carver and Slichter in solid-state [61–63].

1.4.1 MAS-DNP NMR

For some time, DNP was only used to polarize nuclear spins in the solid targets for nuclear physics experiments [64–66]. In the early nineties, DNP was made compatible with modern high resolution magic angle spinning (MAS) solid-state NMR spectrometers [67–69]. MAS-DNP NMR instruments integrate the conventional MAS setup with hardware designed for spinning the sample at low temperature and a high-power microwave source (most commonly a gyrotron or extended interaction klystron) tuned to the EPR frequency of the polarizing agent at the magnetic field of the spectrometer [70]. MAS-DNP is usually performed at 100 K with liquid nitrogen as cryogen. A commercial MAS-DNP system (Bruker Biospin) offers up to 400-fold NMR signal enhancements in the solid state [71]. However, sample temperatures lower than 100 K further increase the electron spin polarization, decrease nuclear spin relaxation rates, and therefore significantly improve the MAS-DNP enhancement [72]. An innovative low-temperature MAS-DNP probe (commercially available from Revolution NMR LLC) uses nitrogen to spin and helium to cool the sample to 25 K [73]. MAS-DNP systems that use helium for both spinning and cooling the sample (60 and 36 K) have also been developed [74,75]. Despite its shortcomings, in particular strong line-broadening effects due to cryogenic temperatures and paramagnetic polarizing agents, MAS-DNP has become a very useful tool for the NMR studies of various biomolecules and without a doubt will continue to have a big impact on solid-state biomolecular NMR [70,76–81]. An interesting extension of MAS-DNP, termed in situ temperature jump DNP, was reported in 2006 [82]. DNP samples of ^{13}C-urea, Na[1,2-$^{13}C_2$,$^{2}H_3$]acetate, [U-^{13}C,$^{2}H_7$] glucose, and L-[U-$^{13}C_5$]-proline were prepared and polarized as in MAS-DNP experiments, in this particular case, with a 140 GHz gyrotron at 90 K and 5 T in the presence of TOTAPOL in a partially deuterated DMSO-water or glycerol water mixture. TOTAPOL is a TEMPO-based biradical polarizing agent operating via the cross-effect. It is often used for MAS-DNP because of its high efficiency [83]. ^{1}H to ^{13}C cross polarization (CP) was used to transfer ^{1}H polarization to the ^{13}C spins at 90 K. At this point, the experiment deviates from traditional MAS-DNP because, rather than recording the NMR signal at cryogenic temperature, the frozen sample is rapidly melted with a pulse of 10.6 μm infrared laser light, and the spectrum is acquired in the liquid state. Freezing and melting of the sample took about 60–90 s and less than 1 s, respectively, while the buildup of ^{13}C polarization required 15–40 s. Solid-state enhancements about 290 times the thermal equilibrium signal were achieved at 90 K and 5 T, which translates into about a 1000-fold enhancement in the liquid state at 300 K and 5 T. A great advantage of the method is that it does not involve dilution, and after spectrum acquisition, the entire process (freezing, DNP, melting, and signal recording) can be repeated as many times as needed to improve SNR by signal averaging. Temperature-jumped DNP was primarily intended for analytical NMR spectroscopy.

1.4.2 Dissolution dynamic nuclear polarization

For biomedical spectroscopy/imaging applications, the breakthrough came in 2003, when it was demonstrated in groundbreaking experiments that the compounds hyperpolarized in the solid state could be dissolved and transferred into an NMR magnet for spectrum acquisition with negligible loss of polarization during the transfer process [84–86]. Dissolution DNP is based on the assumption that dissolution of the frozen polarized sample to produce a solution of hyperpolarized compounds and the transfer of the solution to an NMR spectrometer can be performed within a time period that is short on the nuclear spin-lattice (T_1) relaxation timescale. Theoretically, solid-state DNP works for all NMR active nuclei, including the quadrupolar ones. However, as will be discussed later in more detail, the hyperpolarized magnetization is not persistent but decays to thermodynamic equilibrium by spin-lattice (longitudinal or T_1) relaxation. This means that when the microwaves are turned off, the hyperpolarized magnetization will relax both in the solid state and in solution during dissolution and transfer, and no signal enhancement will be observed if the nucleus has a very short T_1. The typical

dissolution and transfer time for dissolution DNP polarizers is about 8−12 s. As enhanced NMR signals can be observed for about 5 T_1 time periods, no significant signal enhancement would be detected if the T_1 value is less than a couple of seconds. Successful dissolution DNP studies were done with ^{1}H, ^{6}Li, ^{13}C, ^{19}F, ^{29}Si, ^{15}N, ^{31}P, ^{77}Se, ^{133}Cs, ^{89}Y, and $^{107,109}Ag$.

Experimentally, dissolution DNP-NMR is performed in an isotropic glass matrix that involves doping the sample with paramagnetic species, most commonly a stable organic free radical, cooling it to around 1 K in a strong magnetic field (3−10 T), and irradiating the frozen sample with microwaves near the EPR frequency. Under these conditions, the underlying mechanism of polarization transfer is the SE, TM, or a combination of the two. The frozen hyperpolarized sample is then rapidly dissolved in a superheated solvent to produce solutions of hyperpolarized compounds at or near room temperature. Dissolution DNP polarizers are commercially available but can also be home-built.

1.4.2.1 Dissolution DNP polarizers

Although there are several variants described in the literature, the basic structure of DNP polarizers consists of a high-field superconducting magnet housing a cryostat/variable temperature insert (VTI), a DNP insert with the microwave cavity, sample holder, and NMR coil, a microwave source and waveguide, and a dissolution system (Fig. 1.2).

1.4.2.1.1 The HyperSense

The first commercial dissolution DNP polarizer, the HyperSense, was developed by Oxford Instruments. It is now discontinued, but several instruments around the world remain operational. The main hardware components include an actively shielded 3.35 T superconducting magnet, a pumped 4He cryostat, a cryostat insert containing the microwave cavity and the solid-state NMR coil, the sample holder insert, the microwave source, a waveguide, and the sample dissolution and transfer system. The HyperSense uses the commercially easily available W-band (95 GHz) microwave frequency source, and the operating magnetic field of the superconducting magnet (3.35 T) was chosen to match this EPR frequency ($\gamma_e = 28{,}000$ MHz/T). The sample cup containing the DNP sample is immersed into the liquid helium in the cryostat, and during DNP, it is irradiated with microwaves. When the solid-state polarization reaches a plateau, the polarized frozen sample is rapidly dissolved with superheated dissolution liquid to produce a room-temperature solution of the hyperpolarized compound. A step-by-step operational guide for the HyperSense is available in the literature [89]. The HyperSense uses a cryostat with a pumped liquid helium bath in which the sample is immersed. To reach an operating temperature of around 1 K, the vapor pressure above the liquid helium surface is dynamically reduced by using a high-flow dry mechanical vacuum pump. Such a cryostat has an inherently large helium consumption, which is a major disadvantage. The HyperSense uses separate inlet and outlet lines for the transfer of the dissolution liquid and hyperpolarized solution. The sample cup of HyperSense in the original design has a maximum capacity of 400 μL, but usually less than 200 μL sample volumes are polarized. It is worth noting, however, that the system can be modified to allow the polarization of larger samples (up to 1 g in 10 mL of solvent) with an increase in dissolution liquid volume and temperature [90]. While the Hypersense is not manufactured any more, its basic design has inspired the development of newer generation commercial instruments (SPINlab and SpinAligner) as well as a number of custom-built dissolution DNP polarizers operating at higher fields.

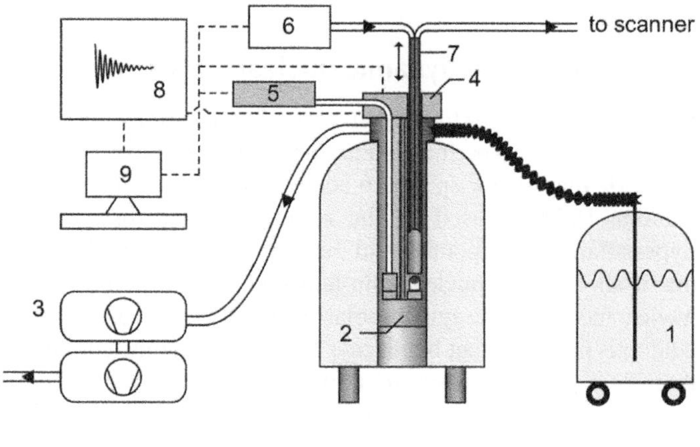

FIGURE 1.2 The generic structure of dissolution DNP polarizers [87] (1) liquid-He supply, (2) cryostat/variable temperature insert (VTI), (3) vacuum pumps, (4) DNP insert with the microwave cavity, sample holder and NMR coil, (5) microwave source and waveguide, (6) Dissolution system including the solvent boiler reservoir, with temperature and pressure transducers, (7) dissolution wand, (8) NMR console, (9) computer and software [88]. *Reproduced from Ref. [88] with permission.*

1.4.2.1.2 SPINlab [91]

The SPINlab is a DNP polarizer developed by GE Health Care for clinical applications. It retains the basic design of HyperSense but operates at a higher field (5 T) and lower temperature (0.8 K), can handle larger sample sizes (up to 2 mL), and can simultaneously polarize up to four samples with mechanically controlled sample positioning. The SPINlab has a closed cryogenic cycle to regenerate the liquid helium (complete regeneration takes about 12 h). The SPINlab uses closed, preassembled fluid paths for the dissolution and transfer of the hyperpolarized sample to maintain a sterile environment during the course of the polarization and dissolution process (Fig. 1.3). In contrast to the HyperSense, the fluid path in the SPINlab consists of two concentric tubes; and the dissolution liquid is injected into the sample vial with a syringe via the inner line, and the hyperpolarized solution passes through the outer tube. The fluid path technology eliminates the need for an external dissolution wand, thereby reducing the heat load during sample dissolution [92]. The fluid path for human applications is assembled as a sterile disposable unit and contains all the necessary components (sample vial, coaxial tube, dissolution syringe, receiver vessel with QC appendages, 0.2 μ sterile assurance filter, and administration syringe) required for DNP, dissolution, and subsequent QC and provides a sterile environment for the chemicals used in these processes [93]. A custom-made fluid path was designed to incorporate a LOD-ESR (longitudinal detected electron spin resonance) probe that allowed the acquisition of the ESR spectrum and the measurement of the electron spin-lattice relaxation time (T_{1e}) of the polarizing agent in the frozen DNP sample under the actual DNP conditions (1.15 K and 5 T) [94].

For human applications, the sample vial (cryovial) is filled with the ^{13}C-labeled substrate/"electroparamagnetic additive" or "electron paramagnetic agent (EPA)" mixture and sealed to the outer tubing of the coaxial line with a single radial laser weld in a sterile environment. During the regular sample preparation procedure, the vial containing the DNP sample is frozen in liquid nitrogen, followed by a pressure check and helium purging, which are necessary when the vial is cooled below 1 K. The sample vial is introduced into the sample pot gradually in multiple steps over 15 min to avoid an excessive increase in the helium temperature. At the end of the DNP period, the sample is moved out of the liquid He bath to a location still within the main magnetic field. The sample dissolution process begins when the inlet/outlet valve automatically switches to the open position and the heated (400 K) dissolution medium is ejected at 1.74 MPa through the inner tube, providing a flow rate of 5−10 mL/s. When pyruvic acid is polarized, the trityl-free radical polarizing agent (EPA) precipitates out of the hyperpolarized dissolution at pH values below four and is removed with a polyethylene frit in-line filter. The acid solution is collected and neutralized in the receiver. The neutralized solution is then transferred into the administration syringe. The solution must pass a QC test before the release of patient injections. The SPINlab has an automated QC module, which has been optimized for the analysis of hyperpolarized [1-^{13}C]pyruvate. The QC module measures the ^{13}C polarization, pH, EPA, temperature, volume, and pyruvate concentrations. The acceptance criteria for the hyperpolarized pyruvate injection are: (1) Polarization $> 15\%$; (2) pyruvate concentration between 220 and 280 mM; (3) EPA concentration < 3.0 μM; (4) pH between 5.0 and 9.0; (5) temperature between 25.0 and 37.0 C; (6) volume > 38 mL. In addition, the 0.2 μ sterilizing filter must also pass the bubble point test for integrity at 50 psi. The next-generation clinical polarizer, SPINlab 2.0, is currently under development. It will operate at a higher field (7 T), will be capable of performing ^{1}H to ^{13}C CP, and is expected to provide over 60% liquid state ^{13}C polarization.

1.4.2.1.3 The SpinAligner

Motivated in part by the high cost of liquid helium, the latest generation dissolution DNP polarizers use variable-field cryogen-free superconducting (dry) magnets, which can be ramped up to about 10 T. These magnets are cooled by an integrated cryo-cooler. These systems still use liquid helium, which is generated by condensing helium gas in the sample space. The commercial successor to the HyperSense polarizer, the SpinAligner (Fig. 1.4), is also based on dry magnet technology [95]. The output frequency of the microwave source is 94 GHz, and it uses a doubler and a tripler to generate microwave frequencies of 188 and 282 GHz. The microwave source has a maximum continuous output power of 100 mW and can deliver constant as well as frequency-modulated irradiation. The dry magnet can be ramped to 3.35, 6.7, and 10.05 T to match these EPR resonance frequencies. The commercially available version has a slightly different configuration from the one published in this paper [96]. The SpinAligner can be operated in a fully automated mode, and it does not need liquid cryogens (unlike the SPINlab, the helium is recycled in a continuous manner). The polarizer can handle a wide range of sample sizes. It uses the fluid path technology, which was initially developed for the SPINlab clinical polarizer, and a sample holder that is easily connected to the fluid path (Fig. 1.4). A detailed step-by-step operational guide and performance and reproducibility data for the SpinAligner have recently been published [96]. It is capable of producing over 60% ^{13}C polarization with [1-^{13}C]pyruvic acid. While the SpinAligner has been primarily designed for research purposes, it can potentially be further developed for clinical applications.

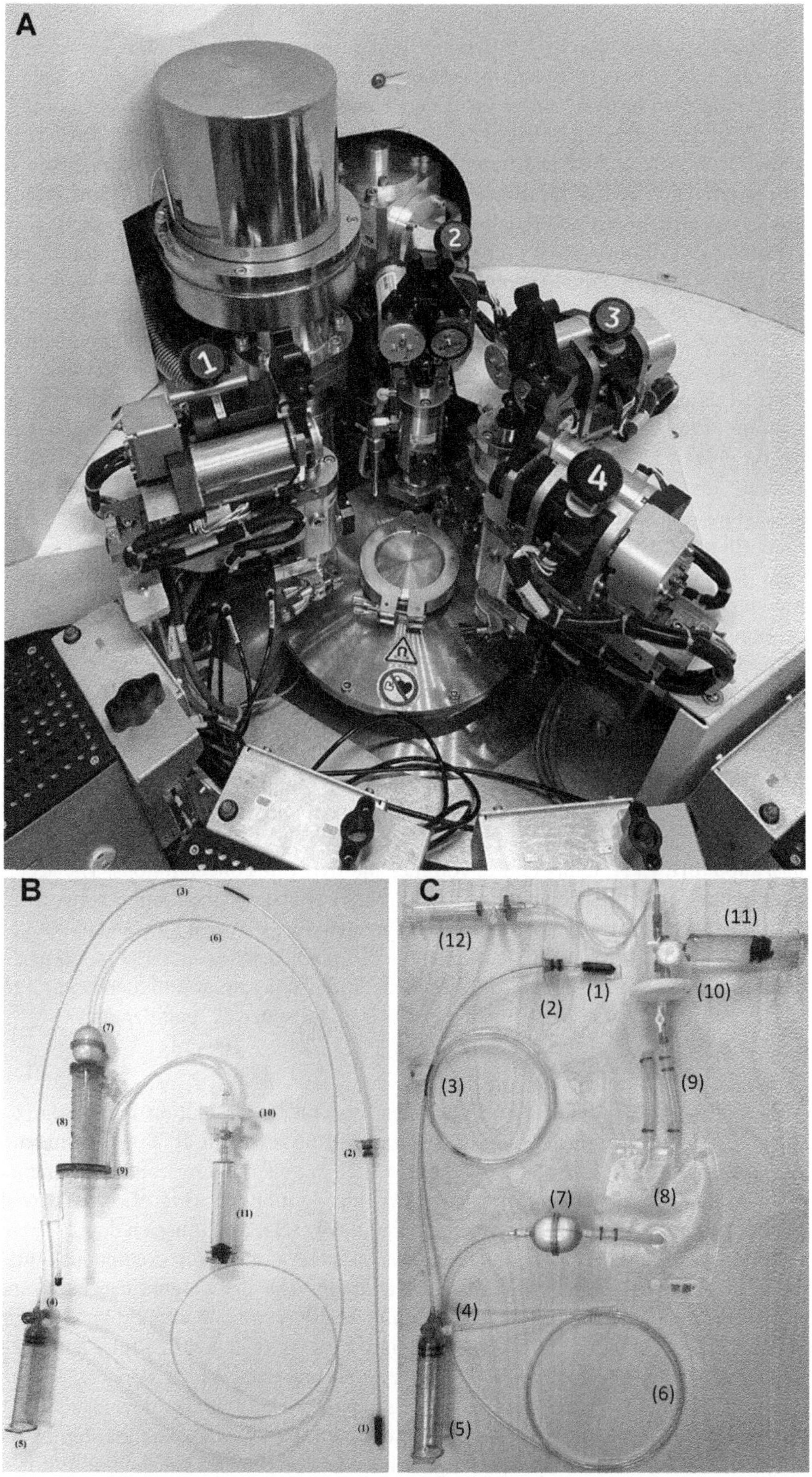

FIGURE 1.3 (A) Top view of the SPINlab showing the four sample bays. (B) The preassembled fluid path for the dissolution and transfer of the hyperpolarized sample. Original design. Part A (loads into the polarizer): (1) vial, (2) dynamic seal, (3) coaxial tube, (4) valve, (5) dissolution syringe. Part B: (6) transfer tube, (7) EPA filter, (8) receiver vessel, (9) QC appendage, (10) sterile assurance filter, (11) Medrad 65 mL MR syringe (left). C) Redesigned fluid path for the new multi probe QC (MPQC) unit. Part A (loads into the polarizer): (1) vial, (2) dynamic seal, (3) coaxial tube, (4) valve, (5) dissolution syringe. Part B: (6) transfer tube, (7) EPA filter, (8) receiver bag, (9) QC appendage, (10) sterile assurance filter, (11) Medrad 65 mL MR syringe, (12) QC syringe.

FIGURE 1.4 The SpinAligner DNP polarizer (left picture). The top view (center picture) shows the dissolution module and the fluid path with the sample being polarized inside the magnet. The sample holder (right picture) can be screwed onto the fluid path. The sample holder is manufactured from the high-performance thermoplastic polymer PEEK (polyether ether ketone) and has a maximum filling volume of 400 μL.

1.4.2.1.4 Bruker Biospin AlphaPolarizer

Bruker is currently developing dissolution DNP instrumentation. A prototype polarizer has been installed at least at one location. The polarizer operates at 7.05 T and has a variable-temperature helium bath cryostat (from 1.2 to 4.2 K). The microwave source has an output of 125 mW and is capable of triangular frequency modulation [97,98].

1.4.2.1.5 Custom-built DNP polarizers

There are several custom-built DNP systems described in the literature. These systems usually operate in higher fields than HyperSense and include extra or improved features. Earlier home-built instruments used conventional superconducting magnets. [57–59,99–103]; later designs utilize dry magnet technology [104–106]. These instruments can provide high nuclear polarization. For example, an efficient cryogen-free DNP system is based on a dry magnet operating at 9.4 T. The polarizer also has an integrated VTI, two NMR coils (tuned to ^{1}H and ^{13}C), and a helium recycling unit. Solid-state ^{1}H and ^{13}C polarizations up to 60% and 50%, respectively, after adiabatic CP were achieved with TEMPOL radical at 1.5 K [105].

1.4.2.1.6 Multi-sample polarization

Some DNP polarizers (including the SPINlab) were designed to accommodate more than one DNP sample [103,106–108]. These instruments generally have a mechanical sample changer and only one dissolution system. The multi-sample polarization feature can be useful when several hyperpolarized samples are required within a short period of time, for example, for consecutive in vivo experiments. The SPINlab is capable of polarizing up to four samples at the same time, while a custom-built polarizer can polarize two samples, allowing the production of two separate hyperpolarized solutions within a few minutes. A revolver-style design in another home-built instrument allows the loading of up to six samples in such a way that one of the samples is positioned in the microwave cavity and another one is in the dissolution port [107].

1.4.2.1.7 Bullet and dual magnet DNP systems

Two ingeniously designed dissolution DNP systems aim at alleviating some of the problems of conventional sample transfer and dissolution (e.g., high dilution and slow transfer rate) by rapidly transferring the frozen hyperpolarized sample into another magnet before dissolution. In bullet DNP [109,110], the hyperpolarized sample is very rapidly transferred as a frozen bullet (up to 80 μL in volume) from the polarizer into a spectroscopy magnet with pressurized helium gas and dissolved only at the point of use. With He gas pressure of 10 bar, the bullet could be shot into another magnet in about

70 ms, which compares very favorably with the 10 s transfer time of the HyperSense. Another advantage of this design is that it eliminates the need to dissolve the hyperpolarized sample in the polarizer, which is one of the limitations of current dissolution DNP technology. As the dissolution does not happen in the polarizer, the solvent is not exposed to cryogenic environments. This means that a significantly reduced dissolution liquid volume is needed to dissolve the sample, yielding much more concentrated hyperpolarized solutions. The HyperSense polarizer uses 4 mL of superheated liquid to dissolve the frozen DNP sample; in contrast, in bullet DNP, only 700 uL is needed to completely dissolve an 80 μL frozen bullet. If the polarizer and the secondary magnet have the same polarity, then the field direction must be reversed along the transfer path. Any change in field direction should occur adiabatically, meaning that the Larmor frequency ω should always be substantially more significant than the frequency of the rotation of the field as seen from a frame fixed with respect to the moving bullet. This was achieved using two pairs of Helmholtz coils around the plug valve at the top of the sample tube. Sample dissolution can be done by directly shooting the bullet into an NMR tube or into a solvent reservoir located above the NMR tube.

The dual-magnet DNP instrument consists of two individually shielded vertical-bore superconducting magnets on top of each other [111]. The upper magnet is charged to 3.35 T, contains a continuous flow cryostat capable of reaching 1.5 K, and is used to perform the DNP. The lower, 9.4 T magnet is dedicated to high-resolution NMR spectroscopy. The two isocenters are separated by 85 cm, with a minimum magnetic field of 200 mT between them. After DNP, the polarized solid sample is rapidly moved into the lower magnet with an electro-mechanical actuator. The spectroscopy magnet is equipped with a dissolution dock that attaches to the bottom end of the VSI. The dissolution dock produces the superheated solvent (2 mL), delivers it to the sample, and transfers the hyperpolarized solution into a 5 mm NMR tube. Due to the proximity and vertical arrangement of the two magnets, the transfer of the solid sample between the two magnets takes only 300 ms with negligible loss of spin polarization. Dissolution and transfer to the NMR tube take another 700 ms, bringing the total time of sample handling to about 1 s [112].

1.4.2.2 Multisubstrate and multinuclear polarization

It is possible to simultaneously polarize two or more substrates in one DNP experiment. Multi substrate polarization was demonstrated in both the HyperSense and SPINlab polarizers [113]. The DNP sample for multisubstrate polarization can be prepared by dissolving two or more compounds at the same time in the glassing matrix if there is no adverse reaction between the components. Alternatively, the substrates can be added to the sample holder in separate steps as frozen pellets or layers frozen in liquid nitrogen after each addition. In the latter case, the use of the SPINlab polarizer was advantageous because the relatively small sample volume of the HyperSense polarizer limited the final concentrations of the hyperpolarized substrates in the dissolution liquid. For example, a SPINlab DNP sample for copolarization of [1-^{13}C]pyruvic acid and [1,4-$^{13}C_2$]fumaric acid was prepared by first loading the vial with pyruvic acid and freezing it in liquid nitrogen. Fumaric acid solution in DMSO was then added as a separate layer on top of the frozen pyruvic acid, followed by freezing the sample again in liquid nitrogen. Copolarization of four ^{13}C labeled substrates ([1-^{13}C]pyruvic acid, sodium [^{13}C] bicarbonate, [1,4-$^{13}C_2$]fumaric acid, and ^{13}C urea) was performed in the HyperSense polarizer, potentially allowing for the simultaneous monitoring of metabolism, necrosis, pH, and perfusion in vivo [114]. It is also possible to polarize a compound labeled with two isotopes or a mixture containing compounds labeled with different isotopes. When the DNP is performed in an isotropic glassing matrix and the mechanism is TM, both nuclei will achieve a common spin temperature, and each nucleus will be polarized to a level determined by this common spin temperature regardless of being in the same or separate chemical entities [113–116].

1.4.2.3 Polarization build-up time

When the microwaves are turned on, the solid-state nuclear spin polarization increases over time, reaching a plateau after about 1–4 h, depending on the DNP conditions (magnetic field strength, polarizing agent concentration, glassing matrix) [117]. The polarization buildup time can be determined by measuring the solid-state polarization over time. The buildup of nuclear magnetization is usually described by a monoexponential function. However, a biexponential model (Eq. 1.13) may provide a better fit for some samples [118,119].

$$M(t) = \Pi_1\left(1 - e^{-\frac{t}{\tau_1}}\right) + \Pi_2\left(1 - e^{-\frac{t}{\tau_2}}\right) \tag{1.13}$$

Where the build-up time constants τ_i (I = 1, 2) are the build-up time constants and Π_i (i = 1, 2) are the plateau values (for a monoexponential fit, $\Pi_2 = 0$) [118]. A biexponential buildup could be due to different polarization processes, i.e., rapid direct polarization transfer from electrons to nuclear spins in the proximity of the radical and slower spin diffusion via homonuclear dipolar interactions throughout the sample. It may also be an indication of a microcrystalline sample with nonuniform radical distribution. Regions with higher radical concentrations have faster buildup than sample regions in which the radical is present at a lower concentration [118,120].

1.4.2.4 Sample dissolution and transfer

When the solid-state polarization reaches a plateau, the microwaves are turned off, and when the dissolution is initiated, the DNP sample is lifted above the liquid He bath, and the polarized sample is quickly dissolved in a superheated dissolution liquid to produce a solution of the hyperpolarized agent [84,85,121]. For biological studies, the dissolution liquid typically contains a buffer (tris(hydroxymethyl)aminomethane (TRIS) or PBS), a neutralizing agent (if needed), and a chelating agent (ethylenediaminetetraacetic acid, EDTA). EDTA chelates paramagnetic metal impurities, which would otherwise cause a more rapid loss of polarization. The buffer and neutralizing agent is used to produce hyperpolarized solutions with a physiological pH of 6–8. For example, [1-^{13}C]pyruvate is most often polarized as neat [1-^{13}C]pyruvic acid and is dissolved in a mixture of TRIS buffer and NaOH. Sample dissolution and transfer of the hyperpolarized solution into a receiving vessel take on average about 10 s in the HyperSense commercial polarizer and some home-built polarizers. The obtained solution can then be manually transferred into an NMR tube or used for biological studies such as magnetic resonance spectroscopy (MRS)/MRI. Since the T_1 decay is exponential, approximately 63% of the initial polarization is lost after one T_1 period. This means that substantial relaxation of hyperpolarized magnetization occurs when the T_1 value is comparable to or less than the transfer time. For nuclei with a T_1 value of a few seconds, this can result in a total loss of hyperpolarized magnetization. To overcome this issue, various technological improvements have been implemented to shorten the transfer and injection times [122–125]. Most custom-built automated transfer and injection devices use an inert gas to drive the sample and can cut the transfer times to less than 3 s. Some of these can also neutralize the free radical polarizing agent and measure the polarization level inside the imaging magnet [123]. In addition, automated injection systems improve the reproducibility of sample administration for in vivo experiments by providing a constant flow rate [58,100,126]. The HyperSense, the SpinAligner, and most custom-built preclinical polarizers have been designed to produce about 4 mL of hyperpolarized solution. In vitro experiments are, therefore, usually performed in a 10 mm NMR tube, which can accommodate this volume. However, some NMR measurements, such as high-resolution NMR spectroscopy of proteins, are preferably performed in a 5 mm NMR tube. Unfortunately, transferring aqueous hyperpolarized solutions from the polarizer into 5 mm NMR tubes is typically accompanied by the formation of bubbles that interfere with high-resolution spectrum acquisition. To avoid this problem, a sample injection apparatus designed for the rapid delivery of the hyperpolarized solution into a 5 mm NMR tube has been developed. This computer-controlled device uses pressurized (1.8 MPa) N_2 gas to transfer the first few hundred microliters of the hyperpolarized solution, in which the concentration of the dissolved sample is highest. The device adjusts the pressure of N_2 to create sufficient turbulence to obtain a homogeneous solution and keeps the MR tube pressurized to prevent the formation of gas bubbles in the sample, thereby allowing the acquisition of high-resolution DNP-NMR spectra of compounds with short T_1 relaxation times [127]. This high-pressure dissolution system was redesigned to make it more stable and reliable. The injection of a preset sample volume via a two-state valve into the NMR tube is triggered by an optical flow detector [128,129]. The injection device suppresses bubble formation by pressurizing the sample in the NMR tube.

Liquid-driven transfer and injection are less commonly used, although high-pressure liquid-driven injection into an NMR tube or flow cell can offer some advantages for high-resolution dissolution DNP NMR spectroscopy work [130]. In particular, liquid-driven transfer systems reliably deliver bubble-free samples, which is essential to achieving high resolution. A sophisticated liquid-driven dissolution and transfer system that uses isopropanol to reproducibly transfer and inject 540 μL of hyperpolarized solution into a 5 mm NMR tube was recently reported. The system consists of two air-actuated 10-way valves, an optical sensor, a high-pressure liquid pump to push the hyperpolarized sample into the spectrometer, and a magnetic tunnel to minimize low-field relaxation. This setup is capable of dissolving and transferring the hyperpolarized sample in less than 5 s with high speed (5 m/s) via a 9.5 m capillary tube without contaminating the sample with the driving liquid or gas bubbles [98].

One of the drawbacks of dissolving DNP is that during the dissolution process, when a 100 μL frozen DNP sample is dissolved with 4 mL of solvent, the hyperpolarized sample will be diluted by a factor of 40. This dilution factor can be reduced by using an immiscible mixture of water and organic solvent in which the substrate is insoluble. The organic phase should have a similar boiling point and heat capacity to water, and it should separate rapidly without forming an emulsion. As much as 75% of the water in the dissolution liquid can be replaced with heptane, toluene, or some fluorinated solvents, yielding a more concentrated hyperpolarized solution (Fig. 1.5) [129,131,132]. The use of an immiscible organic solvent may also have the added benefit of removing the free radical polarizing agent via extraction during the sample transfer.

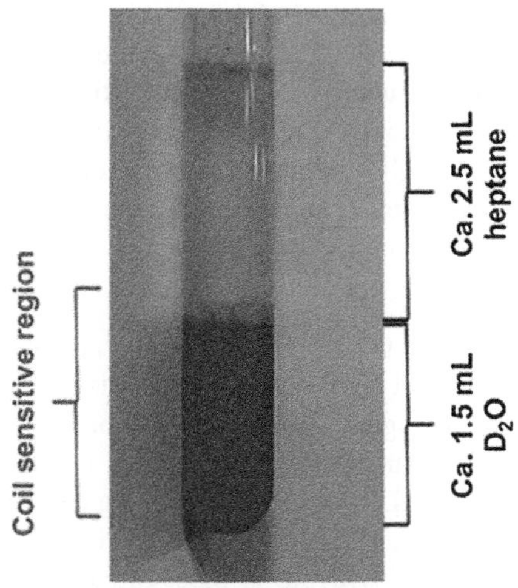

FIGURE 1.5 Dissolution of a frozen water/DMSO sample containing a food dye in the HyperSense with a mixture of heptane water (2:1). 1.5 mL of the dissolution liquid was transferred into a 10 mm NMR tube [129]. *Reproduced from Ref. [129] with permission.*

A fast transfer and infusion system for in vivo studies in small animals was also developed. It consists of an infusion pump integrated with a liquid-gas phase separator made of plastic. The whole system is placed inside the bore of the imaging magnet. After the injection of hot dissolution liquid, the hyperpolarized solution is rapidly transferred into the separator/infusion pump with high-pressure inert gas (6 bars of He). The gas separator is open to the atmosphere, allowing inert gas to escape. The solution is collected at the bottom of the pump, which is located inside interleaved ^{13}C and ^{1}H coils. These coils will enable the acquisition of ^{13}C and ^{1}H NMR signals in the hyperpolarized solution. The separator/infusion pump can be preloaded with a solution containing reagents that can be used to neutralize the free radical polarizing agent or to perform a chemical reaction with the hyperpolarized substrate. After settling and homogenization (approximately 1 s), the piston of the infusion pump is pushed down by applying hydrostatic pressure to a physiological saline solution via a computer-controlled pump. The outlet is monitored with a fiberoptic bubble detector, which stops the infusion in case bubbles are detected in the solution. The hyperpolarized solution is infused into the femoral vein through a catheter. The entire process from dissolution to injection takes about 3 s [58,123].

1.4.2.5 Measuring the polarization

The most straightforward way of measuring the liquid state polarization is by comparing the signal intensity immediately after dissolution with the signal intensity detected when the sample reaches thermal equilibrium after complete relaxation of the hyperpolarized magnetization. This is usually not a problem with ^{13}C. However, detecting the thermal signal may take a very long time for ultralow gamma nuclei such as ^{15}N or ^{89}Y. In these cases, a reference solution containing a known concentration of the nucleus of interest may be used, for example, 2 M $^{89}YCl_3$. For certain compounds that contain at least two nuclear spins that are coupled to each other (e.g., ^{13}C coupled to ^{13}C or ^{15}N, at least one of the coupled nuclei must be enriched), the polarization can be estimated from the asymmetry of the multiplets [99]. For hyperpolarized [^{13}C]urea, the ^{13}C polarization was calculated as the normalized difference of partial integrals of the natural abundance of ^{15}N doublet J-coupled to the ^{13}C-enriched carbon [84]. This approach has been applied to other hyperpolarized compounds; however, for the accurate determination of polarization, cross-relaxation must be negligible and nonfirst-order coupling effects must be taken into account. Some examples include hyperpolarized [1,2-$^{13}C_2$]pyruvate, [1-^{13}C]pyruvate, [1-^{13}C, 3-$^{2}H_3$]pyruvate, and [1,2-$^{13}C_2$; 2-$^{2}H_3$]acetate. The method has also been extended to the real-time determination of polarization in vivo [133]. The theoretical considerations are rather complex, and the time evolution of multiplet asymmetries is not entirely understood yet [134].

1.4.2.6 Performance characteristics of DNP polarizers

The dissolution-DNP technique has been successfully used to generate hyperpolarized agents for in vitro and in vivo molecular MRI applications. Since the publication of the prototype dissolution DNP device, it has been continually being modified and improved, generally with the goal of achieving higher solid-state spin polarization. The most important operating parameters of the DNP instrument are the B_0 magnetic field and the sample temperature [112]. The field in

custom made and commercial DNP polarizers ranges from 2.5 to 10 T. As the electron spin polarization reaches nearly 100% at 1 K above 3.35 T, one would not expect dramatic increases in the nuclear spin polarization with further increases in the magnetic field for a given polarizing agent. There were also concerns that some of the DNP mechanisms may lose their efficiency at high fields [59]. However, experimentally measured ^{13}C enhancement data collected with homebuilt and commercial polarizers operating at around 1 K and up to 10 T clearly show that the efficiency of DNP increases with increasing B_0 field [112]. This has been demonstrated for both trityls and nitroxides [99,112,135]. The net solid-state polarization enhancement in the DNP sample is determined by the rate of polarization transfer, which generates the hyperpolarization, and the nuclear spin-lattice relaxation rate, which characterizes the decay of the polarization. Interestingly, both processes are governed by the electronic spin-lattice relaxation time, T_{1e}. Both the nuclear and electron spin relaxations become slower at lower temperatures. The magnetic field dependence of the electronic spin-lattice relaxation rate ($1/T_{1e}$) of the trityl radical OX063 was measured at 20 K. At lower fields (9.4—95 GHz), the electronic relaxation rate did not change appreciably, but at higher fields (240—360 GHz), it exhibited a sharp increase [136]. This trend is expected if the relaxation occurred by direct process, in which case $1/T_{1e}$ is proportional to B_0 [4,137]. In addition, the relaxation effect of the free electrons on the nuclear spins decreases as B_0 increases. Altogether, the magnetic field dependence of T_{1e} strongly suggests that the efficiency of the DNP should increase in higher magnetic fields. As a rule of thumb, the attainable polarization for low-gamma nuclei, such as ^{13}C, shows more or less linear dependence on B_0/T at commonly encountered B_0 fields and sample temperature ranges (2.5—10 T and 0.8—1.4 K) [59]. Interestingly, the stronger the magnetic field, the higher the optimal radical concentration: 15 mM at 3.35 T, 30 mM at 6.70 T, and 45 mM at 10.10 T [92]. It should also be noted that the enhanced polarization at higher fields and lower temperatures comes at the expense of longer polarization buildup times. In general, the polarization builds up slower at higher B_0 and lower sample temperatures, but becomes faster as the radical concentration increases [135].

1.4.2.7 How to improve DNP

The performance of a dissolution DNP experiment can be greatly improved by optimizing the sample preparation, radical concentration, elimination of low-filed relaxation effects, and by applying microwave frequency modulation and CP techniques [89,115,122,138,139].

1.4.2.7.1 Microwave frequency modulation

Frequency modulation of the microwave irradiation during DNP was found to afford substantial gains in ^{1}H polarization under dissolution DNP conditions with the nitroxide radicals [140,141] The improvement was much more pronounced in samples with relatively low radical content (10—25 mM) when the EPR line of the radical concentration was inhomogenously broadened. The optimal radical concentration without frequency modulation was 50 mM in the DNP sample. Microwave frequency modulation thus allows performing the DNP with a lower radical concentration without compromising the polarization. Low radical concentration has the advantage of reducing the unwanted nuclear T_1 shortening effect of the paramagnetic polarizing agent both in the solid and liquid states [140,142]. Microwave frequency modulation (applied at a rate of 1 kHz and amplitude of 20—50 MHz) also improved the ^{13}C polarization significantly (2—3 fold) with trityl radicals at 15 mM and 10.1 T. At higher radical concentrations, the effect of modulation was much less pronounced [95].

1.4.2.7.2 Cross-polarization

It was demonstrated in very elegant experiments that hyperpolarized ^{1}H magnetization can be transferred to ^{13}C in the frozen DNP sample under the Hartmann-Hahn condition. This entails that the nutation frequency (precession around B_1) of the ^{1}H and ^{13}C spins be equal in their own rotating frame of reference, that is, $\gamma_{1H}B_1(^{1}H) = \gamma_{13C}B_1(^{13}C)$, which can be brought about by simultaneous rf irradiation at both resonance frequencies. When this condition is met, then energy exchange (polarization transfer) can occur between the dipolarly coupled spins [143,144]. CP combined with ^{1}H DNP can advantageously be used to enhance ^{13}C polarization, as ^{1}H DNP with nitroxide radicals at high field is much more efficient than ^{13}C DNP. In the DNP setting, the polarization transfer is performed in multiple CP steps. This technique is remarkably efficient. ^{1}H spins could be polarized to nearly 100% with TEMPO radical (50 mM) within just 20 min at 6.7 T and 1.2K. Polarization transfer from ^{1}H to ^{13}C by CP at low temperatures afforded over 70% solid-state ^{13}C polarization [145]. CP from ^{1}H to other low gamma spins is also possible. It was suggested that ^{1}H to X nucleus CP could be applied to ^{129}Xe, ^{6}Li, ^{15}N, and ^{89}Y [146]. However, successful CP requires a custom-built coil that can be tuned to both nuclei and can provide strong B_1 fields at both resonance frequencies [145—152].

1.4.2.7.3 Magnetic tunnel

The fringe field in a room accommodating superconducting magnets is dependent on the arrangement of the magnets and their shielding. The fringe field can drop below 1 mT between magnets, even when they are unshielded. Low magnetic

fields can amplify certain relaxation processes that can lead to rapid loss of polarization [153]. Loss of magnetization can also occur during the transfer of the hyperpolarized sample from the polarizer to the spectrometer or scanner if the sample passes through a fringe field gradient and the rate of field change matches the Larmor frequency of the hyperpolarized nucleus (nonadiabatic sample transfer) [154]. These unwanted effects can be eliminated using a magnetic tunnel encompassing the transfer line. The design and construction of a magnetic tunnel built from permanent magnets arranged in the form of a Halbach array have been reported. The Halbach arrangement maximizes the magnetic field inside the tunnel [155]. Using strong NdFeB magnets, a magnetic field close to 1 T could be achieved at the center of the tunnel. The effectiveness of magnetic tunnels in maintaining polarization is probably influenced by various factors, such as the compound being used and the laboratory conditions, as well as different relaxation mechanisms and magnetic field variations. A portable transport vessel for hyperpolarized media consisting of a chamber inside an electromagnet has also been patented. The battery-operated electromagnet can generate magnetic fields up to 100 Gauss [153,156].

1.4.2.7.4 Addition of paramagnetic metal ions

The presence of paramagnetic metal ions such as Gd^{3+}, Ho^{3+}, and Mn^{2+} in the DNP sample can dramatically improve the solid-state polarization [157]. The magnitude of the enhancement depends on several factors, such as the nature of the free radical polarizing agent, magnetic field strength, and the concentration of the paramagnetic metal [158–160]. The metal ion is usually present in the DNP sample at the low millimolar concentration range, most commonly at around 2 mM. Gadolinium(III), in the form of the well-known MRI contrast agent ProHance, is often used with trityl OX063. At 3.35 T, the effect of Prohance on the ^{13}C polarization is quite remarkable: up to a 300% increase in the solid-state ^{13}C polarization of sodium [1-^{13}C]pyruvate was reported in the glycerol-water glassing matrix. However, the effect was much weaker with 1,3-bisdiphenylene-2-phenylallyl (BDPA) and 4-oxo-TEMPO [161]. At higher fields, the effect of added Gd(III) is less clear, and the data are somewhat contradictory. Earlier papers reported that the presence of Gd(III) did not significantly improve solid-state polarization [57,99,162,163]. However, more recent work demonstrated that the ^{13}C polarization of uniformly ^{13}C-labeled and perdeuterated glucose could be doubled by the addition of Prohance at 6.7 T with a trityl radical (AH111501, a.k.a. EPA) [163]. Interestingly, a gadolinium containing endohedral metallofullerene ($Gd_2@C_{79}N$) was shown to significantly improve the solid-state ^{13}C polarization with 4-oxo-TEMPO at 5 T, which was attributed to the high-spin ground state (S = 15/2) brought about by the strong exchange coupling between the two encapsulated Gd(III) ions (S = 7/2) and the $C_{79}N^{6-}$ heterofullerene radical ion (S = ½) [164]. While Gd-complexes are the most popular, Ho(III) gives similar enhancements with the added benefit that its liquid-state nuclear T_1 shortening effect is less pronounced. From a theoretical point of view, paramagnetic ions enhance the DNP by modifying the EPR properties of the free radical polarizing agents by shortening the electron spin-lattice relaxation time and narrowing the EPR linewidth [165]. The nuclear T_1 is not affected appreciably [159].

1.4.2.7.5 Glassing matrix deuteration

Deuteration of the glassing matrix can produce significant improvements in the ^{13}C polarization when the EPR linewidth of the polarizing agent is larger than the ^1H Larmor frequency. These agents include nitroxides, galvinoxyl, and DPPH radicals. This is because the ^2H Zeeman reservoir has a lower heat capacity than the ^1H system. However, for the same reason, deuteration of the glassing matrix is not recommended when narrow ESR linewidth-free radicals such as BDPA or trityl OX063 are used because, in this case, the proton Zeeman reservoir has no contact with the electron dipolar reservoir. Since the Larmor frequency of deuterium will be less than the EPR linewidth, the ^2H spin system would increase the heat load in the nuclear Zeeman system, thereby leading to lower ^{13}C polarization. The effect of glassing matrix deuteration on the ^{13}C enhancement can be quite significant. The ^{13}C polarization increased more than 100% with nitroxide radicals but decreased 50% when the narrow linewidth BDPA radical was used as a polarizing agent at 3.35 T. Similar results were obtained at higher fields, although the effect was not quite as dramatic [108,166–169].

1.4.2.8 DNP sample preparation

The composition of the frozen sample is a key parameter for DNP [170]. In general, efficient DNP requires an isotropic glass matrix in which the orientation of the molecules is uncorrelated (the DNP mechanisms are discussed in detail in Chapter 2). The DNP sample is a vitrified homogenous mixture containing the sample to be polarized, the polarizing agent(s), and a glassing agent (if needed). The various commercial and homemade DNP polarizers can accommodate a few hundred μL samples of (the maximum sample size is 2 mL in the SPINlab); however, the sample volume is usually less than 200 μL. For in vivo studies, the concentration of the hyperpolarized substrate in the administered solution should be in the 20–200 mM range (about 250 mM for human studies). As the volume of dissolution liquid is generally 4–6 mL for in vitro or small animal studies, the concentration of the substrate in the DNP sample must be one or two orders of magnitude higher to meet this requirement, generally in the molar or tens of molar range. Some compounds, such as neat

pyruvic acid or 2-ketoisocaproic acid, form glass at cryogenic temperatures and can thus be polarized without adding a vitrifying agent. However, the majority of substrates need to be dissolved in a glass matrix for DNP. The most commonly used biocompatible matrix for in vivo work is glycerol-water, but DMSO-glycerol, DMSO-water, or ethanol-water can also be used [118,166,171]. For non-in vivo studies, methanol-water mixtures are also applicable. Concentrated aqueous or DMSO (melting point 19°C) solutions will often form a glass at low temperatures when rapidly supercooled without the addition of glycerol [121,172].

A potential disadvantage of glycerol is that it can enter both glucose and triglyceride metabolism, thereby potentially interfering with metabolic studies, although the injected amount is usually not very high. Therefore, trehalose (α-D-glucopyranosyl-(1 → 1)-α-D-glucopyranoside) was suggested as a nonmetabolizable glassing matrix component [173]. Trehalose is not metabolized when administered intravenously because the enzyme that breaks down the 1−1 alpha glycoside bond (trehalase) is present only in the small intestine. Trehalose effectively lowers the critical temperature for homogeneous ice nucleation (T_N) in an aqueous solution [174]. It was demonstrated that a glassing matrix consisting of 15% trehalose in water not only affords ^{13}C spin polarizations comparable to glycerol-containing mixture formulations but also offers faster polarization build-up rates [173].

Poor solubility in glassing mixtures is often a problem. The solubility of ionic compounds is dependent on the lattice energy versus the solvation energy, and the solubility may be increased by switching to a different counterion [175]. A good example is the use of highly water-soluble cesium bicarbonate rather than sodium salt in making up the DNP sample for bicarbonate polarization [176]. The use of counterions such as protonated tris(hydroxymethyl)aminomethane (TRIS) and meglumine can also improve the solubility of the substrate and/or the vitrifying properties of the DNP sample because these cations can participate in hydrogen bonding interactions with water due to the presence of OH and NH_2 groups [177]. Acetate was polarized in the form of TRIS salt, which provided slightly higher ^{13}C polarization than the sodium salt, although the absolute concentration of [1-^{13}C]acetate was lower in the TRIS sample [24,171].

Choline forms highly soluble salts with carboxylic acids, and it is also a commonly used component in deep-eutectic solvents [178−180] Choline salts and deep eutectic mixtures were proposed for DNP sample formulations. Choline [1-^{13}C] acetate and choline [1-^{13}C]pyruvate were prepared by mixing equimolar quantities of choline bicarbonate (80% in water) and acetic or pyruvic acid. DNP samples were prepared by adding OX063 radical (15 mM) and Dotarem (1 mM) to these mixtures. Deep eutectic mixtures of choline and acetic acid, pyruvic acid, and urea were prepared by mixing choline chloride and the substrates in a 1−2 molar ratio. These formulations proved to be superior to conventional DNP sample preparations. Both the choline salts and the deep eutectic mixtures had excellent glassing properties without the need for additional glassing agents and could easily be dissolved with superheated water due to their low viscosity. They afforded high substrate concentrations and higher liquid-state ^{13}C polarizations than the conventional sample preparations [181].

The glassing properties of the substrate formulations can be visually tested before the DNP by adding droplets of the substrate solution in the glassing matrix into liquid nitrogen (Fig. 1.6). Well-glassing mixtures will form transparent amorphous pellets (Fig. 1.6 left), while poorly vitrifying formulations appear opaque (Fig. 1.6 right) [182]. Although the liquid DNP sample can be placed into the sample holder and inserted into the cryostat of the polarizer, it is often prefrozen in the form of small (2−3 mm in diameter) pellets before loading it into the polarizer. This is commonly performed by dripping the solution into liquid nitrogen [166]. An improved procedure for vitrification employs isopentane as a heat

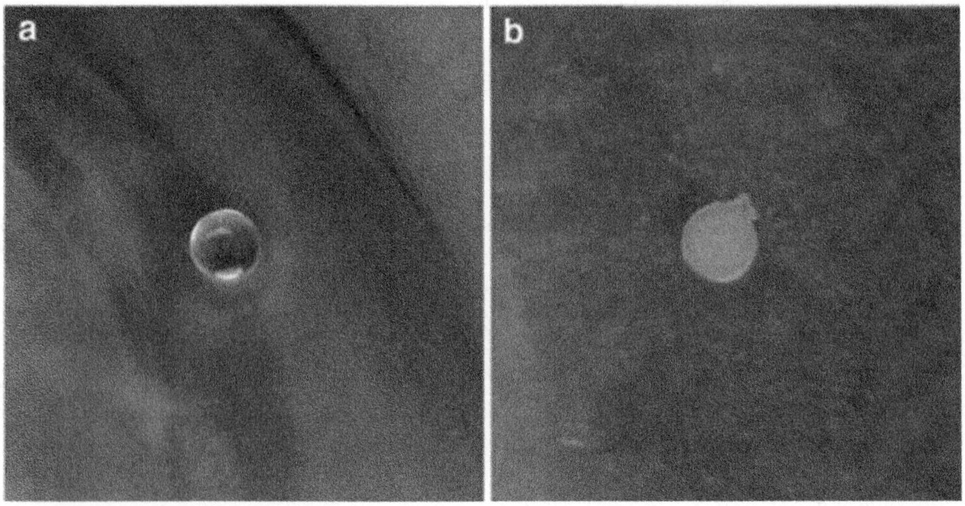

FIGURE 1.6 Visual testing of vitrification of DNP samples in liquid nitrogen [182]. *Reproduced from Ref. [182] with permission.*

transfer medium. The liquid DNP sample is sprayed into isopentane and cooled to 130 K with a liquid nitrogen bath. The frozen sample is then transferred into the sample cup with a spatula [183].

1.5 Polarizing agents

Polarizing agents are paramagnetic substances that supply the free electron spins for DNP. They are diverse materials, including stable (persistent) free radicals, paramagnetic transition metal complexes, in situ-generated transient radicals, or the photoexcited triplet state of pentacene (Fig. 1.7).

1.5.1 DNP mechanisms

Four polarizing mechanisms have been described for DNP in frozen matrix, including the SE, the CE, the TM, and the OE [47,53,184–188]. In general, in a DNP experiment, the relative sizes of three parameters will provide information on the plausible mechanism involved. These are: (i) The homogeneous EPR linewidth (δ) of the polarizing agent; (ii) the inhomogeneous breadth (full width of the EPR signal) (Δ) associated with the g-anisotropy; and (iii) the nuclear Larmor frequency (ω_{0I}). When δ and $\Delta < \omega_{0I}$, the SE is the main polarization process, and the maximum and minimum enhancements are observed at $\omega_{0S} \pm \omega_{0I}$., where ω_{0S} is the EPR resonance frequency. When $\Delta > \omega_{0I}$, the CE is the dominant mechanism of polarization; it is based on electron-electron-nuclear three-spin processes and is most efficient if the ω_{0I} matches the difference of the Larmor frequencies of two electrons. TM is another mechanism operating when $\Delta > \omega_{0I}$ and is dominant when many electrons and nuclei interact strongly at low temperatures (typically $T < 10$ K). TM is best described using spin temperature formalism involving three energy reservoirs: The electron Zeeman system, the electron dipolar system, and the nuclear Zeeman system. Off-center irradiation of the EPR line creates a nonequilibrium polarization gradient across the EPR line, effectively decreasing the spin temperature of the electron dipolar system, which in turn decreases the spin temperature of the nuclear Zeeman reservoir via thermal contact [43,188]. The CE and TM mechanisms can be differentiated by whether the EPR spectrum is either inhomogeneously ($\Delta > \delta$) or homogeneously ($\Delta \approx \delta$) broadened, respectively. The observation of OE in insulating solids was unexpected because the dynamic processes required for electron-nuclear cross-relaxation observed in liquid solution are significantly reduced at cryogenic temperatures. Indeed, OE requires a hyperfine coupling that oscillates on the same time scale as the electron Larmor frequency. OE has been observed in insulating solids when using BDPA as a radical, and the electron hopping rate might

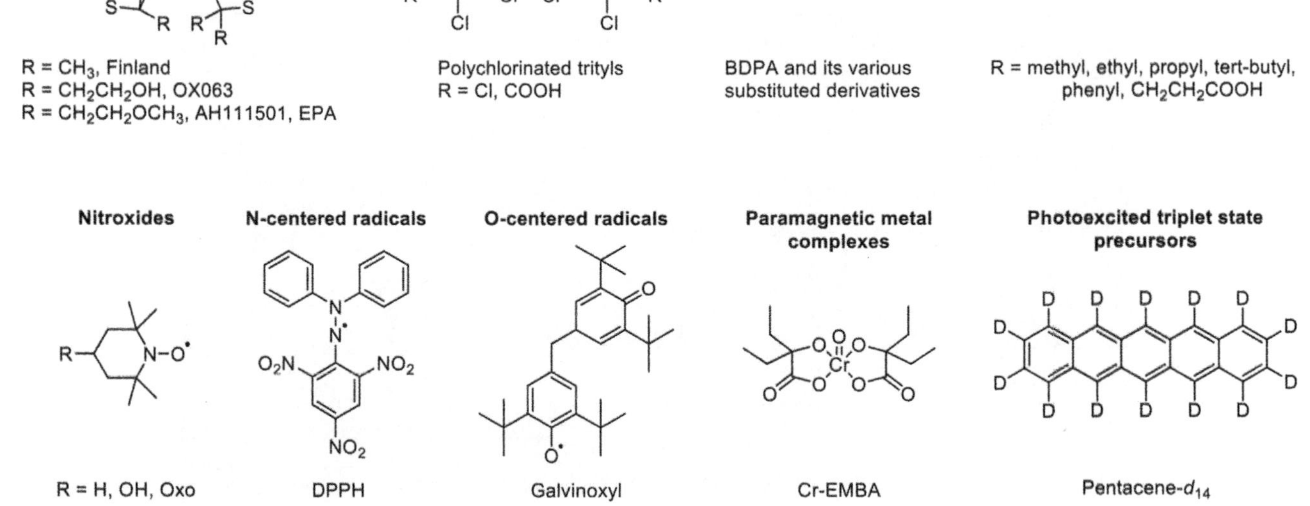

FIGURE 1.7 Examples of paramagnetic species used as polarizing agents in low temperature dissolution DNP.

be responsible for the time dependence of the interaction [189,190]. Nevertheless, at temperatures around 1 K and high fields, the dominant DNP mechanism is TM for most nuclei.

When choosing a polarizing agent, the following points should be considered: nuclei to be polarized, solubility, toxicity, chemical compatibility, removability, response to Gd(III), EPR properties (lineshape and width, electron relaxation, hyperfine couplings), temperature regime, and applied magnetic field. The electron paramagnetic resonance (EPR) properties of the unpaired electron have a great influence on the DNP. Carbon-centered radicals with narrow EPR linewidth are better suited for the direct polarization of ^{13}C and ^{15}N than nitroxides, which work better for ^{1}H DNP (see a more detailed discussion on this topic in Chapter 2).

1.5.2 Stable organic free radicals

Among the paramagnetic species shown in Fig. 1.7, the trityl-based, stable tetrathiatriarylmethyl free radicals are the most frequently used polarizing agents for dissolution DNP because of their efficiency, stability, and wide substrate compatibility [158]. These are substituted derivatives of the parent triphenylmethyl (trityl) radical, also known as Gomberg's radical. The substituents were designed to eliminate the hyperfine coupling of the unpaired electron to the aromatic proton spins, to increase the chemical stability, and to improve not only the solubility but also the removability by precipitation. Of these, the most commonly used ones are the OX063, Finland, and AH111501. The last one is used in the SPINlab to polarize pyruvic acid and is known as EPA ("electroparamagnetic additive" or "electron paramagnetic agent" in the SPINlab terminology). These radicals have the same hydrophobic core but differ in the 12 peripheral groups on the methylene carbons between the S-atoms. These groups have a great influence on solubility and toxicity. OX063, with its 12 hydroxyethyl groups, is water-soluble in a wide pH range and has low toxicity (LD50 = 8 mmol/kg); consequently, it is the preferred choice for in vivo experiments [191]. The Finland radical is much more hydrophobic and has higher toxicity, likely due to its stronger interaction with proteins [192]. In general, tris(p-carboxyltetrathiaaryl)methyl radicals slowly decay in tissues with half-lives ranging from 1 to 24 h, depending on the radical and the conditions [193,194]. Under aerobic conditions, the radicals undergo oxidative decarboxylation by NADPH, or glutathione, and O_2 to produce quinone-methide derivatives, while under anaerobic conditions they are reduced to triarylmethanes [195,196]. Trityl AH111501 is closely related to OX063, but its solubility is very low in aqueous acid solutions (less than 3 μM). It is used for human studies with pyruvate and precipitates out of the hyperpolarized pyruvic acid solution below pH 4 and can be removed by filtration. One limitation of these very efficient DNP polarizing agents is that, in spite of the several synthetic procedures published in the open literature [195,197–201], the synthesis of these radicals is rather costly, and consequently, they have limited commercial availability.

OX063 self-assembles into dimer capsules held together by hydrogen bonding at neutral pH levels above 1–2 mM concentrations. These can form supramolecular inclusion complexes guests, such as tetramethylammonium cation and choline, but not pyruvate. The formation of an inclusion complex with choline enhanced the nuclear relaxation rates, but this unfavorable effect was reduced in the presence of tetramethylammonium chloride, which decreased choline binding to OX063 due to competition. This was also reflected in the improved ^{15}N polarization of choline in the presence of tetramethylammonium chloride [202].

Polychlorinated trityl radicals, also named perchlorotriphenylmethyl (PTM) radicals [203], can also polarize ^{13}C, although they seem to be somewhat less efficient than the commonly used tris(p-carboxyltetrathiaaryl)methyl radicals [204]. It was suggested that electron spin polarization is first transferred to chlorine nuclei, followed by spin diffusion-mediated polarization transfer from the chlorine to ^{13}C [205]. It is also interesting to note the g-tensors of PTM radicals have larger anisotropies than TAM radicals, with $g_x = 2.0034$ and 2.0041, $g_y = 2.0030$ and 2.0023, and $g_z = 2.0023$ and 2.0017 for Finland and PTMTC radicals, respectively [206].

Another class of carbon-centered radicals, BDPA and its derivatives are much more easily available than the trityls. BDPA is an intriguing polarizing agent [207]. Considering its EPR linewidth, it should provide superior results to trityls in DNP. This is, unfortunately, not always the case. BDPA is commercially available as a 1:1 complex with benzene, and its purity must be evaluated before use. Although BDPA can produce the same or slightly higher ^{13}C and ^{15}N polarizations as trityl OX063, the optimal concentration is 2–3-fold higher than it is for the trityls. In addition, BDPA does not respond as well to Gd(III) or Ho(III) doping [208]. The parent radical BDPA is insoluble in water and most polar solvents, including pyruvic acid, but dissolves sufficiently in the dipolar aprotic solvent sulfolane for DNP (over 100 mM). Sulfolane is a crystalline solid at room temperature (m.p. = 27.5°C), but it is glass-forming at cryogenic temperatures when mixed with other solvents. BDPA has been used as a polarizing agent for a wide variety of nuclei in sulfolane-based glassing matrices [101]. Substituted BDPA radicals bearing sulfonate, carboxylate, or quaternary ammonium groups display improved solubility in aqueous media, while hydroxymethyl and aminomethyl substituted BDPA are soluble in neat pyruvic acid

without having to add sulfolane [117,209–212]. The aminomethyl derivative also offers the advantage of a very fast polarization buildup time versus BDPA [117].

Despite their large EPR linewidths, nitroxides as well as other radicals such as DPPH and galvinoxyl, as well as certain paramagnetic metal complexes, can also be used to directly polarize ^{13}C and other nuclei [166,168,213–217]. Nitroxides have the merits of being cost-effective, commercially available, and having low toxicity. However, because of their broad-spectrum linewidth (more than four times the trityl one), nitroxides are poor polarizing agents for direct polarization of low-gamma nuclei such as ^{13}C or ^{15}N. Compared to trityl radicals, they produce a lower polarization percentage but in much shorter buildup times. On the other hand, nitroxides are good agents for ^{1}H polarization, and the CP technique can be used to transfer the large polarization of ^{1}H to ^{13}C [146,149]. Of note, a polychlorinated trityl – TEMPO conjugate biradical was shown to be more efficient in polarizing [2-^{13}C]acetone at 3.35 T and 1.4 K than OX063 or the physical mixture of the two components, likely due to dipolar interactions between the unpaired electrons [218,219]. EPR measurements of the BDPA-TEMPO conjugate revealed a strong J-coupling interaction between the two radical moieties, and it was proposed that this biradical may polarize 13C via the CE [219].

1.5.3 DNP via photoexcited triplet states

Nonthermalized electron spin polarization of photoexcited triplet states of aromatic molecules such as fluorine and pentacene can be transferred to nuclear spin polarization by microwaves. This approach can offer some advantages over conventional DNP [220,221]. The electron spin polarization in photoexcited electronic triplet states is determined by the selection rules for intersystem crossing (ISC) of the excited singlet state to the lowest triplet state and not by thermodynamic equilibrium at a given temperature and magnetic field [222,223]. Therefore, large nuclear polarization (up to 80%) can be achieved by the microwave-driven transfer of the electron spin polarization of the triplet state to nuclear spins at ambient temperatures and low magnetic fields. Because the paramagnetic state exists only during photoexcitation, the nuclear spin relaxation is not affected by the presence of paramagnetic species. Most commonly, pentacene-d_{14} dopant in naphthalene, p-terphenyl or p-terphenyl-2′,3′,5′,6′-d_4 crystal, is used for photoexcited triplet state DNP experiments [224–227]. A highly polarized photoexcited triplet state of deuterated pentacene is generated by a pulsed laser, and the electron polarization is transferred to ^{1}H spins present in the naphthalene or p-terphenyl by applying a microwave pulse [225,227–229]. For practical applications, these highly ^{1}H spin-polarized crystals must be dissolved, and the polarization must be transferred to other molecules. The feasibility of dissolving the polarized crystals and transferring the polarization to target molecules has recently been demonstrated [230]. Optically polarized naphthalene crystals (20%–25% ^{1}H spin polarization) were rapidly dissolved in $CDCl_3$ solutions containing propargyl acetate, 1,1,2,2-tetrachloroethane, dichloromethane acetone, or THF. ^{1}H enhancements of the solutes ranging from −200 to −1730 at 1.45 T were observed as a result of the intermolecular nuclear OE. This method is not limited to ^{1}H but can conceivably be extended to polarizing ^{13}C and ^{15}N via direct polarization transfer or the application of an appropriate pulse sequence such as INEPT [231,232].

1.5.4 Removable polarizing materials

The paramagnetic polarizing agents speed up the spin-lattice relaxation of the hyperpolarized magnetization, notably during the transfer step leading to severe loss of polarization and can have other unwanted effects such as toxicity. Therefore, it is often desirable to remove them before injecting the hyperpolarized solution. For example, the removal of the EPA is required for human applications. Removal of the polarizing agent is generally performed during the dissolution step. This can be achieved by several methods. Hydrophobic radicals can be employed and then extracted into the organic phase when solvent mixture such as water/toluene is used [132]. Other radicals such as trityl, BDPA, DPPH and galvinoxyl can be removed by filtration through an in-line filter after they precipitate out of the dissolution liquid. Nitroxides, and possibly other radicals, can be chemically quenched with L-ascorbate during the dissolution step [233]. A trityl derivative (EPA) has been specifically designed to extensively precipitate while pH changes during the dissolution step.

An alternative approach is based on separating the polarizing agents from the solute to be polarized. The method involves the use of hybrid polarizing solids (HYPSOs), in which the radical (TEMPO, 2,2,6,6-tetramethyl-piperidine-1-oxyl) is homogenously distributed and covalently linked within the pore channels of SBA-15 (Santa Barbara Amorphous-15) mesostructured silica. This type of ordered mesoporous silica has a honeycomb-like structure made of 8–10 nm diameter channels. The nitroxide polarizing agents are linked to the silica wall through a propylamide (HYPSO-1) or 1,2,3-triazole-propyl linker (HYPSO-2) (Fig. 1.8). Later, SBA-16 type mesoporous silica was also used as a solid matrix (HYPSO-3). This type of silica has a three-dimensional cubic network of interconnected pores [234,235]. HYPSO-3 was synthesized by functionalizing the silica with TEMPO via a 1,2,3-triazole-propyl linker. This preparation produced a

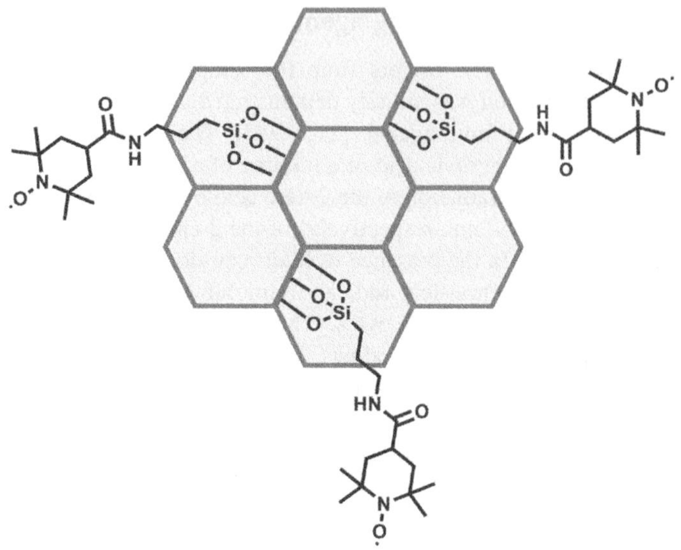

FIGURE 1.8 Structure of HYPSO-1 (hybrid polarizing solid) based on SBA-15 mesoporous silica in which the radical is covalently attached to the silica surface through a propylamide linker.

somewhat higher solid-state ^{1}H polarization than the previous versions, likely due to the more uniform spatial distribution of the radical centers. The three-dimensional network of pores may also allow unrestricted spin diffusion in all three spatial dimensions [236]. Other HYPSO variants with improved performance were also developed; these include HYPSO-W and HYPSO-5. HYPSO-W1 and W2 are mesoporous silica in which the nitroxide radical is incorporated into the wall rather than attached to the surfaces in HYPS 1 and 2 [237]. HYPSO-5 consists of surface-coated silica nanoparticles. It is synthesized by modifying the surface of commercially available mesoporous silica nanospheres (SiliaSphere) with a silica layer incorporating 3-azidopropyl groups. The nitroxide radical is then conjugated to the surface of the nanoparticle using copper-catalyzed alkyne-azide cycloaddition click chemistry. DNP with HYPSOs can be performed conventionally, and the polarizing agent settles out of the hyperpolarized solution after dissolution. HYPSO-5 exhibited the best performance characteristics for DNP among all HYPSO materials. It produced 70% ^{1}H and 51% ^{13}C spin polarizations (after CP) in a 3 M solution of [1-^{13}C]acetate in H_2O/D_2O (2:8) at 6.7 T and 1.2 K [238].

A similar approach uses macroporous polymers (HYPOPs) rather than mesostructured silica as a framework to which the paramagnetic centers are conjugated. These materials are structured epoxy resins synthesized by reacting the diglycidyl ether of bisphenol A, isophorone diamine, and 4-amino-TEMPO in the presence of polypropylene glycol. Depending on the reaction conditions and the mole ratios of the components, various morphologies were obtained with porosities ranging from 50 nm to 1 μm. HYPOP materials were successfully used to polarize [1-^{13}C] glycine [1-^{13}C]sodium formate and 0.9M [1-^{13}C]sodium acetate by ^{1}H to ^{13}C CP in 90% D_2O and 10% ethanol-d_6 glass inside the HYPOP matrix. Modest liquid-state ^{13}C polarizations were recorded after the removal of the agent by filtration [239].

Finally, it is worth mentioning that free radicals covalently conjugated to thermoresponsive polymers and hydrogels were also considered removable polarizing agents. A thermoresponsive hydrogel was prepared by copolymerizing N,N-ethylmethylacrylamide (thermoresponsive unit), 4-methacryloyloxybenzophenone (photocrosslinkable unit), and methacrylic acid (spin labeling unit), followed by cross-linking with UV irradiation. The polymer was functionalized with amino-TEMPO after activating the methacrylic carboxyl groups in the form of pentafluorophenyl ester, achieving up to 15% spin labeling of the carboxyl groups. This formulation was used to polarize water in room-temperature Overhauser DNP and ^{13}C-labeled *tert*-butanol in low-temperature dissolution DNP. In the swollen state, the hydrogel provided a radical concentration of only 6.8 mM, which is nearly an order of magnitude lower than the optimal concentration. When the DNP sample was brought into contact with superheated water, the hydrogel rapidly collapsed and expelled the polarized *tert*-butanol from the hydrogel network. It should be noted that this particular hydrogel is not compatible with acids or aqueous solutions of salts [240,241].

1.5.5 Photoinduced labile radicals as polarizing agents

Another approach to eliminate the paramagnetic centers from the sample after the DNP process is to generate non-persistent, thermally labile radicals. The method was initially demonstrated using pyruvic acid as a radical precursor [242], and thereafter other precursors have been investigated [243–245] The principle is simple and is based on the UV-irradiation of a frozen sample containing pyruvic acid or a mixture of a ketoacid with an analyte. The photogeneration of radicals from pyruvic acid is based on the irradiation of the 2-keto acid by UV light involving the photoexcitation of the π-π* and n–π* transitions (170 and 300–350 nm, respectively) of the 2-carbonyl group, followed by efficient ISC to an excited triplet state [246,247] (Scheme 1.1). In the presence of hydrogen-donating molecules, the excited carbonyl group abstracts a hydrogen atom and gives rise to transient radicals immobilized in the frozen matrix, which annihilate by dimerization or cross-coupling when the temperature rises. The conditions (pH, solvent, and temperature) strongly influence the course of the reaction and the detectable products. When the irradiation of the 2-ketoacid (neat or in D_2O) is performed at 77 K (liquid nitrogen), the recombination products are exclusively CO_2 and the carboxylic acid derived from the 2-ketoacid. X-band (9 GHz) EPR studies revealed a quadruplet signal indicating that the prevalent transient radical species is very likely the C2 carbon-centered radical (ketyl radical) formed from excited pyruvic acid by hydrogen abstraction [242,248]. This photochemical decarboxylation is general for 2-ketocarboxylic acids, and a plausible mechanism is shown in Scheme 1.1. Studies on the photoinduced decarboxylation of pyruvic acid showed that the electron transfer from the triplet state was mediated via hydrogen bonds with pyruvic acid hydrate [249]. Lactic acid was not produced in significant amounts, indicating that most of the ketyl radical is converted back to pyruvate. In any case, accumulation of lactic acid in the reaction mixture is not expected because H-abstraction occurs very easily from a 2-hydroxy acid. However, other products may also form under different conditions [250,251].

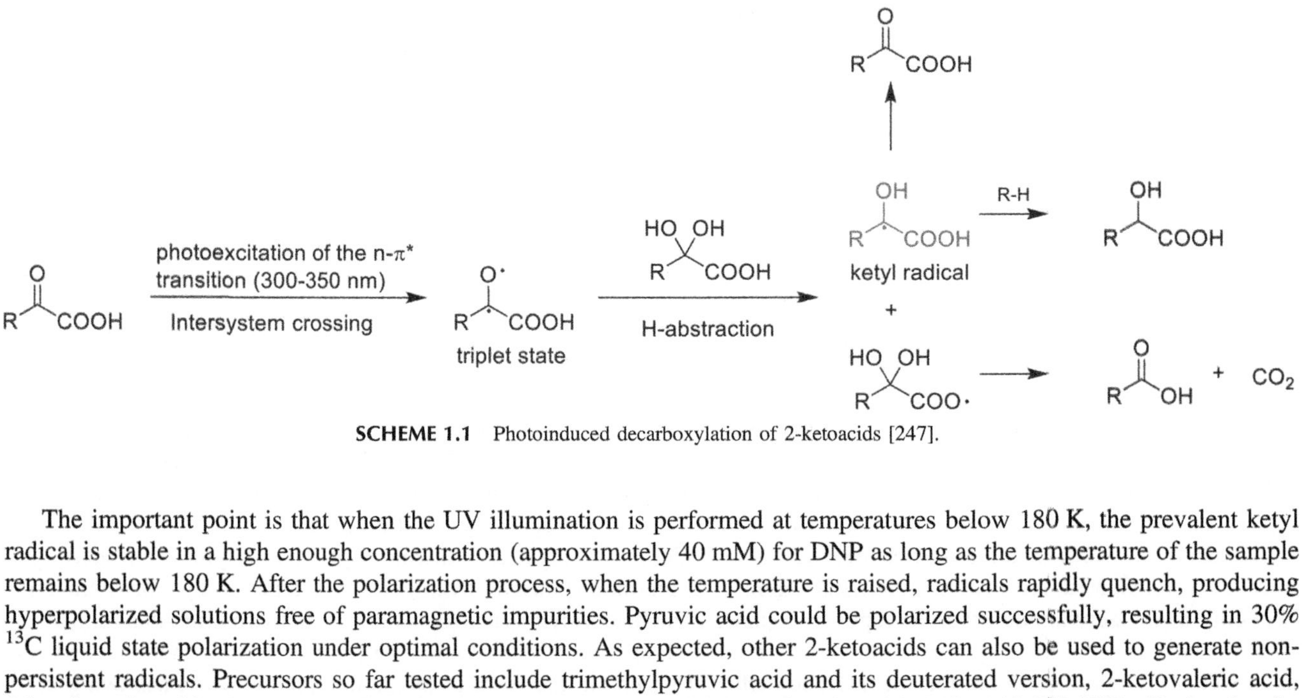

SCHEME 1.1 Photoinduced decarboxylation of 2-ketoacids [247].

The important point is that when the UV illumination is performed at temperatures below 180 K, the prevalent ketyl radical is stable in a high enough concentration (approximately 40 mM) for DNP as long as the temperature of the sample remains below 180 K. After the polarization process, when the temperature is raised, radicals rapidly quench, producing hyperpolarized solutions free of paramagnetic impurities. Pyruvic acid could be polarized successfully, resulting in 30% ^{13}C liquid state polarization under optimal conditions. As expected, other 2-ketoacids can also be used to generate non-persistent radicals. Precursors so far tested include trimethylpyruvic acid and its deuterated version, 2-ketovaleric acid, 2-ketobutyric acid, phenylglyoxalic acid, 2-isocaproic acid, and 2-ketoglutaric acid [243–245,252,253]. Obviously, the most advantageous use of nonpersistent radicals involves the polarization of their precursors because only a negligible amount of decarboxylated product will be present as a contaminant in the room-temperature hyperpolarized solution. For example, when [1-^{13}C]pyruvic acid is polarized with a photogenerated radical, only a trace amount of nonlabeled acetic acid is formed. However, UV-induced radicals can also be used to polarize substrates other than their own precursors. In this case, the precursor has to be added to the DNP sample in sufficiently high concentrations (often in the molar range) to produce 20–40 mM of radicals at 77 K [248,254]. Additionally, efficient hyperpolarization of deuterated [U-$^{13}C_6$] glucose was achieved using trimethylpyruvic acid as a radical precursor, providing 32% ^{13}C polarization in the liquid state after dissolution and transfer [244]. The achievable solid-state polarization seems to be primarily dependent on the radical concentration generated by UV irradiation. Trimethylpyruvic acid and its methyl-perdeuterated derivative were shown to

be superior to pyruvic acid because they offered higher transient radical concentrations and more favorable EPR characteristics upon UV irradiation. Unfortunately, unlike pyruvic acid, trimethylpyruvic acid is not a natural metabolic substrate, and its safety profile has not been investigated. On the other hand, pyruvic acid is safe, but it needs to be present in fairly high concentrations (up to 30% of the DNP mixture), and therefore, it may interfere with metabolism. Trimethylpyruvic acid, however, has the advantage that it is not involved in any metabolic pathway [255]. 2-Ketoglutaric acid proved to be more efficient, capable of producing nearly 150 mM radical concentration in an irradiated DNP sample containing only 300 mM of the precursor in [1-^{13}C]lactic acid:water (80:20 w/w). About 40% solid-state ^{13}C polarization was achieved after 2 h of polarization time at 7 T and 1.35 K.

Inspired by the photoinduction of transient radicals, other methods of generating transient radicals have also been explored. It was shown that radicals can be generated in solid samples of silica, sucrose, mannose, and glucose by exposure to electrical discharge in a dilute atmosphere. Upon dissolution, these radicals are annihilated, yielding the starting materials. These plasma-induced radicals can be used as polarizing agents in DNP [256,257]. Radicals were generated in glucose powder by exposure to low-pressure plasma for a duration of 60 s. DNP was performed at 20 and 50 K and 3.34 T by microwave irradiation. Proton signal enhancements of up to 150-fold in the solid-state were obtained under these conditions [257].

1.5.6 Transporting hyperpolarized materials

The combination of MRI and dissolution DNP for ^{13}C hyperpolarization of metabolites or probes is an excellent technique for tracking tissue-specific abnormal metabolism in real-time and noninvasively. However, dissolution DNP instrumentation is expensive, technically demanding, and requires being installed in close vicinity of a scanner. Large ^{13}C polarization can be achieved and maintained in the polarizer at cryogenic temperature and in the presence of a static magnetic field, even in the presence of radicals. However, radical-induced nuclear spin relaxation becomes prohibitively fast when the sample is in the liquid state and/or in a low magnetic field. Unfortunately, the hyperpolarized samples experience these conditions during the dissolution and transfer steps, and the hyperpolarization lifetime in the sample is only minutes after dissolution and extraction from the polarizer, prohibiting the storage and transport of the hyperpolarized sample. The batchwise operation of current dissolution DNP instruments and the inability to remove polarized samples from the polarizer for storage, transport, and late use have been severe limitations of the technology.

While the T_1 relaxation can be slowed to some extent by deuteration, removal of the paramagnetic polarizing agent, and the proper choice of external magnetic field strength, ^{13}C T_1 values rarely exceed 2 min, and the room-temperature hyperpolarized solutions cannot be transported for longer than about a minute without substantial loss of signal intensity due to the inexorable decay of the hyperpolarized magnetization in the liquid state, making the transport of hyperpolarized solutions unfeasible.

As the solid-state T_1 values at cryogenic temperatures are several orders of magnitude longer than the room-temperature liquid-state T_1s, there have been some attempts to extract the frozen polarized samples after DNP. This approach, however, has not been successful with conventional DNP samples containing free radical polarizing agents (trityls or nitroxides) because the presence of paramagnetic species leads to dramatic shortening of the ^{13}C T_1 in the frozen sample at low field, primarily mediated by electron-nuclear dipolar interactions [258,259]. An interesting experimental approach used TEMPOL (4-hydroxy-2,2,6,6-tetramethylpiperidin-1-oxyl) benzoate as a polarizing agent dissolved in a glass-forming solvent mixture [40 mM in toluene-d_6:THF-d_8:THF (8:1:1)] to overcome this problem and to produce a transportable hyperpolarized form of [1-^{13}C]pyruvate [119]. Before polarization, sodium [1-^{13}C]pyruvate was ground into a microparticulate powder and impregnated with the polarizing solution, followed by DNP of proton spins and then ^{1}H to ^{13}C CP at 1.2 K and 6.7 T. The main advantage of this approach is that the polarizing agent is physically separated from the substrate on a micrometer scale, at which the electron-nuclear dipole-dipole interaction is negligible, thereby creating a "trapped" hyperpolarized ^{13}C state in which the ^{13}C T_1 was significantly increased. This allowed the transfer of hyperpolarized micropowder out of the polarizer and storage in liquid helium for later dissolution. This approach was extended to other hydrophilic substrates that were not soluble in the polarizing solution. Nevertheless, hyperpolarized samples stored at 4.2 K and 1 T for 16 h showed significant loss of ^{13}C polarization (around 1%–2% ^{13}C polarization remaining), which could be ameliorated somewhat by storage in a higher magnetic field (at 6.7 T, the polarization loss in sodium [1-^{13}C] pyruvate was only about 40% over 16 h).

A conceptually different method for transportable hyperpolarized samples takes advantage of the nonpersistent nature of UV-generated transient radicals. Thermal annihilation of the photoinduced radicals above 190 K allows the production of radical-free hyperpolarized frozen samples in which the nuclear T_1 shortening effect of the paramagnetic polarizing agent is no longer present [242,244,248,260,261]. This has been demonstrated for [1-^{13}C]pyruvic acid and

[U-^{13}C, d_7]-D-glucose with deuterated trimethylpyruvic acid as UV-induced radical precursors. The DNP experiment was performed using a custom-made fluid path that was designed to be a reusable, self-contained device. A solution of [U-^{13}C, d_7]-D-glucose (2.2 M) in glycerol:water 1:1 was prepared, and deuterated trimethylpyruvic acid was added (10%). A DNP sample was prepared by freezing 6 μL droplets of this solution. The frozen beads were transferred into a quartz Dewar filled with liquid nitrogen. The beads were then irradiated with a high-power UV light source for 300 s and then transferred into a sample vial. The vial was attached to the custom-designed fluid path and loaded into the polarizer while maintaining cryogenic temperatures. The ^{13}C polarization was performed at 6.5 T under optimal conditions as previously determined [244]. The thermal quenching ("thermalization") of the radicals was carried out by lifting the sample above the liquid He bath and blowing He gas on it. The thermalization parameters (blowing time, blowing pressure, quenching position, and waiting time) were carefully optimized in order to annihilate most (>99%) of the radicals but to preserve as much ^{13}C polarization as possible; the thermalization caused about 20% loss of ^{13}C polarization for glucose. Quenching resulted in a dramatic increase of ^{13}C nuclear spin-lattice relaxation time. However, magnetic fields weaker than 40 mT caused very rapid ^{13}C relaxation in the frozen sample, even in the absence of paramagnetic species. To avoid low-field relaxation effects, a magnetic rail consisting of permanent magnets arranged in Halbach arrays was added to the DNP probe to maintain the magnetic field above 100 mT throughout the extraction step. The sample was quickly removed from the polarizer, placed into a precooled transport device, and stored in liquid nitrogen. It was then transported to an NMR spectrometer about 50 m away, where it was rapidly dissolved in a hot dissolution buffer using a compact dissolution station head to obtain an HP solution of glucose, which was injected into a 5 mm NMR tube. Liquid-state NMR spectroscopy revealed that about 9% ^{13}C polarization remained after transport after optimizing each step. This value represented about 1/3 of the initial ^{13}C polarization (about 30%). As the polarization loss was minimal during the thermalization, the main reason for the loss of polarization is likely the relatively short solid-state ^{13}C T_1 during storage and transport (1 T, 77 K). Under these conditions, the T_1 was estimated to be around 5 min, which is significantly shorter than it is at 4 K and 1 T (about 4 h) [262]. Further improvements included the use of a transportable liquid He cryostat equipped with a 1 T Halbach magnet rather than a liquid nitrogen bath. A [U-^{13}C, d_7]-D-glucose sample polarized with photoinduced radicals generated from deuterated trimethylpyruvic acid could be transported over a 320 km distance from Copenhagen to Aarhus in this liquid He cryostat. The journey took 4 h, over which time the 30% residual solid-state ^{13}C polarization after thermalization dropped to 0.55%. The relatively low ^{13}C polarization was attributed to the presence of nondeuterated trimethylpyruvic acid in the sample, which affected the glucose ^{13}C T_1 values in the frozen sample via methyl-group rotation. These experiments demonstrate that transport and storage of frozen hyperpolarized samples iare feasible under optimized conditions (absence of paramagnetic impurities, cryogenic temperatures, and magnetic fields >1 T). However, all parameters influencing the thermalization, extraction, transport, and remote dissolution must be carefully optimized [263].

1.6 Relaxation of the hyperpolarized magnetization in the liquid state NMR

The nuclear spin system is not isolated from its environment. Once a nonequilibrium spin polarization is generated, either by a change in the magnetic field or temperature, the application of a radiofrequency pulse, or polarization transfer, it will return to the thermodynamic equilibrium state. The process of energy exchange between the nuclear spin system and the environment is called nuclear relaxation, and it involves two principally different relaxation processes [8,264,265].

Spin-lattice (also known as longitudinal, or T_1) relaxation involves energy exchange between the nuclear spin system and its lattice (environment) to reach a state of thermal equilibrium. The major source of longitudinal relaxation in the liquid state is fluctuating local magnetic fields. The change of the longitudinal (along the z-axis) component of the magnetization in time is described by the differential Eq. (1.14) and its solution (Eq. 1.15), showing that longitudinal relaxation involves the exponential change of M_z to reestablish thermodynamic equilibrium. The rate of magnetization change is characterized by the time constant T_1 [266]. The inverse of T_1 is the relaxation rate ($R_1 = 1/T_1$).

$$\frac{dM_z}{dt} = -\frac{M_z - M_0}{T_1} \tag{1.14}$$

$$M_z(t) = M_0\left(1 - e^{-t/T_1}\right) \tag{1.15}$$

As we will discuss it later, the major source of local magnetic field fluctuations is the isotropic tumbling of molecules in solution (Brownian motion). In an analogous manner, the decay or dephasing of the transverse (xy component) of magnetization is characterized by the transverse, spin-spin, or T_2 relaxation time. Spin-spin relaxation is due to temporary and random spin-spin interactions in the transverse plane. The main cause of dephasing is slight differences among the

Larmor frequency of the spins as a result of interactions with the lattice, spin-spin interactions, and magnetic field inhomogeneities. This will lead to slightly different angular velocities in the transverse plane among the preceding individual spins and a loss of phase coherency over time. This dephasing results in the progressive decay of the transverse component of magnetization (M_{xy}) [266]. The relaxation of the transverse component of the magnetization can be described by Eqs. (1.16) and (1.17). In general, T_2 decay is always faster than predicted by molecular mechanisms due to the magnetic inhomogeneity of the B_0 field, which induces a field distortion in the vicinity of the molecules within the field. Overall, this magnetic inhomogeneity effect together with other effects is characterized by the effective spin-spin relaxation time constant (T_2^*) and has great significance in MRI. T_1 and T_2 relaxation occur independently but simultaneously.

$$\frac{dM_{xy}}{dt} = -\frac{M_{xy}}{T_2} \tag{1.16}$$

$$M_{xy}(t) = M_0 e^{-t/T_2} \tag{1.17}$$

1.6.1 Measuring the spin-lattice relaxation time in conventional NMR

There are several methods to measure the spin-lattice relaxation time. Conventionally, T_1 has been measured using inversion recovery [267,268], saturation recovery [269,270], or progressive saturation [271]. Faster Longitudinal relaxation Investigated by Progressive Saturation (FLIP) [272] is a newer approach (Fig. 1.9). In conventional NMR, the sample magnetization in B_0 field is recoverable: after the application of an excitation rf pulse, the M_z magnetization returns to its value determined by the spin polarization in B_0. The recovery of M_z magnetization is an exponential process characterized by a time constant, the spin-lattice relaxation time. Most T_1 determination methods are based on the fitting of the exponential recovery of M_z magnetization over time [272,273]. The standard technique is inversion recovery, which uses

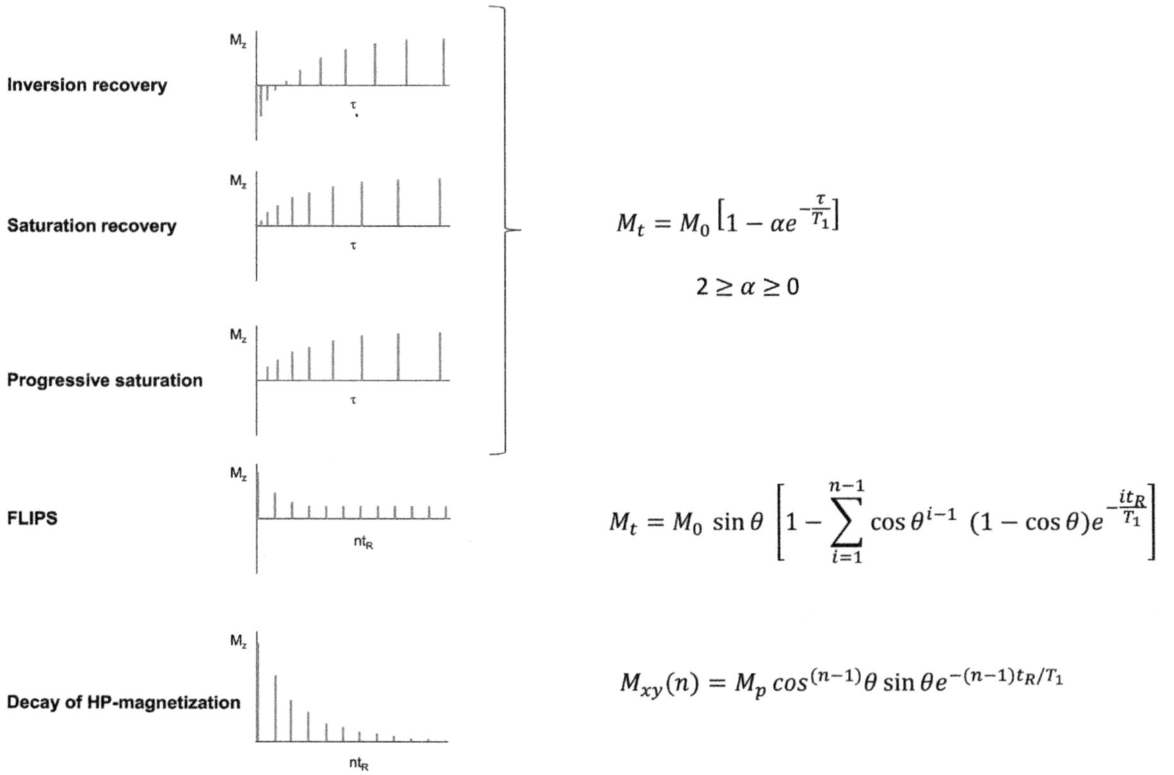

FIGURE 1.9 The evolution of magnetization in various methods used for T_1 estimation. Note that the intensities are not to scale. The hyperpolarized magnetization is generally several orders of magnitude stronger than the thermally polarized one. M_0 and M_p denotes thermal equilibrium and hyperpolarized magnetization, respectively. M_{xy} is the transversal magnetization. The NMR signal intensity is proportional to M_{xy}. Note that when θ flip angle pulse is applied to M magnetization ($M = M_0$ or M_p), the transversal and longitudinal magnetization will be Msinθ and Mcosθ, respectively. *Adapted from Ref. [272] with permission.*

repeated RD (relaxation delay) - $180°-\tau-90°$ pulse sequences. The 180 degree pulse inverts the M_z magnetization. Over τ time, the M_z magnetization is allowed to recover toward its equilibrium value and sampled with a 90° pulse. The experiment is repeated with increasing τ values [268]. A major disadvantage of inversion recovery is that the repetition time (relaxation delay) between consecutive 90-degree pulses must be at least 5 times the T_1 to allow complete relaxation of the spin system before the application of the next 180-degree pulse. The relaxation time T_1 is calculated by fitting the recovery of the magnetization versus time data. Although inversion recovery works well for measuring short T_1 values, it is very time-consuming to apply for spins that have a long T_1. Subsequently developed methods do not wait for the reestablishment of equilibrium magnetization, thereby eliminating the need for long relaxation delays. The saturation recovery method first saturates the magnetization by applying multiple 90 degree RF pulses. Then, after τ delay, a 90-degree pulse is applied, and the signal is recorded. This is repeated with increasing τ values. The relaxation time T_1 can be calculated from the recovery of M_z as a function of time until it reaches its equilibrium value [269]. Progressive saturation applies a repetitive sequence of 90° pulses separated by τ to reach steady-state magnetization, and then the FID is recorded. This is repeated with increasing τ values, and the effect of the 90-degree pulse trains on the signal intensity is determined to obtain a recovery curve from which T_1 is extracted [274]. The recently reported FLIPS method adopts a slightly different approach [272]. Rather than sampling the magnetization recovery curve by performing separate experiments with different τ, in FLIPS, the FID is recorded after each excitation pulse (90 degrees or smaller), which is repeated with increasing t_R repetition time. This results in a steady-state magnetization whose magnitude is determined by the flip angle, t_R, and T_1. The relaxation time can be extracted from the initial and steady-state signal intensities. It should be noted that the FLIPS technique requires complete dephasing of the M_{xy} component during the repetition time t_R. In addition, if signal averaging is used to improve the sensitivity, a relaxation delay longer than $5T_1$ must be included to ensure that the initial magnetization is fully recovered between consecutive acquisitions. Although the FLIPS method is about 10 times faster than inversion recovery and particularly well suited to estimate long T_1 relaxation times, it still takes several hours to measure T_1 values that are on the order of minutes.

1.6.2 Obtaining T_1 from the decay of hyperpolarized magnetization

Although in hyperpolarized NMR the magnetization decays while in conventional NMR the magnetization recovers after a rf pulse, both phenomena involve changes of magnetization in the z-direction and, therefore, are characterized by the same spin-lattice relaxation time. The lifetime of hyperpolarized magnetization in the liquid state is directly related to T_1 relaxation time. After the dissolution of the polarized sample, the hyperpolarized magnetization spontaneously returns to the thermal equilibrium level at a rate determined by the T_1 relaxation time. Virtually no HP signal can be detected after about 5 T_1. The spontaneous decay of HP magnetization offers a very convenient way to measure T_1, and in light of the above discussion, the rapidity of determining long T_1 values this way should be greatly appreciated. The T_1 can be estimated by sampling the decaying magnetization with low-flip-angle pulses. It should be kept in mind that the hyperpolarized magnetization is nonrecoverable, and the applied pulse destroys some of it as determined by the flip angle. This effect must be taken into account in the fitting. The signal intensity at each time point is given by Eq. (1.18) and the decay of signal intensity over time can then be fitted to extract the T_1 (Fig. 1.9)

$$I(n) = I(n-1)\cos\theta e^{-t_R/T_1} \tag{1.18}$$

Where $I(n)$ is the signal intensity after each pulse and $I(n-1)$ is the signal intensity before the pulse, t_R is repetition time, and θ is the flip angle of rf pulse used to monitor the hyperpolarization signal. The term $\cos\theta$ refers to the magnetization destroyed by pulsing and $e^{-(n-1)T_R/T_1}$ denotes the loss of magnetization by T_1 decay [275–277]. To obtain accurate results, it is important to use a small flip angle to ensure that the signal decay is largely due to T_1 relaxation. In addition, care should be taken not to overfill the coil volume with hyperpolarized sample in order to avoid the diffusion of solution that was not exposed to the rf pulses into regions in the coil volume. It goes without saying that the flip angle must be carefully calibrated. It takes about 5 T_1 to collect the data (not counting the DNP time), which is less than 10 min for most ^{13}C spins. In addition, the hyperpolarized solution can be divided into two or more samples, thereby allowing simultaneous T_1 measurements to be performed in different fields if the instruments are available. An example of hyperpolarized magnetization decay is shown in Fig. 1.10.

1.6.3 Nuclear relaxation mechanisms

Spin-lattice relaxation is an energy exchange phenomenon between the spin system and its environment (lattice) as the spin system returns to thermodynamic equilibrium. The relaxation is brought about by weakly fluctuating magnetic fields in the

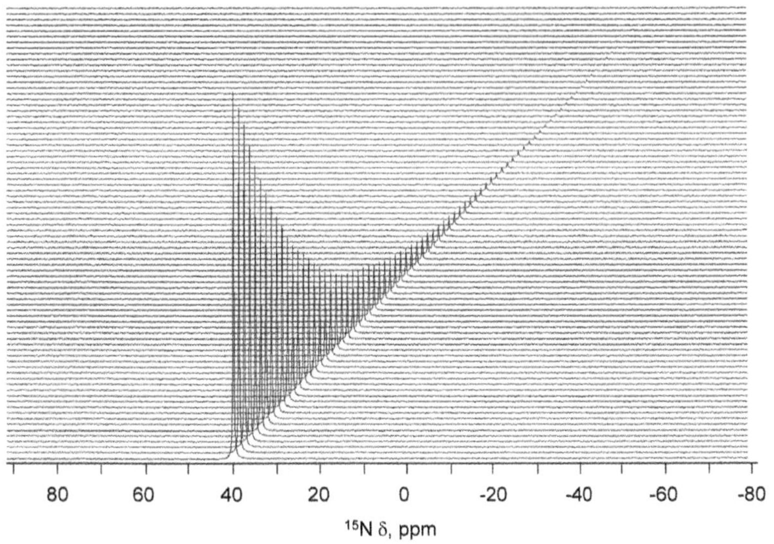

FIGURE 1.10 Sequential ^{15}N NMR spectra of a hyperpolarized sample showing polarization decay. Spectra were collected every 5 s using a 5 degrees pulse at 9.4 T. The fitting of the decay afforded a T_1 of 70 s.

lattice. There are several mechanisms that can generate fluctuating local fields. These include moving nuclear (or electron) dipoles and electric charges. A common feature of these effects is that they all generate local, transient B_{xy} magnetic fields that can rotate the magnetization vector. The fluctuations in the magnetic field are random, but because they are caused by molecular motions, they are a continuous function of time. There is an upper limit to how fast B_{xy} can change. Over very short time periods, B_{xy} can be considered to be nearly constant. Over longer periods, however, the B_{xy} values are uncorrelated. Statistically, the correlation of B_{xy} values at two time points separated by t time interval can be described by an autocorrelation function. Generally, the correlation decays exponentially following $e^{(-\tau/\tau c)}$, where τ_c is the correlation time characteristic for the molecular motion that generates the fluctuating magnetic field components [278].

The spectral density function obtained by the FT of the time—correlation function gives the probability of finding a B_{xy} time varying random magnetic field fluctuating at a given frequency [279–281]. The reduced (or normalized) spectral density function that is not dependent on the strength of the local fields is given by Eq. (1.19)

$$J(\omega) = \frac{2\tau_c}{1 + (\omega\tau_c)^2} \tag{1.19}$$

$J(\omega)$ has central importance in NMR relaxation theory as it expresses the contribution of random magnetic field fluctuations as a function of Larmor frequency (ω) and correlation time τ_c. $J(\omega)$ attains its maximum value when the frequency of the local fluctuating magnetic fields matches the Larmor frequency when ($\tau_c = \omega$).

1.6.3.1 *The dependence of* T_1 *and* T_2 *on molecular motions*

Theoretical treatment of nuclear relaxation using spectral density functions gives the following general formulas for the spin-lattice (longitudinal) and spin-spin (transversal) relaxation rates for homonuclear spins [4,7,8,10,281].

$$R_1 = \frac{1}{T_1} = C\left(\frac{2\tau_c}{1 + (\omega_L\tau_c)^2} + \frac{8\tau_c}{1 + (2\omega_L\tau_c)^2}\right) \tag{1.20}$$

$$R_2 = \frac{1}{T_2} = C\left(3\tau_c + \frac{5\tau_c}{1 + (\omega_L\tau_c)^2} + \frac{2\tau_c}{1 + (2\omega_L\tau_c)^2}\right) \tag{1.21}$$

Where C combines all field and temperature-independent factors (gyromagnetic ratios, vacuum permeability, spin numbers, Planck's constant). For heteronuclear spins, the formulas are similar but more complex.

A plot of T_i ($i = 1, 2$) versus τ_c using Eqs. (1.20) and (1.21) is shown in Fig. 1.11.

The different dependence of T_1 and T_2 relaxation times on the correlation time is very characteristic and has been known since the early days of NMR. The rationalization of nuclear spin relaxation in terms of molecular motions is known as the Bloembergen, Purcell, and Pound theory of nuclear spin relaxation [264,282]. Some of the predictions of this theory

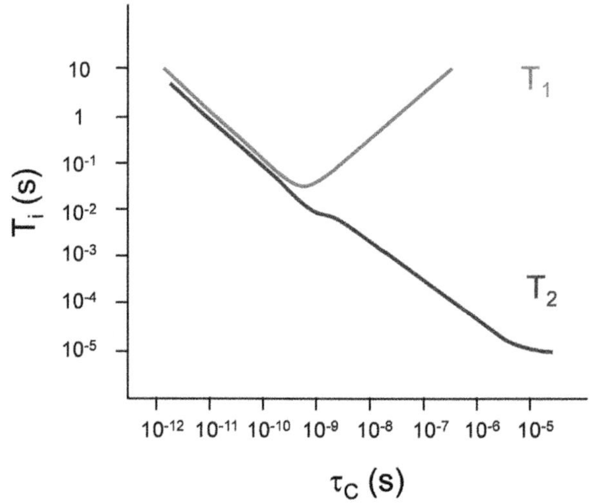

FIGURE 1.11 The dependence of T_1 and T_2 relaxation on the τ_c correlation time for homonuclear spin 1/2 nuclei relaxing by the dipole-dipole mechanism [265].

are discussed in the following: The τ_c is very short for small molecules in nonviscous solutions (when the tumbling rate is very fast) so that $(\omega\tau_c)^2 << 1$ (extreme narrowing limit). Under these conditions, $J(\omega) = 2\tau_c$, and $1/T_1$ is proportional to τ_c. On the other hand, at the opposite end of the time scale, when molecular motions are slow (τ_c is long), the spectral density function results in less efficient relaxation, and $1/T_1$ is proportional to $\omega^2\tau_c$. $1/T_1$ reaches a minimum in the intermediate region, where ω is comparable to $1/\tau_c$ [264]. In general, small molecules and macromolecules have longer T_1 than medium-sized molecules because the molecular rotation frequency of medium-sized molecules is more likely to match the Larmor frequency at commonly encountered magnetic fields. As the name implies, the common notion about T_2 (spin-spin) relaxation is that it only involves interactions between the spins and proceeds without energy exchange between the lattice and the spin system. However, the left side of the graph in Fig. 1.11 clearly shows that in the extreme narrowing regime, T_1 and T_2 follow the same trend, which strongly suggests that in this regime, both T_1 and T_2 relaxation are brought about by the same mechanism, i.e., by random local fields. In the extreme narrowing limit, T_2 can be nearly as long as T_1. As τ_c increases and becomes comparable to and then larger than the Larmor frequency on the right side of the graph, the effects of rotating fields due to the Larmor precession of neighboring nuclear spins increase. This causes loss of phase coherence and occurs most efficiently when the molecular rotation is very slow or nonexistent, as in the solid state [10,265]. Thus, T_1 relaxation time attains a minimum value at a particular rotational correlation time, while T_2 monotonously decreases with increasing τ_c [264].

1.6.3.2 Protein binding

The dependence of T_1 on the correlation time implies that the relaxation time can become very short when the hyperpolarized compound interacts with medium-sized molecules, and this can result in rapid loss of polarization. This is particularly relevant to interactions with plasma proteins because the hyperpolarized agent is directly injected into the bloodstream. The most abundant plasma protein is albumin. The main role of albumin is to regulate osmotic pressure and to transport fatty acids, steroids, and other compounds. Compounds that can reversibly interact with albumin include fatty acids, amino acids, and various aromatic and heteroaromatic "drugs" [283]. Albumin has two main drug-binding sites (Sudlow site I and Sudlow site II), which bind aromatic molecules with high affinity (K_a approximately 10^4 to $10^6\ M^{-1}$) [284,285]. The rotational correlation time (τ_c) of albumin is around 20 ns [286], which falls into the range where T_1 attains its minimum value in commonly encountered magnetic fields. Shortening of the T_1 and increased linewidth of the ^{13}C label in hyperpolarized [8-^{13}C]aspirin due to binding and exchange with albumin was observed both in vitro and in vivo (Fig. 1.12) [287]. These adverse effects were slightly diminished in the presence of a 3-fold excess of unlabeled acetylsalicylic acid.

The binding constant of aspirin to albumin is moderate ($K_a = 3.3 \times 10^6\ M^{-1}$ at 20°C) [288]. Stronger binding may affect the signal intensity even more dramatically. A good example is the Zn^{2+} sensor HP-[^{15}N]-tris(2-pyridylmethyl) amine (TPA). The partially deuterated molecule has a favorably long and largely field-independent T_1 value (about 70 s). The albumin binding constant of this compound is about an order of magnitude higher than that of salicylate, and in the

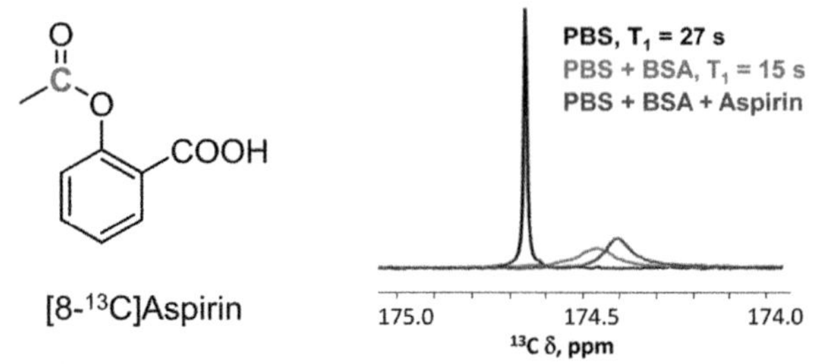

FIGURE 1.12 The effect of protein binding on the ^{13}C T_1 and T_2 relaxation of ^{13}C-labeled Aspirin (acetylsalicylic acid) at 7 T. Nonhyperpolarized ^{13}C-NMR spectra of ^{13}C-labeled Aspirin in PBS (black), PBS + BSA (red [light gray in print version]), and PBS + BSA with 3-fold excess unlabeled Aspirin (blue [gray in print version]) [287]. *Adapted from Ref. [287] with permission.*

presence of albumin, the hyperpolarized ^{15}N magnetization decayed very rapidly. However, the concept of competitive inhibition [289] was successfully applied to restore the HP-^{15}N signal (Fig. 1.13) [290]. Iophenoxic acid was used as a competitive displacer because it has an extraordinarily high binding constant to albumin ($K_a = 5.6 \times 10^7\ M^{-1}$) [291]. An equivalent amount of iophenoxate nearly completely restored the hyperpolarized ^{15}N signal of HP-[^{15}N]-TPA-d_6 in the presence of HSA or in human plasma (Fig. 1.13).

Perhaps it is worth mentioning here that increasing the τ_c of a gadolinium-based ^1H MRI contrast agent via albumin binding results in a dramatic increase in ^1H relativity and has been exploited in the design of angiographic agents and Zn^{2+} sensors [292,293].

1.6.3.3 Spin-lattice relaxation mechanisms

For hyperpolarized NMR, the most important relaxation mechanisms that lead to the relaxation of hyperpolarized magnetization include the dipole-dipole relaxation (R_d), paramagnetic relaxation (R_{para}), chemical shift anisotropy (R_{csa}), spin-rotation (R_{sr}), and scalar relaxation (R_{sc}) of the first, second, and-low field TM [294–296]. Each mechanism can be defined by a time-dependent Hamiltonian operator that denotes the probability of spin transition caused by the specific mechanism (Eq. 1.11) [295–300]. The relaxation rates are additive, as shown in Eq. (1.22).

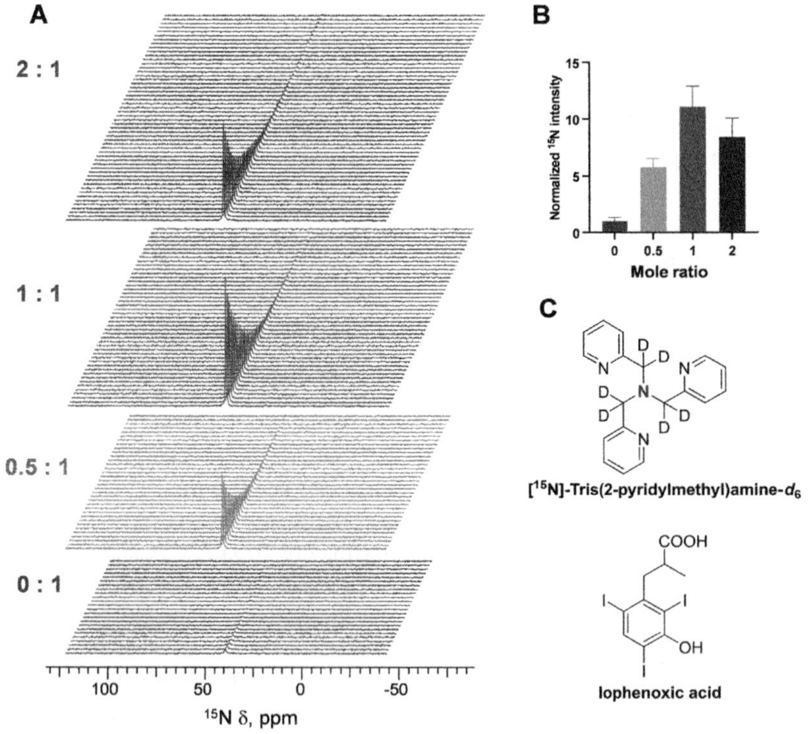

FIGURE 1.13 Iophenoxic acid as competitive displacer restores the ^{15}N NMR signal of HP-[^{15}N]-tris(2-pyridylmethyl)amine-d_6 in human blood plasma (HSA concentration is approximately 0.45 mM) at 9.4 T. (A) Dynamic ^{15}N NMR spectra recorded every 5 s with a 5 degrees flip angle. (B) A plot of ^{15}N signal intensity of the first spectrum of each array versus the mole ratio of iophenoxate to [^{15}N]-tris(2-pyridylmethyl)amine-d_6. (C) Structure of iophenoxic acid and [^{15}N]-tris(2-pyridylmethyl)amine-d_6. The ^{15}N atom is shown in blue (dark gray in print version). *Reproduced from Ref. [290] with permission.*

$$R_1 = \frac{1}{T_1} = \sum_i R_i = R_d + R_{\text{para}} + R_{\text{csa}} + R_{sr} + R_{Sc} \tag{1.22}$$

The individual contributions are temperature- and/or field-dependent. R_d is dependent on the temperature via the variation of the molecular correlation time, while R_{csa}, R_{para}, and R_{sc} are dependent on the applied external magnetic field. The chemical shift anisotropy (CSA) term increases at higher fields, whereas the paramagnetic relaxation rate shows a marked increase at lower fields. A complete quantum mechanical description of a general formalism of spin-lattice relaxation is also available [301].

If the paramagnetic relaxation mechanism is excluded, the relaxation of a hyperpolarized signal is predominantly driven by a dipole-dipole interaction and CSA at fields commonly encountered in spectrometers and scanners (higher than 1 T) [302]. The dipole-dipole relaxation is the magnetic interaction near nuclei with large magnetic moments, mostly ^1H in the molecules. The dipolar relaxation of a nucleus in a molecule depends on the distance between the nucleus and nearby nuclear spins, as described by Eq. (1.12). The dipolar relaxation is due to local oscillating magnetic fields generated by nuclear spins as the molecule tumbles in solution. The dipolar contribution can be determined from the nuclear Overhauser enhancement (NOE) [280,297,303−306].

$$1 / T_{1,dipolar} = \tau_c \gamma_i^2 \gamma_j^2 \hbar^2 / r^6 \tag{1.23}$$

where τ_c is the correlation time for rotational motion (molecular reorientation), and r is the distance between molecular probe nuclei (i) and adjacent dipolarly coupled nuclei (j). ^1H spins have a very strong dipolar effect on directly bonded ^{13}C and ^{15}N nuclei, so quaternary ^{13}C and ^{15}N atoms have longer T_1 values than protonated C−H or N−H atoms [307].

Another major mechanism of relaxation is the CSA, which is proportional to the molecular reorientation time and also dependent on the magnetic field [4,308]. The CSA contribution arises when the shielding tensor is anisotropic. The random molecular motions within the static filed B_0 can lead to a fluctuation in the local magnetic field that promotes the relaxation of the nucleus. CSA contributes to longitudinal relaxation in proportion to the magnetic field (B) and CSA ($\Delta\sigma$) (Equation 1.13) [309]. This effect is more dominant for nuclei with a low gyromagnetic ratio and wide chemical shift ranges. To minimize the CSA effect on longitudinal relaxation, the hyperpolarized probes need to have a more symmetrical structure at the atom of interest since symmetrical structures have lower contributions from CSA. The CSA contribution can be estimated from the dependence of the spin-lattice relaxation rate on B_0^2 [306,310].

$$1 / T_{1,CSA} = 2\tau_c \gamma_i^2 B_0^2 (\Delta\sigma)^2 / 15 \tag{1.24}$$

$$\Delta\sigma = \left[2\sigma_{zz} - \left(\sigma_{xx} + \sigma_{yy}\right)\right] / 3 \tag{1.25}$$

Considering dipole-dipole interactions and CSA effects on T_1 relaxation, low γ nuclei (such as ^{15}N and ^{89}Y) that do not have directly bound protons in highly symmetrical molecules generally exhibit very long spin-lattice relaxation times, occasionally over 1000 s.

1.6.4 Low-field relaxation mechanisms

After the dissolution of the hyperpolarized sample, the hyperpolarized solution is transferred to another magnet for signal acquisition. During transfer, the sample will be exposed to variable and/or low magnetic fields, which may accelerate the decay of the hyperpolarized magnetization via several relaxation mechanisms.

1.6.4.1 Relaxation effect of the free radical

The paramagnetic polarizing agent in the hyperpolarized solution can shorten the T_1 value in low fields. The relaxation effect of unpaired electron spins is dominated by translational motions in the proximity of the paramagnetic center. This effect increases below a critical magnetic field threshold (approximately 100 mT). A significant increase in the relaxation rate of ^1H and ^{13}C spins was observed in the presence of nitroxide radicals. However, trityl radicals seem to affect the T_1 to a lesser degree, presumably due to their larger size, which renders the distance of the closest approach between the electron and nuclear spins about twice as large compared to the nitroxides. This has a substantial effect on the paramagnetic relaxation because the paramagnetic contribution to the spin-lattice relaxation rate is inversely proportional to the cube of this parameter [311−313]. As one might expect, the formation of a supramolecular complex between the radical and the substrate increases the nuclear relaxation rates, as was demonstrated with choline and OX063 [202].

1.6.4.2 Low-field thermal mixing

When the magnetic field is low enough so that the difference between the Zeeman energies (Larmor frequencies) of two dipolarly coupled nuclear spins (e.g., ^{1}H and ^{13}C) becomes smaller than the resonance linewidths, the resonances will overlap, and energy can be exchanged between the two Zeeman energy reservoirs. This process is known as a low-field TM and involves mutual spin flips with energy conservation [34,314]. As a result, the nuclear spins establish a common spin temperature and reach polarization levels accordingly. From a thermodynamic perspective, in low-field TM, the cold (highly polarized) spin reservoir is in heat exchange with the hot (poorly polarized) spins and attains a common spin temperature. When one of the participating spins is ^{1}H and the other one is a low-gamma nucleus such as ^{13}C or ^{15}N, ^{1}H will dominate the TM because of its large specific heat (proportional to g2). This process can be used to transfer high ^{1}H polarization to low-sensitivity nuclei. However, it can also drain the polarization of the low-gamma nucleus if the ^{1}H undergoes rapid relaxation. Low-field TM is described in detail in the classic book of Abragam [4].

1.6.4.3 Scalar coupling as a source of T_1 relaxation

Scalar relaxation is a T_1 relaxation mechanism brought about by stochastic (time-dependent) modulation of scalar (J) couplings due to random transitions between the spin states of the coupled nucleus. Scalar relaxation of the first kind occurs as a result of chemical exchange, for example, protonation of the nucleus of interest. Examples include the protonation of ^{15}N in amines and ^{13}C carboxylate. The relaxation of a ^{13}C nucleus by the exchange of a coupled spin-active nucleus is given by Eq. (1.26), where J is the coupling constant and the τ_{sc} correlation time is the average lifetime of the chemical bond involved in the exchange [4].

$$R_{1,sc} = \frac{8\pi^2 J^2}{3}\left(I_x(I_x+1)\right)\frac{\tau_{sc}}{1+(\omega_{13C} - \omega_x)^2\tau_{sc}^2} \tag{1.26}$$

Scalar relaxation of the second kind occurs when the nucleus of interest is coupled to another nucleus that relaxes very rapidly, most often due to quadrupolar relaxation. Well-documented examples involve ^{13}C spins scalar-coupled to ^{79}Br, ^{81}Br, and ^{14}N nuclei [153,315–317]. The magnitude of scalar relaxation of the second kind is given by a formula that is identical to Eq. (1.27) except that the correlation time in this case is the T_1 of the rapidly relaxing nucleus. It is important to note that the contribution of scalar relaxations is significant only at low field, where the difference between the Larmor frequencies is small. This can be a problem for hyperpolarized NMR when the polarized sample passes through a low-field region. Fast liquid-state ^{13}C relaxation of hyperpolarized [5-^{13}C]glutamine was reported when the sample passed through a low-field (<0.8 mT) due to the quadrupole effect of ^{14}N coupled to the ^{13}C. This undesirable relaxation effect of the quadrupolar ^{14}N could be eliminated by either transporting the sample in a higher magnetic field or by replacing the spin 1 ^{14}N nucleus with the nonquadrupolar spin ½ ^{15}N [153].

1.7 Various nuclei subjected to DNP

1.7.1 ^{1}H

Protons can be polarized to a high degree using DNP or PHIP. Unfortunately, the T_1 of ^{1}H in most compounds is too short for imaging applications. However, as discussed earlier, high ^{1}H polarization can be transferred to other nuclei using CP techniques in the solid state. In some compounds where the ^{1}H spins are isolated from other protons, the ^{1}H T_1 can be surprisingly long, up to 2–3 min (in deuterated solvents), as shown in Table A.1 (Appendix). Examples include partially deuterated pyridine, pyridazine, pyrimidine, and pyrazine derivatives [318,319]. These compounds were polarized extremely well by signal amplification by reversible exchange (SABRE).

Pure water has a T_1 of around 3 s [320,321]. In D_2O, this can be extended to about 30–40 s [129,322], which makes it feasible to use hyperpolarized water for in vivo imaging. and the high ^{1}H polarization can be transferred to other nuclei using CP techniques. Room-temperature Overhauser DNP with nitroxide radicals is commonly used to produce hyperpolarized water, which can then be injected into animals, followed by imaging in a standard MR scanner [323–326]. It is also well suited for the continuous-flow ^{1}H NMR signal enhancement of water [323]. The method is very rapid (<1 s); however, the signal enhancement (ε) is usually low. Theoretically, in Overhauser DNP, ε is given by Eq. (1.27) [327,328].

$$\varepsilon = 1 - \rho f s \frac{|\gamma_s|}{\gamma_e} \tag{1.27}$$

Where $p, f,$ and s are the coupling, leakage, and saturation factors, respectively, and γ_S/γ_e is the ratio of the electron and nuclear gyromagnetic ratios, which is 659 for the proton. Overhauser DNP usually mediated by dipole-dipole interactions, and in this case, the value of p is 0.5 (or close to it). This brings the theoretically attainable enhancement to -328.5. However, the saturation factor s depends on the EPR characteristics of the radical, including the number of hyperfine lines in the EPR spectrum. For commonly used nitroxide radicals under complete saturation s will be 1/3. Thus, the maximum enhancement obtainable by nitroxide radicals under complete saturation ($s = 1/3$) in a concentrated radical solution (when $f = 1$, meaning that nuclear relaxation occurs only via dipole-dipole interaction with the free radical) will be -108.5.

Overhauser DNP also offers the exciting opportunity to directly hyperpolarize water in vivo after the injection of radicals, a technique known as proton-electron double resonance imaging (PEDRI) [329]. For Overhauser DNP to occur, the EPR transition of the radical must be saturated in vivo by microwaves, which can cause excessive heating in the tissues. To avoid harmful heat deposition in tissues, PEDRI is best performed with field cycling. The idea is based on the fact that the DNP process involves relatively long nuclear T_1 values that allow the increase of the B_0 magnetic field before the MRI signal is recorded [330,331]. In field cycling PEDRI, the EPR saturation and polarization transfer are performed in low field (<20 mT) to avoid tissue heating, followed by increasing the magnetic field for MRI [332]. It was shown that for in vivo studies, when the specific absorption rate cannot exceed a certain level, the optimal value of B_0 for polarization transfer ("evolution field") is around 5 mT [333]. On the other hand, it is well known that the sensitivity and resolution of MRI detection increase with B_0, which, therefore, should be as high as possible. Field cycling PEDRI instruments operating at B_0 detection fields of 59 and 450 mT have been reported [331,332]. While these experiments demonstrate the feasibility of performing in vivo DNP, it remains to be seen if PEDRI can be translated to human applications [334].

Water can be polarized to a much higher degree by low-temperature DNP using polarizing agents with broad EPR line, such as nitroxides, Cr(V)-EHBA, or UV generated radicals [141,145,166,335,336]. At high fields, the DNP of ^1H spins with nitroxide radicals is remarkably rapid and efficient. TEMPO very rapidly polarized ^1H spins to almost 100% spin polarization within just 20 min at 6.7 T and 1.2K. CP at low temperatures yields a polarization $P(^1H \rightarrow {}^{13}C)$ in excess of 70% within 20 min. CP from ^1H to ^{13}C spins afforded higher ^{13}C polarization levels than direct ^{13}C DNP in a very short time [145,146].

Water was efficiently hyperpolarized (over 60% solid-state ^1H polarization) using a polymer linked nitroxide as a polarizing agent synthesized by covalently attaching 4-amino-TEMPO to the terminal carboxyl group of the thermoresponsive polymer poly(N-isopropylacrylamide)-COOH (pNIPAM-COOH). The polarizing agent was termed FLAP (filterable labeled agent for polarization). DNP with FLAP does not require the addition of glassing agents because the polymer can be impregnated with the aqueous solution of the compound to be polarized. The solution must contain 1 M NaCl to keep the polymer in suspension after dissolution so that it can be removed by filtration [337]. The FLAP was also successfully used to polarize ^{13}C-labeled sodium pyruvate.

Hyperpolarized water has been used to rapidly generate angiographic and perfusion images in small and medium-sized animals [322,338]. Another interesting application of hyperpolarized H_2O and HOD in biomolecular NMR is based on the chemical exchange of the hyperpolarized ^1H spins with labile protons (amine and amide) of biomolecules. When the exchange rate k_{ex} is comparable to the ^1H T_1 of the exchange site (about 1 s^{-1}), then strong enhancements of the exchange site signal can be detected. A very interesting feature of these experiments is that the NMR signal of the heteroatom (such as ^{15}N) to which the exchanging proton is bound is also strongly enhanced. This spontaneous polarization transfer is due to the heteronuclear Overhauser effect caused by ^1H-^{15}N cross-relaxation. The maximum ^{15}N signal enhancement is determined by a complex interplay of several parameters, including the exchange rate, the ^1H-^{15}N cross-relaxation rate, the ^1H relaxation times of the water and exchange site protons, and the heteroatom (^{15}N) relaxation time [336,339].

The exchange mediated polarization transfer from hyperpolarized water can also be used to enhance the sensitivity of ^{2}D HMQC and water-ligand observed via gradient spectroscopY NMR experiments [129,340,341]. For example, [^{15}N,^1H]-HMQC experiments with hyperpolarized water were used to study the folding rate of a protein (RNase Sa) under different conditions [129,336,342–344].

^1H hyperpolarized probes can also be used to gain structural information on proteins. Polarization transfer from hyperpolarized ^1H spins of the protein-binding ligand to nearby ^1H and ^{13}C spins in the binding pocket can occur via NOE. Rapid Dissolution DNP NMR experiments can be performed to detect the site-specific resolution of these NOE signals using indirect selection based on the ^{13}C chemical shift. This was demonstrated by identifying the ^{13}C signals of the amino acid residue side-chains in the binding pocket of dihydrofolate reductase in HSQC-NOESY experiments performed after mixing hyperpolarized folic acid with the protein [345].

^1H Polarization transfer via NOE could also be observed from hyperpolarized dodecyl phosphocholine protons to a membrane protein inserted in a micelle upon exchange of the hyperpolarized lipid with thermally polarized micellar lipid components [346].

1.7.2 ^{6}Li

^{6}Li (7.5% natural abundance) is a spin-1 nucleus. However, its quadrupole moment is extremely small (Q < 8.5 kHz), and consequently, in many respects, it behaves like a spin½ nucleus. In particular, its spin-lattice relaxation time is usually very long, several 100 s. Lithium chloride is extremely soluble in water and in a variety of organic solvents. For NMR experiments, usually enriched ^{6}LiCl is used. ^{6}LiCl has been hyperpolarized by dissolution DNP in an ethanol-d$_6$/D_2O (1:2) matrix containing 33 mM TEMPO, a D_2O-glycerol-d_8 (1:1) matrix with 58 mM TEMPOL, and a 3:1 sulfolane-diethylene glycol monobenzyl ether doped with BDPA (40 mM) to achieve ^{6}Li polarization levels in the 6%–8% range [207,347,348]. ^{6}Li can also be hyperpolarized by ^{1}H to ^{6}Li CP [151]. When hyperpolarized ^{6}LiCl (2.5 mM, 120 mM) was injected into rats, a ^{6}Li signal could be detected in the brain over a period of 33 to 5 min. The ^{6}Li T_1 data shown in Table A.2 (Appendix) reveal that the ^{6}Li relaxation time is very sensitive to paramagnetic relaxation effects to such a degree that the presence of 0.5 μM Gd-DOTP could be detected. The T_1 was about 10% shorter in whole oxygenated human blood (oxyhemoglobin is diamagnetic) than in deoxygenated blood (deoxyhemoglobin is paramagnetic). Therefore, hyperpolarized ^{6}Li NMR was suggested as a potentially useful tool to generate tissue oxygenation maps [347].

1.7.3 ^{13}C

The gyromagnetic ratio of ^{13}C (10.7084 MHz T^{-1}) is one fourth of that of ^{1}H, which, combined with the low natural abundance (1.1%), makes the receptivity (sensitivity x natural abundance) of ^{13}C about four orders of magnitude lower than that of ^{1}H. Several nonhyperpolarized NMR techniques have been proposed to overcome the low-sensitivity issue of ^{13}C NMR [349,350]. Of these, probably the indirect detection of ^{13}C via J-coupling in the ^{1}H spectrum is the most useful [350]. In terms of sensitivity gain, none of these methods can compete with hyperpolarized ^{13}C NMR. Hyperpolarization dramatically enhances the sensitivity of direct ^{13}C detection; it was shown that it is possible to detect intermediates present in concentration, as low as 1 μM, possibly even lower, in enzyme-catalyzed reactions by HP-^{13}C NMR [24]. Considering the importance of carbon in biochemistry, organic chemistry, and life sciences [351], it is not surprising that the main application of dissolution DNP is to produce solutions of hyperpolarized ^{13}C spins.

^{13}C spins at natural abundance can be successfully hyperpolarized by DNP, but generally, ^{13}C-labeled compounds are hyperpolarized for in vivo applications because enrichment boosts the sensitivity by an additional two orders of magnitude [84]. Experimentally, ^{13}C-labeled compounds can be polarized, as described earlier in this chapter. In fact, dissolution DNP protocols have been developed primarily for polarizing ^{13}C. The most common glassing matrix is glycerol-water, but other biocompatible glassing agents, such as DMSO, can also be used. Narrow EPR linewidth polarizing agents, such as trityl radicals (most commonly OX063) or BDPA, are best suited for direct ^{13}C DNP [352]. With these radicals, the dominant DNP mechanism is TM. Nitroxides can also be used, but these also polarize ^{1}H, and therefore, ^{1}H to ^{13}C CP is more efficient [353].

1.7.3.1 T_1 relaxation of ^{13}C

As already pointed out, sufficiently long spin-lattice (T_1) relaxation and long T_1 values are necessary for successful biomedical imaging with hyperpolarized ^{13}C tracer agents. In general, ^{13}C nuclei have T_1 relaxation times ranging from a few seconds to about 3 min. The ^{13}C T_1 relaxation time depends on the molecular structure and the molecular weight. Intramolecular dipolar interactions and CSA are the most important relaxation mechanisms for HP-^{13}C-labeled substrates [295,300]. Table A.3 (Appendix) lists ^{13}C T_1 values for several compounds. As seen from this list, the T_1 values are very short when protons are attached to the ^{13}C nucleus. This is due to dipole-dipole interactions with directly attached ^{1}H spins. The longest ^{13}C T_1 relaxation times are observed in functional groups in which the ^{13}C nuclei are not directly bonded to ^{1}H. These include carbonyl, carboxyl, and quaternary carbons (long T_1 positions). The 2-keto-carboxyl functional group, which is present in pyruvate oxaloacetate and 2-ketoglutarate, is of special interest because of its great biochemical significance and the exceptionally long T_1 values of the ^{13}C spins. In the absence of dipole-dipole interactions, the main relaxation pathway for carbonyl carbons is the strongly field-dependent CSA, which scales with B_0^2 [354].

The impact of magnetic fields on pyruvate C1 spin-lattice relaxation time is demonstrated in Fig. 1.14. The low-field relaxation mechanisms have a greater influence when the magnetic field is less than 100 mT. The T_1 reaches its maximum at around 1 T, and as the field increases, CSA becomes more and more prominent, leading to a gradual decrease in T_1 [354–356]. In addition to the B_0 field, the T_1 value of carboxylate is strongly influenced by protonation equilibria, intermolecular hydrogen bonding, and the presence of paramagnetic impurities. In case of pyruvate, T_1 is also affected by the formation of pyruvate hydrate in an acidic solution and the base-catalyzed keto-enol tautomeric exchange (Fig. 1.15) [357].

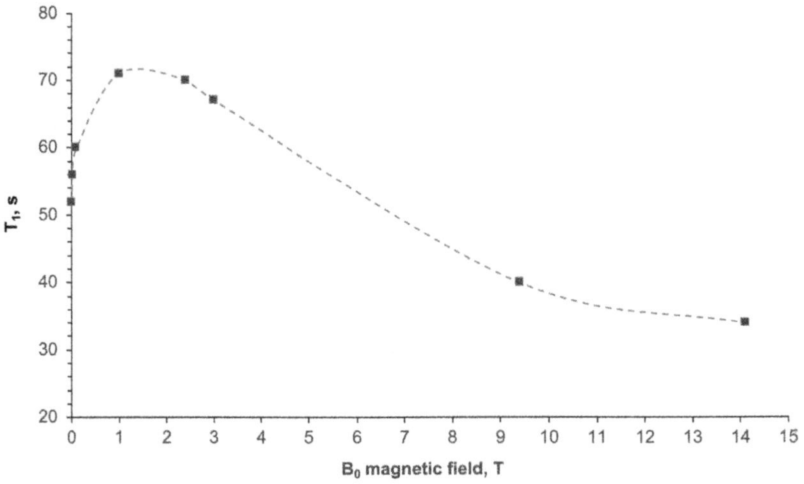

FIGURE 1.14 The field dependence of pyruvate C_1 spin lattice relaxation time in water at pH 7. Note that the data were compiled from different reports and minor differences in the experimental conditions (temperature, pyruvate concentration, free radical concentration) have been ignored [313,355,358–360].

In an analogous manner to the dipolar effect of ^1H, dipolar interactions of directly linked ^{13}C spins also shorten the ^{13}C T_1 values [361]. Although, as expected, the dipolar effect of a ^{13}C spin on the T_1 of an adjacent ^{13}C is weaker than that of ^1H, it is noticeable. For example, the ^{13}C T_1 values of [1,2-$^{13}C_2$]pyruvate (C1 56 s, C2 44 s at 3 T in H_2O) and [1,2-$^{13}C_2$]-2-ketoglutarate are somewhat shorter than the respective T_1 of the singly labeled derivatives [362,363].

1.7.3.2 NMR spectroscopy with HP-^{13}C

The dramatically enhanced sensitivity of hyperpolarized ^{13}C NMR makes it feasible to perform ^{13}C NMR experiments that were either not possible or would have required an unreasonably long acquisition time by conventional NMR. Hyperpolarized ^{13}C was successfully used to identify otherwise nonobservable reaction intermediates [364], allowed the real-time analysis of protein folding [365,366] facilitated the acquisition of 2D ^{13}C-observed HMQC spectra of [1-^{13}C]acetylated amino acids in blood serum [367] and provided valuable insight into cellular metabolic networks with unprecedented sensitivity [25]. However, its greatest impact, arguably, has been on in vivo metabolic imaging. The use of carbon isotopes has a long and successful history in biochemical research. Because metabolic activities involve chemical transformations of organic molecules, carbon isotopes are excellent tools for metabolic studies. The key idea is that the isotope distribution in the metabolic products of the administered tracer encodes the underlying mechanisms. Early work that contributed significantly to our textbook understanding of metabolism used substrates labeled with the radioactive ^{14}C

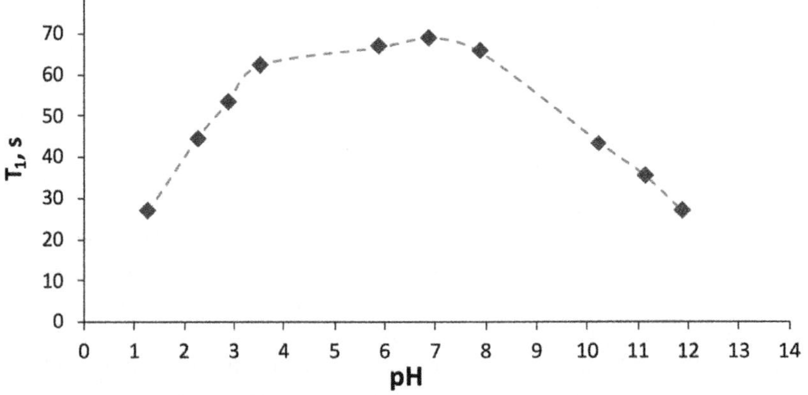

FIGURE 1.15 The pH dependence of T_1 of pyruvate C1 spin lattice relaxation time at 1 T. The T_1 data were collected with hyperpolarized [1-^{13}C] pyruvate. The maximum T_1 is attained between pH 7 and 8 [357].

isotope, followed by the detection and quantification of the presence and distribution of ^{14}C in the downstream metabolic products isolated from tissue extracts or present in the expired air as $^{14}CO_2$ [368,369]. Although the detection and quantification of the radioactive ^{14}C is straightforward, the use of ^{14}C involves ionizing radiation and is time-consuming because the metabolites must be separated, and the determination of the label position requires carbon-by-carbon degradation of the carbon skeleton [370]. Due to improvements in the sensitivity and resolution of the analytical methods used for the detection of ^{13}C and other isotopes, stable ^{13}C has largely replaced ^{14}C for metabolic profiling and metabolomics studies. In particular, ^{13}C labeling patterns and fractional enrichments in specific positions of downstream metabolic products can be established by high-resolution NMR, MS, or the combination of the two analytical techniques [284,371–373]. ^{13}C isotopomer analysis by ^{13}C magnetic resonance spectroscopy generally relies on the ^{13}C chemical shifts as well as the ^{13}C-^{13}C homonuclear coupling patterns to discern the fate of the ^{13}C-labeled tracer and provides a very convenient way to estimate fluxes through specific enzyme-catalyzed steps [374]. Mathematical models of varied complexity are employed to interpret the spectroscopic data and to identify metabolic flux via distinct pathways [349,350,375–377]. Although the poor sensitivity of ^{13}C largely precludes the in vivo detection of ^{13}C by conventional MRI technology, hyperpolarized ^{13}C boosts the sensitivity by several orders of magnitude [84], thereby enabling the in vivo observation of the ^{13}C NMR signal provided that the ^{13}C T_1 is sufficiently long. Conventional ^{13}C isotopomer analysis can be used to validate the HP-^{13}C data [363]. Since isotopomer analysis relies heavily on ^{13}C–^{13}C coupling patterns, the uniformly or multiply labeled versions of metabolic substrates serve best for ^{13}C isotopomer studies [24,378,379].

1.7.4 ^{15}N

Nitrogen-15 (0.37% natural abundance) has a favorable nuclear spin of ½ and a wide chemical shift range (900 ppm), yet ^{15}N NMR is much less widely used as a tool in biological systems than ^{13}C and ^1H because of its extremely poor sensitivity. ^{15}N is a low-sensitivity nucleus with a gyromagnetic ratio about one 10th of that of ^1H. Its low natural abundance makes the direct detection of ^{15}N very time-consuming without enrichment. However, ^{15}N-enriched compounds are relatively easy to analyze by conventional NMR spectroscopy. Nitrogen is one of the four essential elements in living systems. Nitrogen utilization is fundamentally important for cell proliferation, and therefore, it is tightly regulated. Altered nitrogen metabolism is characteristic of cancer cells [380]. The ^{15}N nucleus has several favorable properties for hyperpolarized NMR: narrow linewidths, often very long T_1 values, and a much larger chemical shift range than ^{13}C. Unfortunately, the low sensitivity of conventional ^{15}N NMR precludes its use to monitor nitrogen metabolism in vivo. ^{15}N spins can be hyperpolarized by DNP under standard conditions with various polarizing agents (nitroxides, trityl, or BDPA radicals) [381]. ^{15}N-labeled compounds can also be hyperpolarized using parahydrogen-based techniques and most importantly, SABRE. This requires a dynamic, coordinated interaction between the ^{15}N-substrate and the iridium catalyst that mediates the polarization transfer from the parahydrogen to the substrate [382,383]. Compounds that can successfully be polarized via SABRE include a wide range of ^{15}N-labeled amines, especially heterocyclic compounds [384]. HP-^{15}N-labeled compounds are particularly well suited for NMR sensor design because of the high sensitivity of the ^{15}N chemical shift to the environment. For example, several hyperpolarized ^{15}N pH and metal ion sensing probes have been reported in which the protonation or metal ion coordination causes a large (up to about 100 ppm) ^{15}N chemical shift change [26,385,386]. ^{15}N T_1 values range from a few seconds up to about 1200 s, depending on the molecular structure and symmetry, magnetic field, solvent, and temperature. The distinction of having the longest liquid state ^{15}N T_1 of any small molecule goes to [^{15}N, $^2H_{14}$]trimethylphenylammonium iodide, which was reported to have a T_1 of 1128 s at 14.1 T and 1177 s at 9.4 T at 30°C in D_2O. The main relaxation mechanisms for ^{15}N include dipole-dipole interactions and CSA for larger molecules [297,387]. The contribution of spin rotation relaxation can be significant for small ^{15}N-labeled molecules and ions such as $^{15}NH_3$, nitrate, and nitrite [388,389]. As with ^{13}C, deuteration can significantly slow the ^{15}N T_1 relaxation [26,297,331,390]. Table A.4 (Appendix) lists ^{15}N of T_1 values for several compounds.

1.7.5 ^{19}F

The spin ½ of the ^{19}F nucleus has a high gyromagnetic ratio and 100% natural abundance. The main relaxation mechanisms for ^{19}F spins are dipole-dipole interactions, CSA, and spin rotation, which result in relatively short spin-lattice relaxation times (on the order of seconds) [391]. Nevertheless, in a similar way to ^1H, ^{19}F spins can be polarized with nitroxide radicals in the deuterated glassing matrix, although the short T_1 necessitated the development of a rapid dissolution system [392]. The chemical shift of ^{19}F is quite sensitive to its local environment. Consequently, fluorinated compounds have been used as ^{19}F NMR probes to study protein binding. In addition to changes in the ^{19}F chemical shift,

the binding of fluorinated ligands to proteins can cause strong line broadening due to chemical exchange. The relatively strong CSA of ^{19}F can also lead to line broadening when the rotational correlation time increases upon binding [392]. Although the ^{19}F of T_1 is too short for in vivo studies, hyperpolarized ^{19}F probes have been used in protein-binding experiments. The enhanced ^{19}F signal allows the binding experiments to perform over a much wider concentration range than conventional thermally polarized ^{19}F NMR. Fluorinated reporter probes were also used to study protein-nonfluorinated ligand interactions in competitive binding experiments [393].

1.7.6 ^{29}Si

Natural silicon has one NMR active isotope, the spin ½ ^{29}Si (4.7% natural abundance). The spin-lattice relaxation time of ^{29}Si can be fairly long in small molecules T_1 values up to 48 and 88 s were reported [207,272]. Although the feasibility of the DNP of ^{29}Si spins with carbon-centered radicals was demonstrated [207,394], the potential of hyperpolarized ^{29}Si for HP-sensor design has remained largely unexplored despite the favorable T_1 values. ^{29}Si nuclei in silicon and silicon carbide micro- and nanoparticles could also be hyperpolarized by DNP at cryogenic temperatures. The polarization transfer occurs from the unpaired electrons in surface and lattice defects, and therefore, there is no need to add external polarizing agents. The spin-lattice relaxation time of ^{29}Si at natural abundance in silicon particles can be extremely long. Relaxation occurs through surface relaxation, and the T_1 increases with particle size leveling off at about 5 h for large particle sizes. In vivo imaging with hyperpolarized silicon nanoparticles was demonstrated in mice [395–398].

1.7.7 ^{31}P

The ^{31}P nucleus (100% natural abundance) has a gyromagnetic ratio of 17.235 MHz T^{-1}. ^{31}P spins primarily relax through dipole-dipole interactions, with varying contributions from CSA and spin rotation, depending on the molecular structure and experimental conditions [399]. In general, ^{31}P T_1 values in water are short, less than 30 s, which limits the applicability of hyperpolarized ^{31}P compounds in biomedical imaging [400]. Only a few examples of ^{31}P dissolution DNP can be found in the literature. Trimethyl phosphate was polarized using BDPA in a sulfolane-diethylene glycol monobenzyl ether matrix [207]. Inorganic phosphates and phosphocreatine were polarized in glycerol-H_2O and glycerol-D_2O, respectively (Fig. 1.16) [400].

The dominant mechanism for ^{31}P DNP with carbon-centered radicals is the SE, with some contribution from TM [394]. Phosphate participates in several fundamental biochemical processes. Phosphorylation of proteins is the most common posttranslational modification, and it plays a critical role in cell signaling. Therefore, hyperpolarized phosphate would potentially be very useful for visualizing biological processes involving the phosphorylation of various biomolecules. In addition, the chemical shift of inorganic phosphate is pH sensitive, and hyperpolarized phosphate could be used as a ^{31}P NMR pH probe. Unfortunately, the in vivo application of hyperpolarized phosphate and other phosphorus compounds is

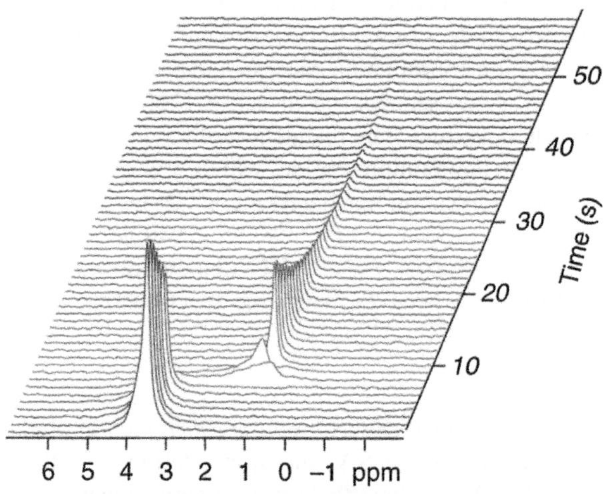

FIGURE 1.16 Hyperpolarized phosphate as a ^{31}P NMR pH probe. KH_2PO_4 was polarized trityl OXO63 in glycerol-water matrix. After dissolution, the pH was 8.1. The pH was then rapidly adjusted to 4.8 by the addition of citrate-Tris buffer upon which the ^{31}P peak exhibited a 3 ppm upfield shift. The ^{13}P NMR spectra were recorded at 5.8 T with a repetition time of 1 s and a flip angle of 10 degrees [400]. *Reproduced from Ref. [400] with permission.*

hampered by the short T_1. While fully deprotonated phosphate is symmetric, all protonated phosphate species are anisotropic, and besides dipolar interactions, CSA and spin rotation are expected to strongly contribute to the relaxation [401]. To fully explain the contributions of CSA and spin rotation relaxation to the overall spin-lattice relaxation of the phosphate ion, the concept of quasi-rotation was proposed [402]. Phosphate is considered a "structure-making" ion because it forms strong H-bonded networks with water, thereby increasing the local order of water. Proton conduction in water occurs by proton exchange via the H-bonded network of water molecules. Partially protonated phosphate can efficiently participate in this process by simultaneously accepting and donating a proton. This process involves a rapid shift of proton (and the negative charge) from one phosphate oxygen to another, which will reorientate the chemical shift tensor and the magnetic moment created by the electrons (spin rotation). Overall, the process has the same effect as if the PO_4 tetrahedron would physically rotate (quasi-rotation). Table A.5 (Appendix) lists ^{31}P of T_1 values for several compounds.

1.7.8 ^{77}Se

There is only one report of the dissolution of DNP with selenium compounds [403]. It was shown that ^{77}Se DNP is feasible with commercially available hardware and polarizing agents. The relaxation of ^{77}Se spins is dominated by CSA. Interestingly, dipole-dipole relaxation is negligible even in molecules that contain Se—H bonds. Therefore, the ^{77}Se T_1 values are expected to be long enough for dissolution DNP experiments only in highly symmetrical molecules.

1.7.9 ^{89}Y

Yttrium is a rare earth metal that forms diamagnetic Y^{3+} cations. Y^{3+}, although it lacks f-orbitals, has very similar properties to those of the trivalent lanthanide ions; therefore, it is considered a pseudo-lanthanide. Yttrium consists of only one natural isotope, the spin ½ ^{89}Y. This is an ultralow-sensitivity nucleus with a gyromagnetic ratio about one 20th of that of ^1H. Consequently, at room temperature, the thermal equilibrium ^{89}Y spin polarization is very low ($P = 1.578 \times 10^{-4}\%$ at 9.4 T and 298 K). The low γ imparts long nuclear spin-lattice relaxation times to Y^{3+} and its complexes. The yttrium aqua ion in strongly acidic solution has an extraordinarily long T_1 value (2200 s) [404]. To the best of our knowledge, this is the longest liquid-phase spin-lattice relaxation time for any nucleus ever reported. Unfortunately, the combination of low thermal polarization with long T_1 also means that direct detection of ^{89}Y by conventional NMR is rather time consuming. For these same reasons, however, ^{89}Y is well suited for hyperpolarization [115,116,405]. Dissolution DNP under standard conditions (glycerol water matrix and trityl OX063, BDPA, or nitroxide radical polarizing agent) works well for various ^{89}Y-compounds and complexes, affording strong ^{89}Y signal enhancements. DNP-NMR made it easy to measure the chemical shifts of a variety of ^{89}Y complexes. These studies revealed that the ^{89}Y resonance frequency is quite sensitive to the chemical environment of the $^{89}Y^{3+}$ ion. The strong signal enhancement combined with the sensitivity of the ^{89}Y chemical shift to environmental effects offers the possibility of designing ^{89}Y NMR probes to monitor physiological parameters such as pH, temperature, redox, and other biomarkers in vivo [406]. In addition, ligand systems that have been developed for lanthanide-based MR contrast agents (Gd^{3+} and Eu^{3+}) can directly be used to safely chelate Y^{3+} for biomedical applications. The chemical shift in Y-poly amino carboxylate complexes covers about a 250 ppm range (Fig. 1.17) [407]. An empirical equation was developed that can accurately predict the ^{89}Y chemical shift based on the number and type of donor atoms bound to the Y^{3+} ion and can be used as a guide in the design of ^{89}Y NMR sensors. Because the shielding of hydroxyl, carboxylate, and water oxygen donor atoms is different, the ^{89}Y chemical shift changes upon the substitution of carboxylates and/or hydroxyls for the Y-bound water molecules. For example, the two inner sphere water molecules in $Y(EDTA)(H_2O)_2$ can be replaced by lactate, which binds in a bidentate fashion. This results in about a 10 ppm downfield shift, demonstrating that this complex could be used for lactate sensing [407]. Hyperpolarized ^{89}Y complexes in which the ligand can undergo protonation demonstrated the feasibility of using ^{89}Y NMR probes to detect pH (Fig. 1.18). For example, in Y complexes with polyaminophosphonate-type ligands such as DOTP or EDTMP, the ^{89}Y chemical shift changes ~10 ppm over the pH range 5—9 due to protonation of the noncoordinated phosphonate oxygen atoms in the complex. HP-^{89}Y NMR spectroscopy has also been used to monitor the rate of complex formation [408].

Although ^{89}Y signal could be observed in vivo after the injection of hyperpolarized YDOTA into rats in preliminary experiments, in vivo spectroscopy/imaging remains challenging despite the large signal enhancements and long T_1 values. Table A.6 (Appendix) lists ^{89}Y of T_1 values for several compounds.

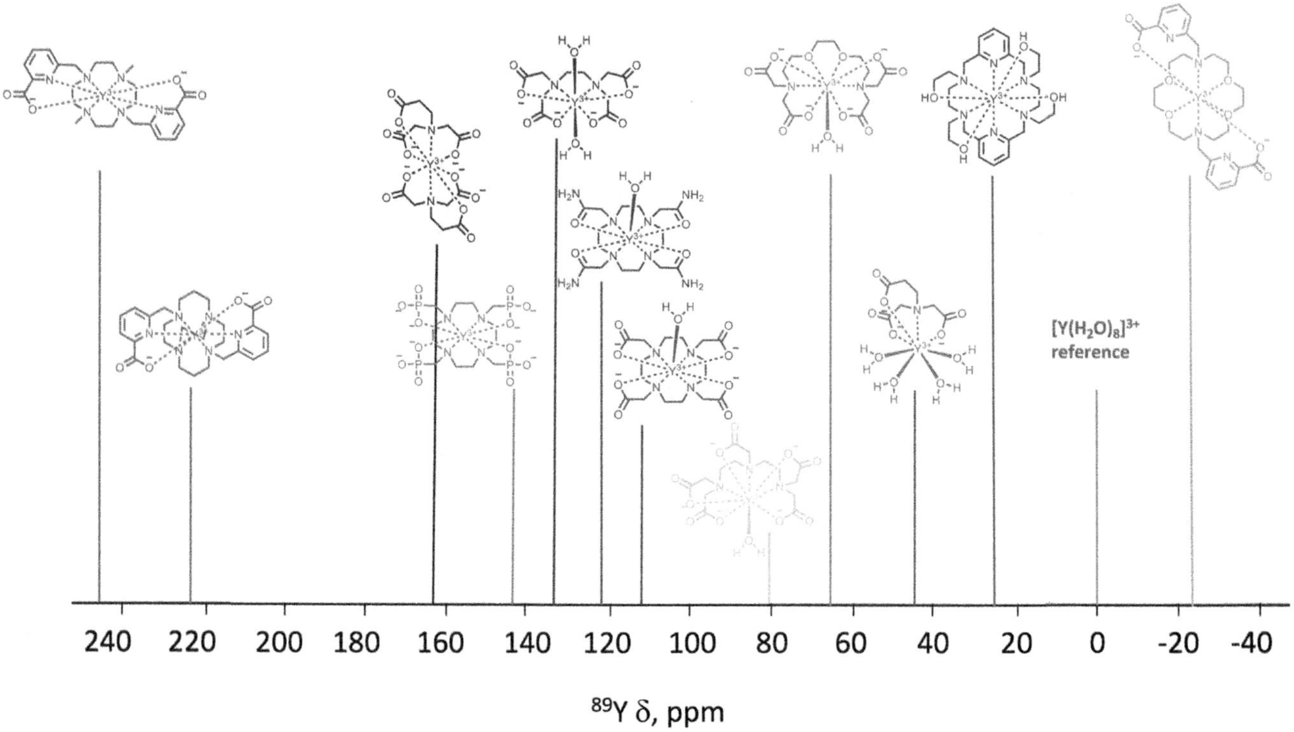

FIGURE 1.17 The ^{89}Y chemical shifts in various yttrium complexes. Relatively minor changes in the ligand structure result in large differences in the ^{89}Y chemical shifts.

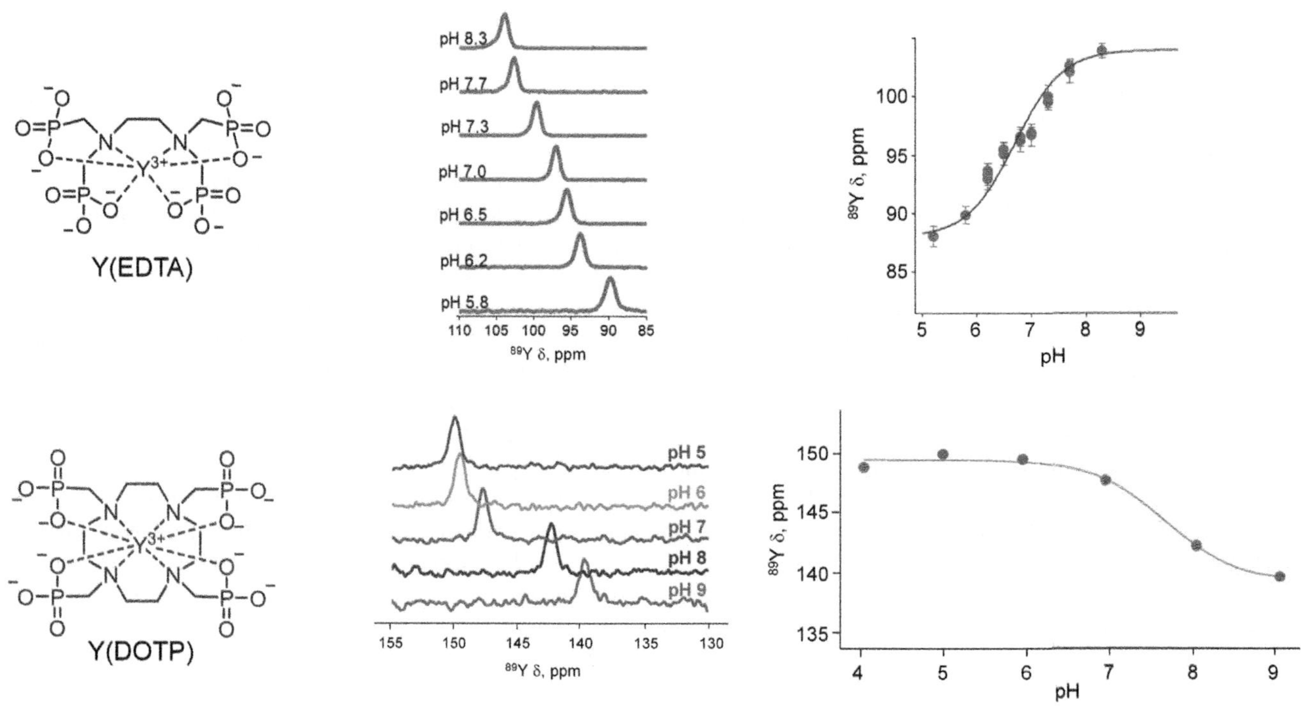

FIGURE 1.18 Protonation of the noncoordinating oxygens in yttrium polyaminophosphonate complexes imparts pH sensitivity to the ^{89}Y chemical shifts. *Adapted from Refs. [405,408] with permission.*

1.7.10 ^{107}Ag and ^{109}Ag

Silver is unique in that it is composed of a nearly 1:1 mixture of two spin I = 1/2 isotopes with very similar NMR properties (^{107}Ag 51.8%, ^{109}Ag 48.2%). Both of these isotopes have an extremely low gyromagnetic ratio and consequently very long spin lattice relaxation times, thereby posing a challenge to direct detection by conventional NMR. Various polarization transfer techniques have been applied to improve detection, with limited success [409]. Therefore, these two isotopes, along with ^{89}Y, are among those that could benefit the most from hyperpolarization. So far, there has been only one report on hyperpolarized $^{107,109}Ag$ [410]. Silver trifluoromethanesulfonate (AgOTf) could be polarized via DNP using the free radical BDPA as a polarizing agent in sulfolane-based glassing matrices. It is worth noting that the other two radicals that were tested (trityl OX063 and 4-oxo-TEMPO) failed to give any enhancements to AgOTf or $AgNO_3$, likely because Ag^+ reacts with these radicals. Hyperpolarized Ag-complexes were generated by either direct polarization of the complexes or by reacting hyperpolarized AgOTf with ligands during the dissolution phase. Ligands with soft donor atoms (C, N, P, and S) were chosen to match the soft acid character of Ag^+. Silver(I) complexes are in general very labile, and the complex formation was complete even with the rigid macrocyclic ligands during transfer. Both silver isotopes have very large chemical shift ranges, and the dominating relaxation pathway is CSA. Dipolar relaxation effects of ligand protons are negligible, as evidenced by the nearly identical ^{109}Ag T_1 values of the nondeuterated and deuterated dipyridyl complexes. The two isotopes have very similar DNP spectra, indicating that the polarization proceeds by TM (Fig. 1.19).

As a result, both isotopes are simultaneously polarized, acquiring the same spin temperature, and can be detected independently from each other (Fig. 1.20) potentially allowing the performance of two consecutive NMR experiments with $^{109,107}Ag$-based probes (Fig. 1.19). Table A.7 (Appendix) lists silver T_1 values for several compounds.

1.7.11 ^{129}Xe

Naturally occurring inert gas Xe has 9 main isotopes, of which the most abundant, ^{129}Xe, has a spin ½ [411]. The T_1 of ^{129}Xe in the gas phase is about 3 h. The main relaxation mechanism of ^{129}Xe gas is spin-rotation, but wall relaxation may also contribute significantly. Wall relaxation involves dipolar interactions of ^{129}Xe with nuclear or electronic spins on the surface of the gas container [412]. In solution, the T_1 is much shorter—26 s in distilled water and less than 10 s in plasma [413,414]. Conventionally, ^{129}Xe is polarized by SEOP [415,416]. Close to 100% ^{129}Xe spin polarization can be achieved by SEOP under optimized conditions [417]. Large quantities of hyperpolarized ^{129}Xe can also be produced in a continuous flow manner [418,419]. However, the experimental setup for SEOP is rather involved and requires the handling of metallic rubidium or cesium [420]. Fortunately, commercial dissolution DNP hardware can be adapted for the polarization of ^{129}Xe. Besides being simpler and safer, ^{129}Xe DNP also offers the advantage of a large expansion factor: a small volume of frozen solid DNP sample will produce a large volume of hyperpolarized room temperature gas (the expansion coefficient is about 530) [421]. The greatest challenge in applying dissolution DNP to ^{129}Xe was the optimization of the DNP sample composition. Xenon forms a nonpolar liquid/solid and has a limited solubility in most polar solvents [421,422]. To attain

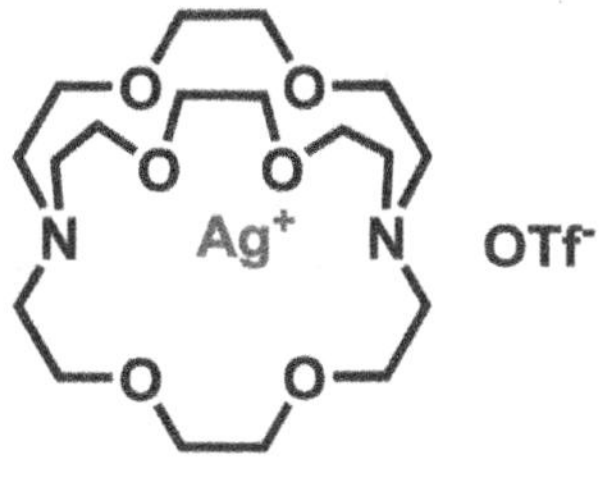

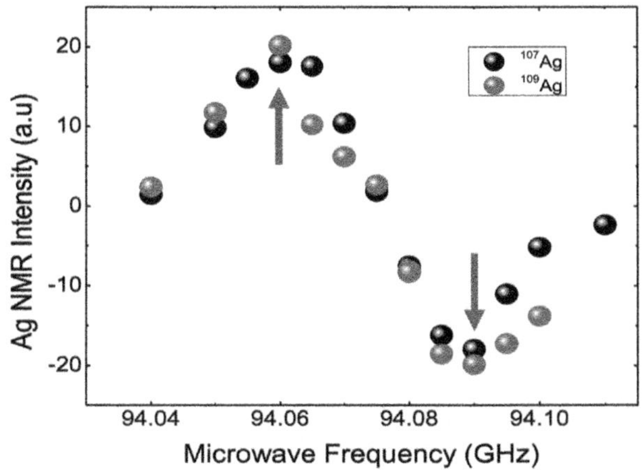

FIGURE 1.19 The DNP spectra of ^{107}Ag and ^{109}Ag isotopes in hyperpolarized Ag(crypt-222)OTf measured by plotting the liquid state signal intensities immediately after dissolution as a function of microwave frequency. The theory of thermal mixing predicts that both silver isotopes are polarized simultaneously and will achieve the same spin temperature. *Reproduced from Ref. [410] with permission.*

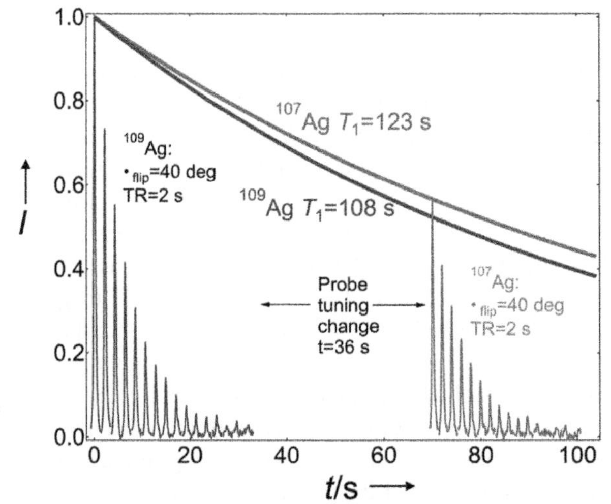

FIGURE 1.20 The sequential detection of hyperpolarized ^{109}Ag and ^{107}Ag signals in $^{107,109}Ag$(crypt-222)OTf at 9.4 T. At this field, there is a 2.4236 MHz difference between the ^{109}Ag and ^{107}Ag Larmor frequencies. First, the ^{109}Ag signal was recorded using 40 degrees pulses. This did not affect the ^{107}Ag magnetization, which could be detected after probe tuning (36 s). The ^{107}Ag signal has lower intensity due to T_1 decay. *Reproduced from Ref. [410] with permission.*

maximum solid-state ^{129}Xe polarization, the solvent should have a low freezing temperature (near or lower than the triple point of xenon) and should be able to dissolve a large amount of xenon and the radical. The frozen DNP sample is usually a heterogeneous mixture of xenon and the polarizing agent dissolved in the glassing matrix and clusters of pure frozen xenon. Therefore, the ^{129}Xe chemical shift difference between the dissolved ^{129}Xe and pure frozen ^{129}Xe should be small to facilitate efficient spin diffusion, which propagates the spin polarization across the DNP sample. So far, isobutanol (melting point 165 K) and ethanol (melting point 159 K) proved to be the best choices for the matrix, as xenon has relatively high solubility in these solvents, and at 500 kPa there is about a 30 K temperature range at which both xenon and the solvent are liquids [423]. The DNP sample is prepared by condensing xenon gas into a solution of the polarizing agent in the matrix at 170 K and 100–500 kPa. The sample is then frozen in liquid nitrogen and transferred into the sample holder of the polarizer. The solid-state polarization can be improved by pulverizing and annealing the sample [424]. Pulverization involves crushing the frozen DNP sample into small pellets, which increases the surface area. During annealing, the DNP sample is removed from the liquid helium bath and kept at temperatures above 120 K for a few hours before DNP. This process was shown to dramatically increase the solid-state ^{129}Xe T_1 [424]. When the DNP is done, the sample is lifted above the liquid helium bath, and the hyperpolarized xenon gas is sublimed from the frozen polarized sample by treating it with hot water vapor and helium gas (440 K and 1 MPa) using a custom-built insert. The hyperpolarized gas is then separated from the solvent and water and recondensed in liquid nitrogen in a magnetic field of 0.1 T. Two important points about the sublimation DNP of ^{129}Xe must be emphasized. First, the solvent does not have to be biocompatible because it is separated from the xenon gas during the sublimation step. Second, the ^{129}Xe T_1 is stable for several hours at temperatures below 120 K and magnetic fields higher than 0.05 T [425]. This allows the transport and/or storage of hyperpolarized ^{129}Xe under these conditions. ^{129}Xe was successfully polarized with both nitroxide and trityl polarizing agents as well as nonpersistent radicals generated from pyruvic acid by UV irradiation. Maximum solid-state ^{129}Xe polarization levels up to about 30% have been reported, which is quite comparable to polarizations achieved with SEOP [423].

It is worth mentioning that the sample preparation technique described above was unsuccessful in achieving DNP for other gases like ethylene, propane, and butane, apparently because a mixture of these compounds and ethanol did not form a homogeneous glass at cryogenic temperatures. However, mesoporous hybrid organosilica materials (hyperpolarizing solids, or HYPSOs), originally developed for the dissolution DNP of pyruvate and other metabolic substrates, were successfully adapted to the DNP of these hydrocarbons. HYPSOs contain surface-bound polarizing agents (TEMPO here) that are homogeneously distributed on the surface of the mesopores [426].

The potential of hyperpolarized inert gases (^{3}He and ^{129}Xe) for lung imaging is obvious [427–431]. However, xenon is more suitable for dissolved phase imaging because it is, unlike helium, fairly soluble blood and has sufficiently long T_1 in vivo to preserve enough ^{129}Xe magnetization to image other organs [432,433].

1.7.11.1 *HP-^{129}Xe biosensors and HP-^{129}Xe hyperCEST*

Xenon has an easily polarizable electron cloud, which makes the 129Xe chemical shift very sensitive to its environment. For example, in the brains of animals and humans, up to five ^{129}Xe peaks can be resolved after hyperpolarized 129Xe inhalation, corresponding to xenon dissolved in different tissues [434,435]. This sensitivity has been exploited in the design of HP-^{129}Xe NMR based biosensors. These sensors employ a cage molecule with a hydrophobic cavity that can encapsulate Xe reversibly but preferably in slow exchange with the dissolved bulk Xe on the NMR timescale, thereby affording two separate ^{129}Xe signals [436,437]. With an appropriately functionalized host molecule, these systems can detect the presence of a target molecule, a binding event, or a chemical reaction based on the ^{129}Xe chemical shift [438–441]. Most commonly, the xenon-binding host is a cryptophane moiety, which has nearly optimal binding affinity and exchange kinetics with Xe [442–444]. These sensors have the disadvantage that only a small percentage of the dissolved ^{129}Xe is bound in the host molecule, which translates into relatively poor sensitivity [445]. A very innovative application of HP-^{129}Xe, which greatly improves the sensitivity, combines chemical exchange saturation transfer with HP-^{129}Xe NMR. HyperCEST is based on the reversible encapsulation of Xe in the cavity of supramolecular cages [446]. The noncovalent interaction between the xenon atoms and the host molecule causes a very large chemical shift difference (up to 170 ppm) between the bound and free ^{129}Xe (Fig. 1.21) [444].

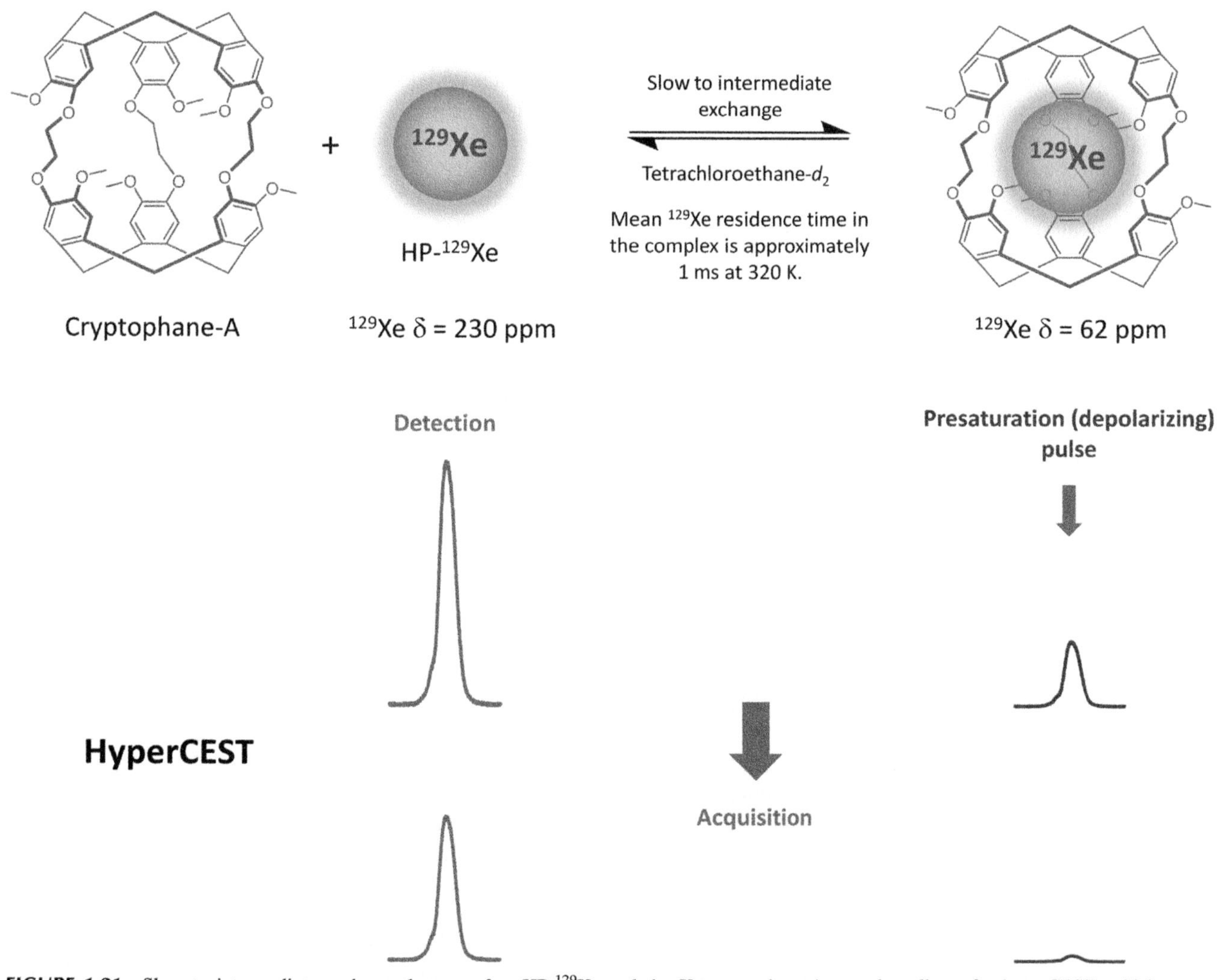

FIGURE 1.21 Slow to intermediate exchange between free HP-^{129}Xe and the Xe-cryptophane-A complex allows for hyperCEST, which uses a depolarizing saturation pulse at the resonance frequency of the xenon-cryptophane complex at 62 ppm. The exchange of saturated ^{129}Xe spins decreases the free ^{129}Xe signal intensity at 230 ppm.

When a saturation pulse is applied at the encapsulated Xe resonance frequency (62 ppm), chemical exchange between the bound and free xenon pools transfers the saturated (depolarized) ^{129}Xe spins into the free HP-^{129}Xe pool (230 ppm) thereby decreasing its signal intensity. The hyperCEST spectrum is obtained by plotting the free ^{129}Xe signal intensity versus presaturation frequency. ^{129}Xe hyperCEST takes full advantage of the signal amplifying effect of CEST as many Xe atoms are exchanged over the duration of the saturation pulse. The wide chemical shift range, long relaxation times and the favorable (slow to intermediate) exchange rate between the host-bound and free xenon makes hyperpolarized ^{129}Xe an excellent candidate for the design of CEST sensors [437,447]. Suitable hosts include functionalized cryptophanes, cucurbiturils, liposomes, microemulsions, paramagnetic metal organic capsules and genetically engineered proteins [436,441,448–456] The sensitivity of the technique is remarkable: some of the host molecules can be detected in the picomolar range [457]. A rather ingenious way to design responsive ^{129}Xe hyperCEST sensors is based on cleavable cucurbituril rotaxanes [458]. A rotaxane is a supramolecular construct in which a molecular axle is threaded through the cavity of a macrocylic host. The two ends of the axle are capped with bulky moieties that prevents the rotaxane from disassembling. Cucurbiturils are often used as the macrocyclic component in the construction of rotaxanes [459]. The axle threaded through the cucurbituril cavity prevents the xenon from binding. However, when the blocking groups are removed, the axle diffuses out of the cavity allowing the cucurbituril to interact with Xe. Cucurbituril based rotaxanes in which the capping group is cleaved in response to an external stimulus such as changes in pH, redox state or enzymatic activity have been used in the design of responsive ^{129}Xe hyperCEST agents [454,460,461] Table A.8 (Appendix) lists ^{129}Xe of T_1 values for several compounds.

1.7.12 ^{133}Cs

^{133}Cs is a spin 7/2 nucleus whose gyromagnetic ratio is roughly 1/8 that of ^{1}H. ^{133}Cs is 100% naturally abundant and relatively easily visible in vivo by standard solution-state NMR [462]. Generally, ^{133}Cs T_1 values are much shorter than those reported for 6Li. Although the quadrupolar moment of the ^{133}Cs nucleus is small, the predominant ^{133}Cs spin-lattice relaxation mechanism is largely quadrupolar, as reflected by the insensitivity of the T_1 to solvent deuteration [463,464]. Cesium exists as Cs^+ in aqueous media. Cs^+ is a large cation, and its electron cloud is easily polarizable. Consequently, the ^{133}Cs chemical shift is sensitive to the chemical environment. Cs^+ acts as a surrogate for K^+ and accumulates intracellularly via the transmembrane Na^+/K^+ (ATPase) pump. A particularly interesting feature of in vivo ^{133}Cs NMR is that the resonances of ^{133}Cs nuclei in the intracellular space are a few ppm downfield shifted from the extracellular $^{133}Cs^{+}$ signal (Fig. 1.22). This effect was attributed to the high concentration of various biomolecules, which disrupts the hydrogen bonding network of water molecules in the cytosol. The weak binding of Cs^+ to intracellular biomolecules such as 2,3-diphosphoglycerate and ATP may also contribute to the observed shift [465]. Despite its relatively short T_1 (16 s at 9.4 T), ^{133}Cs in the form of CsCl or Cs-acetate could be hyperpolarized to a high degree under standard dissolution DNP conditions in a water-glycerol (1:1) matrix with trityl OX063 (20 mM) and Prohance (1.5 mM) [464,466]. While the detection of ^{13}3Cs signals by conventional NMR is relatively easy, hyperpolarized ^{133}Cs allowed the monitoring of Cs^+ transmembrane flux into red blood cells with 1 s time resolution (Fig. 1.22) [466].

1.8 A general overview of the applications of hyperpolarized probes

As discussed in this chapter, hyperpolarization can significantly increase NMR signals of low-γ nuclei, thus allowing for the otherwise insensitive nuclei to be detected in real-time by NMR and MRI with relatively short acquisition times. The improved MR sensitivity has introduced new venues for the development of MR molecular imaging probes, a task traditionally hindered by the poor imaging sensitivity of several MR-active nuclei. Several hyperpolarized metabolic substrates, primarily labeled with ^{13}C, have been proposed and evaluated since the inception of dissolution DNP [298,467–470]. Other less conventional nuclei, such as ^{15}N, ^{89}Y, ^{6}Li, and $^{107,\ 109}Ag$, have also been tested, further expanding the scope of HP-NMR for molecular imaging applications. Since in hyperpolarized MR spectroscopy and spectroscopic imaging, the frequency information enables the simultaneous detection of more than one molecular species at the same time and in *real-time*, achieving a reasonably large chemical shift separation between the substrate and downstream metabolic product or a large chemical shift response upon sensing is essential. Because of the intrinsic properties of their chemical shielding tensor, most low-γ nuclei possess a large chemical shift range, e.g., ~200 ppm for ^{13}C and ~900 ppm for ^{15}N, making them highly suitable candidates for molecular MR imaging applications.

Based on the design of the HP imaging probes, there are three main approaches HP nuclei can be utilized for molecular imaging applications. The first and most common method involves HP molecules that can be converted to a new chemical species via a chemical or an enzymatic reaction. This approach allows for chemical reaction rates or enzymatic fluxes to be

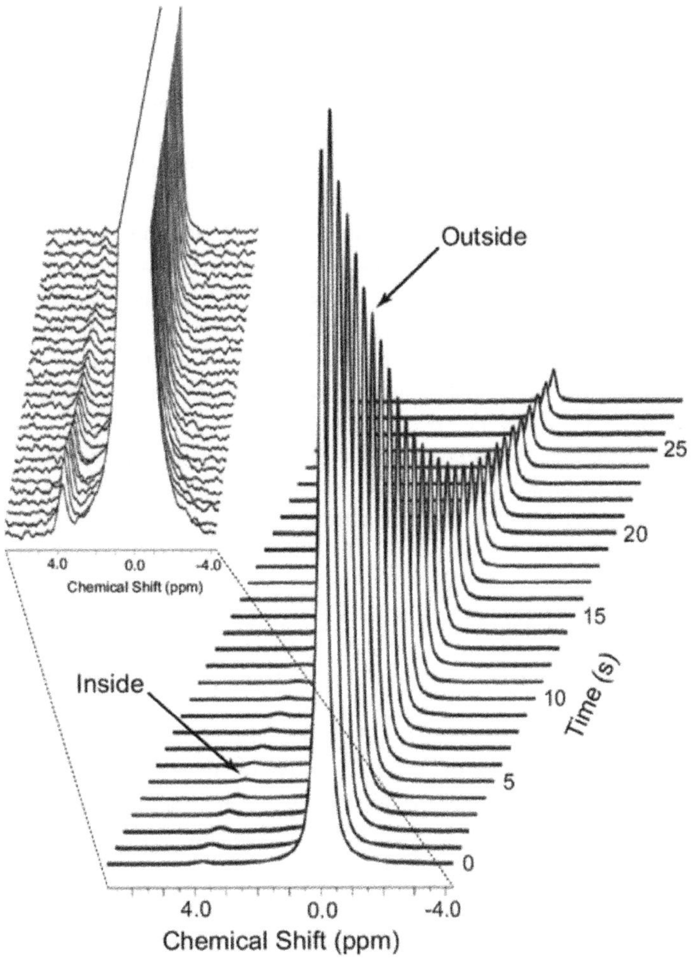

FIGURE 1.22 Monitoring of Cs^+ transmembrane flux by hyperpolarized ^{133}Cs-NMR. Red blood cells (RBCs) were treated with a mechanosensitive cation channel activator (yoda1). The RBCs were then exposed to hyperpolarized $^{133}Cs^+$ (3.3 mM initial extracellular concentration) at 37°C and 9.4 T. The first-order rate constants characterizing the fluxes were obtained from the statistical analysis of the DNP data [466]. Table A.9 (Appendix) lists ^{133}Cs of T_1 values for several compounds. *Reproduced from Ref. [466] with permission.*

estimated from the signal intensities and formation rates of the products. Another approach involves HP probes, in which the chemical shift of the label is sensitive to some physiological parameter such as pH, redox or the local concentration of some chemical species of interest (Ca^{2+}, Zn^{2+}, etc.). In this approach, changes in the chemical shift of the HP-nucleus allow the monitoring of biomarkers or the quantification of analytes. The third and somewhat less common application relies on the direct imaging of HP probes as perfusion markers for angiography. In this approach, the HP signal of the tracer is imaged following intravenous injection into a subject. Nonmetabolizable agents with long T_1 have been designed for this purpose, although long T_1 is not an absolute requirement because hyperpolarized water was also used to generate angiographic images. All three approaches, along with examples of HP probes specific to each method, will be discussed in the following sections.

1.8.1 HP probes for quantifying metabolic processes

The main application of hyperpolarized ^{13}C NMR is the real-time detection and quantification of metabolites in various tissues or organs following the injection of a hyperpolarized ^{13}C-labeled substrate [377,471–474]. Because ^{13}C MRS/MRI can provide spectral information (^{13}C chemical shift), the substrate and its downstream metabolites can be observed at the same time, independently from each other (each compound has a distinct ^{13}C NMR) spectrum. In addition, ^{13}C MRI offers ^{13}C spectral information with spatial encoding, allowing the monitoring of metabolic processes in various organs. In this context, the lack of spectral information in positron emission imaging, which is the current standard clinical imaging

modality for monitoring metabolism, is a big disadvantage, which severely limits the obtainable information in comparison with ^{13}C MRS. Soon after its introduction, hyperpolarized NMR/MRS/MRI was mostly seen as a novel imaging tool for characterizing metabolic activities in diseases due to the much-improved sensitivity combined with the rich spectroscopic information this technology is capable of providing [475–479]. HP probes in the earlier stage of technology development were almost exclusively ^{13}C-enriched metabolic substrates that, upon undergoing metabolism, were at least partially converted to a completely new chemical species. As the hyperpolarized magnetization decays by T_1 relaxation, the HP-signal can be observed for $5T_1$ at best [312]. However, the spectral information contained in the ^{13}C NMR data also implies that complete conversion of the injected substrate within this time window is not necessary. Successful monitoring of specific enzyme-catalyzed steps only requires sufficient labeling of the downstream metabolite pools to observe the HP-^{13}C NMR signals of the products. Signal intensities of the HP metabolites are generally normalized to the total signal intensity or the signal of another HP species, e.g., the injected parent HP probe, another downstream metabolite, or a coinjected nonmetabolizable reference probe, to minimize potential errors in signal quantification arising from the inevitable variations in polarization levels across HP exams. Conversion rates or metabolic fluxes can then be estimated by analyzing the signal intensities of the products, i.e., downstream metabolites, as a function of time. Imaging HP signals of the metabolites also allows for metabolite concentrations to be visualized and quantified in tissues of interest, such as malignant tumors, the heart, and other metabolically active tissues. Ideal imaging probes for this approach are metabolic substrates that can be rapidly metabolized to produce downstream metabolites with chemical shifts resolvable from those of the original HP probe. Even though a large number of ^{13}C-labeled compounds have been hyperpolarized, so far most in vivo experiments have been performed with HP-[1-^{13}C]pyruvate. Hyperpolarized pyruvate remains the only hyperpolarized tracer tested in humans. This is because ^{13}C-labeled pyruvic acid polarizes extremely well by DNP, and the ^{13}C label at C1 or C2 positions has favorably long T_1 values. Pyruvate is rapidly transported into cells, and it lies at a key intersection in intermediary metabolism. Pyruvate, the end product of glycolysis, connects the glycolytic metabolism of glucose to the *tricarboxylic acid cycle (TCA cycle, also known as citric acid or Krebs cycle)* via PDH. It is converted to lactate via lactate dehydrogenase (LDH) and to alanine via alanine transaminase (ALT) [471,473]. Thus, these metabolic processes can be monitored by measuring the production of HP ^{13}C-bicarbonate, [1-^{13}C]lactate, and [1-^{13}C]alanine from [1-^{13}C]pyruvate. Imaging the metabolic downstream products of pyruvate could therefore provide useful information for detecting and characterizing metabolic abnormalities associated with several diseases.

Up to date, the most important application of HP [1-^{13}C]pyruvate is the metabolic characterization of cancer [376,480,481]. The LDH-catalyzed rapid HP [1-^{13}C]pyruvate to HP [1-^{13}C]lactate exchange, which proceeds with negligible de novo production of lactate, gives rise to a lactate signal whose intensity is proportional to the size of the preexisting lactate pool [482]. The lactate/pyruvate ratio obtained after the injection of HP-pyruvate has been shown to be a good biomarker for detecting the Warburg effect [467,483]. The presence of a large lactate pool in tumor tissues due to the Warburg effect is a well-known phenotype of fast-growing tumors (Fig. 1.23). Other metabolic products of HP [1-^{13}C] pyruvate have also been imaged as biomarkers for diseases. For example, the production of HP [1-^{13}C]alanine from HP [1-^{13}C]pyruvate was used as a diagnostic marker for liver injury in rats treated with carbon tetrachloride (CCl_4) [484]. The increased metabolism of HP [1-^{13}C]pyruvate to [1-^{13}C]alanine agreed well with elevated ALT levels, which is a marker for liver injury and inflammation. Last but not least, the bicarbonate signal detected from the decarboxylation of HP [1-^{13}C] pyruvate by PDH has been used as a measure of mitochondrial oxidative metabolism in various tissues [485]. These representative examples demonstrate how hyperpolarized ^{13}C-labeled tracers can be used to monitor metabolic states in tissues by detecting the signal of downstream products. A large number of hyperpolarized ^{13}C-labeled substrates have been developed for metabolic imaging [486].

1.8.2 Applications of HP probes with stimulus-sensitive chemical shifts

Several MR-active nuclei possess an extremely large chemical shift window, e.g., 900 ppm for ^{15}N, with chemical shifts highly sensitive to the chemical environment. This unique NMR property can be exploited for the design of molecular imaging probes for sensing applications. A large number of hyperpolarized probes with stimulus-sensitive chemical shifts have been developed to measure local concentrations of certain analytes, such as H^+ and metal concentrations, based on stimulus-induced chemical shift changes. The hyperpolarized tracer in these probes includes ^{13}C, ^{129}Xe, ^{15}N, and ^{89}Y. The changes in chemical shift are generally induced by a chemical or enzymatic reaction, protonation, coordination of a metal ion, or changes in noncovalent interactions, as is the case for ^{129}Xe. Metal ion coordination, or protonation, alters the chemical environment and hence affects the shielding of the HP nucleus. When these processes occur directly on the HP nucleus (e.g., ^{15}N), the effect on the chemical shift is dramatic. However, protonation or metal ion coordination often occurs at a heteroatom that is in close proximity to the hyperpolarized nucleus, for example, a carboxylate or phosphonate

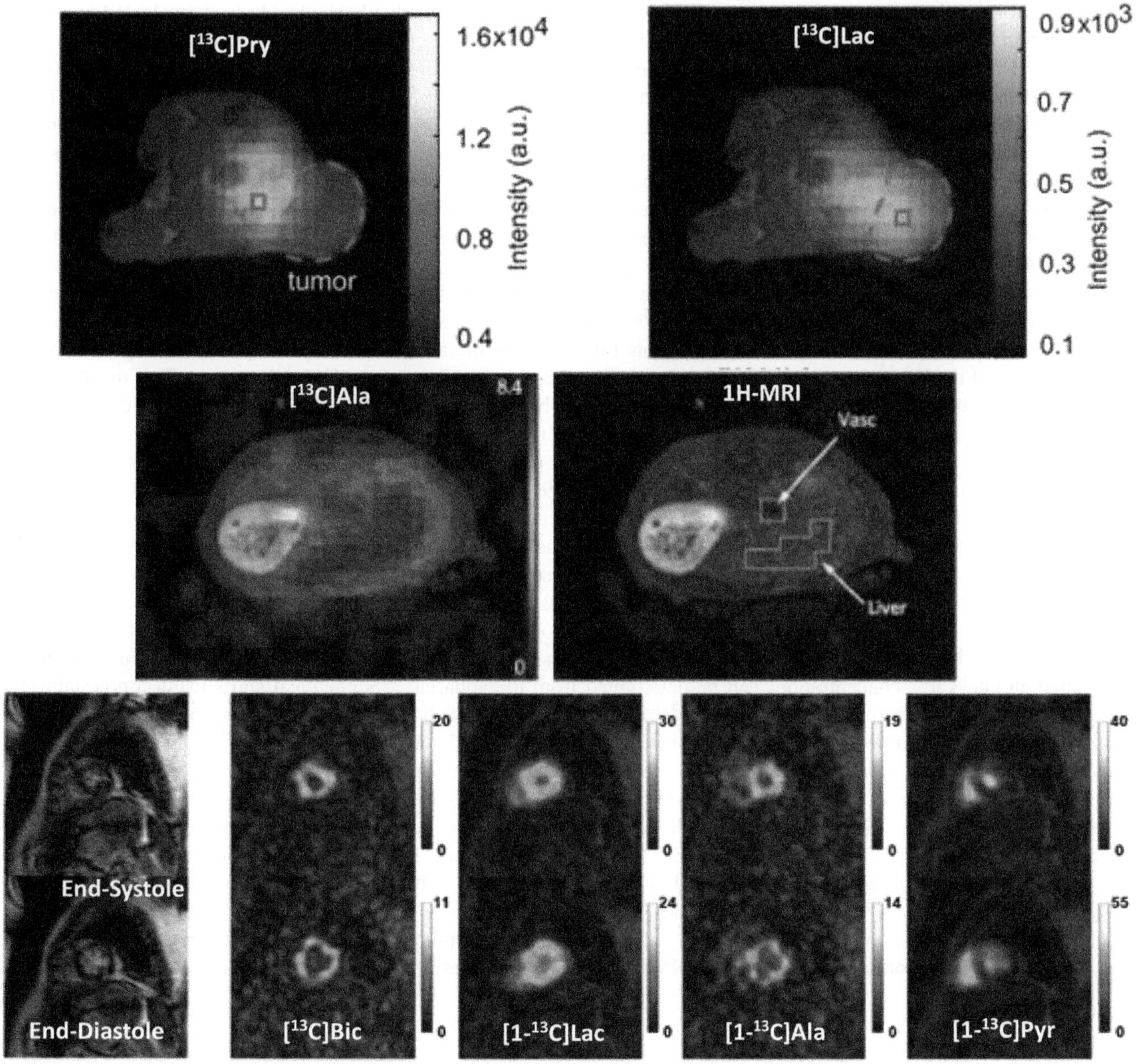

FIGURE 1.23 Metabolic imaging with HP [1-^{13}C]pyruvate. (A) metabolic imaging of LDH activity in a subcutaneous MAT-B-III 13762 breast cancer tumor at the right flank of a rat [487]. (B) imaging of HP-[1-^{13}C]alanine as a marker for liver injury in rats [484]. (C) imaging pyruvate metabolism in the human heart [488]. *Reproduced from Refs. [484,487,488] with permission.*

oxygen. In these cases, the effect on the chemical shift is weaker. Occasionally, long-range effects in which the readout nucleus is three or more bonds away from the sensing atom can also be observed. To determine the local concentrations of the analytes, it is generally necessary to record calibration curves to establish the relationship between the analyte concentration and the apparent chemical shift. Ideal candidates for hyperpolarized MR sensors should contain sensing nuclei that have a long T_1 to preserve hyperpolarization. Their chemical shift should also be responsive to provide reasonable sensitivity. For this reason, the use of non-^{13}C heteronuclei, in particular ^{15}N, is more advantageous. However, spectral resolution depends on the field as well as the gyromagnetic ratio of the nucleus and should be considered in sensor design. Finally, yet importantly, the effect of the sensing process (i.e., coordination or protonation) on the T_1 and T_2 should be minimal. For example, exchange-related line broadening effects may lead to overlapping peaks.

The development and in vivo translation of hyperpolarized probes in this category have not been as well established as those in the other two groups because of the challenges associated with the complexity of the probe design and difficulties

in imaging non-^{13}C heteronuclei. However, there are examples of HP sensors that have already been evaluated in rodents, demonstrating the potential of using HP chemical shift imaging agents for sensing applications. The first examples of HP probes for chemical shift imaging applications included ^{89}Y chelates that showed pH sensitive ^{89}Y chemical shifts [405]. In this probe design, the protonation of the ligand imparted pH-sensitivity to the ^{89}Y in the physiologically relevant pH range. However, the pH response was moderate, 10–20 ppm (Fig. 1.18). While ^{89}Y MRS may not be readily translatable in vivo due to difficulties associated with the detection of such a low-sensitivity nucleus, the results from this study demonstrated for the first time that HP agents could potentially be designed to measure local concentrations of certain physiological biomarkers, such as pH. Other HP ^{89}Y chelates based on the same principle were also developed and tested as pH sensors [489]. However, the majority of the follow-up studies have been focused on more traditional NMR-active nuclei, such as ^{15}N and ^{13}C, that are easier to translate into in vivo applications.

^{15}N is an attractive candidate for chemical shift imaging because of the large chemical shift range and the high chemical shift sensitivity to the chemical environment of the ^{15}N center. Taking advantages of ^{15}N NMR, ^{15}N-enriched pyridine and its derivatives have been tested as pH imaging agents [385]. Unlike the aforementioned ^{89}Y probes, changes in the ^{15}N chemical shifts of the HP ^{15}N-pyridine derivatives are induced by direct protonation of the ^{15}N centers. This results in very large chemical shift differences (~90 ppm) between the free base and fully protonated ^{15}N centers. The data shown in Fig. 1.24 also highlights the disadvantage of this probe design: the protonation of the ^{15}N in the pyridine ring unfavorably affects both the T_1 and T_2 of the label in the pH region near the pK_a. HP ^{15}N probes with chemical shifts sensitive to other analytes, such as metal ions, have also been developed [26,490,390]. An interesting HP ^{15}N labeled metal ion sensing probe design contains a quaternary ^{15}N ammonium center as a ^{15}N NMR readout unit because of its long T_1 [116,390]. This moiety has been proposed as a general platform for the design of hyperpolarized ^{15}N sensors. A Ca^{2+} sensor incorporated a metal binding unit based on the well-known chelator ortho-aminophenol-N,N,O-triacetate, which has good selectivity for Ca^{2+}. Binding to Ca^{2+} ions affects the electron distribution on the ^{15}N center via the aromatic ring, inducing concentration-dependent changes in ^{15}N chemical shifts. Only one ^{15}N signal was detected, indicating that the exchange between the free ligand and the Ca^{2+} complex is fast on the ^{15}N NMR timescale. As with the pH-sensitive probes mentioned earlier, Ca^{2+} concentrations can be determined from the apparent ^{15}N chemical shifts using a preestablished calibration curve. However, the change in the ^{15}N chemical shift was only 1.5 ppm, which may not provide sufficient resolution for in vivo studies. Using a similar probe design, another calcium-sensitive probe, ^{15}N-labeled ortho-aminophenol-N,N,O-triacetate, was developed in which the coordinating N-atom was ^{15}N-labeled. As expected, this sensor provided a much better response (5 ppm upfield shift upon Ca^{2+} binding) because the ^{15}N label is directly involved in the metal ion chelation. The larger chemical shift change window improves the accuracy of the probe for measuring Ca^{2+} concentrations [490]. Hyperpolarized ^{15}N NMR Zn^{2+} sensors were designed based on the well-known Zn^{2+} chelator TPA. The deuterated version of the ligand ^{15}N labeled at the central nitrogen exhibited long T_1 and was successfully used to detect and quantify Zn levels in biological samples. The ligand displayed a sharp ^{15}N resonance that underwent a large, 20 ppm upfield shift upon complexation with Zn^{2+}. Unlike the above mentioned Ca^{2+} chelator, the exchange between the free ligand and the Zn complex is slow on the NMR time scale. As a result, separate signals were observed for the two

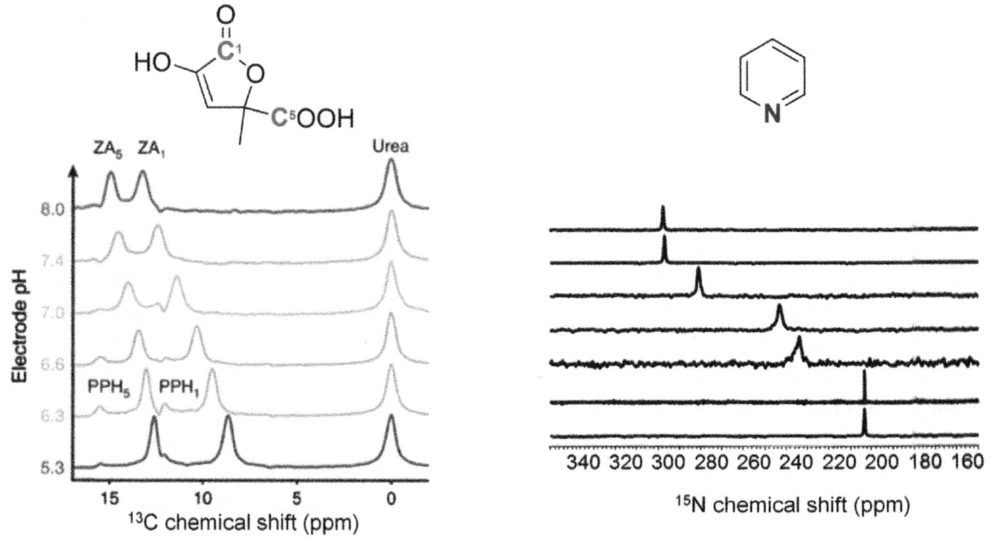

FIGURE 1.24 pH-dependent chemical shift changes: (Left) HP ^{13}C-zymonic acid [491] and (Right) HP ^{15}N-pyridine [385]. *Reproduced from Refs. [385,491] with permission.*

species in the presence of excess ligand, and the Zn^{2+} concentration could be estimated from the ratio of the bound versus unbound ^{15}N signal intensities [26].

^{13}C-containing sensors are not as attractive for chemical shift imaging applications as their ^{15}N counterparts due to the narrower ^{13}C chemical shift window. Moreover, ^{13}C does not possess lone-pair electrons and therefore cannot participate in protonation equilibria or directly coordinate with metal ions in aqueous media (apart from a few exceptions such as acetonyl acetone derivatives). Stimulus-induced chemical shift changes for HP ^{13}C probes are therefore expected to be rather small. Nevertheless, the sharp ^{13}C signals combined with the well-established ^{13}C MRI technology may compensate to some extent for these potential drawbacks. Several ^{13}C-labeled carboxylic acid and amino acid derivatives in which the protonation of a carboxylate or the amino group imparts pH sensitivity to the ^{13}C label have been proposed as hyperpolarized ^{13}C pH sensors. These are *N*-(2-acetamido)-2-aminoethanesulfonic acid (ACES) [492], dialkylated malonates, zymonic acid, and alanine ethyl ester [493]. Although the ^{13}C chemical shift dispersion between 6 and 8 is less than 6 ppm for all these sensors, considering the relatively large gyromagnetic ratio of ^{13}C, this may provide sufficient resolution and accuracy in vivo. In vivo experiments with hyperpolarized, ^{13}C-labeled zymonic acid demonstrated that pH mapping by ^{13}C chemical shift imaging is feasible [491]. In this study, HP ^{13}C-zymonic acid was evaluated as a pH imaging agent (Fig. 1.24). Both carboxylate carbons in this interesting pyruvate derivative were ^{13}C-labeled. The pH-induced chemical shift dispersion curves of the two ^{13}C centers were slightly different, and the chemical shift differences between the two ^{13}C centers were also pH-dependent and could be used for pH imaging. A pH map of a rat tumor generated with HP ^{13}C-zymonic acid is also shown (Fig. 1.24). ^{13}C-based metal-ion sensors have also been reported. For example, the carboxylate group in ^{13}C-labeled EDTA exhibited about a 5 ppm downfield shift upon binding to Ca^{2+} at neutral pH [494,495]. Although this response in ppm is about the same as displayed by the above-discussed ^{15}N calcium sensor, in terms of frequency, the ^{13}C probe outperforms the ^{15}N one by a factor of about 2.5 ($\gamma_{13C}/\gamma_{15N}$) at any given field.

1.8.3 Long-lived, chemically inert HP probes as MR contrast agents

The third class of hyperpolarized agents comprises compounds that can be used to visualize perfusion in various organs, including the heart, liver, or brain. Unlike the previously discussed agents, HP probes in this category do not metabolize, nor do they offer sensing capabilities, but rather provide MR signals for tracing and visualization purposes. This approach involves detecting and imaging the signals of the injected parent HP nuclei [496]. The chemical shifts of these HP nuclei should remain unchanged during the course of the imaging session. Therefore, the ideal hyperpolarized perfusion marker should have a small molecular weight and be nonmetabolizable and chemically inert (it should not undergo protonation, metal ion coordination, or redox reactions).

Interactions with biological macromolecules, e.g., binding to plasma proteins, are also undesirable, as these interactions are known to shorten T_1 and HP signal lifetimes. Lipophilicity versus hydrophilicity (as characterized by the octanol water partition coefficient, K_{ow}) is also an important factor because lipophilic compounds with high K_{ow} are permeable to a lipid bilayer, while hydrophilic molecules with a small K_{ow} are membrane impermeable [497]. Long T_1 values, as for all HP-NMR applications, are preferable, but this is not an absolute requirement because hyperpolarized water was also used to generate angiographic images.

A number of HP probes have thus far been evaluated as potential exogenous perfusion agents (Fig. 1.25). The first one, (1-hydroxymethyl-1-^{13}C-cyclopropyl)-methanol and its deuterated version, was developed specifically for this purpose [498–500]. This HP molecule with a ^{13}C enrichment at the tertiary carbon has a T_1 of 82 s at 2.35 T. Following an intravenous injection of the HP probe into healthy rats, ^{13}C intensity maps revealed that several blood vessels, i.e., the vena cava and hepatic veins, as well as the heart ventricle, were clearly visible in healthy rats [498,500]. The results showed that

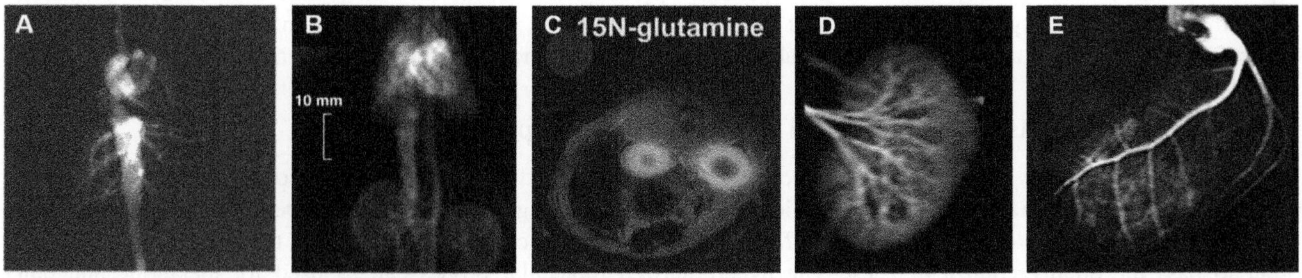

FIGURE 1.25 HP ^{13}C molecules as contrast agents. (A) A ^{13}C MR angiography of a rat following an injection of (1-hydroxymethyl-1-^{13}C-cyclopropyl)-methanol [498] (B) a ^{13}C MR image of a rat injected with [^{13}C]urea [502] and (C) a ^{15}N intensity map showing rat kidneys following an injection with α-tris(deuteromethyl) [^{15}N]glutamine [510] (D) kidney angiography with hyperpolarized water [322], (E) coronary MR angiography in a pig with HP water [511]. *Reproduced from Refs. [322,498,511] with permission.*

a relatively high spatial resolution can be achieved with this ^{13}C probe, with vessels as small as 0.5 mm clearly visible. These seminal studies suggested that HP ^{13}C probes could potentially be used in MR angiography, stimulating a strong interest in the development of chemically inert HP molecules as contrast agents. Currently, two promising ^{13}C-labeled tracers are being investigated with the goal of human translation: ^{13}C-labeled urea and ^{13}C-labeled, perdeuterated *tert*-butanol. These probes differ dramatically in their lipophilicity [501]. Urea is a very attractive candidate for ^{13}C MR angiography because it is a safe, small molecule naturally occurring as a product of protein metabolism in mammals. ^{13}C-enriched urea ([^{13}C]urea) has a relatively long T_1 (e.g., 40 s at 2.35 T) [502]. Enriching both nitrogen atoms with ^{15}N, a spin ½ nucleus, significantly increases T_1 of the carbon center at low magnetic fields [153,503]. The small molecular weight and metabolic inertness make [^{13}C]urea an ideal probe for ^{13}C MR angiography, as the molecule can be transported across tissues via both diffusion and perfusion mechanisms while retaining its chemical integrity and, hence, the ^{13}C chemical shift. Additionally, negligible binding to plasma proteins is expected due to its favorable polarity and hydrophilicity, minimizing an HP ^{13}C signal loss in circulation. ^{13}C MR angiography with HP ^{13}C-urea was first tested in healthy rats [502]. In Fig. 1.25B, a ^{13}C MR image of a rat taken 2 s after an administration of HP [^{13}C]urea clearly shows the presence of HP [^{13}C]urea in the rat vasculature, highlighting an image contrast in multiple blood vessels and organs. Since then, several follow-up studies by the same group as well as others have been reported, demonstrating the feasibility of using this unique HP probe as an imaging agent for ^{13}C MR angiography as well as a probe for evaluating blood flow, tissue perfusion, and renal functions [153,500,501,503–509].

Nonmetabolizable ^{15}N probes with a long T_1 have been explored for potential uses in similar applications [387,510,512–514]. These include ^{15}N-enriched analogs of endogenous substrates such as choline and glutamine. A glutamine derivative in which the alpha amino group was ^{15}N labeled and quaternized with perdeuterated methyls exhibited long T_1 (330 s at 3 T) and was successfully used to image the kidneys [307,510]. (Fig. 1.25C). This tracer offers several advantages over ^{13}C-labeled urea: It has a longer T_1 and less background signal, and because of its zwitterionic structure, it cannot enter cells. This is in contrast to urea, for which there are transporters in various tissues [510,515]. These results show that the ^{15}N-enriched glutamine derivative is a good candidate for renal studies.

In addition to small-molecule HP probes, nanoparticles comprised of MR-active ^{29}Si nuclei have also been evaluated as MR contrast agents. The extremely long T_1 of ^{29}Si signals (about 20 min at 7T) is the main attractive property of these rather unorthodox HP probes. Although in vivo imaging of these probes has been demonstrated in rodents [397,398,516], two major drawbacks of these probes are the poor spatial resolution this system offers and the unknown toxicity profile of silicon nanoparticles, potentially hindering further translations of these probes for biomedical imaging applications.

Finally, two unique perfusion agents should be mentioned: hyperpolarized water and ^{129}Xe (Fig. 1.25D and E). The signal in conventional MRI originates from water present in tissues. Therefore, hyperpolarized water is an obvious candidate as a contrast-enhancing agent for ^{1}H MRI. Water can be hyperpolarized by continuous flow Overhauser DNP or low-temperature dissolution DNP [322,323,517,518]. Perfusion contrast imaging with hyperpolarized water can be performed by continuously infusing it or injecting it as a bolus into the organ of interest through an arterial catheter. High-contrast ^{1}H angiographic images similar to conventional angiography were obtained in the kidney, heart, and other organs of rats and pigs [322,334,511]. Although the spin-lattice relaxation time of water can be over 30 s in D_2O solutions, the T_1 shortens to about a second in blood and tissues. This necessitates the injection of hyperpolarized water via an arterial catheter, which severely limits its application for in vivo MRI. Xenon is an inert gas, and it can be administered by inhalation. It is soluble in blood, and therefore, hyperpolarized ^{129}Xe has potential for angiographic imaging. However, so far, image resolution and SNR in organs other than the lung were rather poor, largely due to the combined effect of the limited transport of hyperpolarized ^{129}Xe and rapid T_1 relaxation in vivo [413,434,519–521].

One could argue that gadolinium-based contrast agents (GBCAs) such as Gadofosveset (sold under the trade names Vasovist and Ablavar, now discontinued) could provide similar, if not better, image contrasts for MR angiography with superior spatial resolution [522]. However, hyperpolarized probes offer an unrivaled advantage over GBCAs technology due to the absence of background signals as well as a greatly reduced risk of toxicity. In addition, hyperpolarized perfusion tracers can be copolarized and injected with metabolic probes for simultaneous metabolic and perfusion imaging [523,524].

1.8.4 Practical aspects of the biological applications of HP probes

For biological studies, the hyperpolarized compound is quickly dissolved in a superheated dissolution liquid to produce a solution of the hyperpolarized agent. For preclinical studies, the polarizing agent OX063 is not removed because it has low toxicity and its relaxation effect is negligible [233]. However, it is critical for human studies to remove free radicals in order to pass the quality control unit test. The hyperpolarized molecular probes can be used for metabolic and other

biochemical studies in various settings (with isolated enzymes, in cell or tissue lysates, cell suspensions, isolated, perfused organs, and in vivo in intact animals) [89,525–527]. These are briefly reviewed below.

1.8.4.1 Isolated enzymes

These studies simply require the addition of an isolated, purified enzyme at an ideal concentration in a physiologic buffer and temperature in an NMR tube compatible with the RF coil, together with the necessary cofactors and ligands [525]. This method is very useful to evaluate the reaction kinetics under controlled conditions as well as to detect if any invisible low-population intermediate species are produced. Besides reaction kinetics, this method is very informative and provides insights into molecular interactions [129] and protein folding [528]. It is important to note that enzymes are frequently present in a variety of isoforms and are produced from diverse sources; thus, care must be taken to choose the correct isoform and species.

1.8.4.2 Cell or tissue lysate

Cell or tissue lysate is a common alternative for enzymes that are either too expensive or just unavailable in purified form. Depending on the origin of the enzyme, the cell or tissue is selected to prepare lysate and optimized for the number of cells. The lysate can be generated either mechanically or chemically to ensure maximum disruption of intracellular organelle membranes. Before HP testing, the soluble supernatant is treated with excess cofactors to account for unavailability due to nonspecific binding to other enzymes present in the lysate. To investigate noncytosolic enzymes, a buffer containing a cocktail of protease and phosphatase was used to inhibit protein breakdown under ice-cold conditions [525].

1.8.4.3 Cell suspension

Testing an HP probe in cell suspension is widely adopted by several published studies [89,529,530]. In a similar way to the previous method, the selection of cells is based on the specific objectives of the study, and the quantity is optimized accordingly. However, the key disadvantage of this approach is the possibility of altering the metabolic status of the cells due to the transition from the frozen state to the NMR tube, assessing metabolism upon separation of adherent cells, and deviation from physiologically relevant temperatures and pH [89]. Conventional ^{1}H, ^{2}H, ^{13}C, ^{23}Na, and ^{31}P NMR have been used to study metabolism in cell cultures since the early 1970s [531]. The improved sensitivity of hyperpolarized NMR greatly facilitated metabolic studies in cell cultures with ^{13}C-labeled substrates. Short T_1 values do not pose such a severe limitation as for in vivo studies. For example, HP-[U-$^{13}C_6$, U-$^{2}H_7$]-D-glucose has successfully been used to study glycolysis in cultures of yeast (*Saccharomyces cerevisiae*), *Escherichia coli*, and human cancer cell lines [467]. Experimentally, live cell experiments can be performed by simply loading a sufficient number of cells (10^6–10^8) into an NMR tube, followed by the rapid injection of the hyperpolarized solution. The metabolism is then monitored by ^{13}C MRS, by using either a series of frequent small flip-angle pulses or carefully timed 90-degree pulses, which provide higher sensitivity but lower temporal resolution [25]. Longitudinal investigations are not possible in this way since the viability of the cells is considerably diminished over time due to ischemia and excessive concentrations of HP agents. This problem can be at least partially overcome by using MR-compatible bioreactors. These provide a continuous supply of oxygen (air, 5% CO_2) and nutrients so that the cells can be kept viable for longer time periods (up to 3 days). A simple design for NMR spectroscopy uses an NMR tube and a perfusion system consisting of polyethylene tubing and a pump for the delivery and circulation of the perfusate and oxygen from a container in the NMR tube [532]. Cells are immobilized either by encapsulation in alginate or agarose beads [533–535] or by growing them on the surface of microcarrier beads such as Biosilon [536]. The beads are then loaded into the NMR tube. A more sophisticated setup may involve the use of hollow fiber bioreactors, which are commonly used for conventional MRS/MRI experiments. These are cartridges containing bundles of hollow fibers, which are capillaries with semipermeable membrane walls (molecular weight cutoff 10–30 kDa) made of a polymer such as polysulfone. They usually have a diameter of 150–200 microns. The cells are grown in the extracapillary spaces, while the culture medium carrying oxygen and nutrients is pumped through the intracapillary spaces [537,538].

1.8.4.4 Isolated perfused organs

Perfused organs, most commonly mouse and rat hearts and livers, have been used to study various metabolic pathways with mass spectrometry (MS) and conventional NMR spectroscopy for a long time, and they are also adopted for HP-NMR studies [539–543]. This is an excellent method for assessing metabolism in both healthy and pathological tissues, and it is

particularly useful to study various models of metabolic diseases. Although an isolated organ is obviously not a perfect model for in vivo conditions, perfused organ experiments have some advantages over in vivo measurements. They allow the assessment of organ-specific metabolism under highly controlled and reproducible conditions without the confounding effects of other organs [89,544]. In particular, they offer greater sensitivity from inherent localization and a high filling factor, allowing the investigation of metabolic intermediates that would otherwise be inaccessible for a better understanding of the biochemical processes. Several HP-^{13}C MRS studies were reported to study pathophysiology and metabolism in perfused isolated organs from rodents, including the heart [545–547], liver [548–550] and perfused isolated organs from porcine models [551]. The usual setup and sequence of steps in a perfused organ HP-^{13}C experiment are outlined in Fig. 1.26. The results of these investigations can be used to assist in the planning and validation of in vivo experiments.

Although sophisticated perfusion systems with electronic pressure and flow rate control are available, the adaptation of the classic vertical glass column-based perfusion apparatus for a vertical bore spectroscopy magnet is easier. The column is equipped with a water jacket to regulate the temperature of the perfusing medium. A continuous supply of oxygen is maintained by aerating the perfusing medium with 95% O_2 and 5% CO_2. The hyperpolarized solution is injected into the perfusion medium directly above the organ via a catheter.

Detailed protocols describing isolated perfused organ experiments are available in the literature and will not be repeated here [89,552,553]. Briefly, excised hearts are generally perfused with constant hydrostatic pressure using Langendorff's method, i.e., in retrograde mode (opposite to normal blood flow), via a cannula in the aorta [554,555]. The perfusing medium is usually a modified Krebs-Henseleit solution, which often contains unlabeled oxidizable substrates such as octanoate [544,556]. The heart rate and diastolic and systolic pressures can be monitored with a water-filled balloon inserted into the left ventricle and connected to a pressure transducer. Global ischemia can easily be induced by blocking the perfusion lines, while regional ischemia can be brought about by the ligation of a coronary artery. Livers are perfused through the portal vein with a modified Krebs-Henseleit medium in a similar way to hearts.

1.8.4.5 Animal models

These are especially critical for safety investigations and are the final stage before clinical translation of any innovative diagnostic or therapeutic procedure [557]. The biological significance of choosing an animal model for HP MR studies to explore biochemical processes transcends all applications [377]. The use of large animals is very close to human biology in terms of organ shape and physiology, but the logistics and disease model availability necessitate the extensive use of

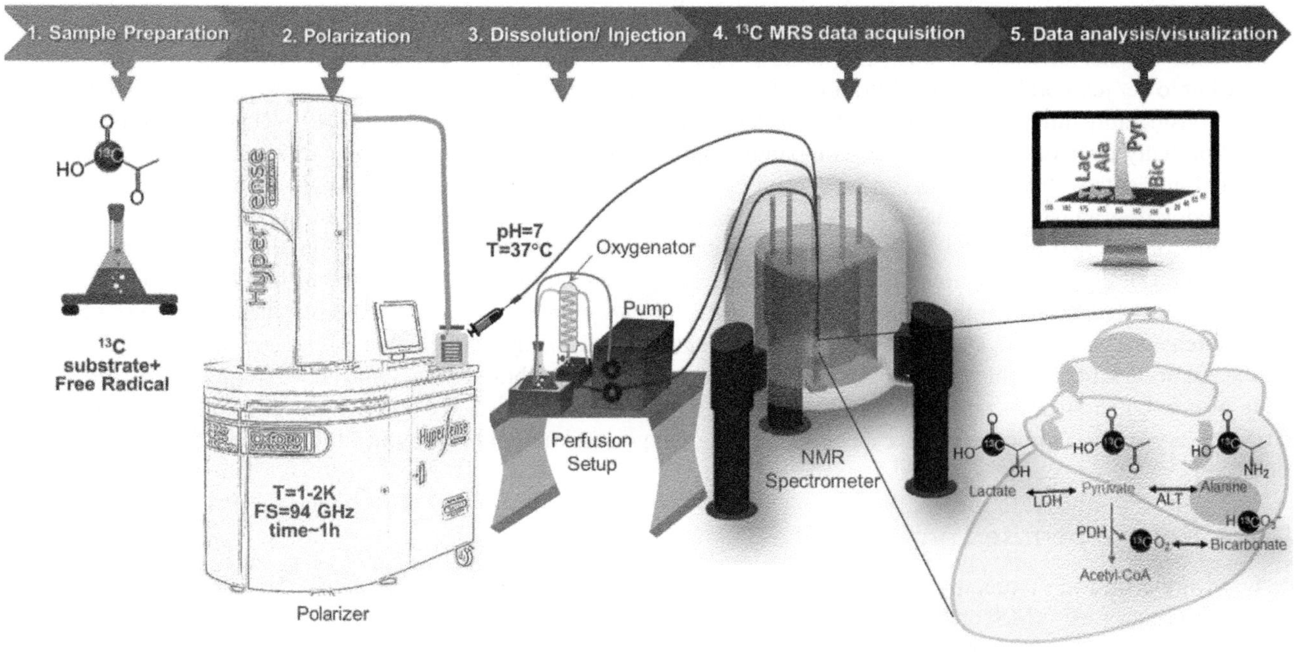

FIGURE 1.26 A typical workflow of a hyperpolarized ^{13}C MR study in isolated, perfused organs.

smaller animal models such as rats and mice. Several factors are typically considered for in vivo preclinical HP studies; the two most important are the physiological and nutritional status of the animals and the concentration of the HP probes.

1.8.4.6 Data analysis and interpretation of HP data

The data analysis and interpretation of HP results are highly dependent on specific goals and the type of study design that was employed. The relevant literature emphasizes the several components that regulate the HP signal as well as the intricacies of the underlying enzyme activity [558,559]. For example, the conversion of HP-pyruvate to HP-lactate is not always considered a function of enzyme activity but rather also depends on the availability of NADH as a cofactor and pool sizes. Several studies also reported fitting models for HP signals and explained the compartmentalization of HP metabolism [560]. The magnetization decay curves can be fitted to various models to extract kinetic parameters (K_M and V_{max}) and quantify fluxes through enzyme catalyzed steps [544,561–570]. For perfusion of isolated organs and preclinical studies, the findings are usually validated by conventional ^{13}C NMR isotopomer analysis of tissue or plasma samples as well as other spectroscopic techniques as appropriate [350,363,373,571–574]. Studies have also used various software, such as tcaCALC [575] and INCA [576] for metabolic flux analysis from NMR data. Taken together, the hyperpolarized ^{13}C studies are very informative and provide novel insights into the pathophysiology and metabolism, especially when combined with other conventional spectroscopy (^1H, ^{13}C, and ^{31}P) and isotopomer analysis.

1.9 Design and synthesis of ^{13}C-labeled compounds for dissolution DNP

Organic synthesis, by providing access to structurally diverse metabolic substrates and intermediates, has played a key role in metabolic studies since the early days of the 20th century. Synthetic advances not only enabled making large quantities of metabolites in shorter routes but also helped identify and characterize the structure of metabolic intermediates before the development of sophisticated analytical techniques such as NMR and MS. A significant milestone in metabolism research occured shortly after the discovery of deuterium by Harold Urey in 1931 [577]. Konrad Bloch, David Rittenberg, and Rudolph Schoenheimer were among the first scientists who introduced isotopes in nutrients as a tagging strategy. With these newly developed molecules, they could determine the origin and fate of intermediary metabolites and perform quantitative measurements.

Metabolic transporters and enzymes are highly selective toward their substrates; therefore, minimal structural modifications should be used for these studies. Isotope labeling produces insignificant changes to the physiochemical properties of metabolites, unlike modifications such as optical reporters. In addition, isotope enrichment provides the means to study parallel metabolic pathways [578]. Isotope labeling, as opposed to regular organic synthesis, requires unique treatment due to the cost and availability of the precursor materials and reagents; therefore, it is limited to certain reactions. Biocatalysis or chemoenzymatic reactions offer new opportunities by enabling stereoselective and stereospecific modification of different types of molecules. Isotope-labeled molecules, particularly with stable isotopes, are less hazardous and can be studied with MS and magnetic resonance. In mass spectrometric studies, isotope dilution provides an accurate measure of the quantity of the isotopes in the analyte [579]. Using hydrogen-deuterium exchange (HDX) in microdroplets, isobaric ions can be distinguished based on their unique MS spectral patterns [580]. Different branches of magnetic resonance utilize the enrichment of MR-visible isotopes to compensate for the inherently low signal, allowing the fate of exogenous molecules to be studied [581]. In addition, isotope labeling allows for investigating ligand-protein interactions [582].

Hyperpolarized substrates for dissolution DNP have thus far mostly been enriched with ^{13}C; however, recent studies have shown that additional enrichment with isotopes such as ^2H and ^{15}N, where applicable, can improve the relaxation properties of the ^{13}C signal. In this chapter, we will discuss some common synthetic strategies that can be used to develop different categories of hyperpolarized probes. We focus on ^{13}C because it is the most important for biomedical imaging. The synthesis of ^{15}N-labeled probes will not be discussed because there are excellent recent reviews on this topic [384,583]. Rather than presenting a comprehensive overview of synthetic methods for isotope labeling, we will focus on synthetic methodologies for ^{13}C labeling from a dissolution DNP perspective. Multiple isotope labelings to improve ^{13}C relaxation properties will also be discussed. We aim to demonstrate the power of organic synthesis and biocatalysis in the development of hyperpolarized probes, with the hope that future investigators can apply these strategies in the synthesis of novel ^{13}C-labeled hyperpolarized probes.

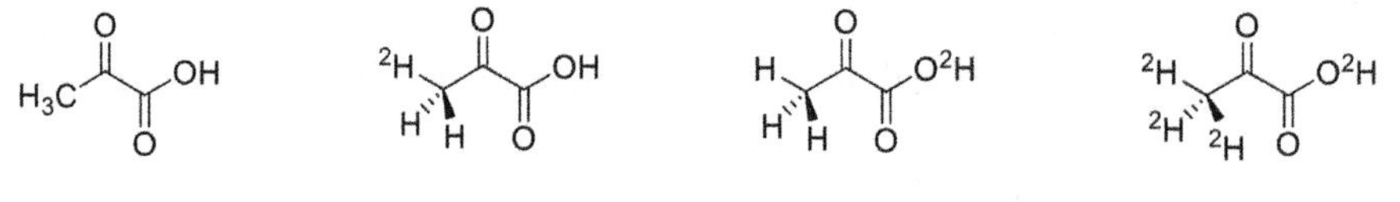

formula	$C_3H_4O_3$	$C_3H_3{}^2HO_3$	$C_3H_3{}^2HO_3$	$C_3{}^2H_4O_3$
name	pyruvic acid	[3-^{2}H]pyruvic acid	[O-^{2}H]pyruvic acid pyruvic [^{2}H]acid	[O,3,3,3-$^2H_4$]pyruvic acid

FIGURE 1.27 IUPAC nomenclature of isotopically substituted pyruvic acid derivatives.

1.9.1 Nomenclature of isotopically labeled compounds

Multiple nomenclature systems have developed in parallel for different fields and applications. As science has become increasingly interdisciplinary, it is important to have a common naming convention to avoid ambiguities and conflicts in nomenclature. Section H of the International Union of Pure and Applied Chemistry (IUPAC) Rules on the Nomenclature of Organic Compounds provides an unambiguous naming convention for isotopically modified compounds. In this section, we introduce some of the IUPAC nomenclature rules relevant to d-DNP probes [584].

An *isotopically modified compound* has a composition in which the isotopic ratio of nuclides for at least one element varies from natural abundance. When an isotope is used to modify a compound, this is described using the atomic symbol of the nuclide, with the isotope being shown as a left-hand superscript, as in ^{2}H, ^{13}C, and ^{15}N. Isotopically modified compounds can be categorized as either *isotopically substituted* or *isotopically labeled.* In an isotopically substituted compound, all molecules are enriched with the indicated nuclide(s) at the designated position(s). An isotopically labeled compound is a mixture of isotopically unmodified molecules and one or more analogous isotopically substituted compound(s). Thus, isotopically labeled compounds are mixtures of different chemical entities. A specifically labeled compound is a mixture containing a unique isotopically substituted compound called the isotopically unmodified compound.

The formulae of isotopically modified compounds are written in the usual way, except that the appropriate nuclides' symbols are used. To define the labeling site, the position and nuclide symbol are placed in the relevant brackets immediately before the name of the labeled portion of the compound. Parentheses are used for an isotopically substituted compound, and square brackets are used for an isotopically labeled compound (i.e., a mixture). Note that there is no hyphen or space between the bracket and the name (Fig. 1.27).

Note that some commercial sources may still designate deuterium as *d* and tritium as *t*, as shown below for labeled pyruvic acid and its corresponding sodium salt (Fig. 1.28).

When there are two otherwise identical chains, the group bearing the label should be incorporated into the stem and not treated as a substituent. In a name consisting of two or more words, the isotopic descriptor is placed before the word corresponding to the labeling site (Fig. 1.29).

The numbering of an isotopically modified compound should be the same as that of the unlabeled compound. The locants (numbers and letters defining the position of functional groups and substituents) are assigned following the general organic chemistry nomenclature. Essentially, the numbering is based on the main carbon chain (even if there is a

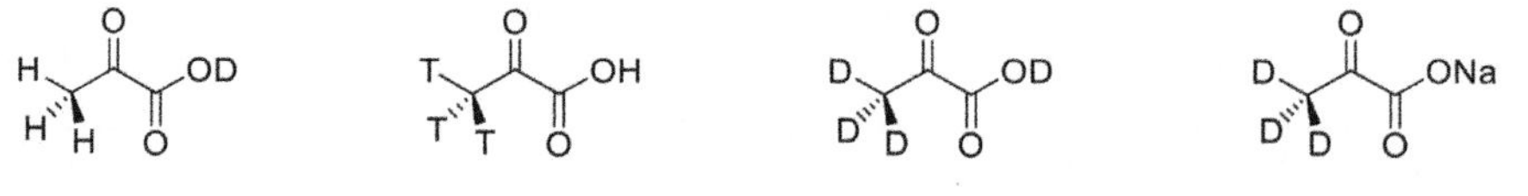

formula	$C_3H_3DO_3$	$C_3HT_3O_3$	$C_3D_4O_3$	$C_3D_3O_3Na$
name	pyruvic acid-*d* [O-*d*]pyruvic acid	pyruvic acid-3,3,3-t_3	pyruvic acid-d_4	sodium pyruvate-3,3,3-d_3

FIGURE 1.28 Examples of older terminology for isotopically labeled pyruvic acid derivatives.

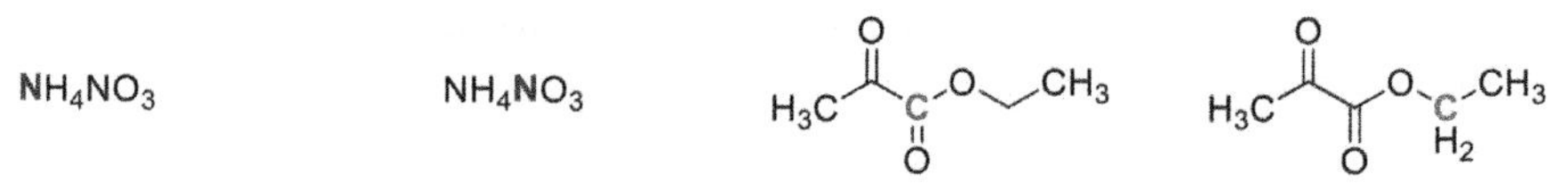

formula	$N^{15}NH_4O_3$	$N^{15}NH_4O_3$	$C_4{}^{13}CH_8O_3$	$C_4{}^{13}CH_8O_3$
name	[^{15}N]ammonium nitrate	ammonium [^{15}N]nitrate	ethyl [1-^{13}C]pyruvate	[1-^{13}C]ethyl pyruvate

FIGURE 1.29 Placement of isotopic descriptor in identical chain of atoms.

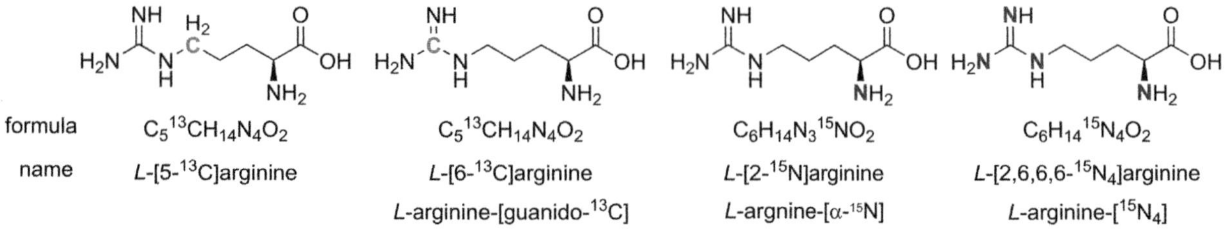

FIGURE 1.30 Examples of numbering of isotopically labeled compounds.

heteroatom in the chain), and the principal functional group should receive the lowest locus. For isotopically enriched heteroatoms attached to the main carbon chain, the numbers correspond to the carbon to which they are attached (Fig. 1.30).

When isotopes of different elements are present as nuclides in a molecule, the nuclide(s) with the lowest locants are written first. When there are more than one isotope of different elements with the same locants, symbols are arranged alphabetically (Fig. 1.31).

A chiral designation (*R*), (*S*), *D*, or *L* precedes the isotopic designation (Fig. 1.32).

The symbol U is used in place of locants to denote uniform labeling when the mixture has an even distribution of labeling across all possible positions with the same isotopic ratio. The symbol G is used in place of locants to denote general labeling when all positions of the designated element are labeled, but not necessarily uniformly (Fig. 1.33). For

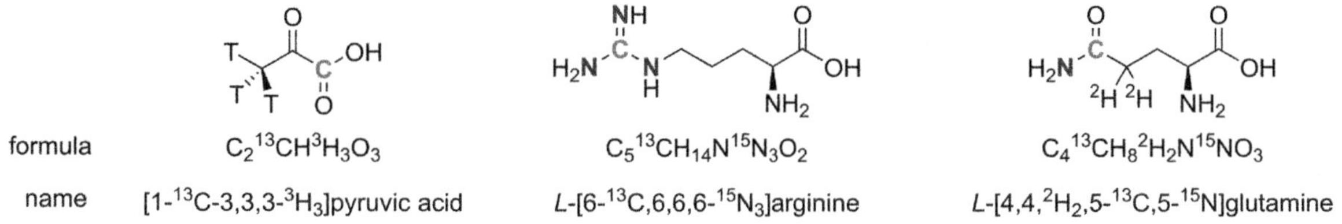

FIGURE 1.31 Nomenclature of compounds labeled with different isotopes.

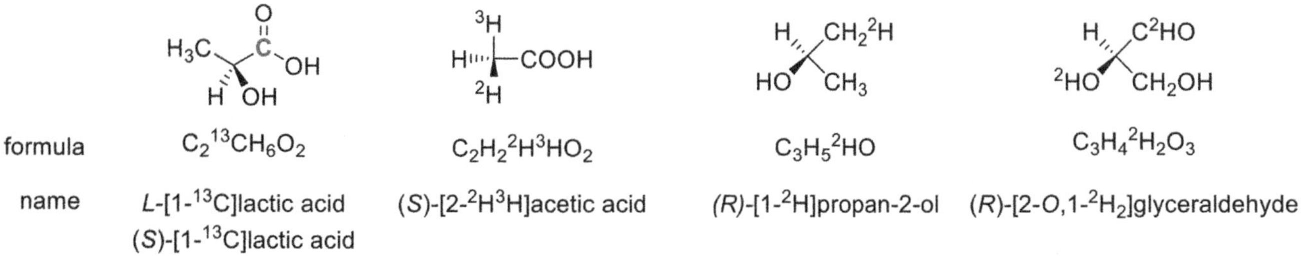

FIGURE 1.32 The chiral designation (*R*, *S*, *D*, *L*) precedes isotopic designation.

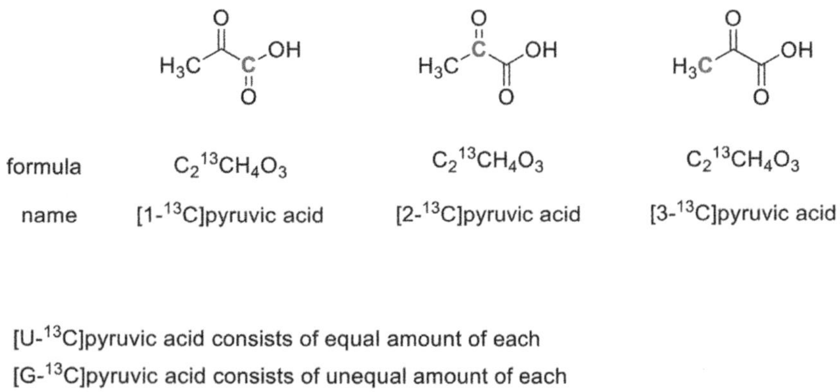

FIGURE 1.33 Uniformly ^{13}C-labeled pyruvic acid is an equimolar mixture of C1, C2, and C3 labeled pyruvic acids. Generally, ^{13}C-labeled pyruvic acid is an unequimolar mixture of C1, C2, and C3 labeled pyruvic acids.

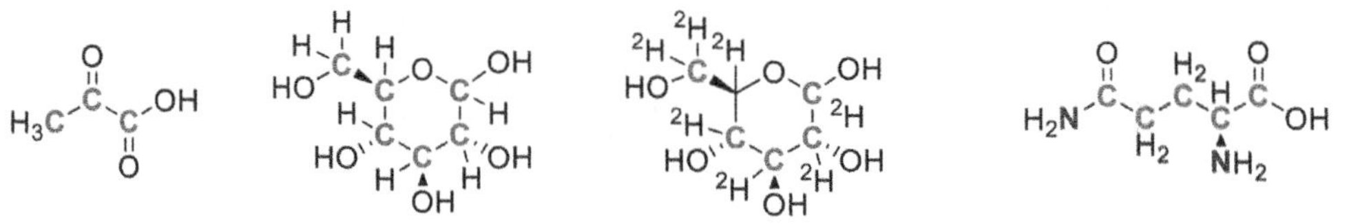

formula	$^{13}C_3H_4O_3$	$^{13}C_6H_{12}O_6$	$^{13}C_6H_5{}^2H_7O_6$	$C_4{}^{13}CH_8{}^2H_2N^{15}NO_3$
name	[UL-$^{13}C_3$]pyruvic acid	*D*-[UL-$^{13}C_6$]glucose	*D*-[UL-$^{13}C_6$,UL-$^2H_7$]glucose	*L*-[UL-$^{13}C_5$,UL-$^{15}N_2$]glutamine

FIGURE 1.34 Universal labeling (UL) means that all positions of the designated element are labeled.

example, tritiation via exposing a compound to tritium gas for prolonged periods (Wilzbach exchange) will afford general labeling [585].

The symbol UL is used in place of locants to denote universal labeling when all positions of the designated element are labeled in a molecule (Fig. 1.34).

When naming chemical entities with labile protons/deuterons (amines, alcohols, carboxylic acids, amino acids, sugars, etc.), the possibility of exchange should be considered. For example, when universally ^2H-labeled acetic acid, urea, or glucose is dissolved in water, the labile ^2H will exchange to ^1H (Fig. 1.35).

Isotopically deficient (depleted) compounds are named by adding the italicized syllable *def* immediately preceding, without a hyphen, the deficient nuclide symbol. If selectively depleted atoms exist, the locants should be specified. When different isotopes of the same elements are present in a molecule, their symbols are arranged in order of increasing mass number (Fig. 1.36).

The IUPAC Rules on the Nomenclature of Organic Compounds cite many other useful examples. It should be noted that the above-outlined rules are seldom, if ever, followed in publications. For example, no distinction is generally made between isotopically substituted and isotopically labeled derivatives. The IUPAC-recommended term "universally labeled" (UL) is rarely used to denote compounds in which all positions of the designated element are labeled, although it can occasionally be seen [586]. Most frequently, "uniformly labeled" is used in that sense. Most of the time, when reading literature that uses nonIUPAC naming conventions, the labeling is unambiguous in the context. However, it is always useful to provide a clear diagram of structural formulae with labeling sites.

It will also be useful to define the meanings of "isotopologue" and "isotopomer." Isotopologues are molecules that have a different isotopic composition, while isotopomers have the same number of isotopes but in different positions [587,588]. For example, [1-^{13}C]pyruvic acid and [1,2-$^{13}C_2$]pyruvic acid are isotopologues, but [1,2-$^{13}C_2$]pyruvic acid and [1,3-$^{13}C_2$] pyruvic acid are isotopomers.

in D_2O solution

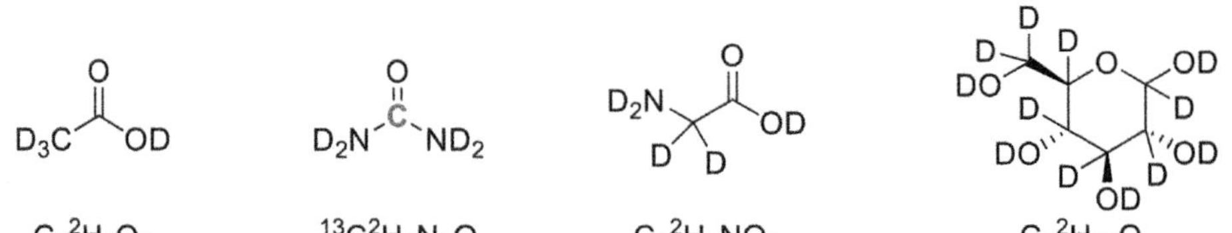

formula	$C_2{}^2H_4O_2$	$^{13}C^2H_4N_2O$	$C_2{}^2H_5NO_2$	$C_6{}^2H_{12}O_6$
name	[UL-$^2H_4$]acetic acid acetic acid-d_4	[^{13}C, UL-$^2H_4$]urea	[UL-$^2H_5$]glycine glycine-d_5	*D*-[UL-$^2H_{12}$]glucose *D*-glucose-d_{12}

in H_2O solution

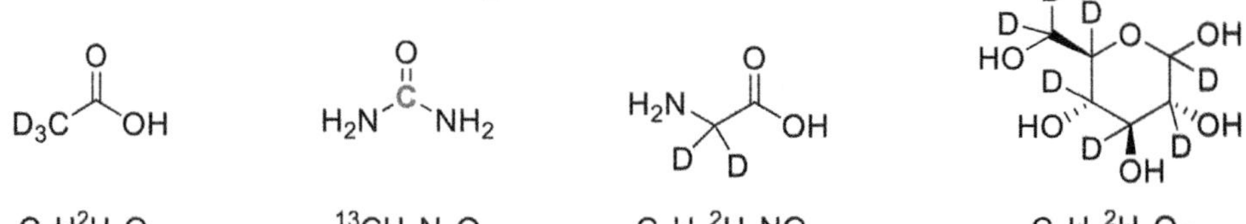

formula	$C_2H^2H_3O_2$	$^{13}CH_4N_2O$	$C_2H_3{}^2H_2NO_2$	$C_6H_5{}^2H_7O_6$
name	[2-$^2H_3$]acetic acid acetic acid-2,2,2-d_4	[^{13}C]urea	[2-$^2H_2$]glycine glycine-2,2-d_2	*D*-[1,2,3,4,5,6,6-$^2H_7$]glucose *D*-glucose-1,2,3,4,5,6,6-d_7

FIGURE 1.35 Labile protons/deuterons will undergo chemical exchange in a solvent that contains labile deuterons/protons (protic solvent).

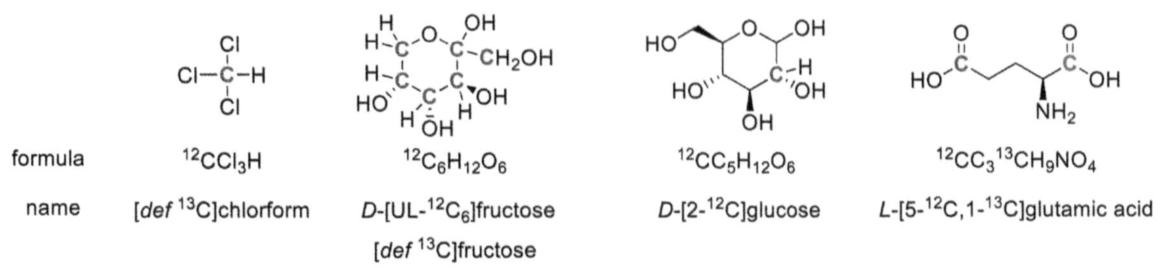

FIGURE 1.36 Nomenclature of isotopically deficient compounds.

1.9.2 Isotope enrichment with ^{2}H and ^{15}N in ^{13}C-labeled probes

Multiple isotope labeling, where more than one isotope is present in a single molecule, is becoming more common among hyperpolarized probes. These modifications not only label molecules with MR-visible isotopes, but also improve the magnetic properties of the probe for dissolution DNP experiments. For example, deuteration limits dipole-dipole interactions and results in longer T_1 relaxation times.

As we have discussed earlier, dipole-dipole interactions with intramolecular ^{1}H spins ($\gamma_{1H} = 42.6$ MHz/T) are a major relaxation mechanism for both ^{13}C and ^{15}N. The undesirable dipolar relaxation effect of protons can be reduced by replacing them with ^{2}H spins, which have a much lower gyromagnetic ratio ($\gamma_{2H} = 6.54$ MHz/T). Theoretically, based on the ratio of the γ values and the spin quantum numbers of ^{1}H and ^{2}H, substitution of a deuterium for a proton directly attached to a ^{13}C nucleus would increase the ^{13}C T_1 by a factor of $3\gamma_{1H}^2/8\gamma_{2H}^2$, approximately 16, if the only relaxation mechanism were T_1 relaxation by the attached proton. In practice, the observed increase is less than that because of the contribution of other relaxation pathways [358,589,590]. Nevertheless, deuteration can be very useful for hyperpolarized NMR studies because it can increase the ^{13}C relaxation time from a few seconds to over a minute in some compounds [26,297,307,358,390,591,592]. Deuteration can be achieved by synthesizing the desired molecule from deuterated starting materials and reagents [26,592,593].

An example is the synthesis of ^{15}N-labeled, deuterated choline (Scheme 1.2). The synthesis started from ethylene glycol-d_4, which was monoprotected with tertbutyldiphenylsilyl chloride in the presence of sodium hydride. The unprotected OH was then activated by converting it into tosylate with *p*-chloro toluenesulfonyl chloride. The amino group was installed in a Gabriel reaction using ^{15}N-labeled potassium phthalimide. Hydrazinolysis of the phthalimide provided the primary amine. Subsequently, permethylation with iodomethane-d_3 followed by acid deprotection of the silyl group afforded the desired final product (Scheme 1.2) [593].

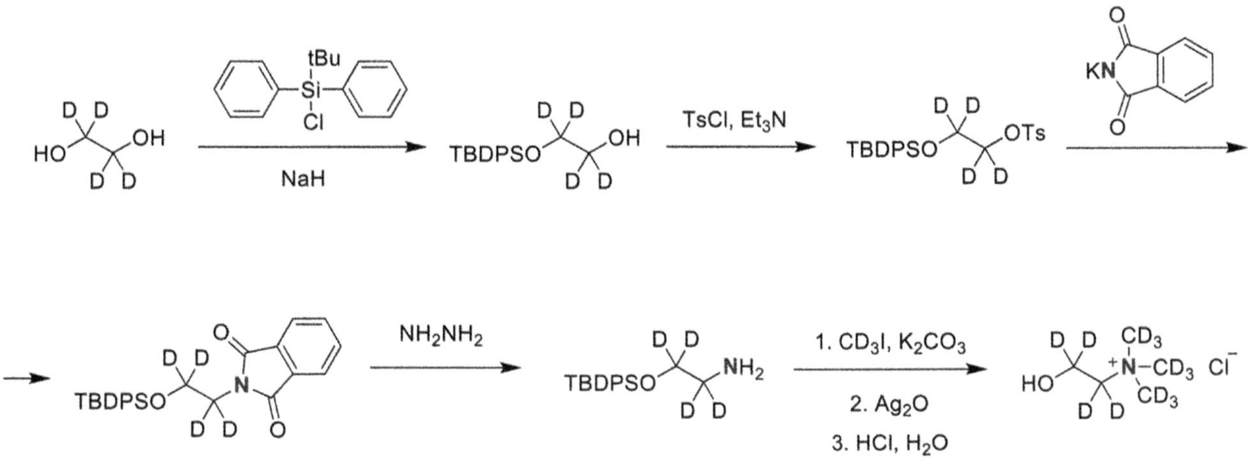

SCHEME 1.2 Synthesis of [^{15}N]choline-d_{13} using deuterated starting materials and reagents.

Another option is functionalizing the target molecule itself through HDX (late-stage deuteration) [594]. Although this strategy would be more convenient, high yielding regio- and stereoselective H/D exchange remains a challenge despite recent progress in the field [595–598]. The development of novel synthetic approaches for H/D exchange is primarily driven by the favorable and often unusual pharmacological properties of deuterated drug candidates [596,599–601]. From a hyperpolarized NMR perspective, the most useful direct deuteration methods include the base-catalyzed deuteration of carbonyl compounds via the keto-enol tautomeric equilibrium [602] and the catalytic H/D exchange of CH bonds adjacent to heteroatoms such as O and N [597,603,604]. These methods usually use D_2O as a deuterium source. For example, pyruvate can easily be deuterated on the methyl group by dissolving labeled pyruvic acid in D_2O and refluxing the solution for 5 h followed by neutralization with $NaHCO_3$, or by treating sodium pyruvate with ammonium carbonate in D_2O at pH 10 for 2 days at 37°C [605–607] Deuteration of pyruvate increases the C3 T_1 to about a minute, enabling the detection of its HP-^{13}C signal, which otherwise cannot be observed due to rapid relaxation [605]. Likewise, 2-ketoglutarate can be easily deuterated at C3 in 0.1 M NaOD [608].

Ruthenium-based catalysts (Ru on C, Ru nanoparticles, and various ruthenium complexes) have been found to be effective catalysts for the stereoselective deuteration of alcohols, amines, amino acids, hydroxy acids, and sugars [603,604,609–615]. These reactions usually proceed with excellent stereoselectivity and reasonably good yields. Very efficient deuterium incorporation (95%–97%) at the alpha position with full retention of configuration was achieved in one step when alpha hydroxyl and amino acids were treated with ruthenium on a carbon catalyst in D_2O (Scheme 1.3) [616].

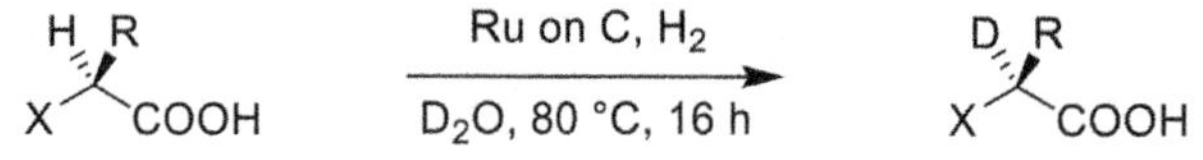

SCHEME 1.3 Deuteration of alpha hydroxy and amino acids with ruthenium catalyst under H_2 atmosphere. The source of deuterium is D_2O.

Deuteration at the alpha position led to a modest (16%–29% at 3 T) improvement of the carboxylate T_1, indicating that the dominant relaxation mechanism is CSA rather than the dipolar effect of the alpha protons [616].

It should be kept in mind that deuteration may change the rates of biochemical reactions due to kinetic isotope effects [617]. Although the deuterium substitution may not change overall flux in most cases such as deuterated glucose in glycolysis [618], at the enzyme level reactions may show a large deuterium kinetic effect [619] Therefore, the reactivities of multiple-isotope-labeled molecules need to be measured and compared to those of natural-abundance molecules to ensure that experimental results can be correctly interpreted and reasonably applied to our understanding of endogenous molecules.

The quadrupolar ^{14}N can affect the T_1 of the coupled ^{13}C nucleus through scalar relaxation of the second kind [153] and can also cause line broadening (T_2 shortening) through the rapid change of its nuclear spin state [503,620]. Since the effect of scalar relaxation by a quadrupole nucleus on the T_1 of coupled spins is inversely dependent on the square of the Larmor frequency difference, it is only noticeable at very low fields (less than 800 μT) [153]. On the other hand, the quadrupolar linebroadenig (T_2 shortening) is apparent at higher fields [621]. The spin ½ ^{15}N is not quadrupolar, so substituting ^{15}N for ^{14}N will eliminate the undesirable quadrupolar effects. This will not influence the T_1 above 100 mT but significantly improve the T_2. The effect on T_2 can be quite remarkable; for example, the ^{13}C T_2 of [^{13}C]urea improves from 0.10 to 34.6 s (in D_2O and at 14.1 T) upon replacing the ^{14}N with ^{15}N [621]. In a similar way, the ^{13}C T_2 in L-[5-^{13}C]glutamine and L-[5-^{13}C, 5-^{15}N]glutamine increased from 0.46 to 4.7 s (in D_2O and at 14.1 T) upon replacing the ^{14}N with ^{15}N in the amide group [622]. Likewise, ^{13}C T_2 in L-[6-^{13}C]arginine and L-[6-^{13}C,$^{15}N_3$]arginine improved from 0.23 to 2.68 s (in D_2O and at 14.1 T) upon substituting ^{15}N for all three ^{14}N in the guanidino group [623]. It should be noted that substituting ^{15}N for ^{14}N will also lead to the appearance of multiplets because of the heteronuclear ^{15}N-^{13}C J-coupling. Interestingly,

these unwanted quadrupolar effects are not seen with ^{2}H, which is also a spin nucleus, because its quadrupole moment is much smaller than that of ^{14}N, and therefore, the relaxation among its Zeeman levels (change in the nuclear state) is slow compared to the ^{2}H-^{13}C J-coupling. Strategies that can be used to introduce ^{15}N in place of ^{14}N include substitution with ^{15}N-labeled cyanide, amidation of carboxylic acids and nitration with ^{15}N labeled reagents. We will show examples for each of these reactions in the following sections.

1.9.3 Isotope depletion

In hyperpolarization, the signal is enhanced by polarization transfer, regardless of labeling. Therefore, the signal of nuclei present in natural abundance in the molecule will also be enhanced and can be detected after dissolution. The enhancement of nonenriched positions becomes problematic when these nuclei have long T_1 and their signal intensity is comparable to that of the enriched sites of the metabolic products. This challenge become particularly pronounced when resonances overlap. This has recently been addressed by depleting ^{13}C from the overlapping nonlabeled position [624]. The benefits of MR-visible isotope depletion were demonstrated with HP-^{13}C-labeled 2-ketoglutarate. 2-Ketoglutarate is converted to 2-hydroxyglutamate by the mutant isocitrate dehydrogenase I (IDH1) enzyme present in various cancers, in particular gliomas. The C5 resonance of 2-ketoglutarate overlaps with the [1-^{13}C]-2-hydroxyglutarate signal at physiological pH. Depletion of ^{13}C at C5 in [1-^{13}C, 5-^{12}C]-2-ketoglutarate significantly improved the detection of 2-hydroxyglutarate and facilitated interpretation of the metabolic data.

1.9.4 ^{13}C-enriched starting materials and reagents

Today, a wide diversity of compounds are available commercially as starting materials and reagents for conventional organic synthesis. However, for isotope labeling, the choices of stating materials are limited due to the availability of the reagents. In addition, the cost of the reagents and starting materials plays a critical role in planning a practical synthetic route for the incorporation of the label in the desired position. This is particularly true for hyperpolarized magnetic resonance probes, where the synthesis of the labeled tracers is often performed on a gram scale. The natural abundance of ^{13}C, the NMR-visible isotope of carbon, is only 1%. ^{13}C is enriched by the cryogenic distillation of carbon monoxide or, less commonly, methane, which is then converted to other ^{13}C-labeled reagents for further synthesis (Fig. 1.37) [625].

The cheapest and most easily available ^{13}C enriched starting materials and reagents include sodium carbonate, potassium cyanide, acetic acid, and methyl iodide. As we will see later, ^{13}C-labeled cyanide is a particularly useful synthon in the synthesis of a variety of ^{13}C-labeled compounds. ^{13}C-labeled potassium or sodium cyanide is obtained from [^{13}C]CH_4 by reacting it with excess ammonia over a platinum catalyst at 1000°C and absorbing the [^{13}C]HCN in KOH or NaOH [627].

1.9.5 Design and synthesis of ^{13}C-labeled compounds for dissolution DNP

Since the lifetime of a hyperpolarized signal is dependent on the spin-lattice relaxation, T_1 plays a very important role in design of HP probes. However, this should not be the only parameter to be considered. If the goal is to monitor enzyme-catalyzed biochemical processes, then the rate of the reactions and the pool sizes of the intermediates must also be considered. In fact, the activity of an enzyme-catalyzed step can be successfully observed with an HP probe whose T_1 is in the range of 10−15 s, provided that the flux of the pathway is high and the ^{13}C polarization is large at the time of injection (this may require a rapid dissolution and injection system). Glycolysis is a good example: it is rapid, and the pool size of lactate is large. Note that strictly speaking, the end product of glycolysis is pyruvate; however, lactate is instantly formed from pyruvate due to high activity of LDH. Thus, glycolytic flux is often characterized in terms of the moles of glucose converted to lactate per unit time [467]. Even though the T_1 of deuterated, ^{13}C-labeled glucose tracers is only about 10−13 s, the production of HP-lactate from glucose via 10 enzyme-catalyzed steps could be observed in the normal mouse brain as well as in tumor-bearing mice in vivo [591,628]. This is quite remarkable, considering the fact that this short T_1 allows for a time window of only about a minute for signal detection. However, flux through many other biochemical processes (e.g., nucleotide biosynthesis) may not be high enough to be assessed with HP-^{13}C NMR. While carboxylate carbons are labeled most commonly for HP-^{13}C NMR studies, there are several other functional groups in which the ^{13}C spins relax relatively slowly ("long T_1" positions). In the following sections, we provide an overview of the various classes of ^{13}C probes used in dissolution DNP experiments.

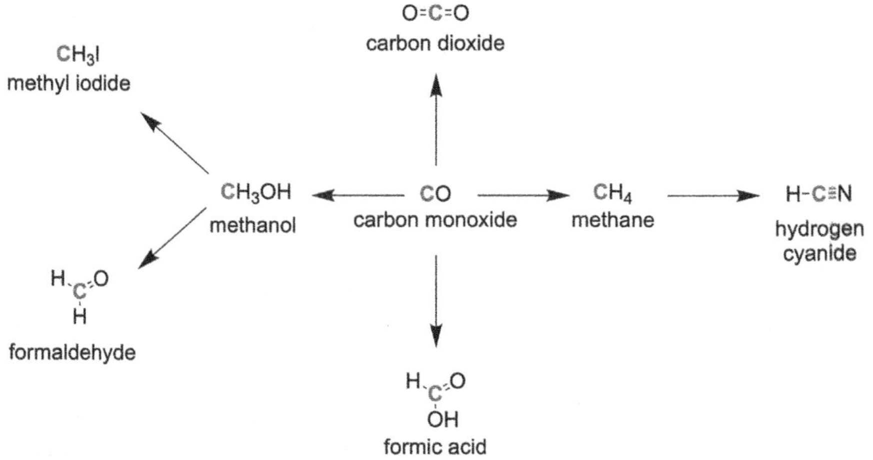

FIGURE 1.37 ^{13}C is enriched by the cryogenic distillation of CO, which is then converted to other commonly used reagents in organic synthesis. *Adapted from Ref. [626] with permission.*

1.9.5.1 Primary carbons

Primary carbon-13 atoms are rarely considered for dissolution DNP, largely because of their unfavorable relaxation properties. The ^{13}C-^{1}H dipole-dipole relaxation shortens the T_1 relaxation time; however, as discussed previously, reasonably long T_1 values in the range of 25–50 s can be obtained by deuteration. HP probes in which the label is a deuterated primary or secondary ^{13}C have been used successfully as metabolic tracers or ^{13}C NMR sensors. The deuterated ^{13}C methoxy group can serve as a new long-lived hyperpolarization unit in HP-^{13}C NMR sensors [592]. For example, HP-[^{13}C, $^{2}H_3$]-*p*-anisidine is a sensitive ^{13}C NMR sensor for hypochlorous acid (HOCl) because it selectively reacts with HOCl via *ipso*-substitution to rapidly produce 1,4-benzoquinone imine and [^{13}C]CD_3OH, which results in a large, about 12 ppm change in the ^{13}C chemical shift (Scheme 1.4). The effect of deuteration is well illustrated by the T_1 values of [^{13}C, $^{2}H_3$]-*p*-anisidine and the nondeuterated compound: 44 and 6 s, respectively, at 9.4 T and 37°C in $H_2O/D_2O = 9{:}1$.

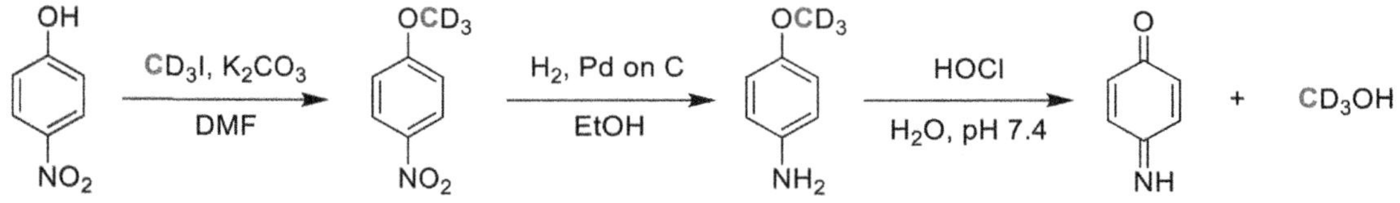

SCHEME 1.4 Synthesis of [^{13}C, $^{2}H_3$]-*p*-anisidine and its reaction with hypochlorite.

1.9.5.2 Secondary carbons

Metabolic tracers that contain HP-[^{13}C]CD_2 moieties include [1-^{13}C, U-$^{2}H_5$] ethanol and ^{13}C-labeled cholines. HP-[1-^{13}C, U-$^{2}H_5$] ethanol was used to assess ALDH2 activity in mouse liver in vivo [629]. Hyperpolarized choline tracers were proposed as probes to monitor choline acetylation and phosphorylation as well as betaine aldehyde synthesis and oxidation [358,513,593,630–632]. Choline has two secondary carbons and is involved in multiple biochemical reactions (Scheme 1.5).

Isotope labeling of choline with ^{13}C and deuterium provides probes with reasonably long T_1: CMP1 (C1 labeling) and CPM2 (C2 labeling) (Fig. 1.38). The T_1 of these probes is not field-dependent, indicating lack of CSA and the dominance of the dipolar relaxation mechanism. The conversion of these molecules to betaine aldehyde and betaine can be assessed with purified enzymes [358,630,631].

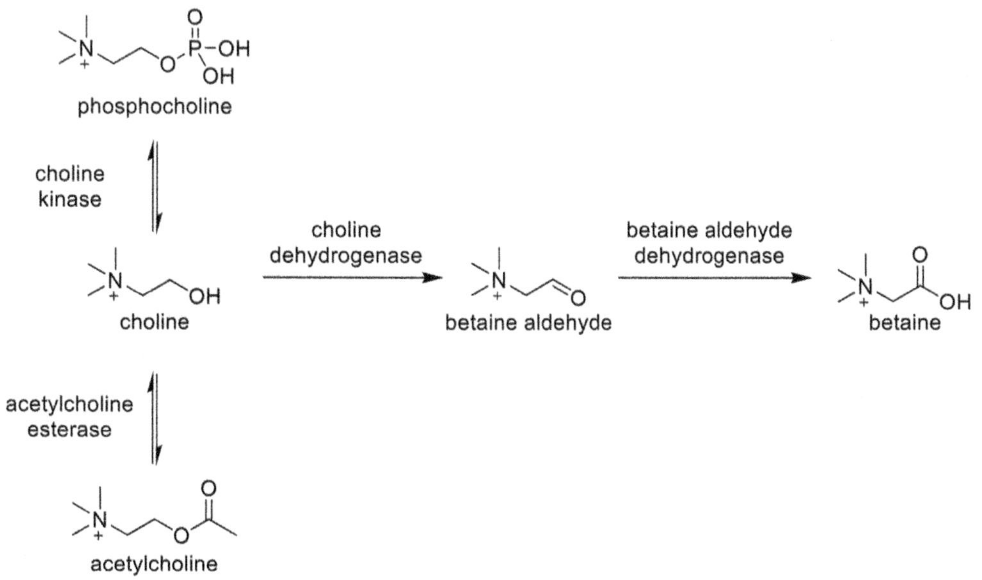

SCHEME 1.5 Enzymatic oxidation of choline to betaine aldehyde and betaine catalyzed by choline oxidase. Choline phosphorylation and acetylation are catalyzed by kinase and esterase respectively.

Later, CMP2 was developed to measure the kinetic rates of choline acetylation. They used commercial acetyltransferase enzymes to show that acetylation can be detected in the time frame of the HP-^{15}N experiment [589]. So far, there has been no report on the observation of choline phosphorylation by HP-^{13}C NMR. However, ^{15}N labeling of choline has been demonstrated to be a useful HP probe for the kinase reaction. ^{15}N exhibits a larger chemical shift range and possibly shows more significant changes even when the reaction site is a few bonds away from the ^{15}N-labeled site. For example, the 15N chemical shift difference between choline (^{15}N $\delta = 43.54$ ppm) and phosphocholine (^{15}N $\delta = 43.37$ ppm) is large enough to allow the separation of the signals, observe the phosphorylation of choline by human choline kinase, and measure the reaction rate, albeit only in vitro [513,593]. The in vivo metabolic utility of choline-derived probes is still under investigation [633].

1.9.5.3 Tertiary carbons

As discussed earlier, the main spin-lattice (T_1) relaxation pathways of ^{13}C are dipole-dipole relaxation and CSA, the latter of which has a greater impact at higher magnetic fields. Tertiary ^{13}C attached to a heteroatom also exhibits long T_1 values due to a lack of directly attached hydrogen. Deuterated, ^{13}C-labeled *tert*-butanol was shown to be a promising perfusion agent in the brain [504]. Although perdeuterated ^{13}C-labeled *tert*-butanol is commercially available, it can also be synthesized starting from the much cheaper [2-^{13}C]acetone, which can be deuterated under basic conditions. [2-^{13}C, U-$^{2}H_6$] acetone has been converted to the desired product with deuterated methylmagnesium iodide (Scheme 1.6) [634,635].

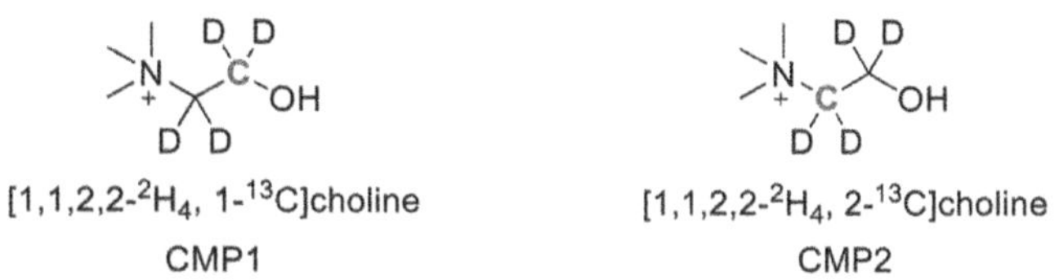

FIGURE 1.38 The structure of ^{13}C-labeled and deuterated choline probes.

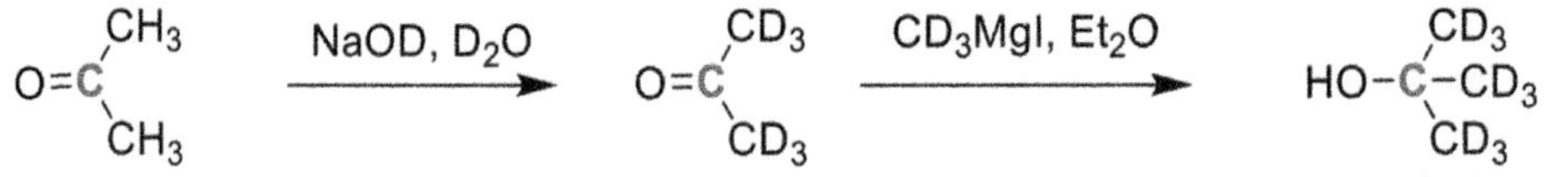

SCHEME 1.6 Synthesis of deuterated, ^{13}C-labeled tert-butanol from [2-^{13}C]acetone.

HP-[^{13}C]-*tert*-butanol-β Gal is an example of an enzyme-responsive hyperpolarized agent. It was used to measure the activity of the enzyme β-galactosidase, which cleaves the glycosidic bond of β-D-galactopyranosides. Since the accumulation of lysosomal beta-galactosidase occurs in senescent cells, this probe was proposed as an HP-^{13}C reporter for cell senescence [636]. [1-^{13}C]-*tert*-butyl-β-D-galactopyranoside was synthesized using the well-established Koenigs-Knorr glycosylation method by reacting 2,3,4,6-tetra-O-acetyl-α-D-galactopyranosyl bromide with ^{13}C-labeled *tert*-butanol in the presence of silver triflate. The basic hydrolysis of the acetate esters affords the final product (Scheme 1.7) [637].

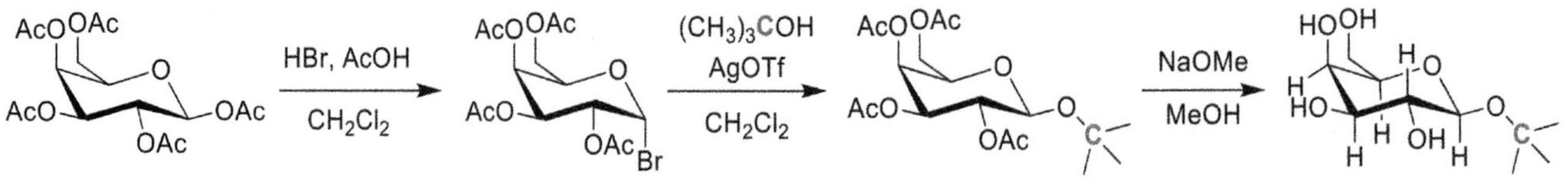

SCHEME 1.7 Synthesis of [1-^{13}C]-*tert*-butyl-β-D-galactopyranoside from [1-^{13}C]-*tert*-butanol.

1.9.5.4 Quaternary carbons

Quaternary carbons lack directly attached hydrogen atoms, which results in the absence of ^{13}C-^{1}H dipole-dipole relaxation. In a similar manner to tertiary carbons, sp^3 quaternary carbons tend to minimize CSA, allowing for relatively long T_1 values. While, the quaternary ^{13}C is typically of limited utility for metabolic studies, the long T_1 makes it attractive for the design of long T_1 scaffolds, nonmetabolizable perfusion agents, and sensors. A prime example of this concept is the uniformly deuterated [^{13}C]-*tert*-butylbenzene and its sulfonated derivative, in which both the dipole-dipole and CSA relaxation are minimized. The tertiary carbon center in these molecules can potentially sense the chemical modification that occurs on the aromatic ring. The synthesis of these derivatives was designed with consideration of the commercial availability of the labeled starting materials and reagents. The synthesis started from the commercially available [$^2H_5$] bromobenzene. This compound can also be prepared by the bromination of [$^2H_6$]benzene [638]. Next, the deuterated bromobenzene was converted to the Grignard reagent. To install the ^{13}C-labeled *tert*-butyl functional group on the benzene ring, fully deuterated benzene was first brominated [638]. Next, the ^{13}C label was introduced via the reaction of the Grignard intermediate with [^{13}C]CO_2. The resulting pentadeuterio benzoic acid (BA) was converted to fully deuterated [^{13}C]tert-butylbenzene with deuterated methylmagnesium iodide in two consecutive steps. Further on, sulfonation with fuming sulfuric acid resulted in the highly water-soluble sulfonic acid with an extremely long ^{13}C T_1 of 141 s at 9.4 T in CD_3OD (Scheme 1.8) [306].

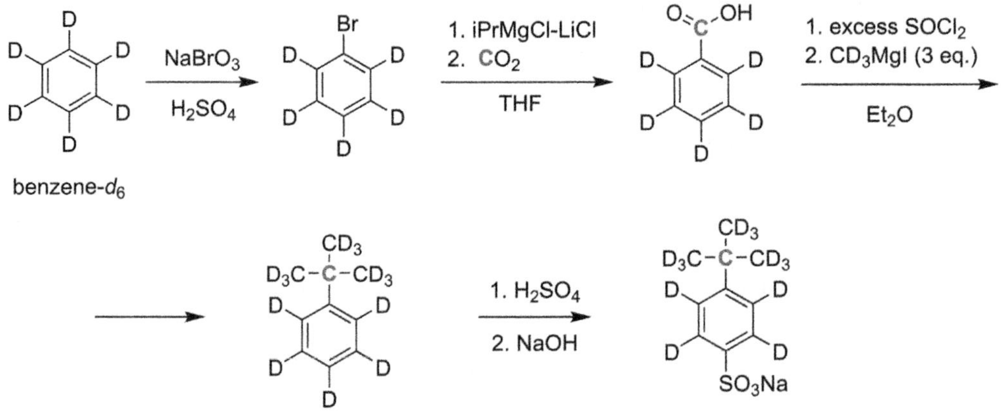

SCHEME 1.8 Synthesis of ^{13}C-labeled, uniformly deuterated p-*tert*-butyl benzenesulfonic acid sodium salt.

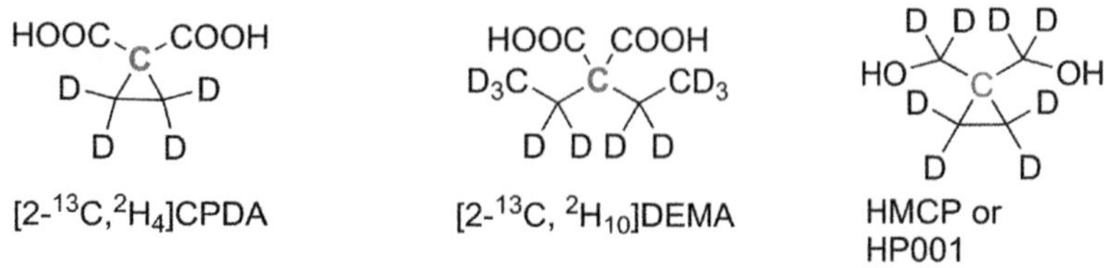

FIGURE 1.39 Deuterated, ^{13}C-labeled dialkylated malonic acid derivatives as pH sensors and perfusion markers.

HP-^{13}C pH sensors containing long T_1 quaternary carbons include various malonic acids dialkylated at C2, such as [2-^{13}C, D_{10}]diethylmalonic acid (DEMA) and [2-^{13}C, D_4]cyclopropane-1,1-dicarboxylic acid (CPDA) (Fig. 1.39). The second pK_a in these dicarboxylates falls into the physiological range. These compounds exhibited moderate ^{13}C chemical shift dispersions (2.5 ppm) of the quaternary carbon between pH 6 and 8.

The preparation of these probes is based on the well-known malonic ester synthesis, in which diethyl [2-^{13}C]malonate is alkylated with deuterated alkyl halides in the presence of a strong base to generate the dialkyl malonic acid ester. Hydrolysis of the ethyl ester with NaOH afforded the desired dicarboxylates (Scheme 1.9) [493]. The ^{13}C label in these compounds has a very long T_1, for example, in [2-^{13}C, U-$^2H_{10}$]DEMA, the ^{13}C T_1 is 106 s at 11.75 T in H_2O (pH 7.1). Reduction of the carboxylates to the alcohol functionality provided by HMCP or HP001. This compound was designed as a perfusion agent. The ^{13}C chemical shift of the label (^{13}C $\delta = 23$ ppm) is pH-insensitive and well separated from naturally occurring lipid signals, making this compound useful as an HP-^{13}C perfusion marker.

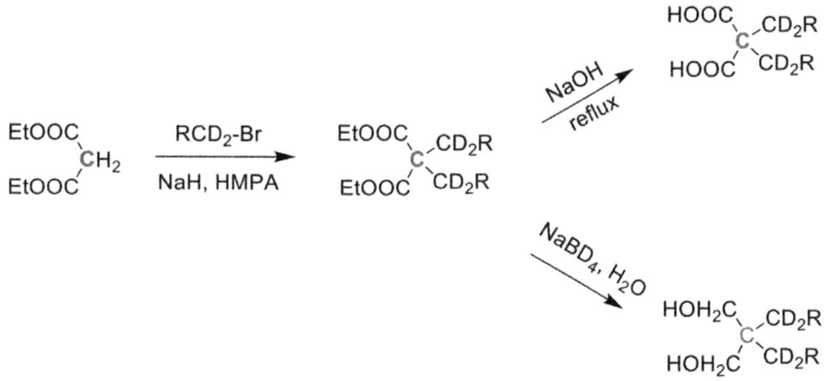

SCHEME 1.9 Synthesis of deuterated, ^{13}C-labeled dialkylated malonic acid derivatives as pH sensors and perfusion markers.

[5-^{13}C, $^2H_9$]dimethyl pyrroline-N-oxide ([5-^{13}C, $^2H_9$]DMPO). Saito et al. polarized [5-^{13}C, $^2H_9$]DMPO and demonstrated the feasibility of in vivo detection. DMPO and similar molecules have been used as spin traps to detect free radical adducts with EPR spectroscopy [639]. Hyperpolarized ^{13}C-labeled DMPO was proposed as a probe to detect diamagnetic end-products formed in reactions with reactive oxygen species, although it remains to be seen whether there is sufficient chemical shift change upon redox reaction for this probe to be used for ^{13}C MRS. The T_1 value of [5-^{13}C, $^2H_9$] DMPO was fairly long, 60 s at 4.7 T in PBS buffer, while the nondeuterated DMPO had a T_1 of 26 s. This probe was prepared starting from the commercially available [2-^{13}C]acetone, which was deuterated in D_2O under basic conditions. Next, reduction of carbonyl with D_2 gas over Ru catalyst gave [2-^{13}C, $^2H_7$]-2-propanol. Bromination followed by nitration afforded [2-^{13}C, $^2H_7$]-2-nitropropane. Aldol condensation with methyl acrylate-2,3,3-d_3, furnished [4-^{13}C, $^2H_9$]-4-methyl-4-nitro pentanoate. The methyl ester group was reduced to the aldehyde with DIBAL-H. Subsequent treatment with zinc powder in acetic acid afforded the desired product as [5-^{13}C, $^2H_9$]DMPO (Scheme 1.10) [634].

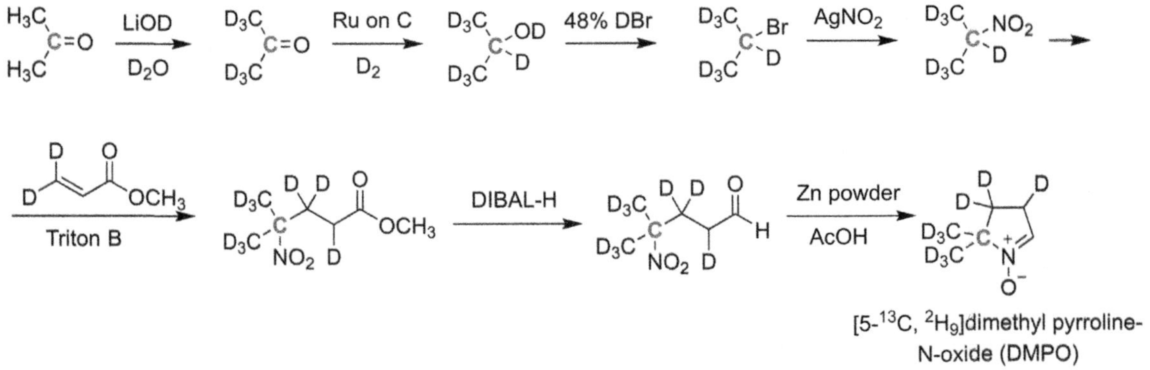

SCHEME 1.10 Synthesis of [5-^{13}C, ^{2}H$_9$]dimethyl pyrroline-N-oxide starting from [2-^{13}C]acetone.

1.9.5.5 Carbonyl compounds

The carbonyl group is a common functionality in organic chemistry. In an aqueous solution, carbonyl compounds usually exist in equilibrium with their hydrated forms. Depending on the complexity of the structure and the presence of other functional groups such as carboxyl or hydroxyl in the molecule, they can dimerize, cyclize intramolecularly, or exist mainly in the tautomeric enol form (Scheme 1.11). The ratio of these species is highly dependent on the experimental conditions, such as pH and temperature. If the conversion between different species occurs slowly, then each species may have a significantly different T_1 relaxation time.

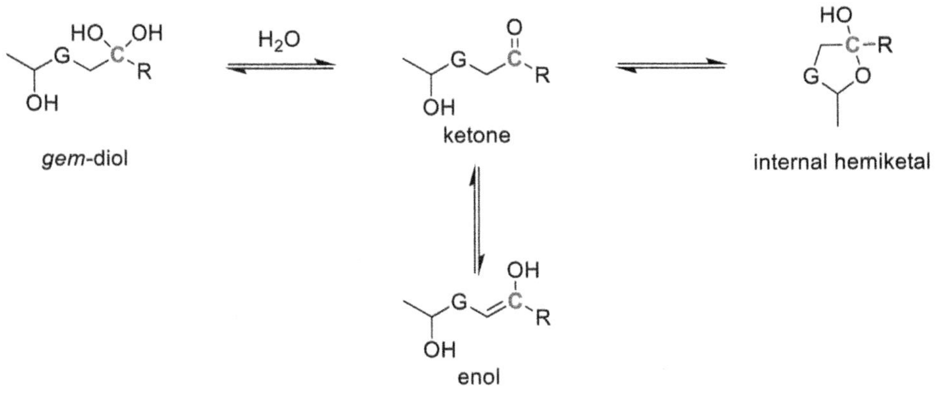

SCHEME 1.11 Carbonyl containing molecules can exist in various interconverting forms depending on the structure.

[2-^{13}C]1,3-Dihydroxyacetone (DHAc) is an intermediate of glycolysis and gluconeogenesis. Hyperpolarized [2-^{13}C]-1,3-DHAc is a promising probe for the detection of hepatic gluconeogenic and glycogenolytic states [640]. Synthesis of [2-^{13}C]-1,3-dihydroxyacetone relies on the condensation of ^{13}C-labeled nitromethane with formaldehyde to give 1,3-dihydroxy-2-nitropropane; the nitro group is converted to a carbonyl group using the Nef reaction (acid hydrolysis of nitronate salts) to afford [2-^{13}C]-1,3-dihydroxyacetone. The product exists in a hemiketal dimer form (Scheme 1.12) [641].

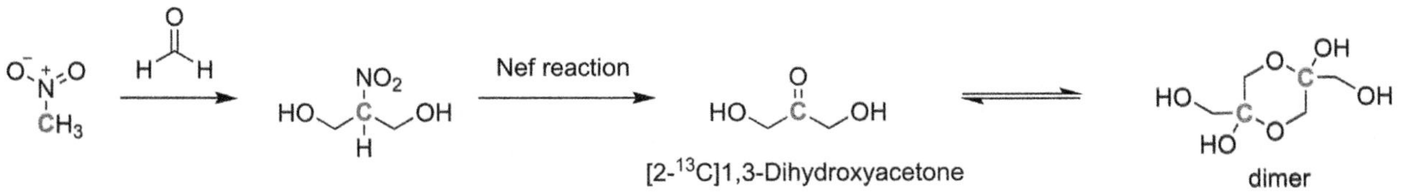

SCHEME 1.12 Synthesis and dimerization of [2-^{13}C]-1,3-dihydroxyacetone.

1.9.5.6 Carboxylic acids

The majority of d-DNP probes developed to date belong to this category, largely due to their great biological relevance and favorable properties for DNP NMR. Although there are numerous synthetic methods for the preparation of carboxylic acids, the ones that can be conveniently adapted for the introduction of a ^{13}C label in the carboxyl group are limited to reactions that

use [^{13}C]CO, [^{13}C]CO_2, or [^{13}C]KCN as labeled reagents. The carboxylation of Grignard reagents and the alkylation of the cyanide ion are classic methods for the synthesis of carboxylic acids. Carbonylation of aryl, vinyl, allyl, benzyl, and alkyl halides requires transition metal catalysts and is currently an active field of research (Scheme 1.13) [642–644].

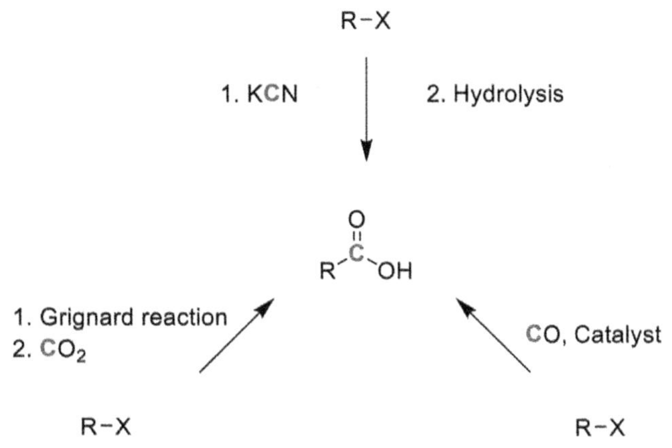

SCHEME 1.13 The most common synthetic approaches to ^{13}C-labeled carboxylic acids use ^{13}C enriched cyanide, carbon dioxide or carbon monoxide.

1.9.5.6.1 Simple carboxylic acids (^{13}C-labeled acetate, propionate, butyrate, octanoate, succinate, and fumarate)

Acetate is a fundamental two-carbon unit and a key metabolic intermediate in the form of acetyl-CoA, an acyl carrier, which participates in carbohydrate, lipid, and protein metabolism [645]. Acetate enters cells rapidly and has a favorably long T_1 (72 s at 1 T in D_2O PBS buffer) [646] so one would expect that HP-[1-^{13}C]acetate would be a very useful metabolic imaging agent. Acetate is rapidly taken up by cells and forms acetyl-CoA via acetyl-CoA synthetase. Acetyl-CoA is then reversibly converted into acetylcarnitine by carnitine acetyl-transferase. Acetylcarnitine can be transported into the mitochondria, where it is converted back to acetyl-CoA, which in turn enters the TCA cycle. Although [5-^{13}C]citrate could be detected in vivo, the signal overlap of [1-^{13}C]acetate with [5-^{13}C]glutamate and [5-^{13}C]-2-ketoglutarate at around 182 ppm hampers its usefulness a a probe for the TCA cycle [363,647,648]. Acetylcarnitine buffers the amount of acetyl-CoA inside the mitochondria. Acetyl-CoA is produced in the beta-oxidation of fatty acids via acetoacetyl-CoA. It is well established that fatty acid β-oxidation is the main energy source of normal myocardium. While [1-^{13}C]acetyl-CoA is very difficult (although not impossible) to detect by HP-^{13}C due to its small pool size, acetyl-carnitine is easily observed in vivo after the injection of HP-[1-^{13}C]acetate [649]. The HP-[^{13}C]acetylcarnitine/[1-^{13}]acetate signal ratio reflects fatty acid metabolism in the myocardium. In an ischemic heart, this ratio is significantly lower than in normal tissue, indicating mitochondrial dysfunction [650].

Unlike pyruvic acid, neat acetic acid does not vitrify well, so hyperpolarized acetate is produced by DNP of the TRIS or sodium salt under standard conditions [glycerol-water matrix, OX063 radical, plus trace amounts of a Gd-chelate such as Dotarem (GdDOTA)] [171,646,651]. Taking advantage of the excellent ^{13}C polarization levels that can be achieved for pyruvic acid, HP-[1-^{13}C]acetate was produced by the oxidative decarboxylation of HP-[2-^{13}C]pyruvate in the presence of hydrogen peroxide (Scheme 1.14). This method afforded somewhat higher ^{13}C polarization for [1-^{13}C]acetate than direct DNP [652].

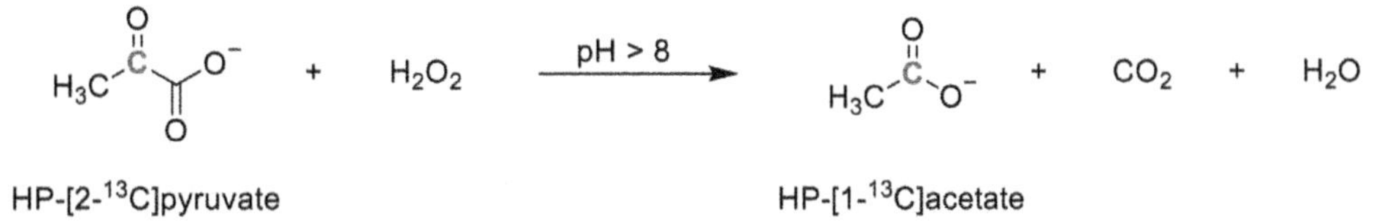

SCHEME 1.14 Generation of HP-[1-^{13}C]acetate by the rapid oxidative decarboxylation HP-[2-^{13}C]pyruvate in the presence of H_2O_2.

HP-[1-^{13}C]propionate was metabolized analogously to acetate to produce HP-[1-^{13}C]propionyl-CoA and HP-[1-^{13}C] propionylcarnitine, but the signal intensities were lower, indicating that propionate is a less preferred substrate for acetyl-CoA synthetase [650].

HP-[1-^{13}C]butyrate is a very useful probe to monitor short-chain fatty acid metabolism [254]. Observed metabolites in isolated, perfused rat hearts included acetoacetate, and β-hydroxybutyrate, as well as metabolites derived from acetyl-CoA (acetyl carnitine, citrate, and glutamate). The unknown peak at around 175 ppm was likely butyryl carnitine [653]. A similar metabolite pattern was also observed in vivo in rat hearts. It was also shown that, as expected, in fasted animals, a significant increase in acetoacetate signal was observed. Pyruvate competition with butyrate for the production of acetyl-CoA was observed when HP-[1-^{13}C]pyruvate and HP-[1-^{13}C]butyrate were coinjected [654]. HP-[1-^{13}C]butyrate can be produced by the DNP of [1-^{13}C]butyric acid in glycerol-water or DMSO water matrices and dissolving the DNP sample in a solution containing NaOH and buffer. DMSO was shown to be a better glassing agent for [1-^{13}C]butyric acid than glycerol. Trityl OX063 was used as polarizing agent, but HP-[1-^{13}C]butyrate was successfully produced by the DNP of sodium [1-^{13}C]butyrate with nitroxide radicals [654]. A mixture of [1-^{13}C]pyruvic acid and [1-^{13}C]butyric acid was also copolarized using photo-induced nonpersistent radicals. The T_1 was 56 s in Tris buffer (pH 7.6) at 1.05 T and 20 s at 11.7 T in Trizma buffer (pH 7.4) [118,653].

Hyperpolarized [1-^{13}C]octanoate, a medium-chain fatty acid, was also tested as a probe for myocardial fatty acid metabolism. [1-^{13}C]Octanoate was polarized in a 4 M solution in DMSO with a trityl radical. The T_1 was about 29 s in buffered D_2O at 9.4 T; however, it was only about 10 s in blood, likely due to interactions with blood proteins such as albumin. Only one metabolite peak, assigned to [1-^{13}C]acetylcarnitine, was observed in rat hearts [655].

Fumarate and succinate are important intermediates of the TCA cycle. [1,4-$^{13}C_2$]Fumaric acid has been used to measure malate dehydrogenase activity in necrotic tissues [656]. Diethyl [1-^{13}C, 2,3-$^{2}H_2$]succinate hyperpolarized via PHIP of [1-^{13}C, 2,3-$^{2}H_2$]fumarate ester was reported as a ^{13}C metabolic probe for the TCA cycle [657]. However, in vivo experiments with HP-diethyl [1,4-$^{13}C_2$]succinate failed to show downstream metabolic products derived from succinate [658].

The classic synthesis of [1-^{13}C]acetate is based on the reaction of methylmagnesium iodide with [^{13}C]CO_2 [659]. The synthesis of [1,4-$^{13}C_2$]succinate and its subsequent conversion to [1,4-$^{13}C_2$]fumarate demonstrate the usefulness of ^{13}C-labeled cyanide as a C1 synthon for the introduction of the ^{13}C-labeled carboxyl group [660,661]. The resulting α,ω-di [^{13}C]nitrile can be hydrolyzed to carboxylic acids (Scheme 1.15). Bromination followed by base catalyzed elimination and ester hydrolysis completed the synthesis of [1,4-$^{13}C_2$]fumarate. Esters can be prepared using the classic Fischer esterification of the carboxylic acid with an alcohol, such as ethanol, in the presence of trace amounts of acid catalyst [662].

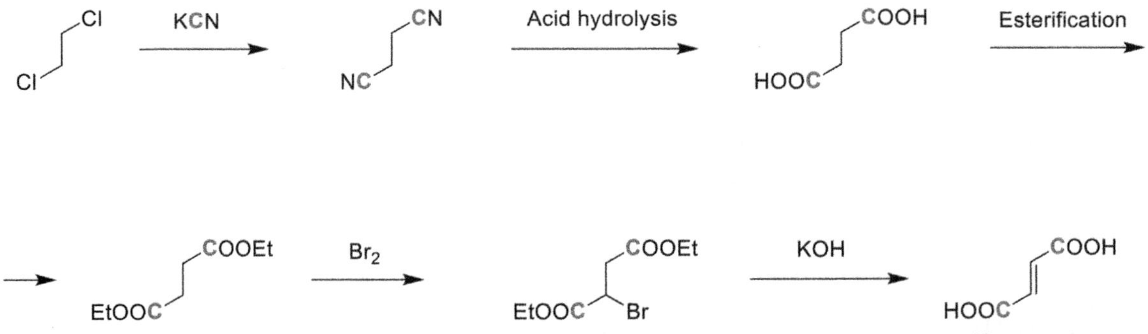

SCHEME 1.15 Synthesis of [1,4-$^{13}C_2$]fumaric acid via [1,4-$^{13}C_2$]succinic acid.

1.9.5.6.2 ^{13}C-labeled keto carboxylic acids

Pyruvate. Keto acids are carboxylic acid derivatives that contain one or more keto (CO) groups in the molecule. 2-Keto carboxylic acids with the keto group positioned adjacent (α) to the carboxyl function play an especially important role in biochemistry because they are intimately involved in such fundamental metabolic processes as the TCA cycle, glycolysis, gluconeogenesis, and transaminations. [1-^{13}C]Pyruvic acid is the most frequently used ^{13}C-labeled tracer for metabolic studies by hyperpolarized NMR/MRS/MRI. Its popularity is due to its essential metabolic role combined with its very favorable properties for DNP. As already discussed, pyruvate rapidly exchanges into the preexisting lactate pool, and therefore, the HP-[1-^{13}C]lactate signal is frequently used to probe the Warburg effect. Another common, if somewhat

incorrect, use of HP-[1-^{13}C]pyruvate is to estimate TCA cycle flux based on the detection of hyperpolarized bicarbonate signals after the injection of HP-[1-^{13}C]pyruvate. The TCA cycle is primarily fed by acetyl-CoA derived from the decarboxylation of pyruvate, catalyzed by pyruvate dehydrogenase (PDH). However, it should be noted that PDH is not a TCA cycle enzyme, and pyruvate is not the only source of acetyl-CoA, as it can also be produced from fatty acids by beta oxidation. [1-^{13}C]Pyruvic acid and its sodium salt are commercially widely available. Unlike [1-^{13}C]pyruvate, [2-^{13}C]pyruvate retains the label after passing through PDH to produce [1-^{13}C]acetyl-CoA, and therefore, it is a promising probe to assess mitochondrial metabolism [663]. HP-[2-^{13}C]pyruvate gives rise to a number of downstream metabolites in the isolated, perfused heart and in vivo. These include [1-^{13}C]acetyl-CoA (202 ppm), [1-^{13}C]acetylcarnitine (175 ppm), [1-^{13}C]citrate (181 ppm), [5-^{13}C]glutamate/[5-^{13}C]-2-ketoglutarate (183 ppm), [2-^{13}C]lactate (71 ppm), and [2-^{13}C]alanine (53 ppm) [664,665]. It was also tested on healthy volunteers [666]. [1,2-$^{13}C_2$]pyruvate can be used to follow more than one downstream metabolic pathway. For example, hyperpolarized [1,2-$^{13}C_2$]pyruvic acid was successfully used to simultaneously probe PDH and mitochondrial flux in a cardiac model, as the C1 is lost when pyruvate is decarboxylated by PDH while the label at C2 enters the TCA cycle as acetyl-CoA [667]. Due to the dipolar interactions of the ^{13}C-^{13}C spins, the T_1 relaxation times and polarization levels for [1,2-$^{13}C_2$]pyruvate were slightly lower than in the singly labeled derivatives.

Although we cannot be sure exactly which processes are used to produce [1-^{13}C]pyruvic acid by the individual companies, the most commonly quoted synthesis uses ^{13}C-labeled cyanide for the introduction of the label at C1. The procedure is based on an older method developed for the synthesis of 2-ketoacids and can be modified for the synthesis of other ^{13}C-labeled pyruvic acids with other labeling patterns. The cyanide reagent is most commonly CuCN, but trimethylsilyl cyanide prepared from KCN and trimethylsilyl chloride can also be used (Scheme 1.16) [668–670].

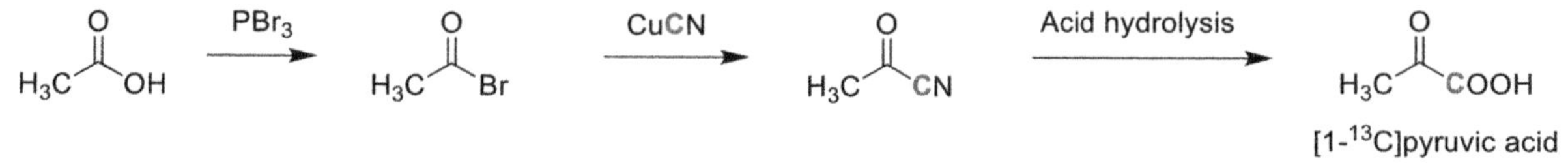

SCHEME 1.16 The synthesis of [1-^{13}C]pyruvic acid from acetyl halide and ^{13}C-labeled cyanide.

Other possible synthetic approaches to ^{13}C-labeled pyruvic acids include the reaction of Grignard reagents and organolithium reagents with diethyl oxalate, the oxidation and subsequent hydrolysis of labeled butyl lactate, and the carboxylation of lithiodithianes with ethyl chloroformate or carbon dioxide (Scheme 1.17) [671–675].

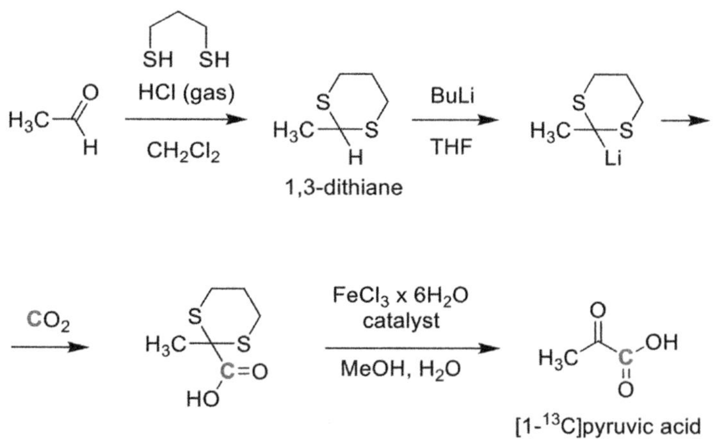

SCHEME 1.17 Synthesis of ^{13}C-labeled pyruvic acid via the carboxylation of the lithium derivative of 2-methyl 1,3-dithiane.

C2 and multiple ^{13}C-labeled pyruvate derivatives can be synthesized by the above methods, starting from the appropriately labeled reagents [676]. An early synthesis of [2-^{14}C]pyruvic acid started from [^{14}C]CO_2, which was converted to [1-^{14}C]acetate with methyl magnesium bromide. The rest of the synthesis followed Scheme 1.17 [677].

Pyruvic acid is in equilibrium with its hydrate form in an aqueous solution. Unfortunately, pyruvic acid and sodium pyruvate solutions are not very stable and cannot be stored for extended periods [678–681]. Over time, pyruvic acid condenses with itself to form a homoaldol product known as parapyruvic acid, which also exists in equilibrium with its hydrate in aqueous media [682]. Parapyruvic acid can undergo further intramolecular cyclization to form a cyclic lactone

enol product, zymonic acid [683–686]. Generally, aqueous solutions of pyruvic acid and sodium pyruvate contain pyruvate hydrate, parapyruvic acid, zymonic acid, their tautomers, and hydrated derivatives as unwanted impurities (Scheme 1.18). Some of these can interfere with metabolism. Parapyruvic acid (4-methyl-4-hydroxy-2-ketoglutaric acid) is a potent inhibitor of ketoglutarate dehydrogenase. It reacts with the enzyme to form a covalently linked citramalyl derivative, causing the accumulation of 2-ketoglutarate in the TCA cycle [687,688]. However, zymonic acid does not have this inhibitory effect. As already mentioned in Section 1.8.2, the chemical shift of the carboxylate in zymonic acid was found to be pH-responsive in the physiological range, $pK_{a2} = 6.90$ (OH at -C2) and is thus [1,5-$^{13}C_2$]zymonic acid is a suitable HP probe for measuring physiological pH with a reasonable T_1.

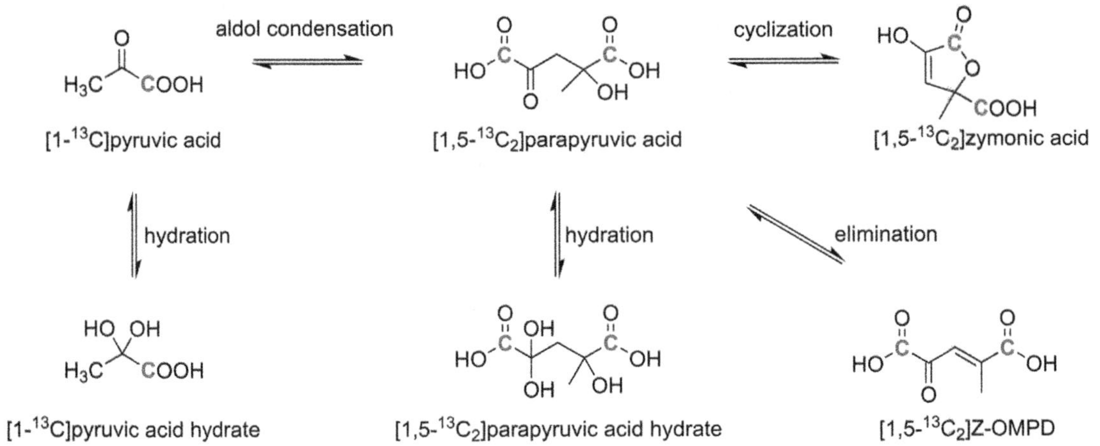

SCHEME 1.18 Pyruvic acid undergoes hydration and self-condensation reactions to form pyruvate hydrate, parapyruvic acid, zymonic acid and Z-OMPD.

Self-condensation of pyruvic acid, pyruvic acid esters, and sodium pyruvate can be brought about by acid or base catalysis [684,689]. The synthesis of [1,5-$^{13}C_2$]zymonic acid was achieved with zinc acetate-catalyzed self-condensation of [1-^{13}C]pyruvic acid to parapyruvate. Next, acid treatment of the parapyruvate intermediate afforded [1,5-$^{13}C_2$]zymonic acid (Scheme 1.19) [491]. The deuterated form, [1,5-$^{13}C_2$, 3,6,6,6-D_4]zymonic acid, was also prepared, and the deuteration resulted in a 39% increase in the T_1 [690]. A related pyruvate derived pH probe, [1,5-$^{13}C_2$]Z-4-methyl-2-oxopent-3-enedioic acid (Z-OMPD) was also synthesized by the base catalyzed self-condensation of ethyl[1-^{13}C]pyruvate. This compound has higher stability in aqueous solution and displays longer T_1 values (138 s for C1 and 119 s for C5 at 1 T in D_2O) than zymonic acid [691].

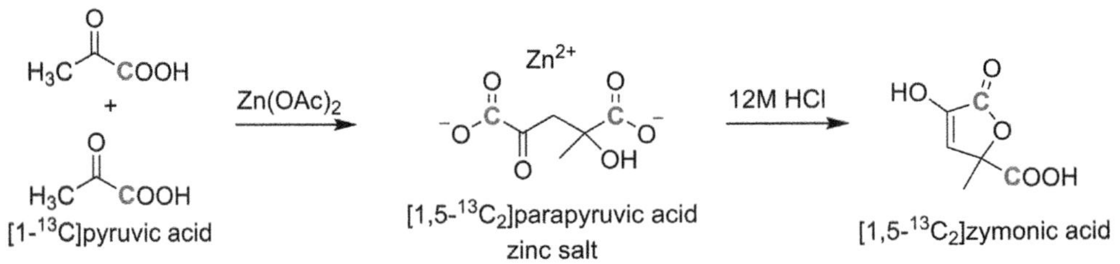

SCHEME 1.19 Synthesis of [1,5-$^{13}C_2$]zymonic acid.

Pyruvate is a monocarboxylate at physiological pH and is rapidly transported across cell membranes by the monocarboxylate transporters. Nevertheless, the neutral and more lipophilic ethyl [1-^{13}C]pyruvate has also been explored to facilitate passive transport. It was shown that hyperpolarized ethyl [1-^{13}C]pyruvate is well suited for brain imaging by ^{13}C MRS in animal models [692]. Esterification of pyruvic acid with ethanol can be performed by the classic acid catalyzed esterification or by activating carboxylic acid with dicyclohexyl carbodiimide in the presence of dimethylaminopyridine (DMAP) as an acyl transfer catalyst to increase the yield of esterification.

As discussed earlier, the oxidative decarboxylation of pyruvate by hydrogen peroxide rapidly produces acetate and CO_2/bicarbonate [652]. Since pyruvate polarizes so well, a new hyperpolarization methodology termed Chemical

Reaction-Induced Multi-molecular Polarization (CRIMP) was proposed, which essentially involves the rapid chemical conversion of a highly polarized starting material into several hyperpolarized agents (Scheme 1.20) [693]. The feasibility of the concept was demonstrated by generating HP-[1-^{13}C]acetate and HP-[^{13}C]CO_2/HP-[^{13}C]bicarbonate from HP-[1,2-$^{13}C_2$]pyruvate, as shown in Scheme 1.20. The mixture of these hyperpolarized agents can be used to simultaneously map pH and assess TCA cycle activity, fatty acid synthesis, and glycolysis (with partial conversion of the sating pyruvate).

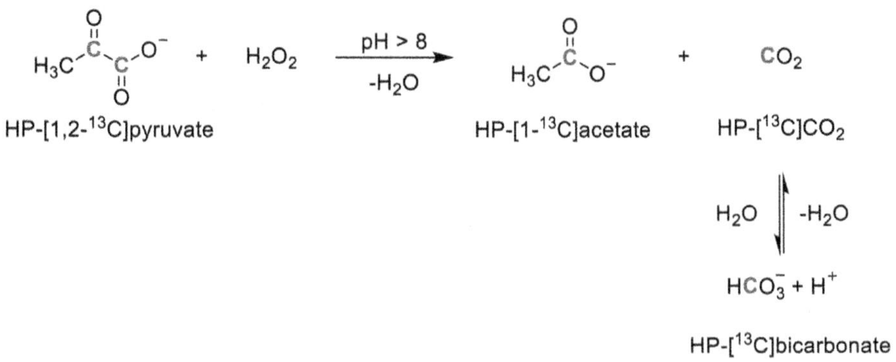

SCHEME 1.20 The concept of CRIMP (Chemical Reaction-Induced Multi-molecular Polarization) involves the rapid chemical production of several hyperpolarized agents from a highly polarized precursor, in this case, [1,2-$^{13}C_2$]pyruvate.

2-Ketoglutarate. 2-Ketoglutaric acid (alpha-ketoglutaric acid) is another ketocarboxylic acid of great biological importance. It is a key intermediate of the TCA cycle, produced via the decarboxylation of isocitrate by the enzyme IDH. The decarboxylation of 2-ketoglutarate to CO_2 and succinyl-CoA is catalyzed by 2-ketoglutarate dehydrogenase (2-KGDH). This is a major control point of metabolic flux through the TCA cycle, as the activity of this enzyme is regulated by both the energy and redox state of the cell. In addition, ketoglutarate is perhaps the most important nitrogen transporter. It can be reversibly transaminated to glutamate. It can also be produced from glutamate by glutamate dehydrogenase, and the TCA cycle can be replenished via 2-ketoglutarate by anaplerotic reactions. Thus, 2-ketoglutarate is intricately involved in both carbon and nitrogen metabolism, and intracellular 2-ketoglutarate levels reflect the metabolic availability of nitrogen and carbon substrates [694]. An interesting and clinically significant metabolic conversion of 2-ketoglutarate occurs in some cancers. About 70%–80% of low-grade II/III gliomas express a gain-of-function mutation in IDH, which will reduce 2-ketoglutarate to 2-hydroxyglutarate (2-HG). 2-HG is considered an oncometabolite as it can trigger oncogenesis via affecting DNA and methylation [695,696]. However, somewhat paradoxically, the presence of mutated IDH predicts longer survival and a better response to temozolomide treatment in low-grade gliomas [697]. The prognostic value of 2-HG levels in gliomas prompted the development of several MRS-based techniques, including HP-^{13}C MRS, for 2-HG detection. Hyperpolarized ^{13}C-labeled 2-ketoglutarate derivatives have been used successfully to detect 2-HG production in orthotopic tumor models, to monitor glutamate production and to probe 2-KGDH activity in the liver in vivo [363,536,698]. 2-Ketoglutarate is a dianion at physiological pH. It exists in equilibrium with the hydrated form and a cyclic lactone formed by intramolecular cyclization (Scheme 1.21) [608,624,699].

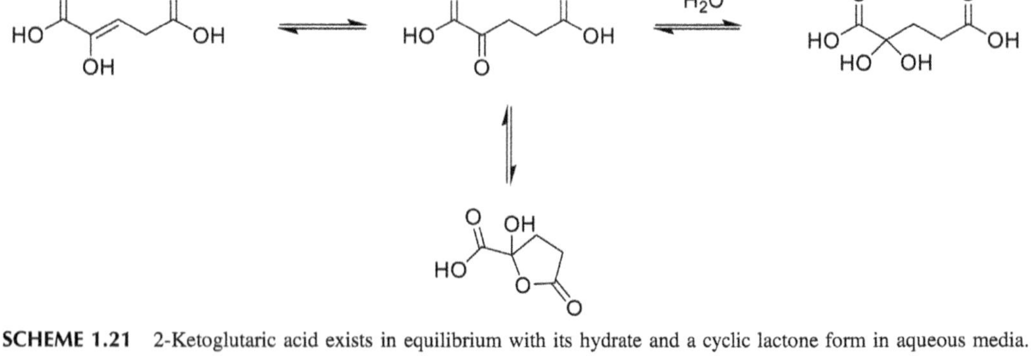

SCHEME 1.21 2-Ketoglutaric acid exists in equilibrium with its hydrate and a cyclic lactone form in aqueous media.

The interconversion of these forms is slow on the NMR timescale, allowing their separate observation and NMR characterization. However, the rate of interconversion is fast on the timescale of the HP-^{13}C lifetime and thus, it does not affect the observation of downstream metabolites. Although there are several methods published in the open literature for the synthesis of 2-ketoglutarate derivatives labeled with ^{13}C at various positions, the synthesis is surprisingly challenging for such a small molecule [624,700]. From the synthetic organic chemistry perspective, ^{13}C-labeled 2-ketoglutarate derivatives are potentially accessible via the Michael addition (1,4-addition of a stabilized carbanion to an a,β-unsaturated carbonyl compound), the Claisen condensation (base-catalyzed condensation of carboxylate esters to β-ketoesters), or the Stetter reaction (1,4-addition of an aldehyde to an a,β-unsaturated compound). The synthesis based on the Michael addition starts from commercially available labeled ethyl bromoacetate, which was converted to the iodoacetate derivative with NaI. Ethyl iodoacetate was then reacted with silver nitrite in diethyl ether to afford ethyl [1-^{13}C]nitroacetate. Next, in the key step of the synthesis, Michael addition of ethyl [1-^{13}C]nitroacetate and ethyl acrylate formed diethyl [1-^{13}C]-2-nitroglutarate. Nef oxidation of the nitronate anion with excess ozone afforded diethyl 2-ketoglutarate, which was subsequently hydrolyzed to the free acid. Subsequent hydrolysis in hydrochloric acid afforded the free acid (Scheme 1.22).

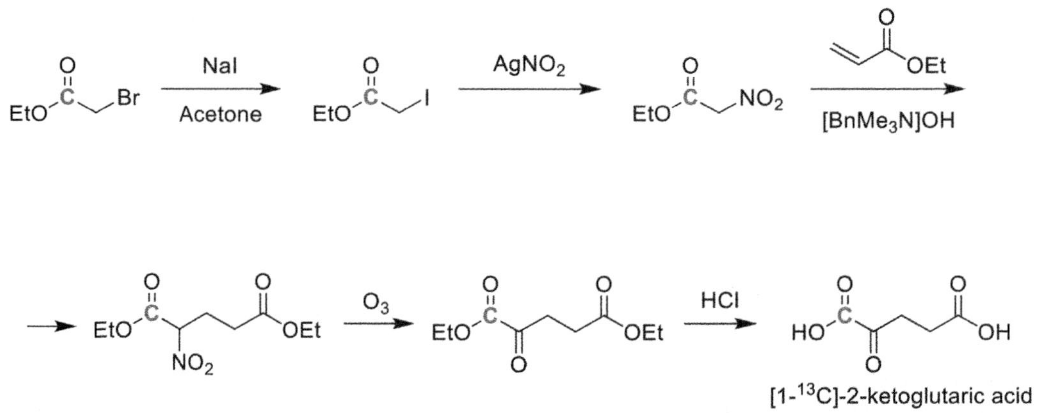

SCHEME 1.22 Synthesis of [1-^{13}C]-2-ketoglutaric acid via Michael addition.

While virtually any labeling pattern is accessible via this approach, it is also low-yielding and difficult to scale up [701]. These disadvantages prompted the development of a synthetic approach based on the Stetter reaction (Scheme 1.23) [624]. The intermediate glyoxylamide was prepared in situ by the oxidation of [1-^{13}C]-2-bromo-morpholino-acetamide with N-hydroxy-4-N,N-DMAP. The Stetter reaction between the glyoxylmorpholide intermediate and benzyl acrylate was catalyzed by 3,4,5-trimethylthiazolium iodide. The Stetter addition product was converted to diisopropyl 2-ketoglutarate by refluxing it in isopropanol in the presence of sulfuric acid. The synthesis was completed by the acid hydrolysis of the diisopropyl ester. The methodology was also adapted for the preparation of [1-^{13}C]-2-ketoglutarate depleted of natural abundance ^{13}C at C5. The use of ^{13}C-depleted [1-^{13}C]-2-ketoglutarate facilitates the observation of [1-^{13}C]-2-hydroxyglutarate formed by the action of the mutant IDH enzyme by eliminating the signal overlap between the natural abundance of ^{13}C at C5 of ketoglutarate (^{13}C $\delta = 184.0$ ppm) and the C1 of [1-^{13}C]-2-HG (^{13}C $\delta = 183.9$ ppm).

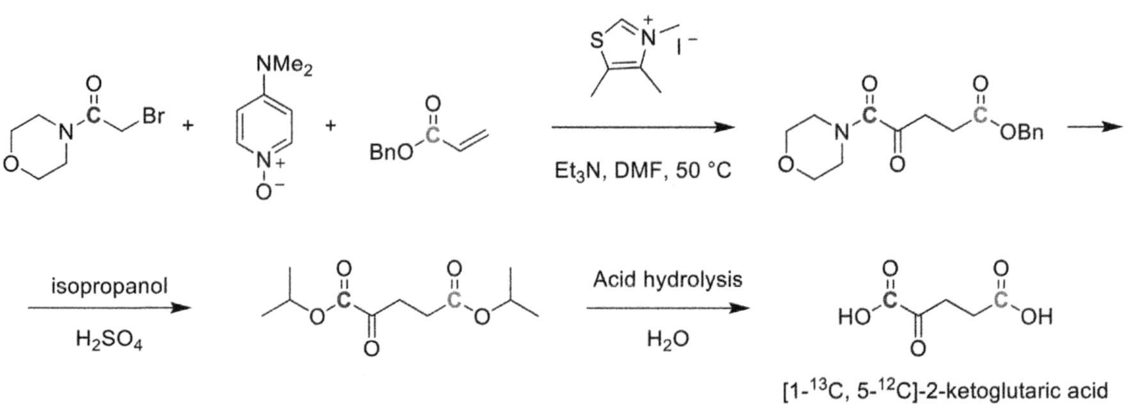

SCHEME 1.23 Synthesis of [1-^{13}C, 5-^{12}C]-2-ketoglutaric acid using the Stetter reaction.

Both the [1,2-$^{13}C_2$] and the [U-$^{13}C_5$]-2-ketoglutaric acid diethyl esters were synthesized in reasonable yields starting from commercially available labeled oxalic and succinic acid by the well-established Claisen condensation, which is frequently used for the synthesis of 2-ketocarboxylic acid esters, including 2-ketoglutarate derivatives [702]. The base-catalyzed condensation of ^{13}C-labeled diethyl oxalate and diethyl succinate afforded oxalosuccinate triethyl ester, which was hydrolyzed and decarboxylated in acid to give ketoglutaric acid. The labeled 2-ketoglutaric acids were converted to ethyl ester derivatives by esterification with acetyl chloride in ethanol (Scheme 1.24) [363].

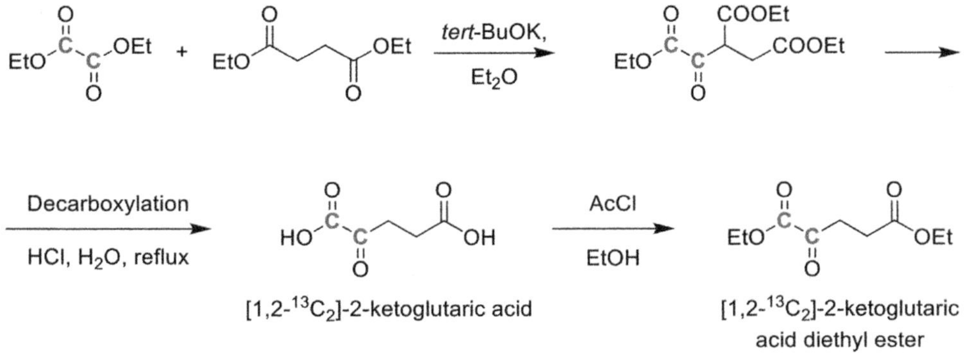

SCHEME 1.24 Synthesis of [1,2-$^{13}C_2$]-2-ketoglutaric acid using the Claisen condensation.

The diethyl ester was used to facilitate the cellular delivery of 2-ketoglutarate. In contrast to cancer cells, uptake of free ketoglutarate in normal cells is slow on the timescale provided by the decay of hyperpolarized ^{13}C magnetization. However, as was demonstrated by conventional ^{13}C isotopomer analysis of liver extracts, the diethyl ester was rapidly taken up and metabolized in the liver to produce glutamate, succinate, glucose, and other metabolites within 3 min [363]. It is worth noting that there is a significant difference in the rate of hydrolysis of the ketoglutarate ester groups. Experiments performed with unlabeled dimethyl 2-ketoglutarate revealed that the C1 ester group hydrolyzed much faster than the C5 ester, which persisted for days [703].

2-Ketoisocaproic acid. 2-Ketoisocaproic acid (αKIC) is a useful probe to study altered branched-chain amino acid (BCAA) metabolism. HP-[1-^{13}C]-2-ketoisocaproic acid can assess both BCAA aminotransferase and branched-chain α-keto acid dehydrogenase complex (BCKDC), which activities could be assessed via the detection of [1-^{13}C]leucine or hyperpolarized bicarbonate in glioblastoma models [704]. Synthesis of [1-^{13}C]-2-ketoisocaproic acid relies on [1-^{13}C] glycine as the starting material, which was reacted with potassium cyanate to form hydantoin. Condensation of hydantoin with isovaleraldehyde and subsequent hydrolysis furnished the desired [1-^{13}C]-2-ketoisocaproic acid (Scheme 1.25) [705].

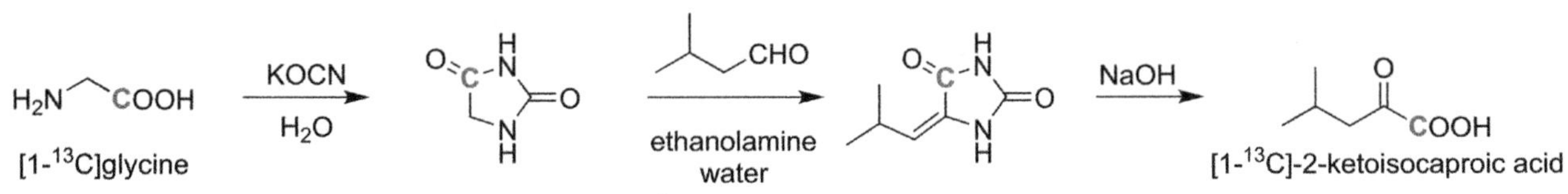

SCHEME 1.25 Synthesis of [1-^{13}C]-2-ketoisocaproic acid starting from C1 labeled glycine.

2-Ketobutyrate. LDH has five isoforms. These are assembled as tetramers of either of the two subunits, M (muscle) and H (heart), expressed by the genes LDH-A and LDH-B, respectively. Skeletal muscle, hepatocytes, and cancer cells almost exclusively contain the homotetramer of M subunits (LDH-5), whereas the heart contains almost exclusively H subunits (LDH-1). Both 2-ketobutyrate and pyruvate can be reduced by LDH enzymes; however, the selectivity constant of 2-ketobutyrate is higher for LDH-1. Therefore, HP-^{13}C-labeled 2-ketobutyrate is the more appropriate probe to study LDH-1 activity in the heart [706]. HP-^{13}C NMR/MRS experiments were performed with the commercially available [U-$^{13}C_4$]-2-ketobutyrate [706]. This probe had a reasonably long T_1 value of 52 s at 3 T. In a similar way to pyruvate, one would anticipate that monolabeling at either C1 or C2 would improve the T_1 and the SNR by eliminating the ^{13}C-^{13}C homonuclear couplings [362,707]. Alternatively, [U-^{13}C]-α-Ketobutyrate can be prepared from[$^{13}C_5$]-valine using an aminotransferase reaction.

[2-^{13}C]benzoyl formic acid (BFA). Hyperpolarized [2-^{13}C]BFA was proposed as a probe to detect hydrogen peroxide (H_2O_2) because it selectively reacts with hydrogen peroxide to produce [1-^{13}C]BA and CO_2 [708]. Carboxylic acids react with (cyanomethylene) triphenylphosphorane to form cyano ketophosphorane ylide. Oxidative cleavage of the ylides affords 2,3-diketonitriles, which can be converted into 2-keto carboxylic acid derivatives. [2-^{13}C]BFA was synthesized by a modification of this method (Scheme 1.26) in which [1-^{13}C]benzoic acid was converted to the cyanosulfur ylide with cyanosulfonium bromide and HATU followed by oxidation with oxone to produce [2-^{13}C]benzoyl formic acid (BFA) [709].

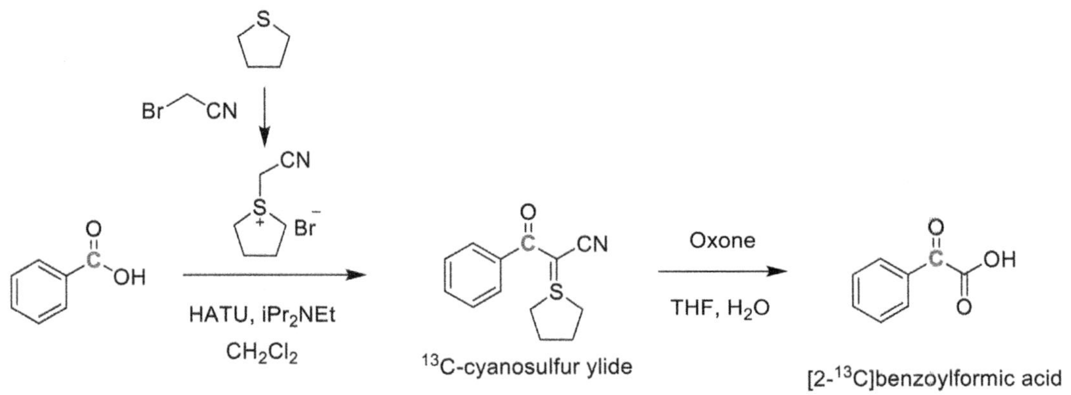

SCHEME 1.26 Synthesis of [2-^{13}C]benzoylformic acid via the ^{13}C-labeled cyanosulfur ylide.

The biochemical relationship between α-amino acids and their respective α-ketoacid analogs is obvious. In biological systems, most amino acids can be derived from their keto acid precursor, and vice versa, α-ketoacids (2-ketoacids) can be produced from their amino acid counterparts. Thus, 2-ketocarboxylic acids ^{13}C-labeled at any position can potentially be synthesized by enzymatic oxidative deamination of the corresponding labeled amino acids, provided they are commercially available (Scheme 1.27) [710]. However, to the best of our knowledge, this method has not been applied yet to the synthesis of hyperpolarized probes.

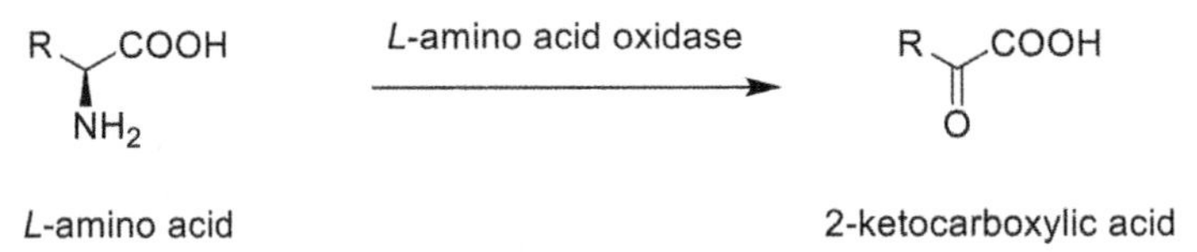

SCHEME 1.27 Synthesis of 2-ketoacids by enzymatic oxidative deamination of amino acids.

^{13}C-labeled acetoacetate derivatives. Acetoacetate is a ketone body (together with β-hydroxybutyrate and acetone) produced by the liver from fatty acids. Ketone bodies are important metabolic intermediates used as energy sources when carbohydrate stores are depleted or glucose uptake is impaired, as in starvation or diabetes. The reversible reduction of acetoacetate is catalyzed by the mitochondrial enzyme β-hydroxybutyrate dehydrogenase enzyme (BDH), and the β-hydroxybutyrate to acetoacetate ratio reflects the mitochondrial redox state (NADH/NAD+) [545]. Thus, hyperpolarized ^{13}C-labeled acetoacetate was used to measure mitochondrial redox via the detection of the β-hydroxybutyrate signal [545,711]. This is analogous to the lactate-to-pyruvate ratio, which characterizes the redox state in the cytosol. In fact, copolarized [1-^{13}C]pyruvate and [1,3-$^{13}C_2$]acetoacetate have recently been used to monitor cytosolic and mitochondrial redox in one HP experiment [546]. Hyperpolarized ^{13}C-labeled acetoacetate was also used to probe ketone body metabolism in diabetic rat hearts [712]. Ethyl [1,3-$^{13}C_2$]acetoacetate is a newly developed probe to assess carboxy esterase activities. Carboxy esterases are a family of promiscuous enzymes, and their activity is decreased in cancer tissues. Metabolic conversion of ethyl [1,3-$^{13}C_2$]acetoacetate to [1,3-$^{13}C_2$] acetoacetate can be monitored based on the C1 chemical shift, as there is a 5 ppm difference between the ester and the free acid C1 signal [713]. The base-catalyzed self-

condensation of ethyl acetate to ethyl acetoacetate is a classic example of Claisen condensation and can be used for the synthesis of the [1,3-$^{13}C_2$], [2,4-$^{13}C_2$] and [U-$^{13}C_4$] labeling patterns (Scheme 1.28) [714,715].

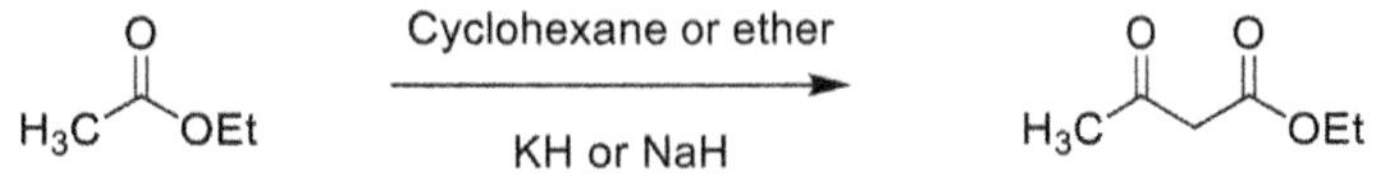

Ethyl [1-^{13}C]acetate — Ethyl [1,3-$^{13}C_2$]acetoacetate

Ethyl [2-^{13}C]acetate — Ethyl [2,4-$^{13}C_2$]acetoacetate

Ethyl [1,2-$^{13}C_2$]acetate — Ethyl [U-$^{13}C_4$]acetoacetate

SCHEME 1.28 Synthesis of ^{13}C-labeled ethyl acetoacetates via the Claisen condensation of labeled ethyl acetate.

The synthesis of other labeling patterns, such as [1-^{13}C] and [3-^{13}C], is more challenging. All possible labeling patterns are available via "crossed" Claisen condensation or the Blaise reaction, as the acetoacetate carbon skeleton is regiospecifically synthesized from two different components. In the crossed Claisen condensation, a strong base is used to fully deprotonate one of the reaction partners to form an enolate salt. This is then reacted with the other component to afford the condensation product (Scheme 1.29) [716].

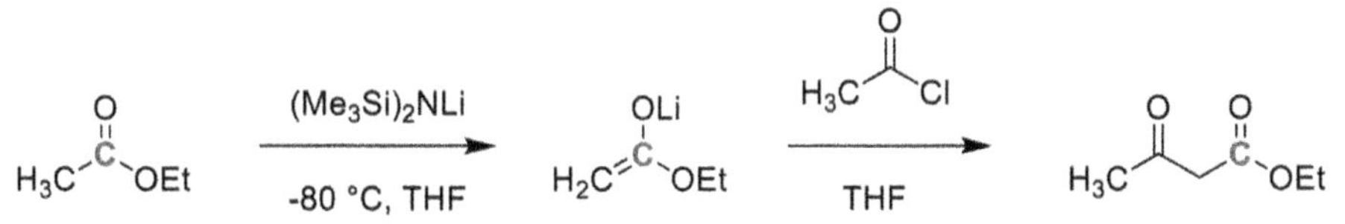

Ethyl [1-^{13}C]acetate — Ethyl [1-^{13}C]acetoacetate

SCHEME 1.29 Synthesis of ^{13}C-labeled ethyl acetoacetates via "crossed" Claisen condensation.

[3-^{14}C]Acetoacetate ethyl ester was prepared in good yield in a related condensation by reacting the magnesium derivative of ethyl-*tert*-butyl malonate with labeled acetyl chloride, followed by the thermal elimination of isobutylene and decarboxylation of the acetyl malonic ester (Scheme 1.30) [717,718].

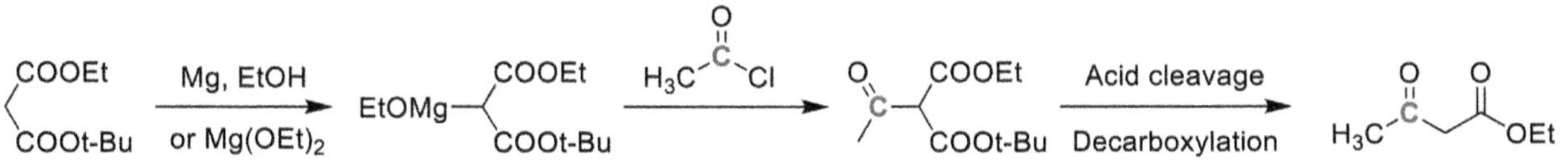

Ethyl [3-^{14}C]acetoacetate

SCHEME 1.30 Synthesis of [3-^{14}C] labeled ethyl acetoacetate via malonic ester synthesis.

The Blaise reaction is a useful synthetic method for the preparation of β-ketoesters from nitriles and alpha-haloesters. The reaction proceeds via an organozinc intermediate, which adds to the CN triple bond of the nitrile [719,720]. The synthesis of [2,3,4-$^{13}C_3$]acetoacetate ethyl ester illustrates the usefulness of this reaction (Scheme 1.31).

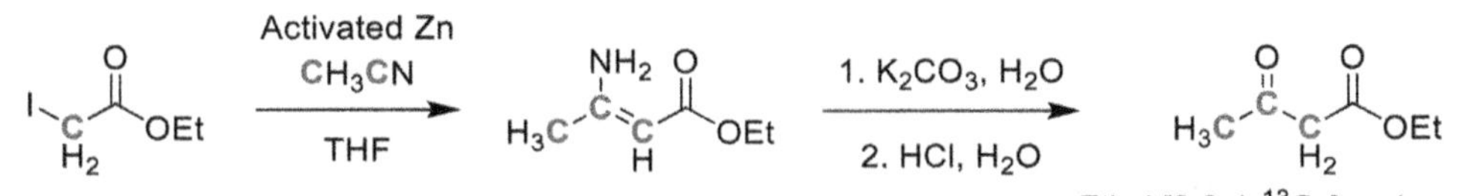

Ethyl [2,3,4-$^{13}C_3$]acetoacetate

SCHEME 1.31 Synthesis of ^{13}C-labeled ethyl acetoacetates via the Blaise reaction.

The ethyl ester can be hydrolyzed to the salt with NaOH or LiOH [545,712]. The ^{11}C-labeled acid was also prepared without going through the ethyl ester by the direct carboxylation of the enolate anion derived from acetone. Although originally developed for the synthesis of [1-^{11}C]acetoacetate, these reactions could potentially be adapted for the preparation of the ^{13}C-labeled derivatives. Acetone enolate can be produced by the reaction of isopropenyl acetate with methyllithium or by the cleavage of the O—Si bond in isopropenyloxy trimethylsilane with CsF (Scheme 1.32). In the next step, the enolate is directly reacted with labeled carbon dioxide to form acetoacetate [721].

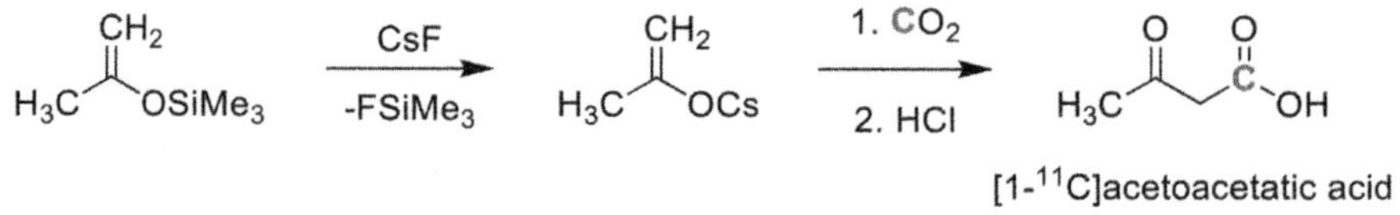

SCHEME 1.32 Synthesis of 11C-labeled acetoacetic acid starting from the enolate anion of acetone.

It should be noted that free acetoacetic acid and acetoacetate anion are not stable. They spontaneously decarboxylate to acetone and CO_2. The decomposition rate of the acid is about 50 times higher than that of the anion at 37 degrees [722.] The decarboxylation is catalyzed by amines via imine formation [723]. In serum, about 40% of acetoacetate decomposes at −20°C within 7 days [724]. In addition to decarboxylation, acetoacetate can undergo retro-Claisen condensation, which produces acetate [545]. Both [2-^{13}C]acetone and [1-^{13}C]acetate have been observed in hyperpolarized [1,3-$^{13}C_2$]acetoacetate. The presence of decomposition products should be considered when the in vivo ^{13}C MRS data are analyzed because they can potentially make the detection of metabolic products confounding.

1.9.5.6.3 Hydroxycarboxylic acids

Hydroxycarboxylic acids are an important class of compounds, some of which (lactate, malate, and 3-hydroxybutyrate) participate in fundamental metabolic processes. The higher-than-normal levels of some hydroxycarboxylic acids are characteristic of certain diseased states and can be used as diagnostic and prognostic biomarkers (e.g., lactate, β-hydroxybutyrate, and 2-hydroxyglutarate) [725–727]. Due to their significance, the synthesis of hydroxy carboxylic acids has been and continues to be the subject of several studies. As most hydroxy carboxylic acids are chiral, numerous chemical and enzymatic approaches have been developed for their enantioselective synthesis (Scheme 1.33) [728–733]. Of these, enantioselective reduction of ketocarboxylic acids and the conversion of amino acids can be easily adapted for the synthesis of ^{13}C-labeled hydroxy acids.

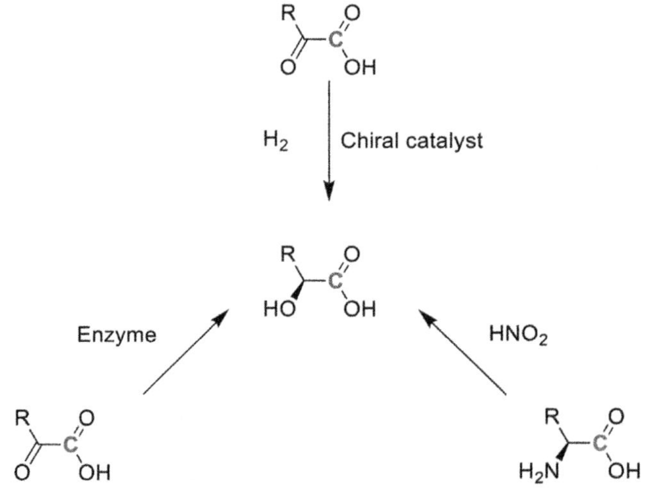

SCHEME 1.33 Commonly used approaches to α-hydroxy carboxylic acids.

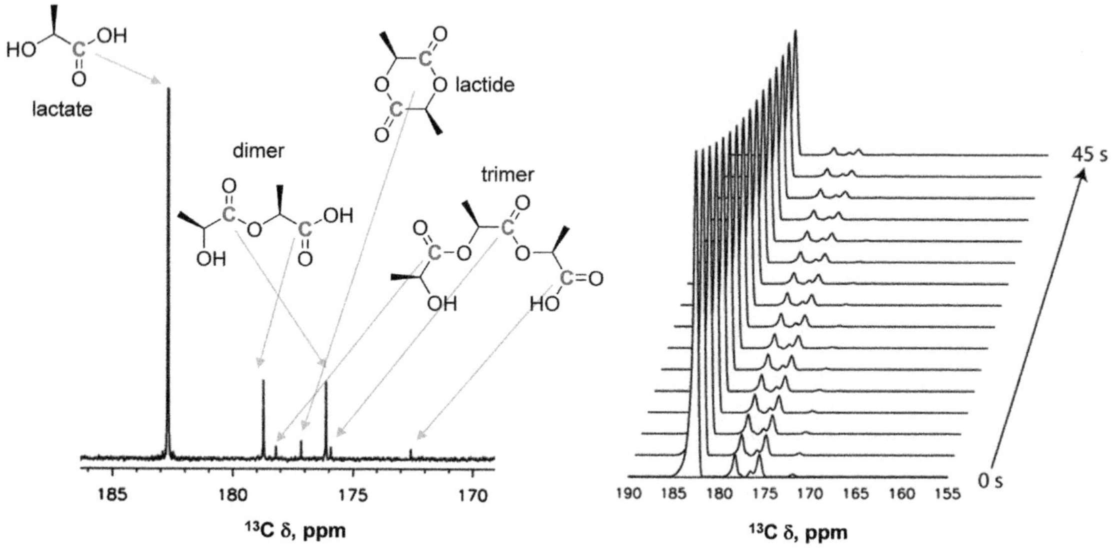

FIGURE 1.40 Thermally polarized ^{13}C NMR spectrum of [1-^{13}C] lactic acid at 11.7 T. The additional resonances were assigned to the dimer, trimer and lactide (left) [743]. Dynamic ^{13}C NMR of hyperpolarized [1-^{13}C]lactate in solution at 3 T (right). *Adapted from Ref. [743] with permission.*

Lactate. Lactate has a distinctly different metabolism from pyruvate. Lactate can be directly utilized as an energy source by the mitochondria in various tissues, such as muscle cells and neurons, especially during physical exercise. In contrast to pyruvate, lactate formation, distribution, and further metabolism play a critical role in regulating intermediary metabolism. Via the Cori cycle, it is the key physiological substrate for hepatic gluconeogenesis. The Cori cycle involves the interorgan recycling of lactate carbons. Lactate is produced in other tissues (e.g., in muscles by glycolysis) and transported by the blood to the liver, where it is used for gluconeogenesis. The glucose is then released into the blood and transported to other tissues for energy production [734]. On a cellular level, lactate participates in the intracellular lactate shuttle. There is mounting evidence that lactate-to-pyruvate oxidation does not occur in the cytosol. Rather, lactate produced in the cytosol by glycolysis is shuttled into the mitochondria and converted to pyruvate. During exercise (accelerated glycolysis), the intracellular lactate shuttle accounts for most of the shuttling of reducing equivalents (NADH) into the mitochondria [735,736]. For these reasons, hyperpolarized ^{13}C lactate is a useful tool to probe metabolism in the liver, heart, muscle, brain, and other tissues [737—741]. An early synthesis of [1-^{13}C]lactic acid started from acetaldehyde, which was converted into cyanohydrin with ^{13}C-labeled KCN. Lactic acid (2-hydroxypropionic acid) contains both a hydroxyl and a carboxyl function and forms oligomeric and polymeric esters by a self-catalyzed esterification process (Fig. 1.40) [742]. The most prominent of the oligomers include the linear dimer (lactoyl lactic acid), and to a lesser extent, the linear trimer and the cyclic dimer (lactide). The carbonyl ^{13}C resonances of these oligomers can overlap and obscure the signals of some metabolites of interest [743]. Commercial lactic acid comes as a solution in water, and the concentration of the oligomer impurities depends on the age and actual concentration of lactic acid in the product. Pure, anhydrous L (or D) lactic acid is a solid with a melting point of 53°C (the racemate DL form melts at around 17°C) [744]. High purity L-lactic acid free from oligomers can be prepared by first distilling the crude product in vacuum, followed by recrystallization from an ethyl isopropyl ether solvent mixture [745].

Glycerate. Glycerate 3-phosphate is an intermediate in glycolysis and gluconeogenesis. Therefore, sodium [1-^{13}C]-D-glycerate has been investigated as an indicator of glycolytic activity in the liver. The tracer has been synthesized starting from [1-^{13}C]acetic acid, which is brominated in acidic conditions with bromine (Br_2) and then esterified with benzyl alcohol. Next, ^{13}C-labeled benzyl acrylate was obtained via triphenyl phosphonium ylide with formaldehyde. Asymmetric

dihydroxylation of the double bond was achieved with AD mix-α (a commercially available osmium tetroxide-containing premixed reagent used for Sharpless asymmetric dihydroxylation reactions) [746]. The benzyl ester was cleaved by catalytic hydrogenation with a Pd/C catalyst to afford the final product (Scheme 1.34) [747].

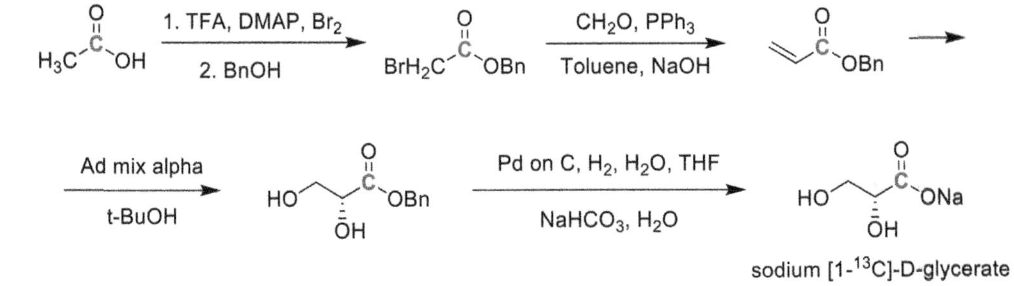

SCHEME 1.34 Asymmetric synthesis of C1 labeled D-glycerate from [1-^{13}C]acetic acid.

2-Hydroxyglutarate. Although ^{13}C-labeled 2-hydroxyglutaric acid has not been used in HP-^{13}C experiments, the uniformly ^{13}C-labeled derivative has been employed in ^{13}C MRS and isotope labeling studies. The L-enantiomer has been synthesized via the treatment of [U-$^{13}C_5$]L-glutamic acid with nitrous acid, followed by the hydrolysis of the resulting lactone (S)-(+)-5-oxo-2-tetrahydrofuran carboxylic acid (Scheme 1.35) [748]. The racemic form was prepared by the sodium tetrahydroborate reduction of uniformly ^{13}C-labeled 2-ketoglutarate in water (Scheme 1.35) [749].

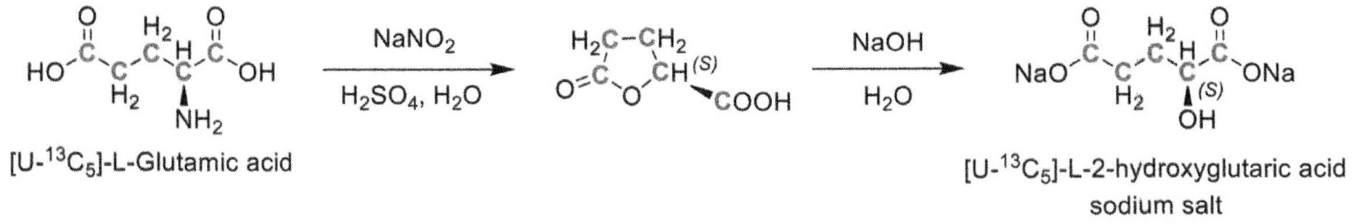

SCHEME 1.35 Synthesis of uniformly ^{13}C-labeled S-2-hydroxyglutarate from L-glutamic acid.

3-hydroxybutyrate. Hyperpolarized [1,3-$^{13}C_2$]-3-hydroxybutyrate has been used to measure BDH (EC 1.1.1.30) activity in the mouse heart [182]. Racemic-labeled 3-hydroxybutyrate derivatives can be prepared by the reduction of the corresponding acetoacetate with sodium tetrahydroborate in water [750]. However, it is more convenient to reduce acetoacetic acid esters (methyl, ethyl, or benzyl) because the separation and purification of the products are much easier. The desired free hydroxybutyrate can be obtained by hydrolysis or hydrogenolysis of the esters [751,752]. Enantioselective chemical reduction of ethyl [U-$^{13}C_4$]acetoacetate was performed via Noyori asymmetric hydrogenation with enantiomer purity in 96% yield with excellent stereoselectivity (99.3% ee) using a chiral Ru(II) catalyst and 85 atm of H_2 gas [753,754]. Racemic ^{13}C-labeled 3-hydroxybutyrate derivatives can be prepared by the reduction of the corresponding labeled acetoacetate derivatives with $NaBH_4$. Enzymatic enantioselective conversion of [3-^{13}C]acetoacetate to the naturally occurring enantiomer, [3-^{13}C]-D-3-hydroxybutyrate has been achieved with 3-hydroxybutyrate dehydrogenase and NADH following a procedure published for the unlabeled compound (Scheme 1.36) [755,756].

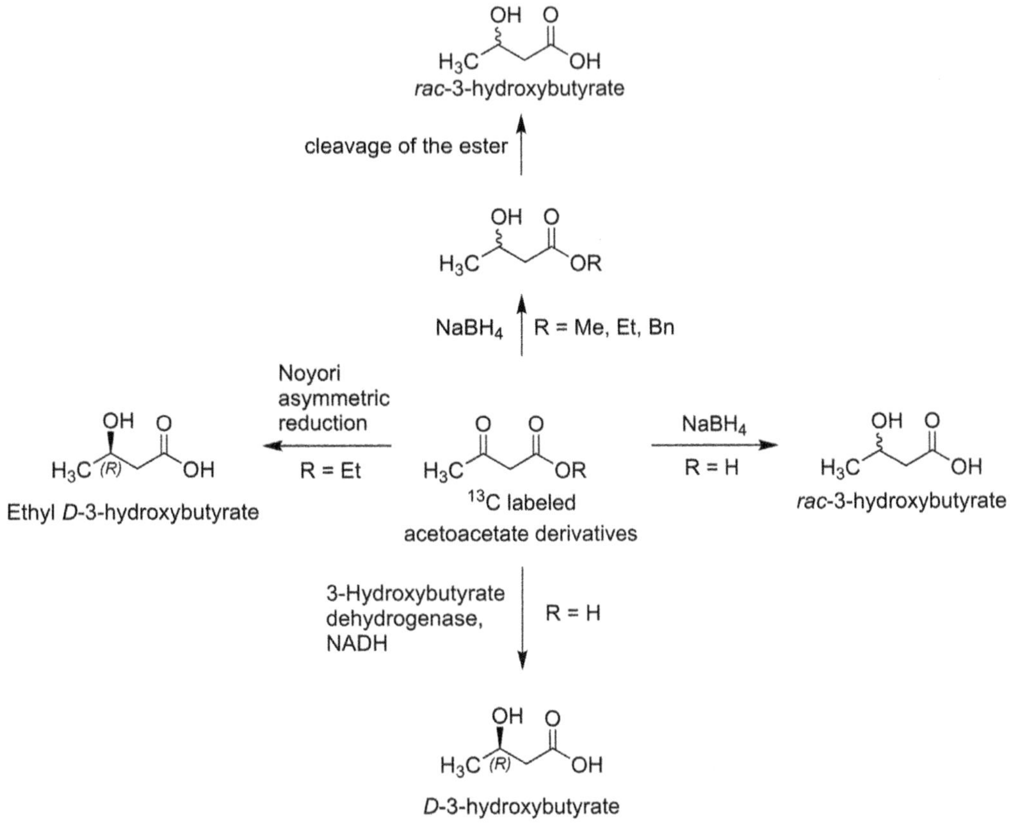

SCHEME 1.36 ^{13}C-labeled acetoacetate derivatives can be reduced to β-hydroxybutyrates chemically or enzymatically.

1.9.5.6.4 CO_2 precursors

The bicarbonate-CO_2 equilibrium system is well suited for measuring physiological pH because the pK_a of bicarbonate is in the physiological range. It has been demonstrated that tissue pH can be assessed in vivo from the ratio of the ^{13}C signal intensities of [^{13}C]bicarbonate and [^{13}C]CO_2 after the intravenous administration of hyperpolarized bicarbonate [176]. The pH was calculated using the Henderson-Hasselbalch equation and assuming a $pK_a = 6.17$ for bicarbonate (Eq. 1.28).

$$pH = pK_a + \log\left(\frac{[HCO_3^-]}{[CO_2]}\right) \tag{1.28}$$

Originally, bicarbonate was polarized in the form of cesium or sodium bicarbonate; however, sodium bicarbonate generally exhibits low solubility in the glycerol-water-based glassing matrices, which makes it difficult to form a concentrated glassing mixture. While cesium bicarbonate has a higher solubility, Cs^+ is toxic and has to be removed by ion exchange before in vivo administration [114]. These issues prompted the development of alternative methods that could be used to generate hyperpolarized bicarbonate/CO_2. One approach involved the rapid oxidative decarboxylation of 2-ketoacids such as pyruvic acid and 2-keto isocaproic acid (Scheme 1.37). Experimentally, the hyperpolarized ketoacid was decarboxylated with hydrogen peroxide, which generated hyperpolarized bicarbonate. Excess oxidant was destroyed by the addition of sulfite [757].

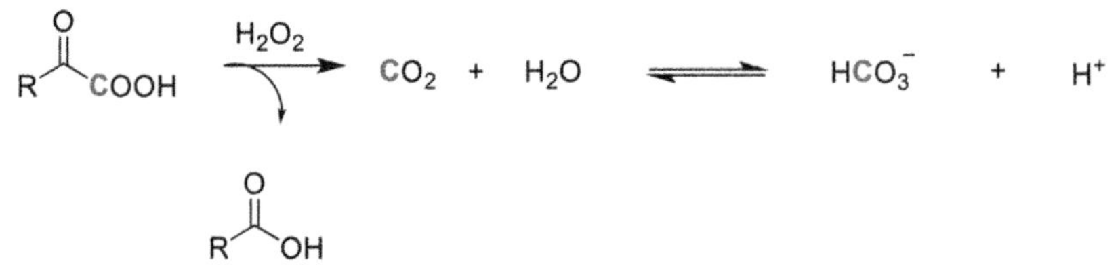

SCHEME 1.37 Oxidative decarboxylation of HP-[1-^{13}C] labeled ketoacids generates HP-[^{13}C]bicarbonate.

Another approach involved the use of ^{13}C-labeled carbonated precursor molecules that could rapidly be cleaved by base-catalyzed hydrolysis (Scheme 1.38). [1-^{13}C]-1,2-glycerol carbonate was found to be the best-performing candidate because of its favorable glassing properties and fast rate of hydrolysis. It was prepared by the transesterification of ^{13}C-labeled dimethyl carbonate with glycerol in the presence of potassium carbonate and magnesium oxide [758]. Alternatively, [1-^{13}C]-1,2-glycerol carbonate may also be prepared by the Pd/PPh$_3$-catalyzed transesterification of unlabeled diallyl carbonate with glycerol in the presence of excess [^{13}C]CO_2 [759].

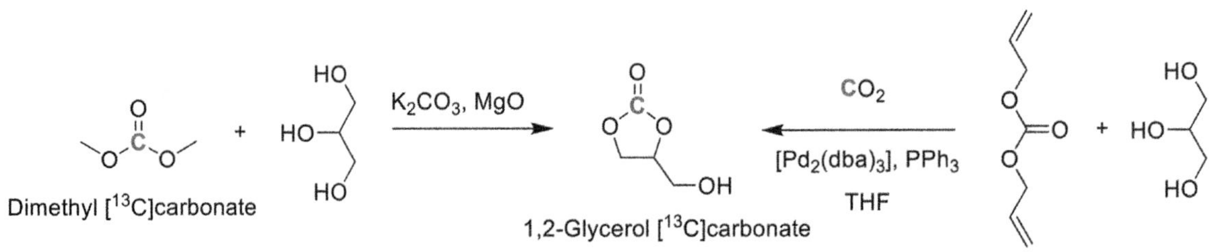

SCHEME 1.38 Synthetic approaches to 1,2-glycerol [^{13}C]carbonate.

Although these bicarbonate precursors generally have favorable properties for DNP, including good glassing behavior and T_1 values longer than bicarbonate, these advantages are largely negated by the decarboxylation or hydrolysis step necessary to generate the hyperpolarized bicarbonate before injection. These aforementioned drawbacks were eliminated by ^{13}C-labeled mixed anhydride tracers that were designed to be rapidly cleaved in vivo by carboxyl esterases to generate highly polarized [^{13}C]CO_2 and [^{13}C]bicarbonate [760]. ^{13}C-labeled ethyl acetyl carbonate was found to be the best-performing candidate. The tracer was prepared by reacting ethyl [1-^{13}C]chloroformate with acetic acid in the presence of N-methyl morpholine as a base in diethyl ether (Scheme 1.39).

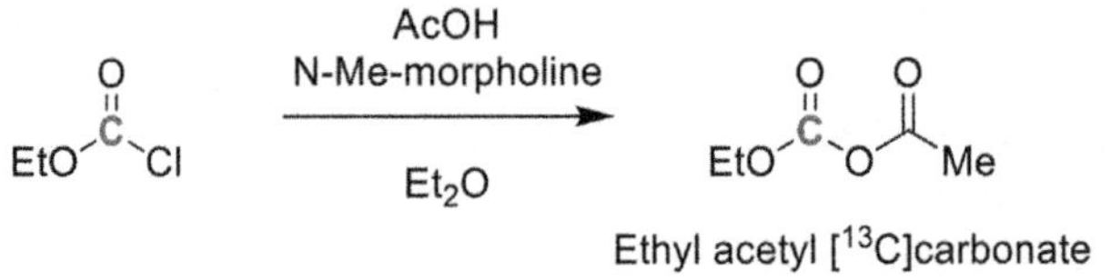

SCHEME 1.39 Synthesis of ethyl acetyl [^{13}C]carbonate from labeled acetic acid.

The CO_2 generation from this tracer takes place in two stages. First, the esterase-catalyzed cleavage of the ethyl ester bond occurs to produce monoacetyl carbonate. This unstable intermediate then undergoes rapid spontaneous

decomposition to release CO_2. As this process occurs in vivo, the newly generated hyperpolarized CO_2 will be rapidly equilibrated to hyperpolarized bicarbonate by carbonic anhydrase.

1.9.5.7 Carbohydrates

Carbohydrates and polyols play key roles in metabolism, cell signaling, and proliferation. Glucose is the most important source of energy in all living cells. In eukaryotes, glucose is first broken down to produce pyruvate during glycolysis. Pyruvate then enters the TCA cycle as acetyl CoA to produce CO_2 and NADH. Subsequently, the reducing equivalents are used to generate the energy carrier molecule ATP in oxidative phosphorylation. Aberrant glucose metabolism is implicated in various pathological conditions. Perhaps the best-known example is the elevated glycolytic rate and increased lactate level often observed in tumors even in the presence of sufficient oxygen supply (Warburg effect). As glucose is the starting material of glycolysis, hyperpolarized ^{13}C-labeled glucose derivatives would be ideal tracers to assess glucose metabolism. However, the carbons in glucose are directly attached to hydrogen atoms, which results in short T_1 relaxation times (<2 s), and nondeuterated [^{13}C]-labeled D-glucose derivatives are unsuitable for hyperpolarized ^{13}C studies. Therefore, deuteration is essential for the development of hyperpolarized ^{13}C-labeled carbohydrate tracers. Deuteration of uniformly ^{13}C-labeled glucose increases the T_1 to about 10–12 s, which represents about a 5-fold improvement. However, it is still short compared to the dissolution and transfer time of commercial dissolution; DNP polarizers and rapid dissolution systems provide significantly stronger signal intensity at the time of injection for in vivo studies. Hyperpolarized [U-$^{13}C_6$, U-$^2H_7$]glucose has been successfully used to visualize D-glucose metabolism in several organisms and various tissues [467]. In particular, the C1 carbon of lactate was observed as a doublet along with signals of a few other metabolites in the tumor after injection of hyperpolarized [U-$^{13}C_6$, U-$^2H_7$]glucose into tumor earing mice using the commercial dissolution system. Although the ^{13}C signals of the unmetabolized tracer were seen in normal tissues, glycolytic activity could not be detected, likely reflecting the lower glucose uptake and glycolytic rate of healthy tissues in comparison to the tumor [591]. However, a custom-made rapid transfer and injection system allowed the observation of glycolytic activity in the brains of healthy mice [628].

[U-$^{13}C_6$, U-$^2H_7$]glucose is derived from biomass in photosynthetic algal cells (Scheme 1.40). The cells can be cultivated under a [^{13}C]CO_2 atmosphere and D_2O in the medium, which results in ^{13}C and ^2H-labeled starch, which is isolated and hydrolyzed to afford uniformly labeled and perdeuterated glucose [626,761].

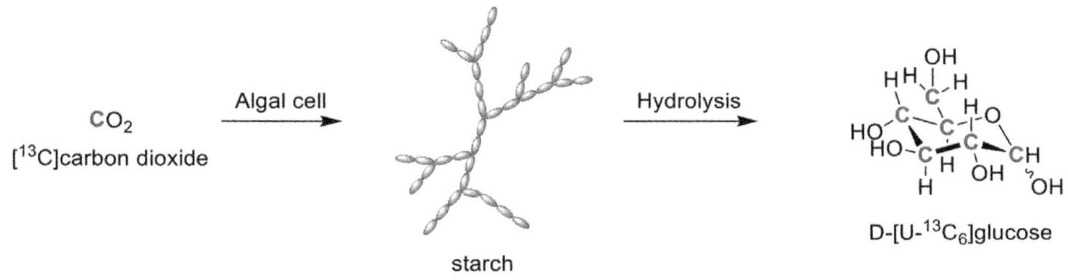

SCHEME 1.40 Synthesis of uniformly ^{13}C-labeled glucose from algal biomass.

Glucose derivatives, either singly or multiply-labeled at specific positions, are generally synthesized by a combination of chemical and enzymatic methods. The introduction of a ^{13}C label at C1 or C6 is relatively straightforward. The synthesis relies on the classic Kiliani-Fischer aldose chain elongation methodology. In the original Kiliani-Fischer synthesis, the aldose monosaccharide is reacted with potassium or sodium cyanide to form a pair of C2 epimeric cyanohydrins, which are hydrolyzed to the corresponding aldonates. The epimers are separated, and the aldonate is converted to the lactone, which is reduced with sodium amalgam [762]. As the original synthesis often suffered from low yields, an improved method, known as cyanohydrin reduction (Scheme 1.41), was developed in which the cyanohydrin is reduced directly rather than converted into lactone. Generally, the cyanohydrins are not isolated but are immediately reduced to the C2-epimeric aldoses with H_2 and a Pd/$BaSO_4$ catalyst at a slightly acidic pH (<4.3), followed by the separation of the epimers [763].

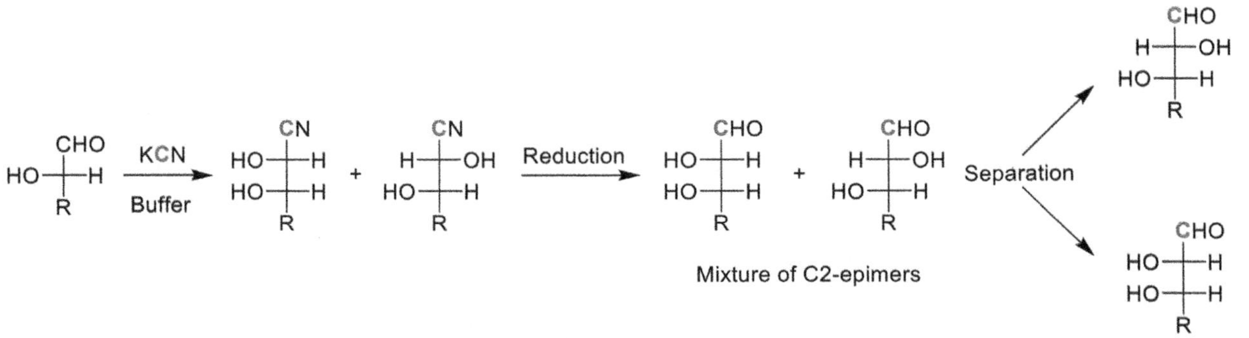

SCHEME 1.41 Synthesis of [1-^{13}C] labeled sugars via cyanohydrin reduction.

Deuteration of the installed ^{13}C label can easily be accomplished by performing the reduction with D_2 gas in D_2O. Alternatively, deuterated formic acid may be used, which breaks down to D_2 and CO_2 in the presence of Pd/$BaSO_4$, which eliminates the need for deuterium gas. The synthesis of [1-^{13}C, 1-^{2}H]-D-glucose involves the reaction of [^{13}C]KCN with arabinose in D_2O at an optimal pH (around 8.5) to form the aldonitrile. When the formation of aldonitrile is complete (about 15 min), the reaction mixture is acidified (pH 4), and excess cyanide is removed by purging the reaction with N_2 gas. The reaction mixture is then reduced with D_2 in a Parr apparatus in the presence of a Pd/$BaSO_4$ catalyst to afford the C2-epimer aldoses. Labeled nitromethane can also be used in place of KCN to introduce ^{13}C in position C1. The nitro-alcohol can then be hydrolyzed into sugar [764].

Labeling aldoses at the inner positions is more challenging. Fortunately, introducing a ^{13}C label at C2 is greatly facilitated by a process known as molybdate-catalyzed epimerization with C1−C2 transposition [765]. This is a very interesting skeletal rearrangement of aldoses in which the CHO group at C1 and the HCOH moiety at C2 change place with each other (Scheme 1.42). The transposition proceeds with epimerization at C2. The reaction occurs in the open-chain form, and the presence of hydroxyl groups at C1, C2, and C3 is necessary. This process provides a very convenient means to shift the label from C1 to C2. For example, D-[1-^{13}C]mannose obtained by the cyanohydrin reduction method can be easily converted to D-[2-^{13}C]glucose.

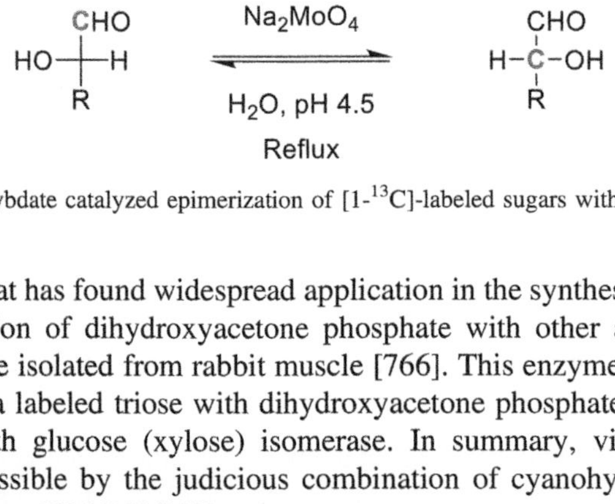

SCHEME 1.42 Molybdate catalyzed epimerization of [1-^{13}C]-labeled sugars with C1−C2 transposition.

Another useful transformation that has found widespread application in the synthesis of ^{13}C-labeled sugar derivatives is the stereoselective aldol condensation of dihydroxyacetone phosphate with other aldehydes catalyzed by the enzyme D-fructose-1,6-bisphosphate aldolase isolated from rabbit muscle [766]. This enzyme allows the introduction of a label at an internal position by condensing a labeled triose with dihydroxyacetone phosphate to a labeled fructose, which can be converted into labeled glucose with glucose (xylose) isomerase. In summary, virtually any ^{13}C labeling pattern of D-glucose and other sugars is accessible by the judicious combination of cyanohydrin reduction, molybdate-catalyzed transposition, and enzymatic reactions [626,767,768].

D-Glucose. Since glycolysis literally means splitting one glucose molecule into two pyruvate units, HP-^{13}C-labeled glucose derivatives would be ideal probes to monitor this fundamental pathway. Unfortunately, the ^{13}C spins in non-deuterated glucose have a rather short T_1 as a result of the strong dipolar relaxation effect of directly attached protons [769]. Therefore, deuteration of glucose is absolutely necessary to improve the ^{13}C T_1 for hyperpolarized ^{13}C MRS experiments. This is true for other carbohydrate molecules as well. While it is certainly possible to introduce the deuterium label at specific positions during the synthesis of the ^{13}C-tracer, for hyperpolarized ^{13}C studies, the fully (or nearly fully) deuterated

molecule is preferred, and it is often more convenient to obtain ^{13}C-labeled sugars commercially and perform the deuteration on these samples. Generally, Ru on C in D_2O is used to catalyze the H/D exchange. A typical example is the deuteration of [3,4-$^{13}C_2$]-labeled D-glucose, as shown in Scheme 1.43. The ^{13}C-labeled D-glucose had to be converted to the methyl glucoside first to avoid extensive decomposition. This was achieved by a treatment with MeOH and H_2SO_4 (Fischer glycosidation) [770]. Next, the methyl glucoside was dissolved in D_2O and treated with a Ru/C catalyst under a hydrogen atmosphere to afford the partially deuterated methyl glycoside. It was necessary to perform the catalytic step 3 times with fresh catalyst to achieve high (>95%) D-incorporation. Note that deuteration of C—H bonds at C1 and C5, which are adjacent to the hemiacetal oxygen or protected OH, did not occur. Finally, acidic hydrolysis afforded the desired [3,4-$^{13}C_2$,2,3,4,6,6-$^2H_5$]glucose. The T_1 values of the ^{13}C labels at C3 and C4 in this partially deuterated glucose increased to about 12 and 11 s, respectively, in water and at 9.4 T. This probe was successfully used to visualize glycolysis in vivo in the mouse brain. It is worth mentioning that the use of the [3,4-$^{13}C_2$] labeling pattern rather than the [U-$^{13}C_6$] one produces [1-^{13}C]lactate, which offers a two-fold increase in the SNR of the C1 lactate by eliminating the carbon-carbon homonuclear coupling [628].

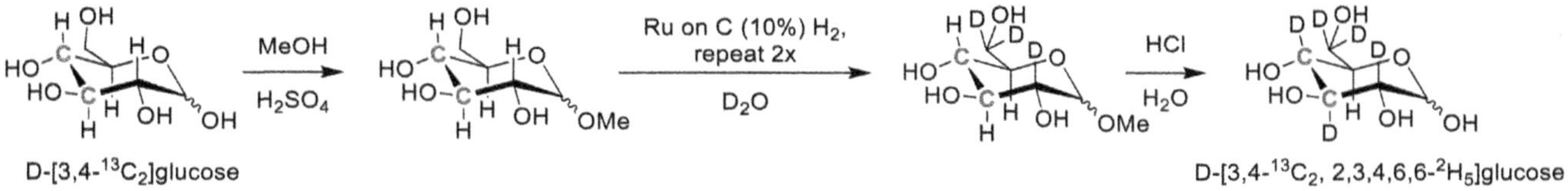

SCHEME 1.43 Ruthenium catalyzed partial deuteration of D-[3,4-$^{13}C_2$]glucose in D_2O.

D-Fructose. D-Fructose is a simple 6-carbon ketose. It is an isomer of glucose with C1. D-fructose exists as an equilibrium mixture of the β-pyranose (75%), β-furanose (21%), α-furanose (4%), α-pyranose (trace), and open chain forms (trace) in D_2O solution at 27°C (Fig. 1.41). The composition of the mixture is strongly solvent- and temperature-dependent [707,771]. However, only the first three are usually observable by ^{13}C NMR. Fructose has a carbonyl group at C2 rather than an aldehyde carbon at C1 and the absence of directly attached protons at this position significantly extends the T_1 value. Even without deuteration, the ^{13}C spins at C2 have reasonably long T_1 values (about 14 s at 3 T and 16 s at 11.7 T). Full deuteration of the molecule extends the T_1 to 34 s, indicating that the spin lattice relaxation is likely dominated by dipolar interactions with ^{1}H spins in the molecule. Unlike many other carbonyl-labeled probes, the ^{13}C T_1 of fructose C2 is not field-dependent, which can be a great advantage for using this probe at higher magnetic fields. Although the T_1 of sp^2 hybridized carbonyl carbons decreases with increasing B_0 field in accordance with the CSA-dominated spin-lattice relaxation, the hemiketal C2 carbon is sp^3 hybridized, and the contribution of CSA is negligible. Consequently, T_1 will be largely insensitive to B_0 [772].

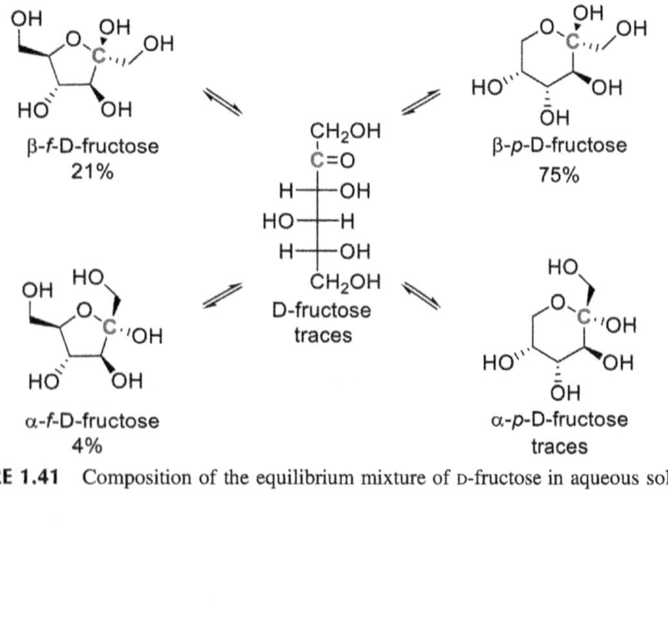

FIGURE 1.41 Composition of the equilibrium mixture of D-fructose in aqueous solution.

The metabolism of fructose differs significantly from that of glycolysis. In particular, fructose uptake and subsequent conversion into trioses are not regulated by insulin. Hyperpolarized [2-^{13}C] fructose shows three peaks that are associated with β-*p*-D-fructose, β-*f*-D-fructose, and α-*f*-D-fructose, and decent signal in mice showed the feasibility of in vivo studies. [2-^{13}C]-D-fructose was deuterated in D_2O in the same manner as [3,4-^{13}C2]glucose, with repeated applications of fresh Ru catalyst (Scheme 1.44). The ^{13}C label in the deuterated [2-^{13}C]-D-fructose had a T_1 of 93 s in D_2O at 1 T. The position of labeling provides chemical shift change upon formation of fructose-1-phosphate; therefore, this probe can serve as a surrogate of ketohexokinase activity. This study shows that in hepatocellular carcinoma, the formation of fructose-1-phosphate is compromised [773].

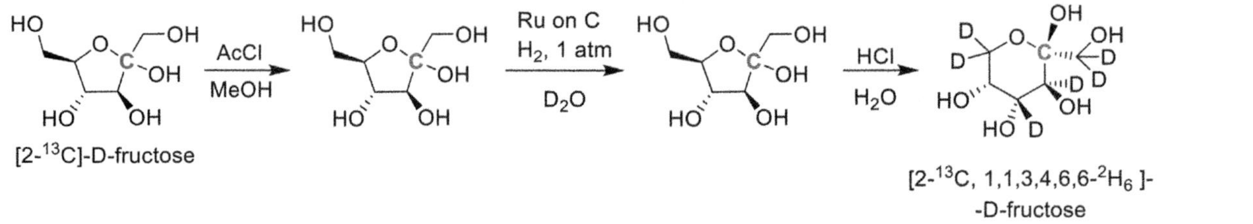

SCHEME 1.44 Ruthenium catalyzed partial deuteration of [2-^{13}C]-D-Fructose in D_2O.

D-[1-^{13}C]glucono-δ-lactone. The pentose phosphate pathway (PPP) branches off glycolysis at glucose-6-phosphate. One of the roles of PPP is to produce NADPH for anaplerosis and to maintain intracellular redox homeostasis. PPP is upregulated in cancer to support cell proliferation. Hyperpolarized D-[1-^{13}C]glucono-δ-lactone has been used as a probe to detect flux through the PPP [774,775]. The probe is phosphorylated at C6 and hydrolyzed to 6-phosphogluconate, which is decarboxylated by 6-phosphogluconate dehydrogenase to produce ribose-5-phosphate and [^{13}C]CO_2 detected as [^{13}C] bicarbonate. D-[1-^{13}C]glucono-δ-lactone was synthesized by the oxidation of [1-^{13}C]-D-glucose with cyclohexanone under anhydrous conditions (Scheme 1.45). D-glucose is an ideal starting material for this probe because it exists almost exclusively in the 6-membered pyranose form, and therefore, it can directly furnish the δ-lactone (6-membered ring). Hydride transfer from the hemiacetal to cyclohexanone was catalyzed by the ruthenium-based Shvo's catalyst. In water, D-glucono-δ-lactone forms an equilibrium mixture consisting of the 5- and 6-membered lactones as well as the open chain gluconic acid.

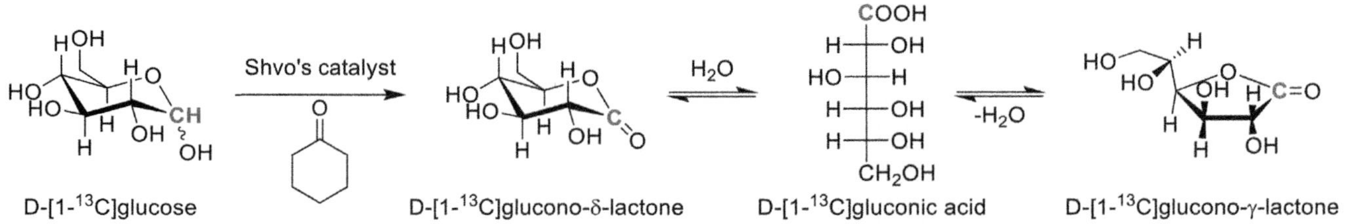

SCHEME 1.45 Synthesis and hydrolysis of D-[1-^{13}C]glucono-δ-lactone.

[2-^{13}C]-*myo*-inositol. Inositols are 6-membered carbocyclic sugars. They occur in various organisms, including mammalian tissues, with *myo*-inositol being the most abundant, particularly in the brain. Hyperpolarized [2-^{13}C,1,2,3,4,5,6-^{2}H$_6$]-*myo*-inositol was developed to assess inositol 3-kinase activity. This enzyme phosphorylates myo-inositol to *myo*-inositol 3-phosphate (Scheme 1.46). The difference in the chemical shifts of C2 in *myo*-inositol and *myo*-inositol-3-phosphate (MI3P) is about 0.5 ppm.

[2-^{13}C] *myo*-inositol was obtained in a multi-step synthesis stating from [2-^{13}C]-D-glucose (Scheme 1.47). Treating [2-^{13}C]-D-glucose with MeOH in an acidic condition gave the methyl glucoside. The primary hydroxyl was then selectively silylated with *tert*-butyldimethylsilyl chloride, and the other hydroxyls were protected as the benzyl ether with benzyl

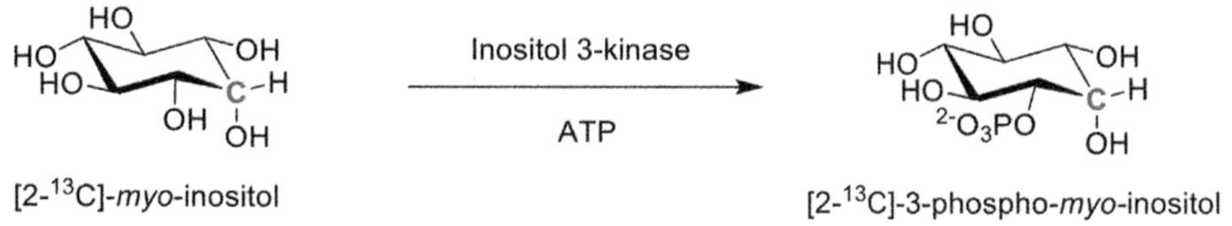

SCHEME 1.46 Phosphorylation of [2-^{13}C]-*myo*-inositol by inositol 3-kinase.

bromide and NaH and purified as a protected α-anomer. Next, deprotection of the silyl group with tetrabutylammonium fluoride, followed by Dess—Martin periodinane oxidation, produced an unstable aldehyde, which was immediately converted to the *Z*-isomer of the enolacetate with acetic anhydride. Next, in the key step of the synthesis, the mercury(II)-mediated rearrangement of the enolacetate in a Ferrier (II) reaction [776] afforded the penta-oxygenated cyclohexanone. Stereoselective reduction of the carbonyl group was achieved with tetramethylammonium triacetoxyborohydride. The acetyl group at C1 was deprotected with guanidinium chloride and sodium ethoxide. Removal of the benzyl groups under a hydrogen atmosphere in the presence of $Pd(OH)_2$ completed the synthesis of [2-^{13}C]-*myo*-inositol. Chemoselective H-D exchange reactions on the carbons adjacent to the hydroxyl groups were performed using the Ru on carbon in D_2O and in H_2 atmosphere. The preparation of [2-^{13}C, 1,2,3,4,5,6-^{2}H$_6$]-*myo*-inositol is a remarkable example of a complex multistep synthesis of a ^{13}C-labeled, deuterated tracer designed for HP-^{13}C MRS applications. The probe was hyperpolarized by dissolution DNP and was successfully used to monitor the myo-inositol 3-kinase reaction in vitro [777].

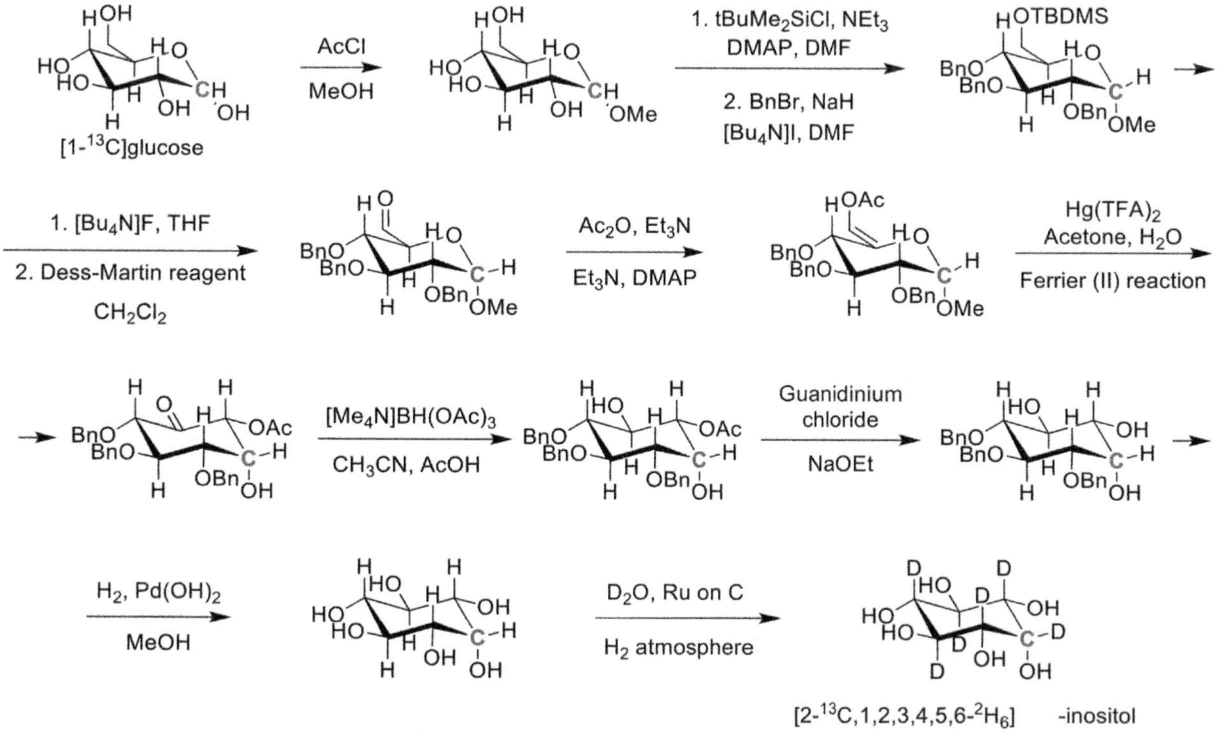

SCHEME 1.47 Multistep synthesis of [2-^{13}C] *myo*-inositol.

[1-^{13}C]-ascorbic acid. Hyperpolarized [1-^{13}C]-ascorbic acid (vitamin C) and its oxidized form, [1-^{13}C]-dehydroascorbic acid (DHAA), have been used as probes to monitor redox state by ^{13}C MRS [778,779]. ^{13}C-labeled ascorbic acid was synthesized by the condensation of labeled KCN with L-threo-pentos-2-ulose followed by the acid hydrolysis of

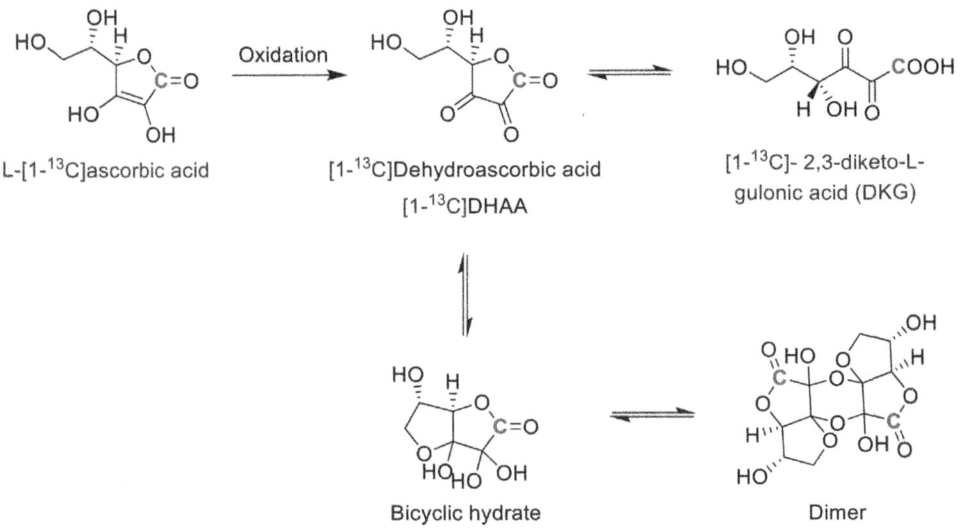

FIGURE 1.42 Various forms of [1-^{13}C]-dehydroascorbic acid.

the imidolactone intermediate [780]. [1-^{13}C] labeled ascorbic acid can be oxidized [1-^{13}C]DHAA by air or oxygen over activated charcoal [778]. Alternate methods use catalytic amounts of copper (II) acetate [781,782], or the Rh(III) complex of a macrocyclic Schiff base to achieve higher yields [783]. DHAA exists as the bicyclic form in a less-polar organic solvent. In the solid-state, it exclusively forms the dimer [784,785] However, in the presence of water it converts to various hydrated forms of the monomer and dimer (Fig. 1.42). LC-MS studies of aqueous solutions DHAA solutions at pH 2 revealed the presence of three dimeric and two monomeric forms [786]. DHAA is hydrolyzed to 2,3-diketo-L-gulonate at higher pH [787]. [1-^{13}C]DHAA has been used as a probe to measure the rate of enzymes maintaining the GSH/GSSG ratio and to assess oxidative stress [779,788]. Unfortunately, it induces transient respiratory arrest, and shows pancreatic toxicity, which will likely limit its in vivo use. Other, less toxic hyperpolarized tracers such as [1-^{13}C]-N-acetyl cysteine, a precursor of glutathione, are currently being developed redox to monitor redox state [789].

1.9.5.8 Amino acids and peptides

1.9.5.8.1 Amino acids labeled at C1

Amino acids are the building blocks of proteins. However, several amino acids participate in many energy-producing and biosynthetic processes, far beyond their role in protein synthesis [790]. For example, most amino acids are glucogenic as they can be broken down into pyruvate or TCA cycle intermediates for gluconeogenesis. Alanine links carbohydrate and amino acid metabolism via reversible transaminase reactions. Glutamate is the starting material for several biologically important molecules including glutathione, N-acetyl-L-aspartyl-L-glutamate, γ—aminobutyrate, and other amino acids (L-proline and L-arginine). Cancer cells have an increased demand for glutamine, which is the most abundant amino acid in the blood, thereby providing an ample source of carbon and nitrogen to cover the biosynthetic and energetic needs of proliferating cells [791]. Lysine and leucine are ketogenic amino acids because they can be broken down to acetoacetyl CoA and acetyl CoA and can be used to generate energy during starvation. Thus, ^{13}C-labeled amino acids are potentially very useful tracers for conventional and hyperpolarized ^{13}C studies of biochemical processes in which amino acids participate. Examples of metabolic studies with hyperpolarized [1-^{13}C] labeled amino acids include [1-^{13}C]-*L*-alanine, which was used to probe transaminase activity and was shown to produce different amounts of pyruvic acid and lactate in the liver and heart of mice due to differences in activities of transaminase and LDH between the tissues [792]. This probe offers some advantages over [1-^{13}C]pyruvate in measuring the cytosolic NAD/NADH ratio because it can more accurately detect the intracellular lactate-to-pyruvate ratio [793,794]. Hyperpolarized D-[1-^{13}C]alanine was used to probe the activity of the peroxisomal enzyme *D*-amino acid oxidase. This enzyme, occurring in the kidney, liver, brain, and polymorphonuclear leukocytes, is responsible for the oxidative deamination of D-amino acids to 2-keto acids [795]. Conversion of hyperpolarized [1-^{13}C]-*L*-glutamic acid into hyperpolarized [1-^{13}C]-L-glutamine was observed in the rat brain after opening the BBB with mannitol, which could be used to assess astrocyte glutamine synthetase (EC 6.3.1.2) activity [796]. The formation of [1-^{13}C]-2-hydroxyglutarate from hyperpolarized [1-^{13}C]-*L*-glutamine via glutamate and 2-ketoglutarate catalyzed

by mutant IDH was observed in mouse models of chondrosarcoma with IDH1 and IDH2 mutations [797]. Hyperpolarized [1-^{13}C]-L-aspartic acid produced HP-[1-^{13}C]oxaloacetate via aspartate aminotransferase (EC 2.6.1.1) in human hepatoma cell suspension [182]. Interestingly, hyperpolarized [1-^{13}C]- and 1,4-^{13}C2]-L-aspartic acid was successfully used to detect renal phosphoenolpyruvate carboxykinase (PEPCK) activity in vivo via the formation of hyperpolarized [1-^{13}C]phosphoenol pyruvate, an important intermediate of glycolysis and glyconeogenesis [550,798]. The conversion of hyperpolarized [1-^{13}C]-*N*-acetyl-*L*-methionine to [1-^{13}C]-*L*-methionine was used to visualize cells that overexpress the *Acy-1* aminoacylase-1 reporter gene. The aminoacylase-1 enzyme (EC 3.5.1.14) cleaves N-acetylated amino acids into acetate and free amino acids. Given the high spatiotemporal resolution of HP ^{13}C MRS, the [1-^{13}C]-*L*-methionine signal can be used to noninvasively track cells that express this gene [277]. Amino acids labeled with ^{13}C in the sidechain are less frequently used for HP ^{13}C studies because, in most cases, the biochemical action happens at C1. However, the glutamine to glutamate conversion catalyzed by glutaminase is important in glutamine metabolism, and hyperpolarized [5-^{13}C] glutamine has been used to monitor the activity of this enzyme in human hepatocellular carcinoma cells in vivo in a rat liver tumor [799,800] as well as in prostate cancer cell lines (PC3 and DU145) before and after treatment [801]. In addition, triple-labeled glutamine in pancreatic tumors showed glutamine flux before and after treatment [622]. Arginase is another hydrolytic enzyme that acts on the arginine sidechain, converting it to ornithine and urea. The activity of this important urea cycle enzyme was monitored with hyperpolarized [6-^{13}C]arginine [623,802].

In addition to monitoring metabolic processes, ^{13}C-labeled amino acids and their derivatives have also been used as ^{13}C NMR sensors of pH, redox, and meta ions (Zn^{2+}, Ca^{2+}) [792,803–805].

As we have seen with ^{13}C-labeled carboxylic acids and sugars, sodium and potassium [^{13}C]cyanide are excellent reagents for the introduction of a ^{13}C label into a carboxylate. An early synthesis of [1-^{13}C]glycine is based on the reaction of [^{13}C]NaCN with N-chloromethylphthalimide, followed by acid hydrolysis of the phthalimide [659]. The classic Strecker synthesis can be applied for the preparation of racemic α-aminonitriles from [^{13}C]HCN, aldehydes, and ammonia. These can then be hydrolyzed into racemic (*DL*) amino acids. Newer methods focus on the enantioselective synthesis of labeled amino acids. The easily available [1-^{13}C]glycine is a good starting material, and there are multiple approaches that allow the introduction of a side chain on the alpha carbon of glycine in a stereoselective manner. Since the alpha C–H bonds are equivalent in glycine, induction of chirality is necessary. Initially, this was achieved by incorporating the glycine moiety into an asymmetric molecule, which required a stoichiometric amount of the chiral auxiliary. The chiral structures used to control the stereochemistry include Schollkopf's bis lactim ethers, Oppolzer's camphor sultams, Xu's tricycloiminolactones, Williams' oxazinones, Seebach's imidazolidinones, and Evans' oxazolidinones (Fig. 1.43) [806–813].

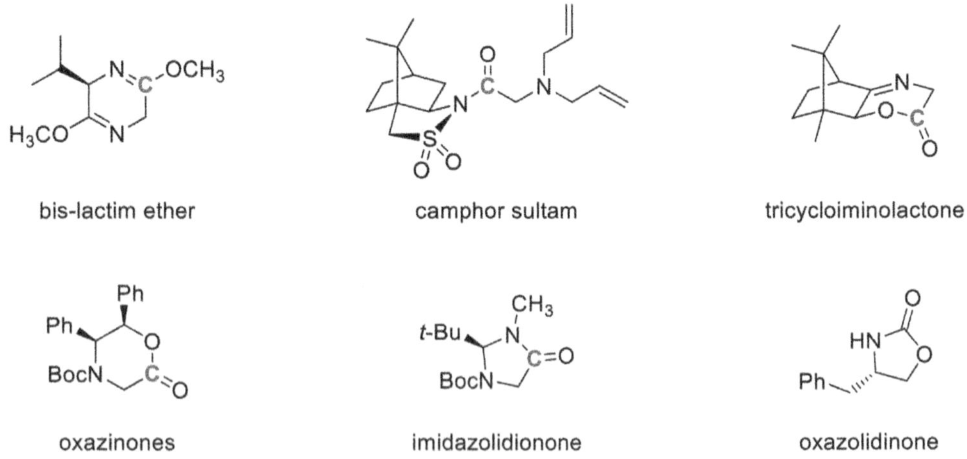

FIGURE 1.43 Asymmetric intermediates designed to control the stereochemistry in the stereoselective synthesis of chiral amino acids. The carboxylate carbon, which generally will be labeled, is highlighted in red (light gray in print version). Note that in Evans' oxazolidinones this carbon is introduced during the functionalization of the ring nitrogen.

These intermediates can be prepared from glycine, and the sidechain can be introduced by alkylation in the presence of a strong base such as LDA or BuLi or by 2,3-Stevens rearrangement, as is the case for the diallylgycine camphorsultam derivative.

Later, in the late 1980s, Martin J. O′ Donnell, inspired by the concept of phase-transfer catalysis (PTC), developed a novel method that greatly facilitated the stereoselective synthesis of isotopically labeled amino acids. The key intermediate in this process is a Schiff base ester that can be prepared from glycine in a few steps [814], and by taking advantage of enantioselective PTC, the chirality at the α-carbon can be induced by chiral catalysts (Scheme 1.48) [815,816].

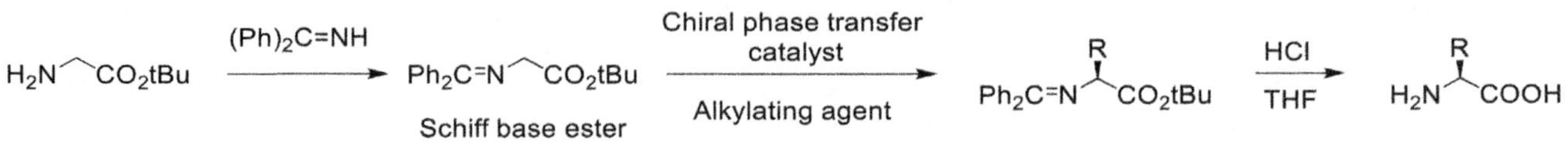

SCHEME 1.48 O'Donnell's method for the stereoselective synthesis of alpha amino acids using biphasic solvent systems and chiral phase transfer catalysts.

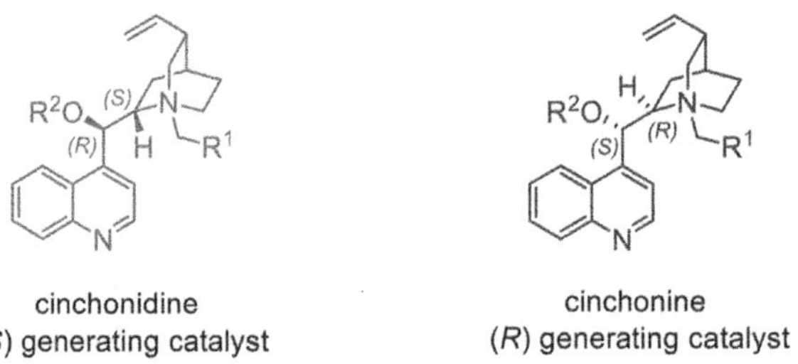

FIGURE 1.44 Phase transfer catalysts based on Cinchona alkaloids provide excellent stereoselectivity in the O'Donnell amino acid synthesis.

The deprotonated Schiff base ester forms an ion pair with the quaternary ammonium salt of PTC at the interface of the phases. The ion pair reacts with the alkylating agent in the organic phase. Originally, the method used *Cinchona* alkaloids as PTCs. The naturally derived form of cinchonidine generates amino acids with *S* stereochemistry, while the diastereomeric form of cinchonine gives rise to the *R* enantiomer (Fig. 1.44). More recently, improved PTCs derived from cinchona alkaloids and binaphthyls have been developed to enhance enantioselectivity [817–819]. The enolate can also be alkylated via Michael addition [820]. In addition to the catalyst, other factors such as base, solvent, temperature, the ester group on the Schiff base, and the type of electrophile will also influence the yield and enantioselectivity of the reaction.

Examples include the synthesis of isotopically labeled homoserine and methionine derivatives via alkylation of *tert*-butyl benzophenone imine glycinate with dimethylallyl bromide and the synthesis of labeled lysine via Michael addition to methyl acrylate in the presence of a chiral phase-transfer catalyst (Scheme 1.49) [821–823].

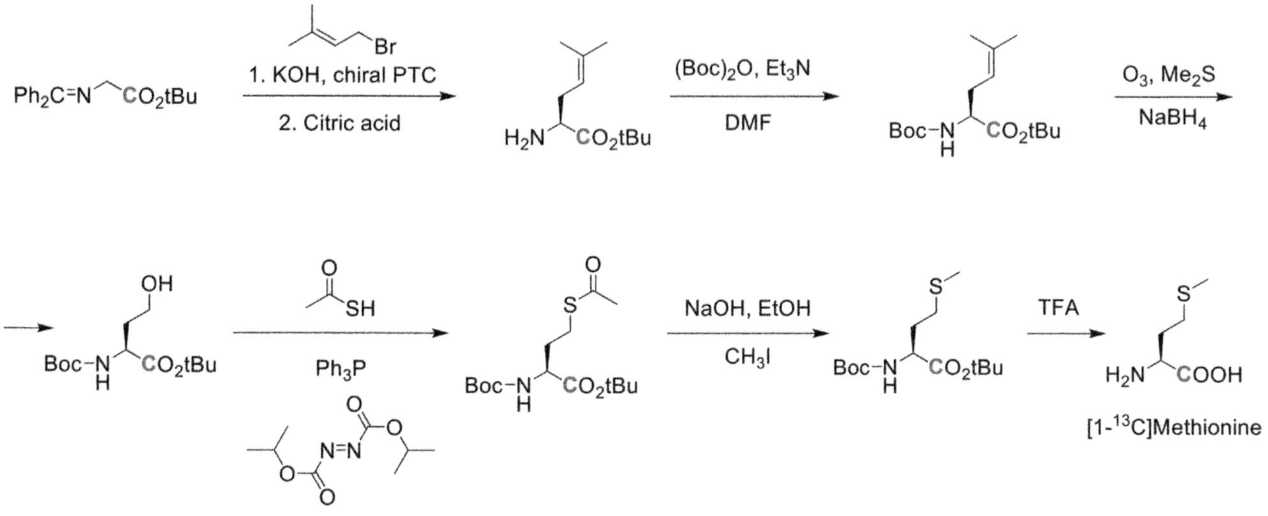

SCHEME 1.49 Stereoselective synthesis of [1-^{13}C]methionine via the O'Donnell method.

Although D-amino acids can obviously be synthesized by the asymmetric alkylation method with the appropriate PTC enantiomer to achieve the desired enantioselectivity, they can also be obtained from the L-isomer through the deprotonated Schiff base enolate, which can be stereoselectively protonated using a chiral proton-donating Brønsted acid (Scheme 1.50) [824]. This is a very convenient way to convert the labeled L-amino acid, which is often commercially available, into the less common D-isomer [825,826].

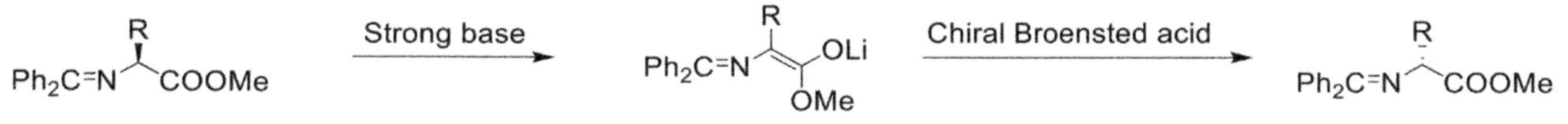

SCHEME 1.50 Synthesis of D-amino acids via the stereoselective protonation of Schiff base enolates derived from L-amino acids.

1.9.5.8.2 Amino acids labeled at the side-chain

The commercially available [5-^{13}C]glutamine has an unfavorably short ^{13}C T_1 (about 15–16 s at 9.4 T), which prompted the development of the C4 deuterated derivative. The key intermediate of the synthesis was C4-deuterated homoserine, which was obtained by the borodeuteride reduction of an N-Boc-L-aspartic acid 1-*tert*-butyl ester. The ^{13}C at C5 was introduced using labeled potassium cyanide after the activation of the homoserine OH by tosylation, followed by conversion to the iodide. Controlled acid hydrolysis of the cyano group and concomitant cleavage of the *tert* butyl protecting groups afforded the desired [5-^{13}C, 4,4-$^{2}H_2$]glutamine (Scheme 1.51). The deuteration significantly improved T_1 from 15 to 33 s at 9.4 T [827].

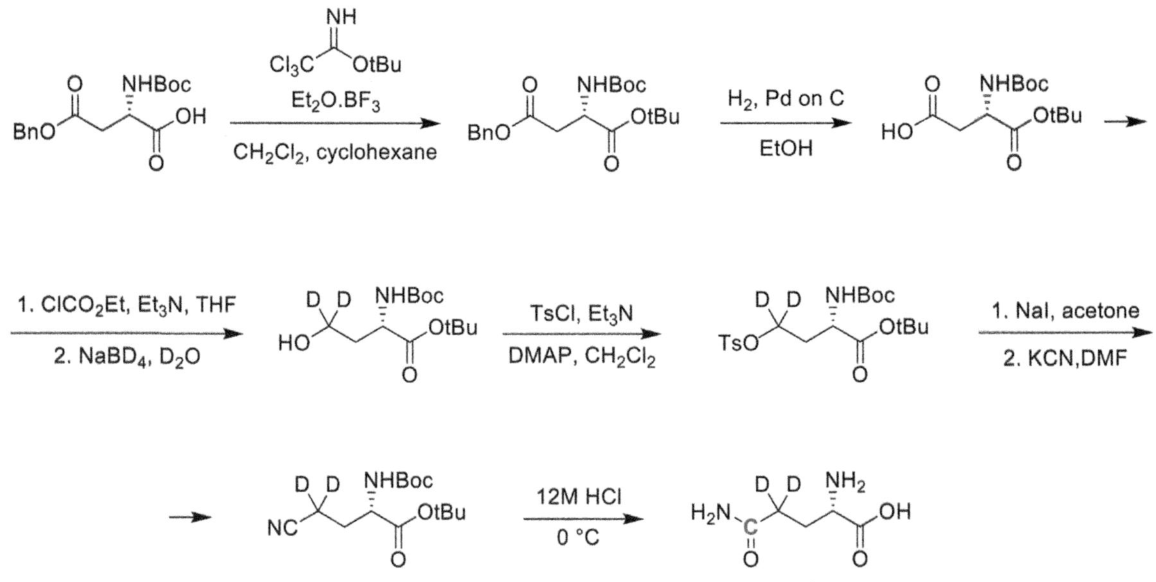

SCHEME 1.51 Synthesis of L-[5-^{13}C, 4,4-$^2H_2$]glutamine.

Eskandari et al. developed **[5-^{13}C, 4, 4-$^2H_2$, 5-^{15}N] glutamine** by introducing ^{2}H and ^{15}N to further improve the relaxation properties of this probe. In this synthesis, the Appel reaction (n-Bu_3P and CCl_4) [828] was used to convert the deuterated homoserine OH to the chloride prior to the reaction with [^{13}C]KCN. Subsequent hydration of the resulting deuterated nitrile to the amide was performed with acetaldehyde oxime in the presence of Wilkinson's catalyst in order to avoid side reactions associated with the hydrolysis of the nitrile with a strong acid [829]. The synthesis of [5-^{13}C, 5-^{15}N, 4,4-$^2H_2$]glutamine was completed by the removal of the protecting group trifluoracetic acid (Scheme 1.52) [622].

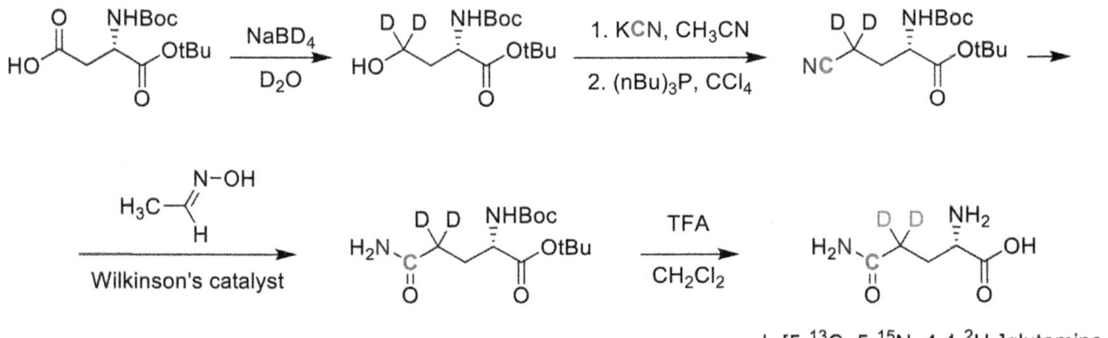

SCHEME 1.52 Synthesis of L-[5-^{13}C, 5-^{15}N, 4,4-$^2H_2$]glutamine.

[6-^{13}C]arginine is a promising probe to monitor arginase activity [802]. Unfortunately, the T_1 of the label in the guanidino moiety is rather short (the T_1 (10 s at 11.7 T and 12 s at 3 T) reflects the quadrupolar relaxation effect of ^{14}N in the guanidino side chain, which results in loss of signal in low magnetic fields). This motivated the synthesis of [6-^{13}C, $^{15}N_3$]arginine. While the replacement of the ^{14}N atoms in the guanidino moiety did not improve the T_1 significantly at high field ($T_1 = 7$ s at 14.1 T), it extended the T_2 from 0.17 to 0.49 s at 14.1 T [623]. [6-^{13}C, $^{15}N_3$]Arginine was synthesized starting from N-Boc-L-aspartic acid 1-*tert*-butyl ester, in which the C4 carboxylate was reduced with sodium borohydride to form the protected homoserine. Next, the hydroxyl group was converted into the tosylate, and the ^{15}N label was introduced in a nucleophile substitution with [^{15}N]KCN to form the protected nitrile. Subsequent hydrogenation over

palladium afforded [5-^{15}N]ornithine, which was converted to N-Boc protected [6-^{13}C, $^{15}N_3$]arginine with ^{13}C and ^{15}N labeled N,N′-Di-Boc-S-methylisothiourea obtained from [^{13}C, $^{15}N_2$]urea. Finally, deprotection of the N-BOC group with acid afforded [6-^{13}C,$^{15}N_3$]arginine (Scheme 1.53) [623].

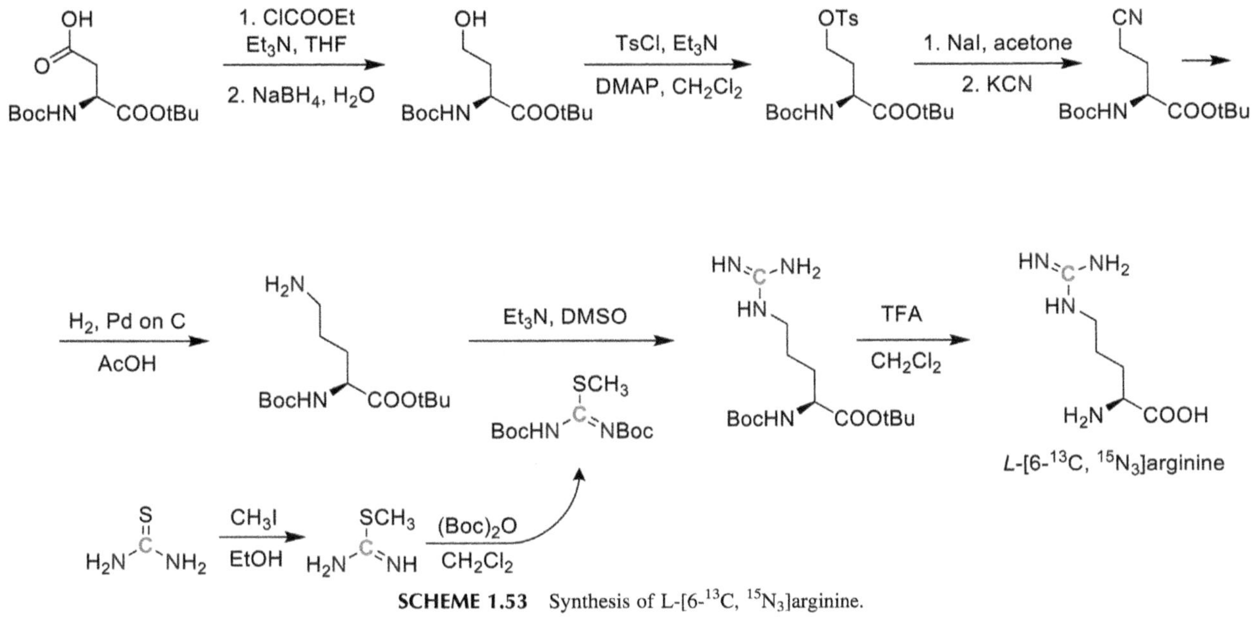

SCHEME 1.53 Synthesis of L-[6-^{13}C, $^{15}N_3$]arginine.

1.9.5.8.3 Esters and amides of labeled amino acids

Amino acids have amino and carboxyl groups and exist in the zwitterionic form in the solid state as well as in an aqueous solution at the isoelectric point. The zwitterionic structures are stabilized by intramolecular hydrogen bonding, which increases the pK_a of the amino group (around 9–10). However, esterification of the carboxylate can drop the pK_a by almost two units, close to the physiological range. Thus, esterification of a [1-^{13}C] labeled amino acid has been performed, bringing the pK_a near 7 in the design of ^{13}C pH probes. For example, the pK_a of aspartate dimethylester ($pK_a = 6.04$), lysine ethyl ester ($pK_a = 7.18$), serine ethyl ester ($pK_a = 7.47$), glycine methyl ester ($pK_a = 7.52$), and alanine ethyl ester ($pK_a = 8.03$) are all in the physiological range [494,803] Esterification may also be a useful modification to facilitate the cellular uptake of an amino acid [805]. Esters of amino acids are easily prepared by saturating the alcoholic solution of the amino acid with dry HCl gas or [805] treating the solution with thionyl chloride to activate carboxylate to form the ester [494]. Conversion to amide is another common modification of amino acid-based probes. Amino acid amides also tend to exhibit lower pK_a values than the corresponding acids (serine amide $pK_a = 7.35$, alanine amide $pK_a = 9.38$, and glycine amide $pK_a = 8.27$) [803]. The ^{13}C and ^{15}N labeled versions of N-(2-acetamido)-2-aminoethanesulfonic acid (ACES) were proposed as HP^{13}C MRS pH sensors because of the large pH-dependent chemical shift of the amide carbon, low toxicity, and lack of interaction with biological cations. The unlabeled form of ACES is a Good's buffer, commonly used in biology [830]. [1-^{13}C, 1-^{15}N]ACES was synthesized from N-Boc-protected [1-^{13}C]glycine. The amino group was activated with carbonyldiimidazole and was converted into the ^{15}N amide. The Boc group protection was removed with TFA to give [1-^{13}C, 1-^{15}N]glycine amide, which was alkylated with sodium 2-bromoethanesulfonate in water and NaOH to afford [1-^{13}C,1-^{15}N]ACES (Scheme 1.54) [831]. The probe exhibited a 3.58 ppm ^{13}C chemical shift dispersion between pH 6.5 and 7.4. It also had a reasonably long T_1 (18 s at 11.7 T and 25 s at 3 T).

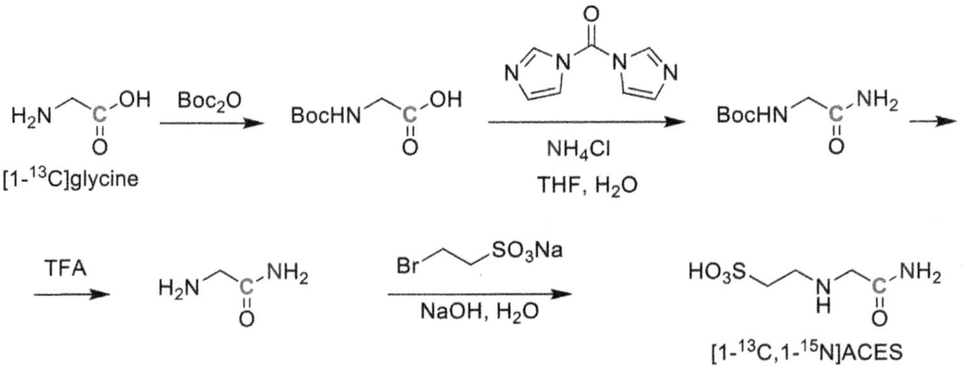

SCHEME 1.54 Synthesis of [1-^{13}C, 1-^{15}N]-N-(2-acetamido)-2-aminoethanesulfonic acid (ACES).

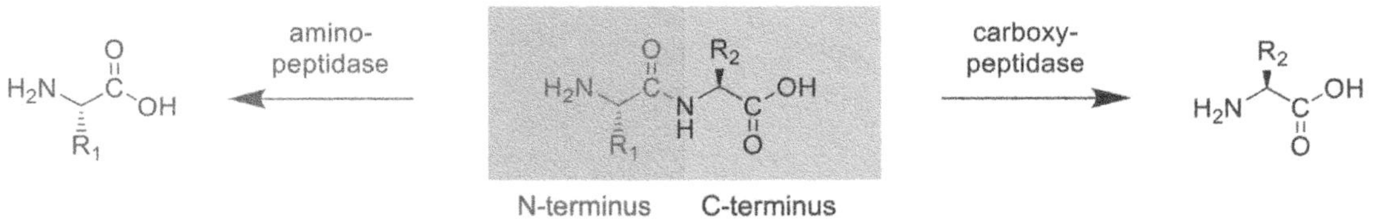

FIGURE 1.45 Aminopeptidases cleave amino acids from the N-terminus while carboxypeptidases cleave amino acids from the C-terminus of a peptide chain.

1.9.5.8.4 Amino acid derivatives as probes for peptidases

Peptidases are hydrolases that can be categorized as either aminopeptidases or carboxypeptidase, depending on from which terminus they catalyze the peptide bond cleavage (Fig. 1.45) [832]. The reaction occurs at a terminal peptide bond; thus, labeling at a peptide carbonyl makes the most sense when designing peptidase probes. Hyperpolarized [1-^{13}C]-L-alanine amide was proposed as a ^{13}C MRS probe to measure aminopeptidase N activity. Aminopeptidase N is an amidohydrolase that acts on an N-terminal alanine ($R_1 = CH_3$) residue.

Hyperpolarized [1-^{13}C]-L-alanine amide was proposed as a ^{13}C MRS probe to measure aminopeptidase N activity. [1-^{13}C]-L-alanine amide was synthesized starting from [1-^{13}C]-L-alanine. The methyl ester was obtained by treatment with $SOCl_2$ in methanol, and the amino group was protected as N-Boc with Boc_2O. The methyl ester was then converted into the amine with amide and ammonium hydroxide. The final product was obtained with acid deprotection of the N-Boc group (Scheme 1.55) [833]. The T_1 was about 25 s at 9.4 T. ^{15}N-labeled ammonium hydroxide may be used in place of the regular one; this might help preserve the ^{13}C polarization when the sample experiences a low magnetic field outside the polarizer. Nevertheless, the probe exhibited sufficiently high reactivity toward the enzyme to allow the observation of the cleavage product [1-^{13}C]-L-alanine before the decay of hyperpolarized ^{13}C polarization in vitro.

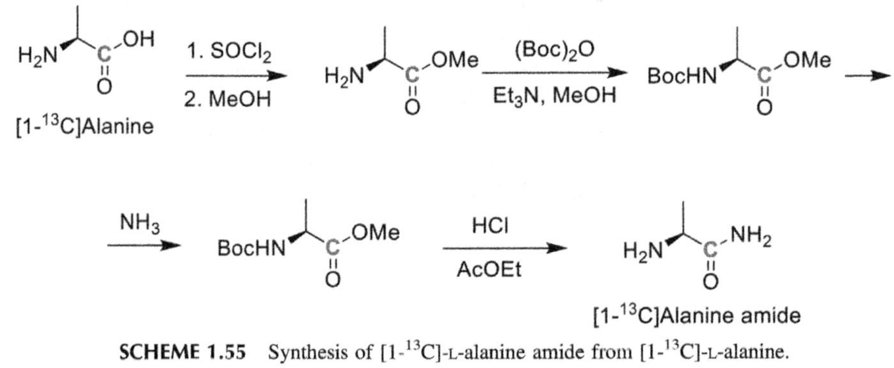

SCHEME 1.55 Synthesis of [1-^{13}C]-L-alanine amide from [1-^{13}C]-L-alanine.

The aforementioned hyperpolarized [1-^{13}C]Ala-NH_2 was not applicable in vivo because of its slow enzymatic conversion and the small chemical shift change in the product. Therefore, further probe development was focused on improving the catalytic efficiency of the enzymatic cleavage and increasing the chemical shift difference between the substrate and the product. As aminopeptidase N rapidly cleaves dipeptides, with Ala being the most preferred N-terminal residue, the dipeptide Ala-Gly was used as a design platform. Among the tested Ala-Gly derivatives, the dimethylamide Ala-Gly-NMe_2 labeled at the Gly carboxamide was identified as the best candidate in terms of rapid cleavage, relatively large chemical shift change (2.4 ppm), and good resistance to cleavage by other peptidases. The Gly methylene was deuterated to improve the relaxation time of the label (57 s in D_2O at 3 T). Ala-[1-^{13}C, 2-$^{2}H_2$]Gly-NMe_2 was synthesized as outlined in Scheme 1.56. The hyperpolarized probe was successfully used to detect aminopeptidase activity in vivo in mice with human pancreatic cancer (MIA PaCa-2) xenografts [834].

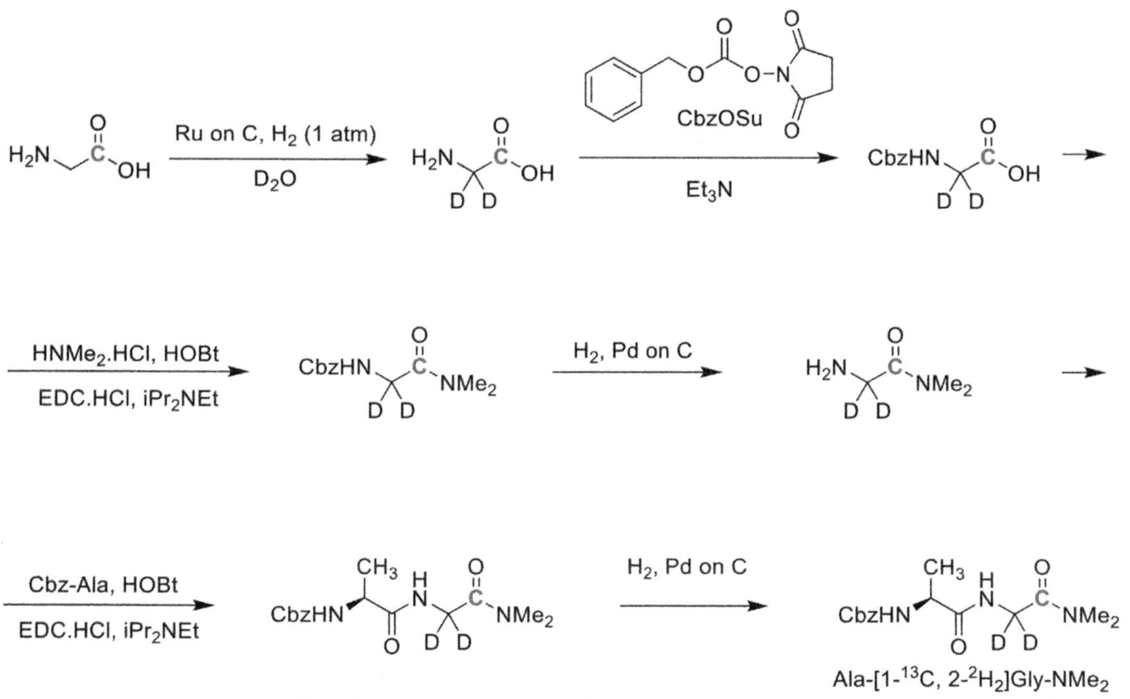

SCHEME 1.56 Synthesis of Ala-[1-^{13}C, 2-$^{2}H_2$]Gly-NMe_2 as an HP-^{13}C probe for aminopeptidase activity starting from [1-^{13}C]glycine.

In this synthesis, hydrogenation of [1-^{13}C]glycine over ruthenium on activated carbon in D_2O forms [1-^{13}C]glycine-d_2. In the basic condition, N-(benzyloxycarbonyloxy)succinimide (CbzOSu) was reacted to protect the amino group. The activated carboxylic group in HOBt and EDCl reacts with dimethylamine hydrochloride. Next, hydrogenation over palladium forms [1-^{13}C]gly-d_2-NMe_2. Reaction with protected alanine(Cbz-Ala) and subsequent hydrogenation over palladium form the final Ala-[1-^{13}C]gly-d_2-NMe_2 [834].

Carboxypeptidase G2 (glutamate carboxypeptidase II) is a bacterial enzyme that acts on a C-terminal glutamate. It also cleaves amide, urethane, and urea bonds between L-glutamic acid and an aromatic moiety. Carboxypeptidase G2 has been used in gene-directed enzyme prodrug therapy for cancer. This is a two-step process that involves the introduction of a foreign gene encoding an enzyme that can convert a prodrug to the cytotoxic drug molecule in the tumor [835–837]. A hyperpolarized ^{13}C reporter probe with a natural abundance of HP-3,5-difluorobenzoyl-L-glutamic acid was proposed to monitor carboxypeptidase G2 activity. It was shown in in vitro experiments that the amide would be the best place for ^{13}C-enrichment because it has the longest ^{13}C T_1 and offers the largest chemical shift change (4.8 ppm) upon cleavage (Scheme 1.57). Unfortunately, the T_1 even in this position was rather short, about 7 s at 11.7 T. Nevertheless, the formation of the cleaved product, 3,5-difluorobenzoic acid, could be observed within the lifetime of the HP-^{13}C magnetization. While in vivo application may not be feasible due to the short T_1, the relaxation properties of this probe could be significantly improved by substituting ^{15}N for the amide nitrogen and performing the experiment in a lower B_0 field.

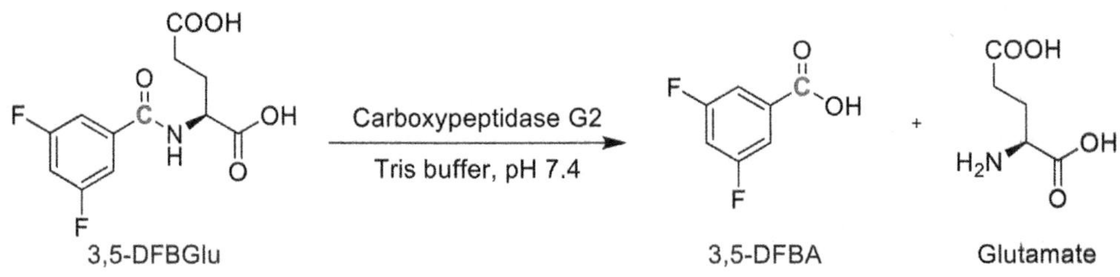

SCHEME 1.57 Natural abundance 3,5-difluorobenzoyl-L-glutamic acid as an HP-^{13}C NMR probe for Carboxypeptidase G2.

Hyperpolarized ^{13}C-labeled γ-glutamyl glycine (γ-Glu-[1-^{13}C]Gly) was developed as a probe for γ-glutamyl transpeptidase (GGT), the key enzyme in the metabolic pathway that regulates homeostasis of glutathione. Glutathione is an unusual tripeptide with a gamma peptide bond between the side chain carboxyl of glutamate and cysteine. GGT catalyzes the cleavage of this gamma-peptide bond. However, the enzyme can accommodate substrates with diverse substituents at the gamma position. The [1-^{13}C]glycine residue offers the advantage of a large chemical shift change (4.3 ppm) upon enzyme cleavage. The probe was synthesized from the N-benzyloxycarbonyl-L-glutamic acid 1-benzyl ester, which was converted to the activated N-hydroxysuccinimidyl ester. The reaction of the activated ester with [1-^{13}C]glycine furnished the protected gamma-peptide. The *N*-benzyloxycarbonyl (Cbz) and *O*-benzyl groups were removed with Pd/C-catalyzed hydrogenative deprotection (Scheme 1.58) [838]. The T_1 of γ-Glu-[1-^{13}C]Gly was 30 s at 9.4 T. This probe was successfully used for the real-time detection of GGT activity in vivo. The deuterated form, γ-Glu-[1-^{13}C, 2-$^{2}H_2$]Gly, was also synthesized, and it exhibited a slightly longer T_1 value (41 s at 9.4 T) [839].

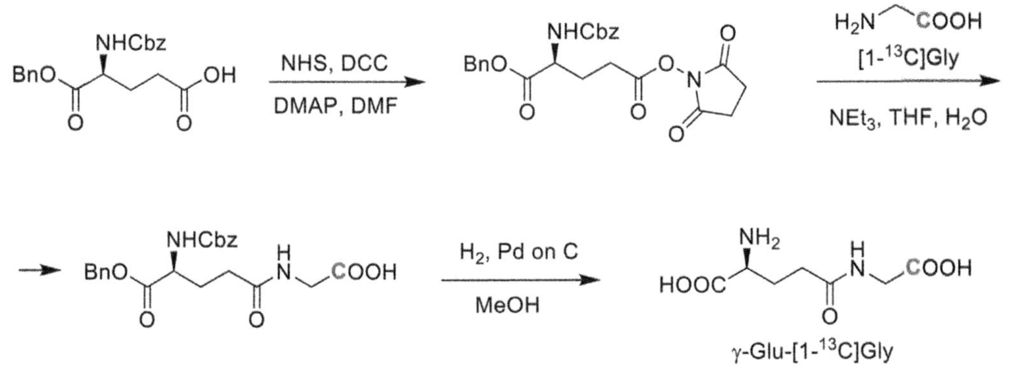

SCHEME 1.58 Synthesis of γ-Glu-[1-^{13}C, 2-$^{2}H_2$]Gly as a probe for γ-glutamyl transpeptidase (GGT).

1.9.5.8.5 Oligopeptides

Peptides are rod-shaped molecules, and depending on the structure, the local rotational freedom of individual amino acid residues, especially when they are in a terminal position, may govern relaxation processes more or less independently from the tumbling of the entire molecule [840,841]. It was recently shown that the carboxylate carbon of a C-terminal [1-^{13}C] Gly is the best location for an HP-^{13}C label in oligopeptides [842]. In glycine oligomers (Gly)x, the T_1 monotonously shortened from 51 to 32 s with increasing molecular weight going from x = 1 to x = 3. After that, the effect leveled off as the T_1 only slightly decreased from 27 s (x = 4) to 24 s (x = 5 and 7). While the presence of a side chain in this residue had an unfavorable effect on T_1 by restricting the rotational freedom of the carboxylate, a side chain on the neighboring residue had little effect. For example, the T_1 of Gly carboxylate in various dipeptides (Xaa-[1-^{13}C]Gly, Xaa = Ala, Val, Leu, Nle,

Phe, Trp, and Glu) was very similar, about 30 s (9.4 T, 10 mM, D_2O, 37°C, pD = 7.4). Deuteration of the Gly methylene further increased T_1 by eliminating the dipolar relaxation effect of these protons. Using these concepts, two oligopeptide-based HP-^{13}C probes of biological importance were designed. Tyr-Pro-Phe-Pro-[1-^{13}C]Gly-d_2 is the ^{13}C-labeled, deuterated version of β-casomorphin-5, which is a β-casein-derived μ-opioid receptor agonist. The peptide was synthesized via solid phase peptide synthesis (SPPS) starting from Fmoc-[1-^{13}C]Gly-d_2. T_1 values were surprisingly long for such a large molecule: 20 s at 9.4 T in D_2O and 24 s at 3 T in H_2O. The second probe was the ^{13}C-labeled, a deuterated derivative of the well-known antioxidant glutathione, γ-Glu-Cys-[1-^{13}C]Gly-d_2. Glutathione is a somewhat unusual tripeptide with a gamma peptide bond between the side chain carboxyl of glutamate and cysteine. The labeled probe had a long T_1 of 52 s at 9.4 T in D_2O and 58 s at 3 T in H_2O. When the HP probes were injected into mice, the formation of HP-[1-^{13}C]Gly-d_2 was observed, thereby allowing the monitoring of enzymatic degradation of the oligopeptides by HP-^{13}C NMR (Scheme 1.59).

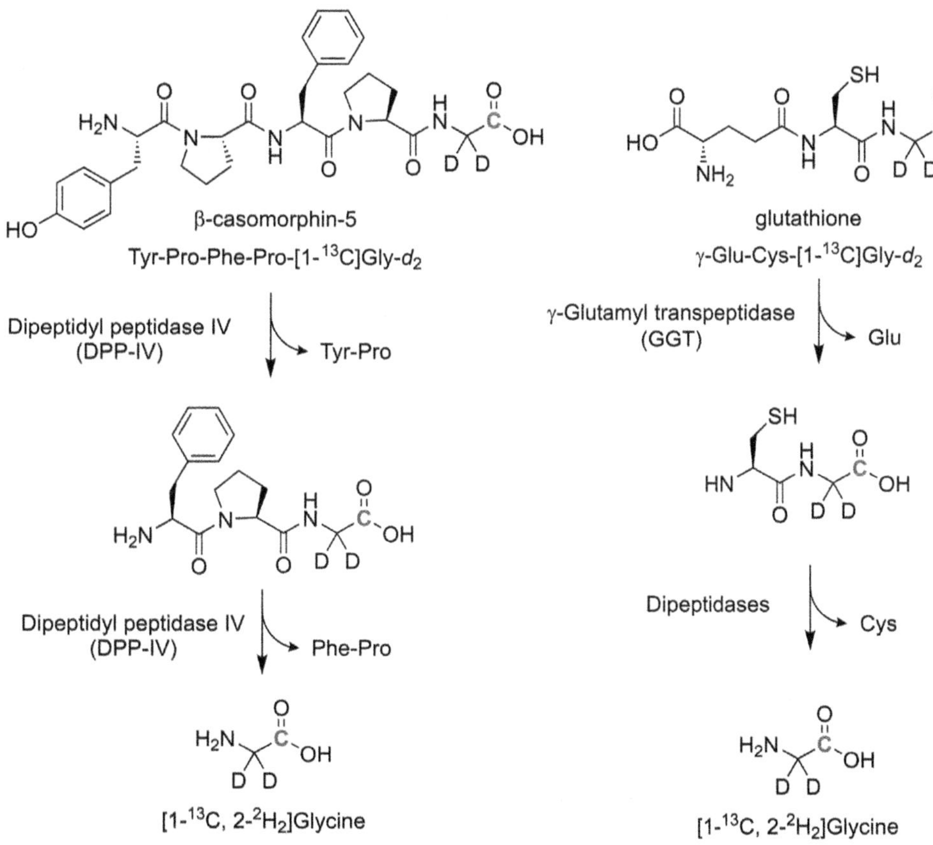

SCHEME 1.59 The formation of HP-[1-^{13}C, 2-$^2H_2$]glycine was observed in vivo after the injection of β-casomorphin-5 (left) and glutathione (right) based HP-^{13}C probes in which the hyperpolarized label was in the terminal deuterated glycine.

1.10 Conclusion and future outlook

HP-NMR/MRI techniques are capable of increasing the sensitivity of conventional NMR/MRI experiments by several orders of magnitude. ^{13}C, by and large, is the most commonly hyperpolarized nucleus, and the most important application of HP-^{13}C NMR/MRI is metabolic imaging in various health, and diseased tissues. Cancer, diabetes, liver, and heart diseases are closely associated with aberrant metabolism. Metabolic profiling already plays an important role in diagnosing and managing these diseases. ^{18}F-FDG-PET has been spectacularly successful in the diagnosis, staging, and treatment response assessment of various cancers, despite the fact that ^{18}F-deoxyglucose only informs about glucose uptake and does not tell anything about downstream metabolism. The lack of spectral information in PET makes real-time metabolic imaging of specific enzyme-catalyzed reactions extremely challenging with PET tracers. On the other hand, in vivo hyperpolarized 13C MRS can detect and quantify tissue metabolites independently since each compound has a unique NMR spectrum. The first human study with hyperpolarized [1-^{13}C]pyruvate in 31 prostate cancer patients was reported in

2013 [843]. Currently, the technology is in clinical trials at several locations around the world [376,377,844]. The prospect of potential clinical applications has driven the synthesis, hyperpolarization, and preclinical testing of a large number of ^{13}C-labeled compounds. Despite these efforts, HP-^{13}C-labeled pyruvate remains the only hyperpolarized tracer tested in humans, which is undoubtedly due to the great biochemical significance and the almost ideal properties of pyruvic acid for dissolution DNP. This is rather unfortunate; for HP-NMR/MRI technology to survive and make a clinical impact, hyperpolarized tracers other than pyruvate must also be developed for clinical applications. A major hurdle in achieving this goal is that the SPINlab clinical polarizer and its quality control module were optimized for the polarization of pyruvic acid. Specifically, the dissolution and subsequent quality assurance processes must be significantly accelerated so that hyperpolarized tracers that have less favorable T_1 values than pyruvate can also be accommodated. In addition, certain technical problems related to imaging coil and pulse sequence development must also be overcome [845].

From a basic science perspective, strongly enhanced hyperpolarized magnetization can give rise to unusual phenomena. For example, under certain conditions, a large negative nuclear polarization can lead to spontaneous emission of coherent rf radiation known as radio-frequency amplification by stimulated emission of radiation (RASER) from the sample. This is analogous to LASER and MASER. Although RASER was first observed as photochemically induced DNP experiment in 1978, it was not until recently that the HP community started to explore it for imaging applications [846,847]. Interestingly, since RASER involves the spontaneous emission of rf waves, excitation pulses are not needed for signal acquisition.

The rapid T_1 decay of the HP signal is perhaps the greatest limitation of HP-NMR/MRI. As discussed earlier in detail, the liquid state T_1 relaxation time of ^{13}C (and other nuclei as well) is largely determined by the molecular size and structure, magnetic field, temperature, solvent, and presence of paramagnetic species. The spin-lattice relaxation time of ^{13}C in small molecules is generally less than 2 min, and one should not expect liquid state ^{13}C T_1 values in the order of hours. While the T_1 decay of the hyperpolarized magnetization is inexorable, long-lived singlet spin states of strongly coupled nuclear spins often have much longer lifetimes, and these systems might be used to store hyperpolarized magnetization [848–856].

In summary, chemistry will continue to play an important role in the further development of this technology by designing and synthesizing novel hyperpolarized tracers, glassing matrices, and polarizing agents, as well as facilitating the clinical translation of new agents by optimizing the dissolution and quality control processes.

Appendix: List of T_1 values

TABLE A.1 ^1H T_1 values of selected compounds.

Name	Chemical structure	Apparent T_1 (s)	B_0 (T)	Solvent	Temp. (°C)	pH	References
Methyl-d_3 nicotinate-4,6-d_2	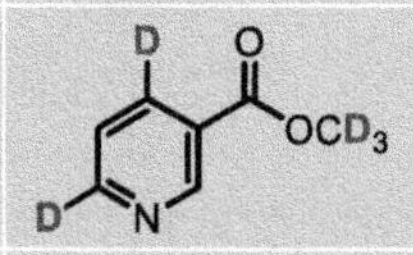	116 (H2) 107 (H5)	9.4	CD_3OD	25		[318]
Pyridazine-3,5-d_2	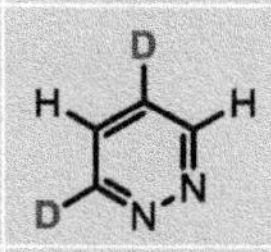	160 (H3) 177(H5)	9.4	CD_3OD	25		[318]
Pyrazine-2,5-d_2	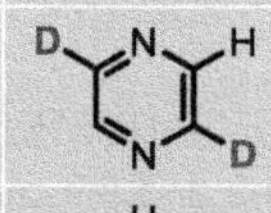	86 (H6)	9.4	CD_3OD	25		[318]
Pyrimidine-2,5-d_2	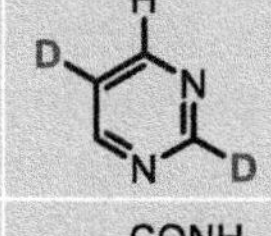	129 (H4)	9.4	CD_3OD	25		[318]
Isonicotinamide-2,5-d_2	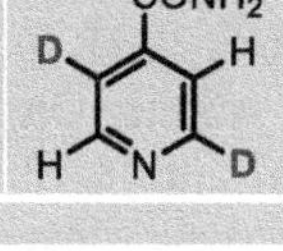	43(H3) 18 (H6)	9.4	CD_3OD	25		[318]

TABLE A.2 ^{6}Li T_1 values of selected compounds.

Name	Chemical structure	Apparent T_1 (s)	B_0 (T)	Solvent	Temp. (°C)	pH	References
Lithium chloride	^{6}LiCl enriched	273	9.4	H_2O			[207]
Lithium chloride	^{6}LiCl natural abundance	1040	1.375	D_2O	80		[857]
		170		H_2O	30		
Lithium chloride	^{6}LiCl enriched	525	9.4	D_2O, not degassed			[858]
		780		D_2O, degassed			
		168		H_2O, not degassed			
		189		H_2O, degassed			
Lithium chloride	^{6}LiCl enriched	506 (0.5 M)	9.4	D_2O			[347]
		177 (130 mM)		Whole human blood			
		157 (130 mM)		Deoxygenated human blood			
Lithium chloride	^{6}LiCl enriched	180 (0.5 M)		0.9 M D_2O, 2 mM TEMPO			[348]
		100		0.9 M D_2O, 2 mM TEMPO, 0.5 mM GdDOTP			
		72		Rat brain			

TABLE A.3 ^{13}C T_1 **values of selected compounds.**

Name	Chemical structure	Apparent T_1 (s)	B_0 (T)	Solvent	Temp. (°C)	pH	References
L-[1-^{13}C]Alanine	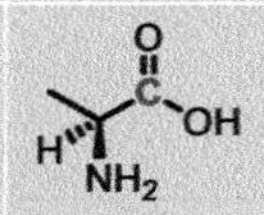	29	9.4	H_2O	37	7.0	[792]
		42	3.0	H_2O		7.3	[793]
		53	3.0	H_2O		7.0	[616]
		66	3.0	H_2O		7.5	[794]
D-[1-^{13}C]Alanine	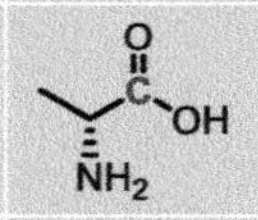	41	9.4	D_2O		7.4	[795]
L-[1-^{13}C, 2-^{2}H]Alanine	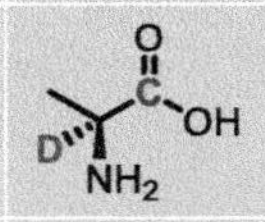	66	3.0	H_2O		7.0	[616]
L-[2-^{13}C]Alanine	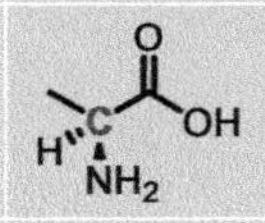	5	3.0	H_2O		7.0	[616]
L-[2-^{13}C,2-^{2}H]Alanine	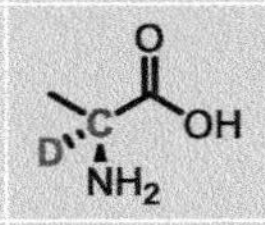	25	3.0	H_2O		7.0	[616]
L-[6-^{13}C]Arginine	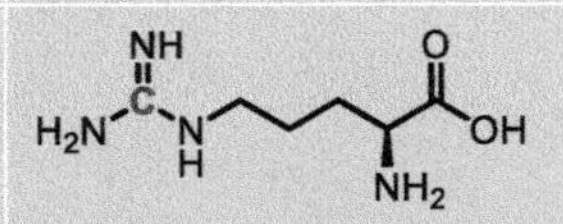	Too short	1.0	H_2O			[621]
		32	1.0	D_2O			
		7	14.1	H_2O			
		10	14.1	D_2O			
		12	3.0	H_2O	37	7.8	[802]
		10	11.7	H_2O			
		25 (C1)	3.0	H_2O			
		13 (C1)	11.7	H_2O			
L-[6-^{13}C,^{15}N$_3$]Arginine	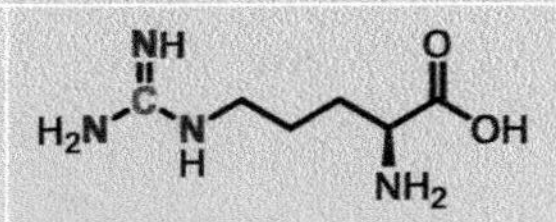	13	1.0	H_2O			[621]
		31	1.0	D_2O			
		7	14.1	H_2O			
		10	14.1	D_2O			
		15	14.1	H_2O			

L-[1-^{13}C]Aspartic acid	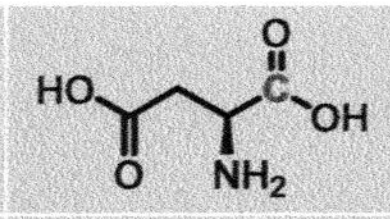	29	9.4	H_2O	37	7.0	[792]
L-[1-^{13}C]Cysteine	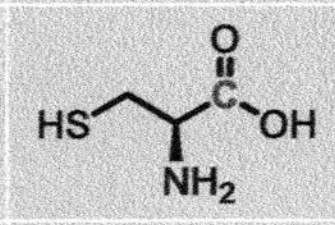	30	9.4	H_2O	37	7.0	[792]
L-[1-^{13}C]Glutamine	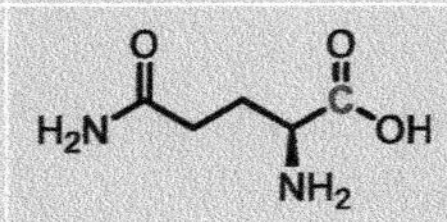	25	9.4	H_2O	37	7.0	[792]
		25	9.4	H_2O	37	7.4	[799]
		31	1.0	H_2O		7.4	[797]
L-[5-^{13}C]Glutamine	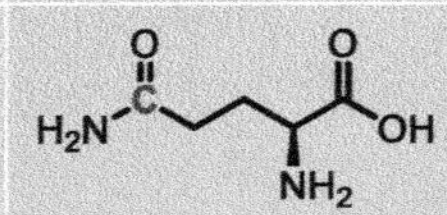	16 (C5)	9.4	H_2O	37	7.4	[799]
		25 (C1)					
		22	1.0	H_2O			[621]
		34	1.0	D_2O			
		11	14.1	H_2O			
		12	14.1	D_2O			[827]
		15	9.4	H_2O	37	7.0	[800]
		20	14.1	H_2O			
		22	1.0	H_2O		5.5	[622]
		32	1.0	D_2O		5.5	
L-[5-^{13}C, 5-^{15}N]Glutamine	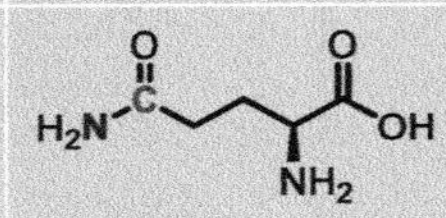	24	1.0	H_2O			[621]
		31	1.0	D_2O			
		24	1.0	H_2O			[622]
		31	1.0	D_2O			
L-[4-^{2}H, 5-^{13}C]Glutamine	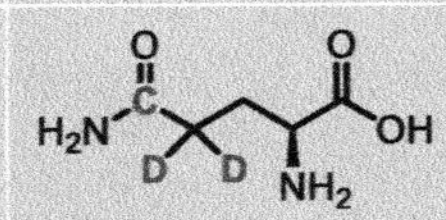	33	9.4	H_2O/Tris buffer			[827]
		37	1.0	H_2O			[622]
		68	1.0	H_2O			

Continued

TABLE A.3 ^{13}C T_1 **values of selected compounds.—cont'd**

Name	Chemical structure	Apparent T_1 (s)	B_0 (T)	Solvent	Temp. (°C)	pH	References
L-[4-^{2}H, 5-^{13}C, 5-^{15}N]Glutamine	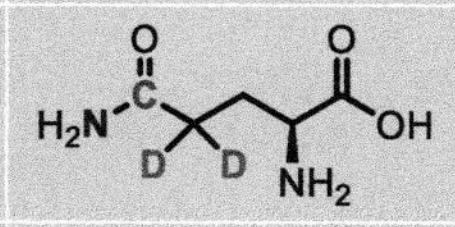	39	1.0	H_2O			[622]
		72	1.0	D_2O			
L-[1-^{13}C]Glutamic acid	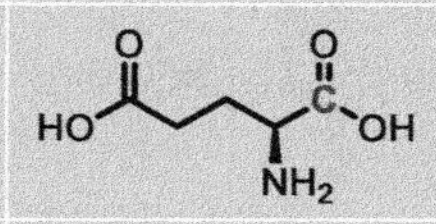	26	9.4	H_2O	37	7.0	[792]
		30	3.0	H_2O		7.4	[796]
L-[5-^{13}C]Glutamic acid	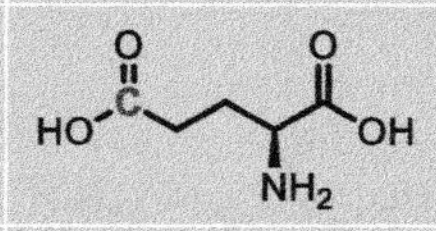	25	14.1	H_2O	37	7.0	[800]
		35	9.4	D_2O	37	7.4	[839]
L-glutamic acid, natural abundance, C5	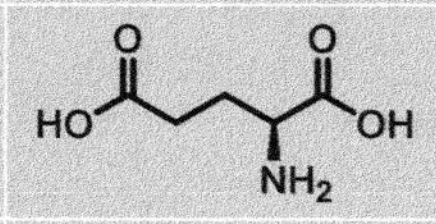	11	11.7	H_2O	30	7.0	[859]
[1-^{13}C]Glycine	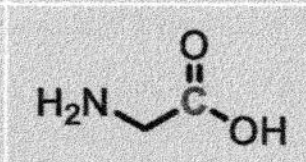	50	9.4	H_2O	37	7.0	[792]
		45	9.4	H_2O	25	7.4	[839,838]
		56	9.4	D_2O	37	7.4	[839]
		52	3.0	H_2O		7.0	[616]
[1-^{13}C, 2-^{2}H]Glycine	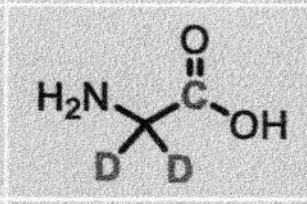	65	3.0	H_2O		7.0	[616]
L-[1-^{13}C]Leucine	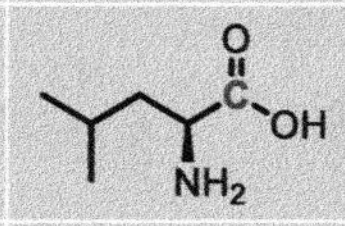	24	9.4	H_2O	37	7.0	[860]
L-[1-^{13}C]Lysine	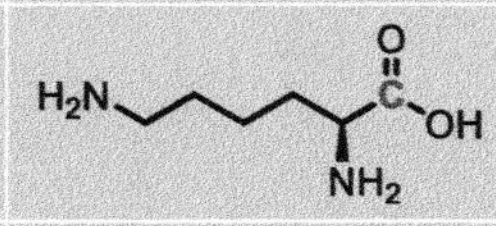	26	9.4	H_2O	37	7.0	[792]
L-[1-^{13}C]Methionine	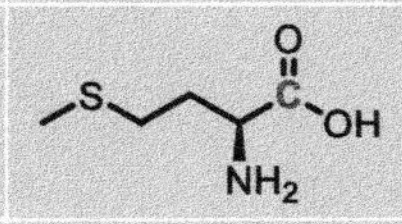	17	9.4	H_2O	37	7.0	[792]
L-[1-^{13}C]Serine	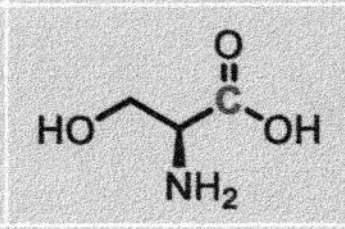	23	9.4	H_2O	37	7.0	[792]
		43	3.0	H_2O		7.0	[616]

L-[1-^{13}C, 2,3-$^{2}H_3$]Serine	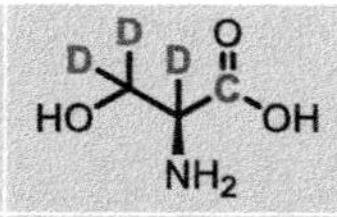	53	3.0	H_2O		7.0	[616]
L-[2-^{13}C]Serine	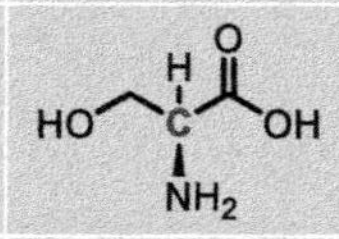	4	3.0	H_2O		7.0	[616]
L-[2-^{13}C, 2,3-$^{2}H_3$]Serine	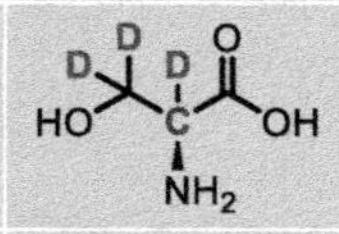	23	3.0	H_2O		7.0	[616]
L-[1-^{13}C]Valine	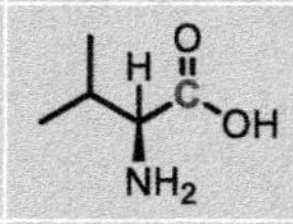	38	3.0	H_2O		7.0	[616]
L-[1-^{13}C, 2-^{2}H]Valine	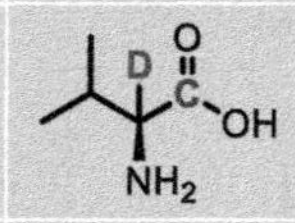	49	3.0	H_2O		7.0	[616]
N-[1-^{13}C]Acetyl-*L*-Alanine	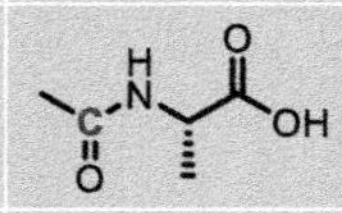	15	11.7	H_2O	37	7.3	[860]
N-[1-^{13}C]Acetyl-glycine	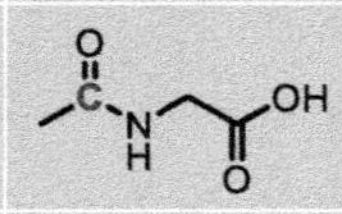	17	14.1	H_2O	27		[861]
		15	11.7	H_2O	37	7.3	[860]
N-[1-^{13}C]Acetyl diglycine	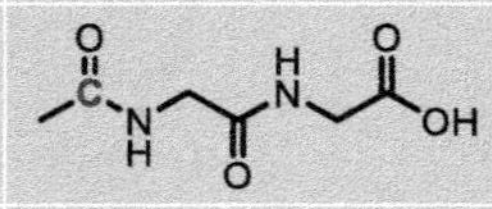	10	11.7	H_2O	37	7.3	[860]
N-[1-^{13}C]Acetyl triglycine	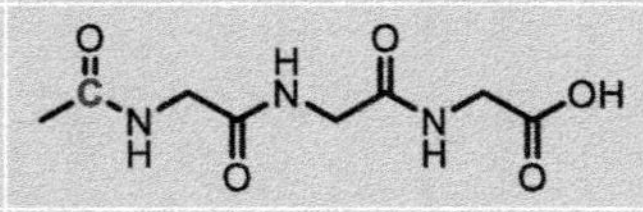	9	11.7	H_2O	37	7.3	[860]
N-[1-^{13}C]Acetyl-d_3 triglycine	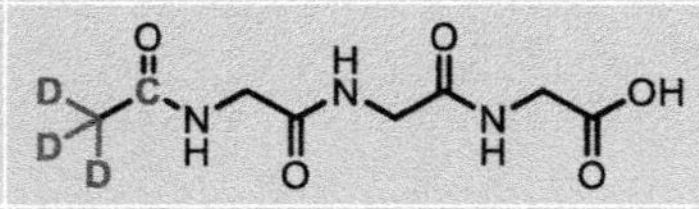	10	14.1	H_2O	27		[861]
L-[5-^{13}C]Pyroglutamate	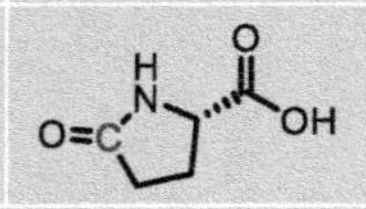	32	14.1	H_2O	37	7.0	[800]

Continued

TABLE A.3 ^{13}C T_1 **values of selected compounds.—cont'd**

Name	Chemical structure	Apparent T_1 (s)	B_0 (T)	Solvent	Temp. (°C)	pH	References
N-[1-^{13}C]Acetyl-*L*-serine	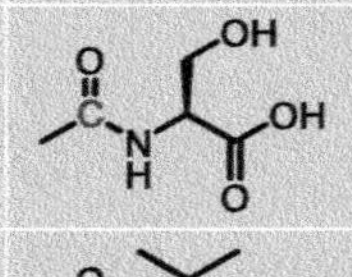	11	11.7	H_2O	37	7.3	[860]
N-[1-^{13}C]Acetyl-*L*-valine	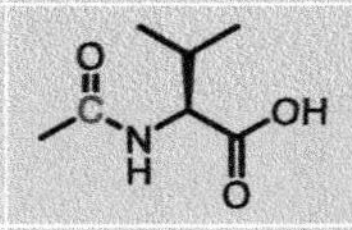	10	11.7	H_2O	37	7.3	[860]
N-[1-^{13}C]Acetyl-*L*-leucine	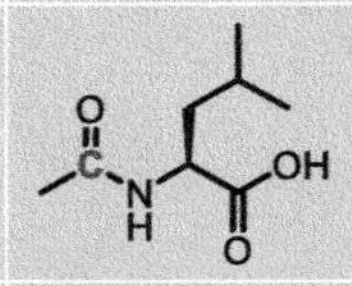	9	11.7	H_2O	37	7.3	[860]
[1-^{13}C]Acetyl-d_3-Arg-Gly-Asp	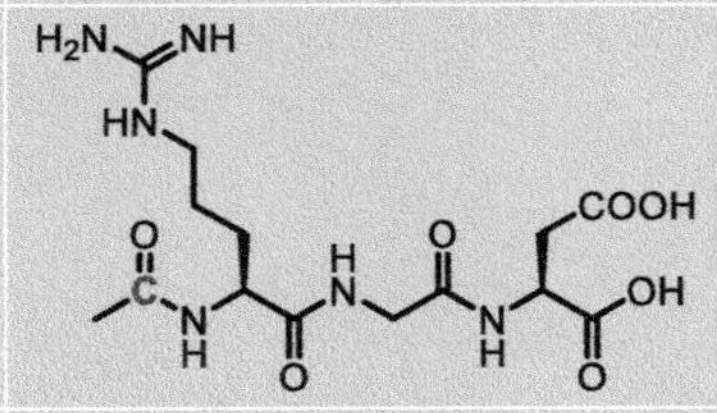	5	14.1	H_2O	27		[861]
N-Acetyl-*L*-[1-^{13}C]methionine	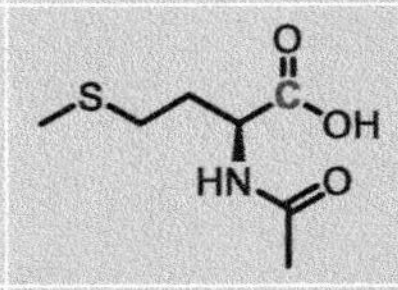	28	3.0	H_2O		7.2	[277]
N^2-acetyl, N^6-[1-^{13}C]acetyl-Lys	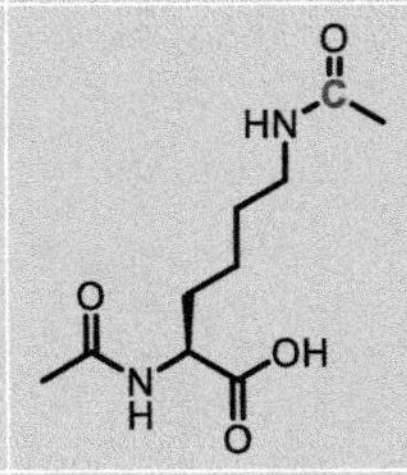	12	11.7	H_2O	37	7.3	[860]
Lys-N-[Acetyl-1-^{13}C] adduct of N-acetyl-α- Melanocyte stimulating Hormone N^2-acetyl, N^6-[1-^{13}C]acetyl-Lys-Pro-Val	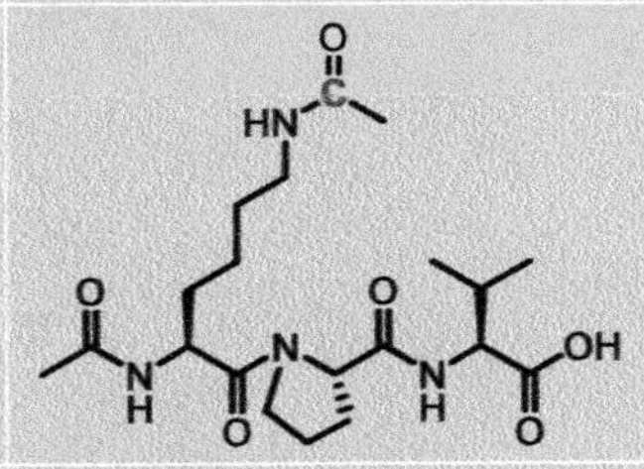	10	11.7	H_2O	37	7.3	[860]
[1-^{13}C]Ala-NH_2	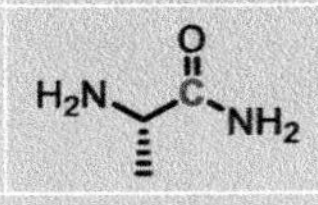	25	9.4	H_2O–D_2O (9:1)	37	7.4	[490]

[1-^{13}C]Ala-Gly	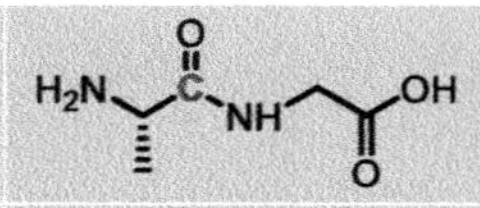	17	9.4	H_2O–D_2O (9:1)	37	7.4	[490]
[1-^{13}C]Ala-Ala	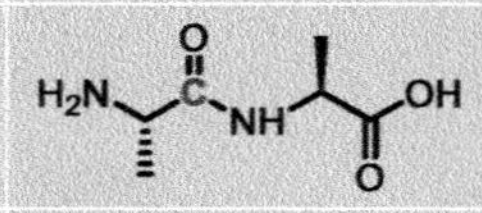	14	9.4	H_2O–D_2O (9:1)	37	7.4	[490]
γ-Glu-[1-^{13}C]Gly	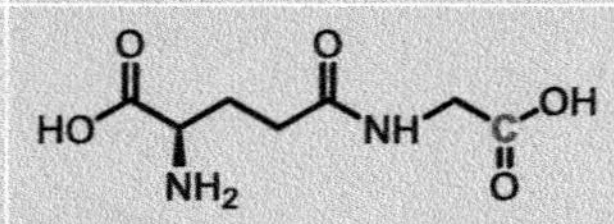	30	9.4	H_2O	25	7.4	[838]
		30	9.4	Serum/10% D_2O	37	7.4	[839]
		31	9.4	D_2O	37	7.4	
		27	11.7	D_2O	37	7.4	
		22	14.1	D_2O	37	7.4	
γ-Glu-[1-^{13}C, 2-$^{2}H_2$]Gly	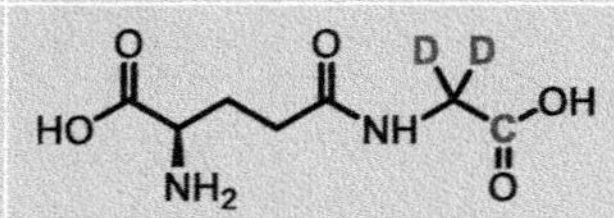	46	9.4	D_2O	37	7.4	[839]
		41	11.7	D_2O	37	7.4	
		29	14.1	D_2O	37	7.4	
		41	9.4	Serum/10% D_2O	37	7.4	
γ-[5-^{13}C]Glu-Gly	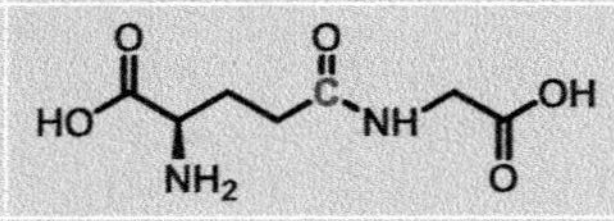	13	9.4	D_2O	37		[839]
3,5-DFB-Glu	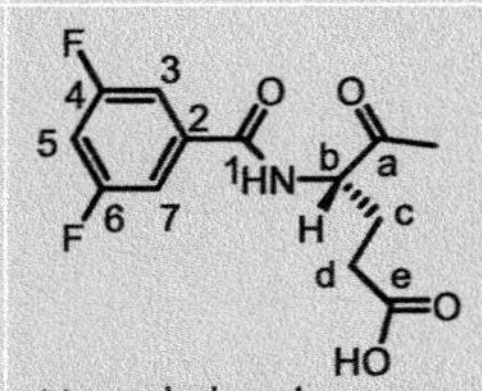 Natural abundance	7 (C1)	11.7	H_2O	30	7.0	[859]
		6 (C2)					
		9 (Ca)					
		8 (Ce)					
L-[1-^{13}C]Lysine ethyl ester dihydrochloride	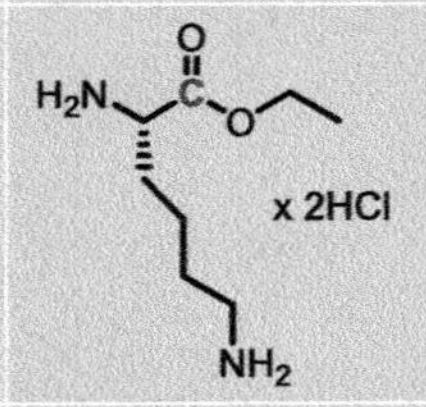	13	9.4	H_2O		7.4	[494]
Ethyl acetyl [^{13}C]carbonate	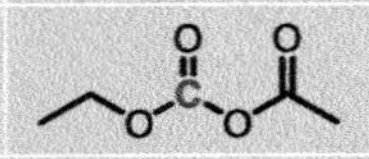	30	9.4	CH_2CH_3OH			[760]
Ethyl [1-^{13}C]acetate	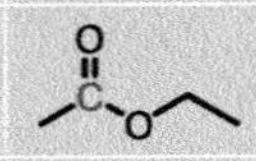	45	9.4	CH_3OH			[216]

Continued

TABLE A.3 ^{13}C T_1 **values of selected compounds.—cont'd**

Name	Chemical structure	Apparent T_1 (s)	B_0 (T)	Solvent	Temp. (°C)	pH	References
[2,3-$^{13}C_2$, $^2H_6$]Diacetyl	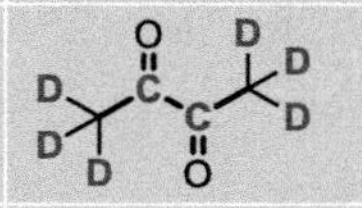	30	7.0	Acetone			[848]
[2-^{13}C]-1,3-Dihydroxyacetone	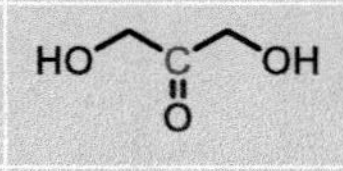	32	9.4	H_2O		7.4	[640]
		39	3.0	H_2O			[845]
N-2-[1-^{13}C, 1-^{15}N]Acetamido-2-aminoethanesulfonic acid	 [1-^{13}C, 1-^{15}N]–ACES	18	11.7	H_2O			[492]
		25	3.0	H_2O			
[1-^{13}C]Ascorbic acid	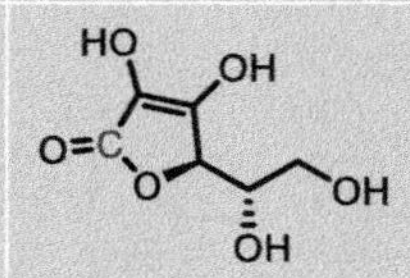	29	3.0	H_2O		7.0	[781]
		16	11.7	H_2O			
		16	9.4	H_2O		7.0	[778]
		9	9.4	In vivo			
[1-^{13}C]Dehydroascorbic acid	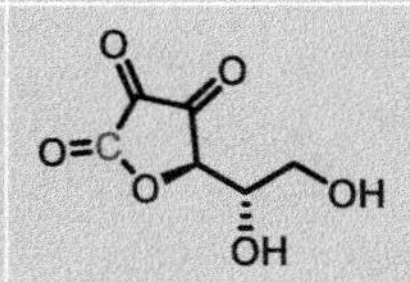	57	3.0	H_2O			[781]
		21	11.7	H_2O			
		21	9.4	H_2O			[778]
		12	9.4	In vivo			
		21	11.7	In vivo			[779]
[8-^{13}C]Aspirin	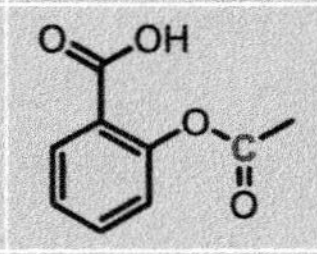	27	7.0	D_2O			[287]
[7,8-$^{13}C_2$]Aspirin	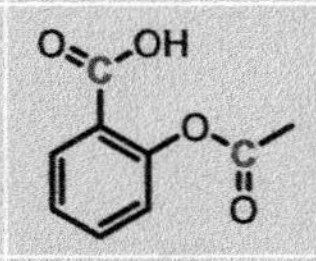	30 (C8)	7.0	D_2O			[287]
		33 (C7)					
[8-^{13}C, 9-$^2H_3$]Aspirin	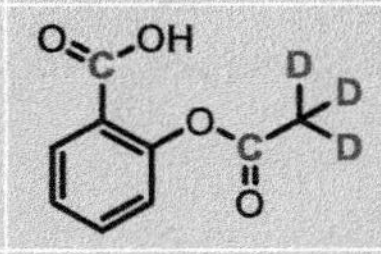	29	7.0	D_2O			[287]
[1,1′-$^{13}C_2$]Acetic anhydride	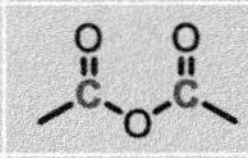	34	11.7	H_2O		7.8	[860]
[1,1′-$^{13}C_2$]Butyric anhydride	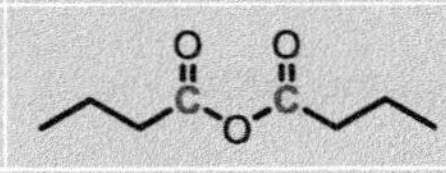	39	14.1	H_2O			[862]
		40	14.1				

[^{13}C, $^{2}H_3$]-*p*-anisidine	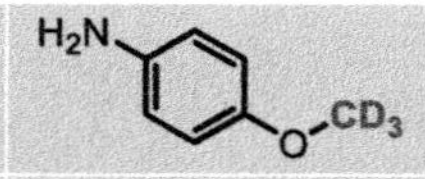	44	9.4	H_2O/D_2O			[605]
Sodium [^{13}C]Bicarbonate	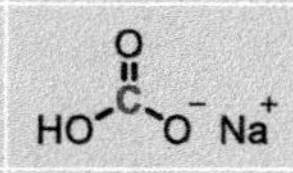	33	11.7	H_2O			[758]
		50	3.0	H_2O			[845]
Sodium [^{13}C]bicarbonate	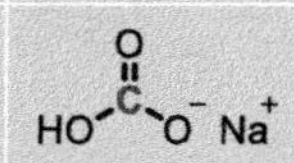	47	11.7	H_2O		7.8	[114]
		50	3.0				
		34	3.0	H_2O			[362]
Cesium [^{13}C]bicarbonate	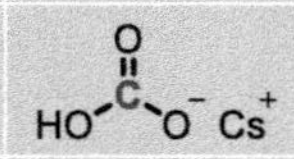	10	3.0	H_2O	37		[176]
[^{13}C]Carbon dioxide	O=C=O	33	11.7	H_2O			[758]
[1-^{13}C]-1,2-Glycerol carbonate	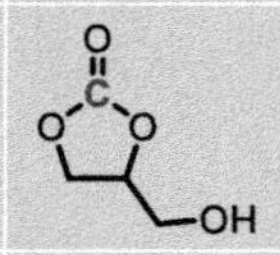	64	11.7	H_2O			[758]
D-[U-^{13}C, U-$^{2}H_{12}$]Glucose	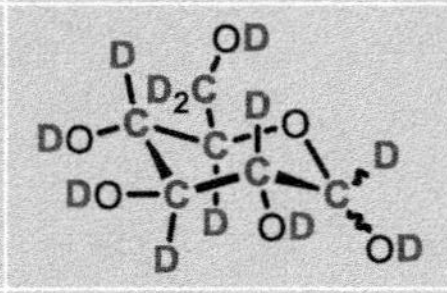	12 (C1–C5) 10 (C6)	14.1	H_2O			[863]
D-[U-^{13}C, $^{2}H_7$]Glucose	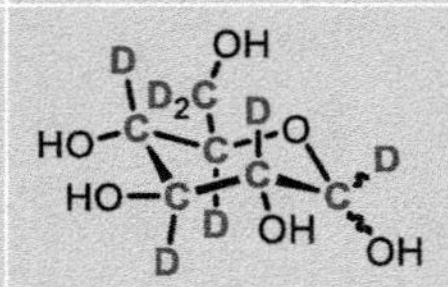	5 (C1–C5) (4.03 M)	11.8	H_2O			[769]
		4 (C6) (4.03 M)					
		17 (C1–C5) (40 mM)					
		11 (C6) (40 mM)					
D-[1-^{13}C]Gluconate	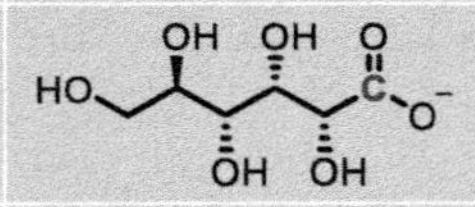	17	9.4	H_2O		7.4	[774]
D-[1-^{13}C]-γ-Gluconolactone	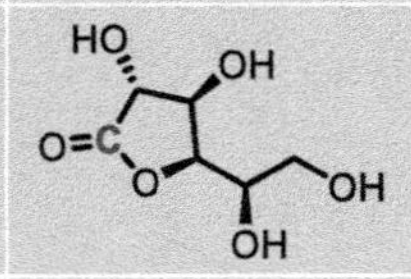	20	9.4	H_2O		7.4	[774]
D-[1-^{13}C]-δ-Gluconolactone	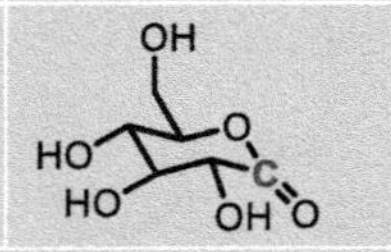	18	9.4	H_2O		7.4	[774]

Continued

TABLE A.3 ^{13}C T_1 **values of selected compounds.—cont'd**

Name	Chemical structure	Apparent T_1 (s)	B_0 (T)	Solvent	Temp. (°C)	pH	References
D-[2-^{13}C]Fructose	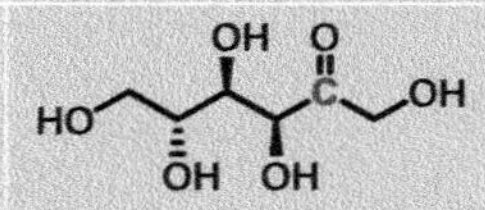	27	14.1	H_2O			[863]
		16	11.7	In vivo			[773]
		19	1.0	H_2O			[773]
[2-^{13}C,1,1,3,4,6,6-$^2H_6$]-β-p-D-Fructose	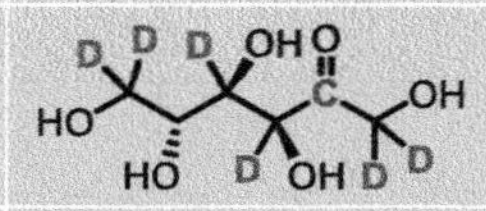	93	1.0	D_2O			[773]
[2-^{13}C]-β-Fructopyranose	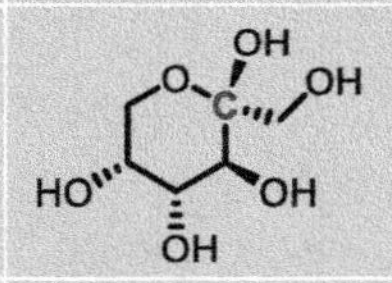	16	11.7	H_2O	37		[772]
		15	3.0				
[2-^{13}C]-β-Fructofuranose	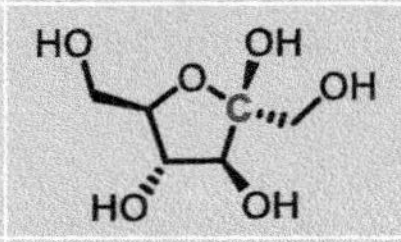	16	11.7	H_2O	37		[772]
		13	3.0				
[2-^{13}C]-α-Fructofuranose	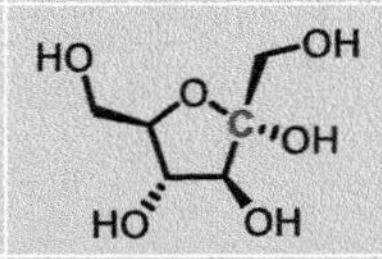	16	11.7	H_2O	37		[772]
		13	3.0				
[$^{13}C_6$]Benzene	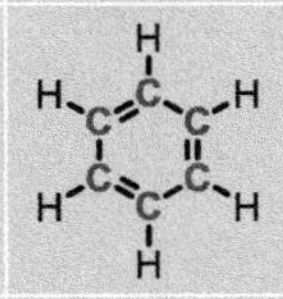	33	9.4	CH_3OH			[207]
Benzoic acid	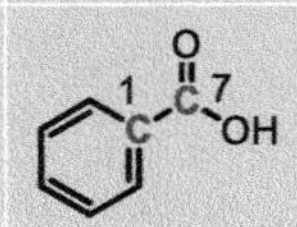	19 (C1)	11.7	H_2O	37	7.8	[864]
		35 (C7)					
		Benzoic acid (36 mM) in the presence of β-cyclodextrin (0.5 mM):					
		16 (C1)					
		19 (C7)					
[^{13}C]-*tert*-butylbenzene	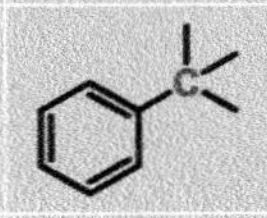	76	9.4	CD_3OD	25		[306]
[^{13}C, $^2H_{14}$]-*tert*-butylbenzene	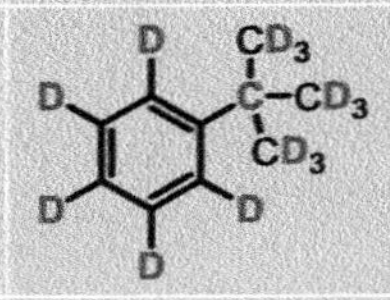	141	9.4	CD_3OD	25		[306]

Sodium [^{13}C, $^{2}H_{13}$]-4-*tert*-butyl-benzenesulfonate	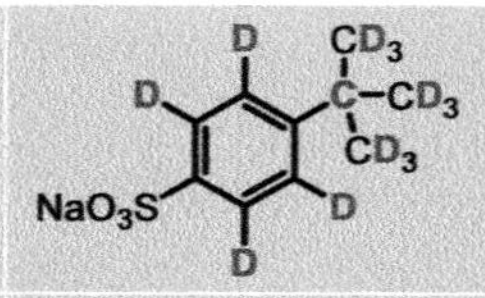	209	9.4	D_2O	22		[306]
		147		90% H_2O			
6-Methoxy-α-methyl-2-naphthaleneacetic acid (naproxen)	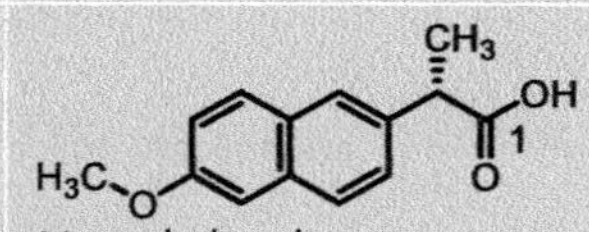 Natural abundance	15 10 (6.5 mM naproxen and 0.5 mM β-cyclodextrin)	11.7	H_2O		7.8	[864]
3,5-Difluorobenzoic acid	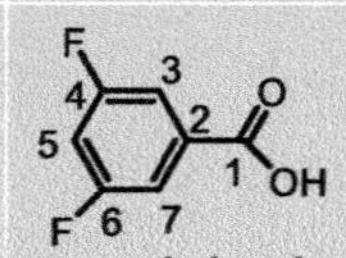 Natural abundance	23 (C1)	11.7	H_2O	30	7.0	[859]
		19 (C2)					
[1-^{13}C]Styrene	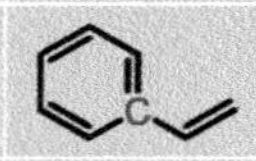	37	9.4	$CDCl_3$			[865]
Naphthalene[a]	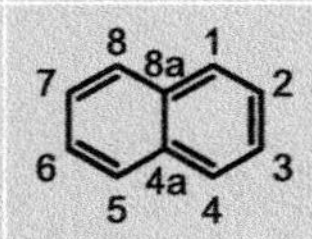	11 (C1, C4, C5, C8)	2.3	$CDCl_3$			[310]
		9 (C2, C3, C6, C7)					
		158 (C4a, C8a)					
		10 (C1, C4, C5, C8)	7.0				
		8 (C2, C3, C6, C7)					
		54(C4a, C8a)					
Phenanthrene[a]	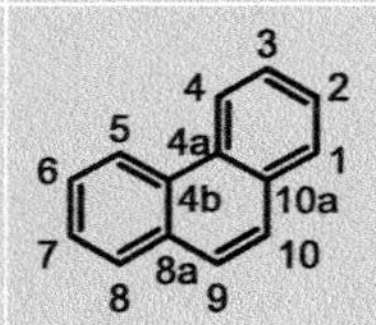	5 (C1, C8)	2.3	$CDCl_3$			[310]
		4 (C2, C3, C6, C7)					
		5 (C4, C5)					
		5 (C9, C10)					
		75 (C8a, C10a)					
		95 (C4a, C4b)	7.0				
		4 (C1, C8)					
		3 (C2, C3, C6, C7)					
		4 (C4, C5)					
		4 (C9, C10)					
		25 (C8a, C10a)					
		29 (C4a, C4b)					

Continued

TABLE A.3 ^{13}C T_1 values of selected compounds.—cont'd

Name	Chemical structure	Apparent T_1 (s)	B_0 (T)	Solvent	Temp. (°C)	pH	References
Pyrene[a]	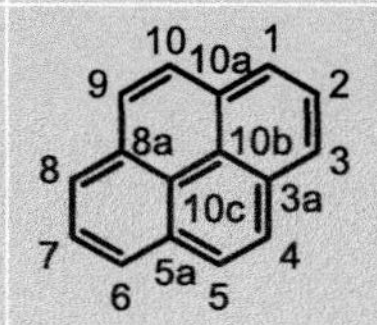	6 (C1, C3, C6, C8)	2.3	$CDCl_3$			[310]
		5 (C2, C7)					
		6 (C4, C5, C9, C10)					
		129 (C3a, C5a, C8a, C10a)	7.0				
		228 (C10b, C10c)					
		5 (C1, C3, C6, C8)					
		5 (C2, C7)					
		5 (C4, C5, C9, C10)					
		27 (C3a, C5a, C8a, C10a)					
		35 (C10b, C10c)					
Acenaphthene[a]	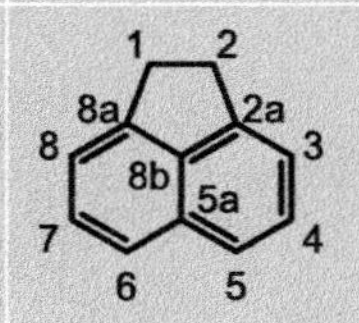	3 (C1, C2)	2.3	$CDCl_3$			[310]
		6 (C3, C8)					
		6 (C4, C7)					
		6 (C5, C6)					
		77 (C2a, C8a)					
		208 (C8b)					
		128 (C5a)					
		3 (C1, C2)	7.0				
		6 (C3, C8)					
		7 (C4, C7)					
		7 (C5, C6)					
		42 (C2a, C8a)					
		66 (C8b)					
		42 (C5a)					

Xanthone[a]	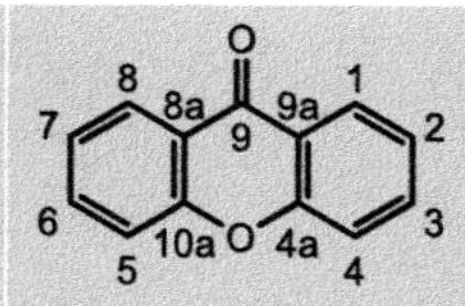	4 (C1, C8)	2.3	$CDCl_3$			[310]
		3 (C2, C7)					
		3 (C3, C6)					
		4 (C4, C5)					
		106 (C8a, C9a)					
		116 (C9)					
		109 (C4a, C10a)	7.0				
		4 (C1, C8)					
		3 (C2, C7)					
		3 (C3, C6)					
		4 (C4, C5)					
		44 (C8a, C9a)					
		32 (C9)					
		45 (C4a, C10a)					
Benzene[a]	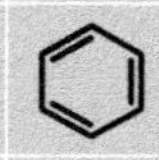	23 (undegassed) 29 (degassed)	2.3	Benzene-d_6	38		[866]
Toluene[a]	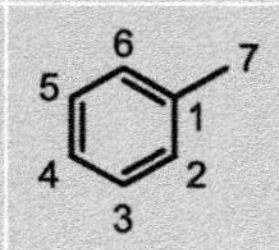	58 (C1)	2.3	No solvent	38		[866]
		20 (C2)					
		21 (C3)					
		15 (C4)					
		16 (Cα)					
tert-Butyl benzene[a]	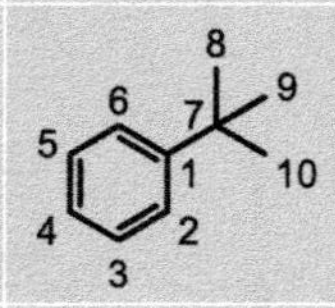	10 (C2)	2.3	No solvent			[866]
		10 (C3)					
		6 (C4)					
Styrene[a]	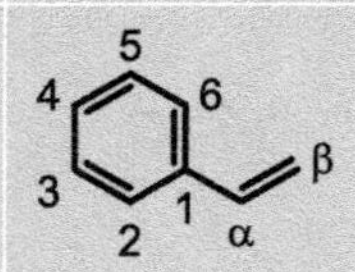	75 (C1)	2.3	No solvent			[866]
		15 (C2)					
		15 (C3)					
		12 (C4)					
		17 (Cα)					
		8 (Cβ)					

Continued

TABLE A.3 ^{13}C T_1 **values of selected compounds.—cont'd**

Name	Chemical structure	Apparent T_1 (s)	B_0 (T)	Solvent	Temp. (°C)	pH	References
Ethynylbenzene[a]	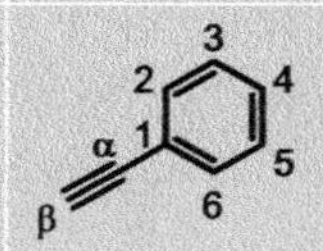	56 (C1)	2.3	Benzene-d_6			[866]
		13 (C2)					
		13 (C3)					
		9 (C4)					
		53 (Cα)					
		9 (Cβ)					
Ethynylbenzene[a]	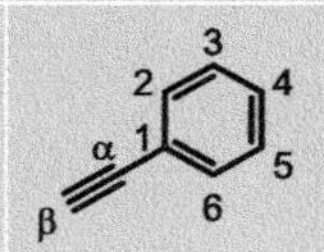	107 (C1)	2.3	Acetone-d_6 degassed			[866]
		14 (C2)					
		14 (C3)					
		8 (C4)					
		132 (Cα)					
		9 (Cβ)					
Phenyl benzene (biphenyl)[a]	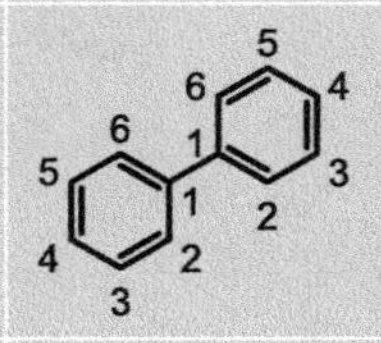	61 (C1)	2.3	Acetone-d_6			[866]
		6 (C2)					
		6 (C3)					
		3 (C4)					
Nitrobenzene[a]	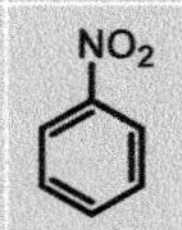	56 (C1)	2.3	No solvent			[866]
		7 (C2)					
		7 (C3)					
		5 (C4)					
Phenol[a]	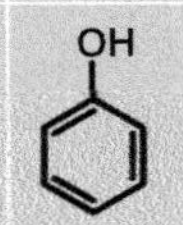	18 (C1)	2.3	CCl_4			[866]
		3 (C2)					
		3 (C3)					
		2 (C4)					
Aniline[a]	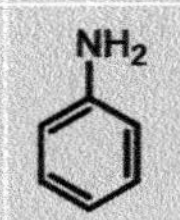	5 (C2)	2.3	No solvent			[866]
		5 (C3)					
		4 (C4)					

(3,4,5-Trimethoxyphenethylamine hydrochloride (mescaline)[a]	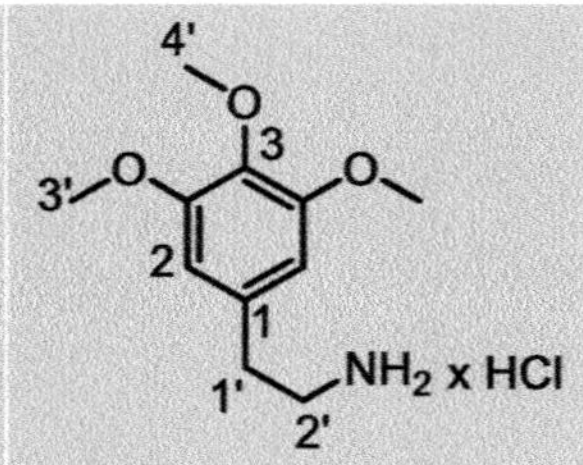	4 (C1)	2.3	D_2O	38		[866]
		0.3 (C2)					
		11 (C3)					
		14 (C4)					
		0.2 (C1′)					
		0.3 (C2′)					
		1 (C3′)					
		2 (C4′)					
Diphenyldiacetylene[a]	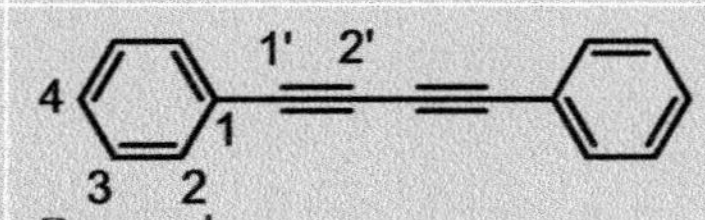 Degassed	5 (C2)	2.3	Acetone-d_6	38		[866]
		5 (C4)					
		1 (C5)					
		1 (C6)					
		8 2(C1′)					
		136 (C2′)					
		15 (C1′)	5.9				
		30 (C2′)					
1,1-Diphenyl- 2-propyn-1-ol	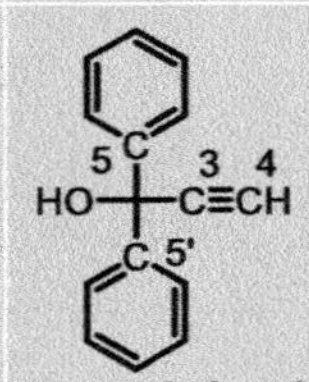 Natural abundance	19 (C3)	9.4	$CDCl_3$			[272]
		14 (C4)					
		20 (C5,5′)					
Phenazone	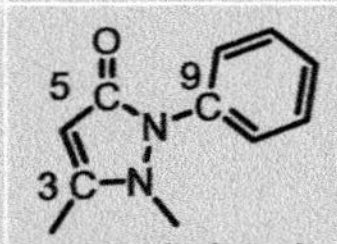 Natural abundance	15.8 (C3)	9.4	$CDCl_3$			[272]
		15.0 (C5)					
		18.5 (C9)					
[^{13}C]-*tert*-butanol	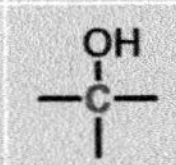	46 (in blood)	3.0	H_2O			[501]
		43 (rat brain)					
		37 (radical free)	14.1	D_2O			[240]
		18 (33 mM TEMPO)					
tert-Butanol	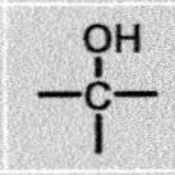	54	9.4	CD_3OD			[306]

Continued

TABLE A.3 ^{13}C T_1 values of selected compounds.—cont'd

Name	Chemical structure	Apparent T_1 (s)	B_0 (T)	Solvent	Temp. (°C)	pH	References
[$^2H_{10}$]-*tert*-butanol	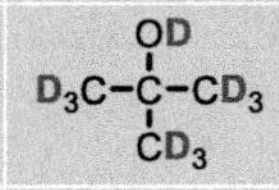	106	9.4	CD_3OD	25		[306]
[$^2H_{10}$,^{13}C]-*tert*-butanol	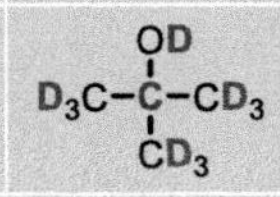	112 (radical free) 29 (33 mM TEMPO)	14.1	D_2O			[240]
[$^2H_9$,^{13}C]-*tert*-butanol	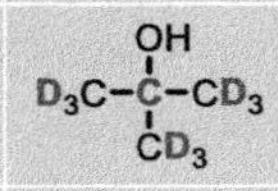	46	9.4	Blood			[504]
Choline chloride	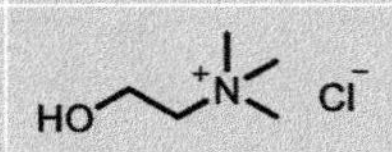	5 (C1)	11.8	H_2O: D_2O (9:1)	25		[589]
		5 (C2)					
		4 (C3)					
[$^2H_{13}$]Choline bromide	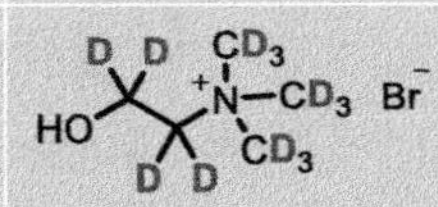	32 (C1)	11.8	H_2O: D_2O (9:1)	25		[589]
		33 (C2)					
		27 (C3)					
[1,1,2,2-$^2H_4$]Choline chloride	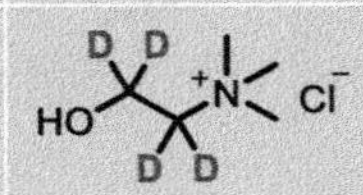	29 (C1)	11.8	H_2O: D_2O (9:1)	25		[589]
		35 (C2)					
		2 (C3)					
[2-^{13}C, 1,2-$^2H_4$]Choline chloride	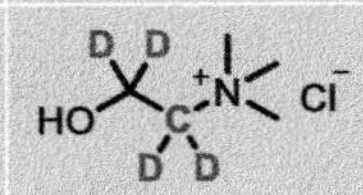	35	11.8	H_2O: D_2O (9:1)	25		[589]
		35		H_2O: D_2O (9:1)	18		
		46		H_2O: D_2O (9:1)	25		
		53		H_2O: D_2O (9:1)	37		[358]
		55		H_2O: D_2O (9:1)	45		
		47	14.1	H_2O	37		
		56	14.1	H_2O	45		
		52	9.4	H_2O	37		
		60	9.4	H_2O	41		

[2-^{13}C, 1,2-^{2}H$_4$]Acetylcholine	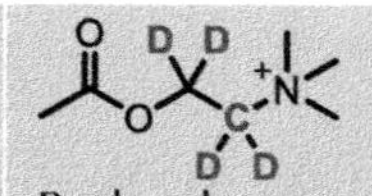 Produced enzymatically from [2-^{13}C, 1,2-^{2}H$_4$]choline	34	14.1	H_2O	37		[589]
Bis-1,1-(hydroxymethyl)-[1-^{13}C, ^{2}H$_8$]cyclopropane	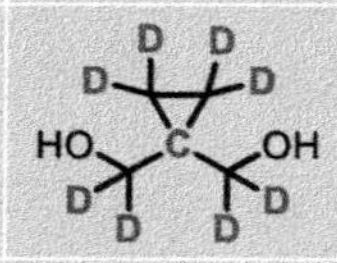	82	2.4	H_2O			[867]
Cyclohexyl isopropyl-1,2-^{13}C$_2$-oxalate-d_{18}	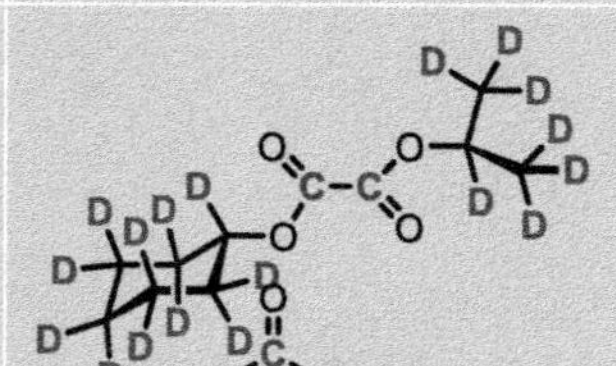	54	4.7	Methanol-d_4	37		[868]
[^{13}C]EDTA	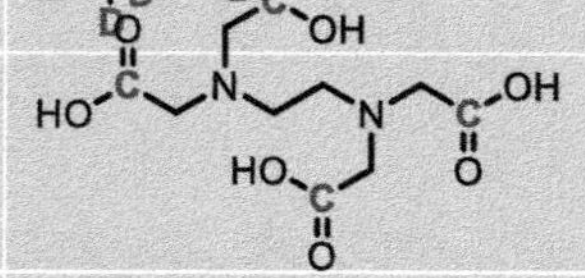	15	1.0	D_2O:DMSO-d6 (9:1)		7.4	[495]
		14	9.4	H_2O	25		[494]
[^{13}C, U-^{2}H$_{12}$]EDTA	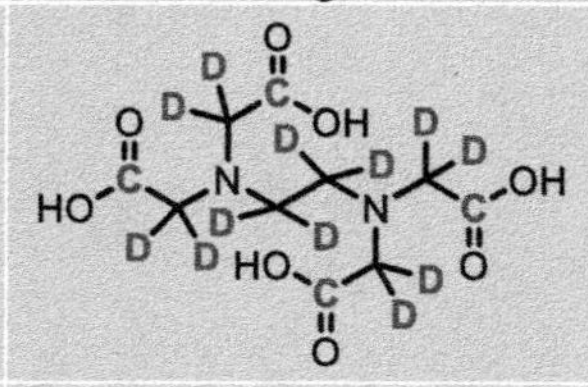	23	9.4	H_2O	25		[494]
Sodium [2-^{13}C]glycerate	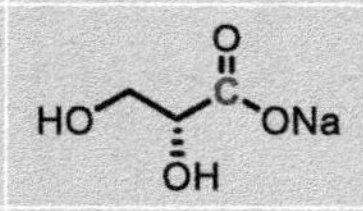	60	3.0	H_2O		7.5	[747]
[2-^{13}C, ^{2}H$_{10}$]Diethylmalonic acid	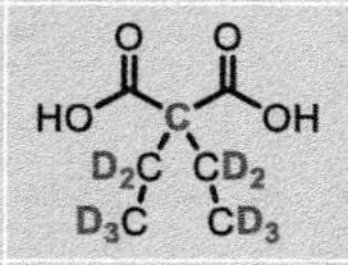	106	11.75	H_2O		7.1	[493]
[1,4-^{13}C]Fumaric acid	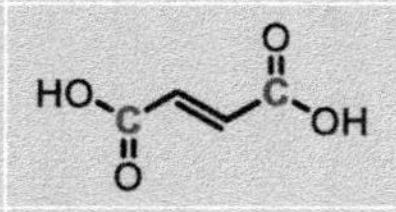	24	9.4	H_2O Cell suspension	37	7.4	[656]
[1,4-^{13}C$_2$]Malate	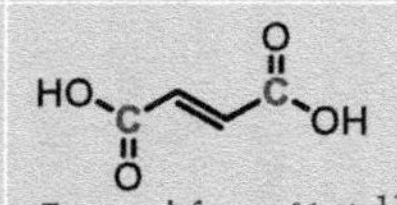 Formed from [1,4-^{13}C]fumarate in cell suspension	18	9.4	H_2O Cell suspension	37	7.4	[656]

Continued

TABLE A.3 ^{13}C T_1 values of selected compounds.—cont'd

Name	Chemical structure	Apparent T_1 (s)	B_0 (T)	Solvent	Temp. (°C)	pH	References
[1,4-^{13}C]Fumaric acid	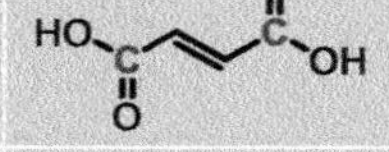	29	11.7	H_2O			[114]
[1,4-$^{13}C_2$]Maleic acid	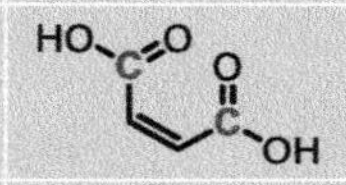	30	9.4	H_2O			[216]
[1-^{13}C]-2-Ketoglutaric acid	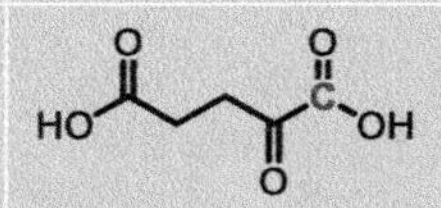	52 (C1)	3.0	H_2O		7.4	[698]
		19 (C1)	11.7				
		41 (C5)	3.0				
		18 (C5)	11.7				
[1-^{13}C-5-^{12}C]-2-Ketoglutaric acid	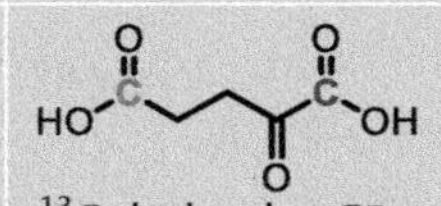 ^{13}C depleted at C5	43	3.0	H_2O		7.4	[624]
[1-^{13}C]-2-Ketoglutaric acid hydrate	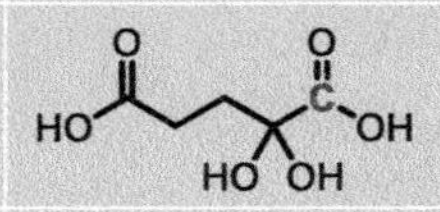	54	3.0	H_2O		7.4	[698]
		16	11.7				
[1-^{13}C]-2-Hydroxyglutaric acid	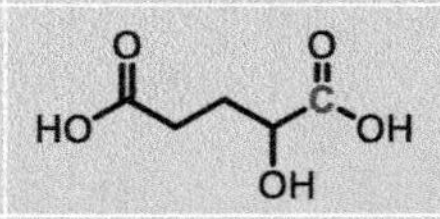	26	3.0	H_2O		7.4	[698]
		17	11.7				
Diethyl [1,2-$^{13}C_2$]-2-ketoglutarate	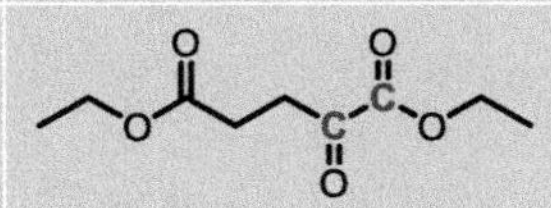	33 (C1)	9.4	H_2O			[363]
		39 (C1)	3.0				
		21 (C2)	9.4				
		23 (C2)	3.0				
Diethyl [1-^{13}C-5-^{12}C]-2-ketoglutarate	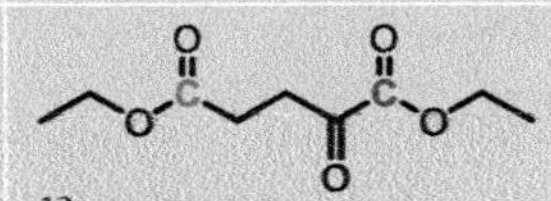 ^{13}C depleted at C5	39	3.0	H_2O		7.4	[869]
Diethyl [1,4-$^{13}C_2$]succinate	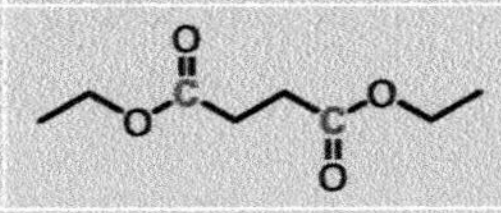	38	3.0	H_2O		7.5	[658]

[1-^{13}C]Pyruvic acid	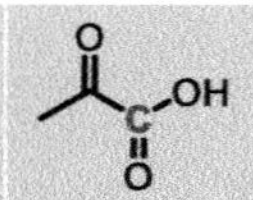	44	14.1	H_2O	27		[861]
		67	11.7	H_2O			[114]
		42	9.4	$NaHCO_3$,	25		[207]
		48	3.0	H_2O	37		[845]
		67	3.0	H_2O			[663]
		63	3.0	H_2O	30	8.2	[870]
		55	1.5	H_2O			
[1-^{13}C]Pyruvate	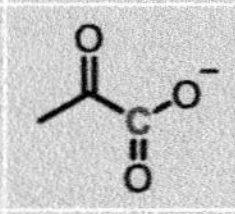	29	3.0	In vivo			[843]
[1,2-^{13}C]Pyruvic acid	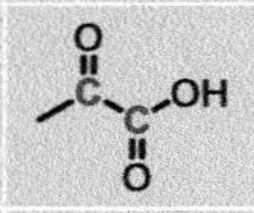	56 (C1) 44 (C2)	3.0	H_2O			[362]
[2-^{13}C]Pyruvic acid	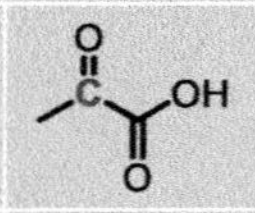	50 39	3.0 3.9	H_2O H_2O			[663] [845]
[2-^{13}C, 3-^{2}H$_3$]Pyruvic acid	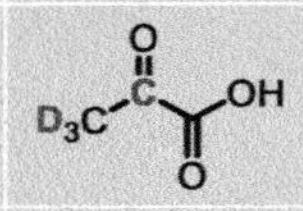	56	14.1	D_2O			[605]
Ethyl pyruvate	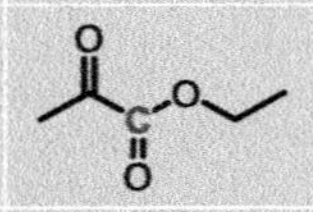	45	3.0	H_2O		6.7	[692]
Methyl pyruvate	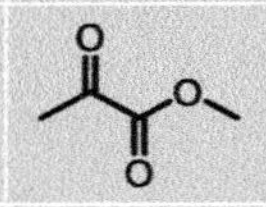	42 (C1) 44 (C2)	9.4	CH_3OH			[216]
Diethyl [1,1′-^{13}C$_2$]oxalate	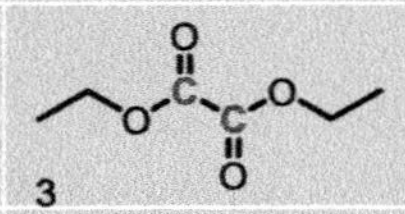	22	8.4	DMSO-d_6			[871]
Tetramethylallene	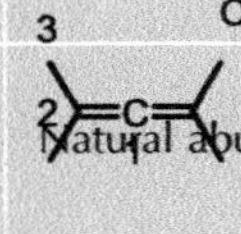 Natural abundance	63 (C1)	9.4	CH_3OH	25		[207]
		59 (C2)					
		8 (C3)					
Ethyl [1-^{13}C]acetate		45	9.4		25		[216]
Diethyl-1,3-acetone-dicarboxylate	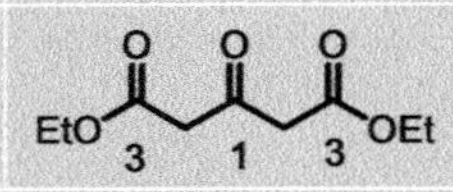	32 (C1)	9.4	MeOH	25		[216]
		34 (C3)					

Continued

TABLE A.3 ^{13}C T_1 values of selected compounds.—cont'd

Name	Chemical structure	Apparent T_1 (s)	B_0 (T)	Solvent	Temp. (°C)	pH	References
[^{13}C]Urea	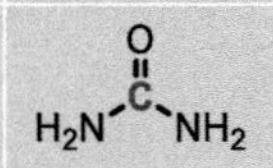	52	1.0	H_2O			[621]
		124	1.0	D_2O			
		31	14.1	H_2O			
		57	14.1	D_2O			
		47	3.0	H_2O			[845,505]
		46	9.4	H_2O			[207]
		44	11.7	H_2O		7.8	
		35	14.1	H_2O	27		
[^{13}C,^{15}N$_2$]Urea	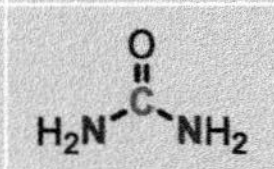	53	1.0	H_2O			[621]
		116	1.0	D_2O			
		32	14.1	H_2O			
		37	14.1	D_2O			
		45	3.0	H_2O			[501]
Zymonic acid	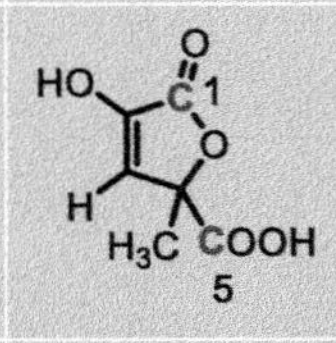	43 (C1)	3.0	H_2O	37	7.4	[491,690]
		51 (C5)					
		44 (C1)	1.0	D_2O			
		70 (C5)					
[1,5-^{13}C$_2$,3,6,6,6-^{2}H$_4$]Zymonic acid	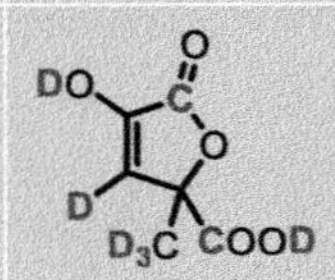	56 (C1)	1.0	D_2O			[690]
		97 (C5)					
		49 (C1)	3.0	H_2O			
		71 (C5)					
[1-^{13}C]-2-Ketoisocaproic acid	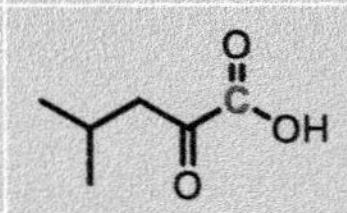	77	3.0	H_2O		7.4–8.2	[704]
							[872]
		55	9.4	H_2O		7.4	

[2-^{13}C]Benzoylformic acid	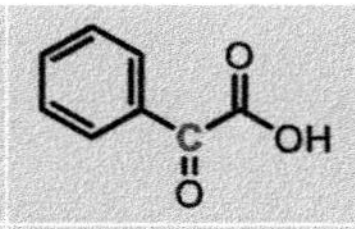	24	11.7	H_2O		7.8	[708]
		19	14.1	H_2O			
Sodium [1-^{13}C]acetate	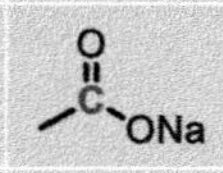	40	9.4	H_2O	27	7.6	[24]
		46	14.1	H_2O			[861]
Sodium [1-^{13}C, 1-$^{2}H_3$]acetate	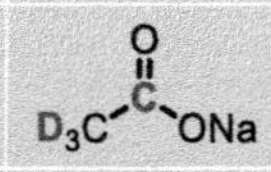	50	14.1	H_2O	27		[861]
Ethyl [1,3-$^{13}C_2$]acetoacetate	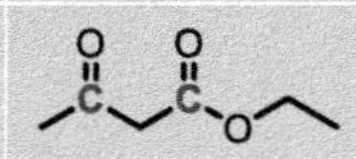	28 (C1)	14.1	H_2O	37		[713]
		31 (C3)					
		44 (C1)	0.5	H_2O			
		45 (C3)					
[1,3-$^{13}C_2$]Acetoacetate	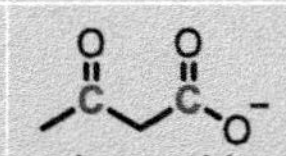 Obtained by enzymatic hydrolysis of ethyl [1,3-$^{13}C_2$]acetoacetate with immobilized carboxyl esterase from porcine liver	25 (C1)	14.1	H_2O	37		[713]
		28 (C3)					
		33 (C1)	0.5	H_2O			
		35 (C3)					
Sodium [1,3-$^{13}C_2$]acetoacetate	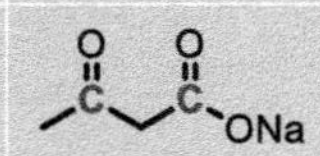	31 (C1)	11.7	H_2O			[711]
		27 (C3)					
		45 (C1)	9.4	H_2O			[545]
		37 (C2)					
		58	3.0	H_2O			[845]
Sodium [1-^{13}C]acetoacetate	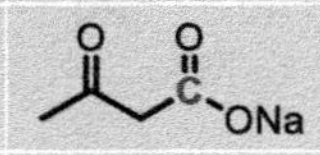	28	11.7	H_2O		7	[873]
[1-^{13}C]Butyric acid	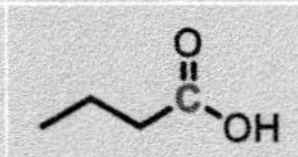	20	11.7	H_2O		7.4	[653]
		56	1.05	H_2O		7.6	[118]
		43	14.1	H_2O		7.4	[862]
[1-^{13}C]-β-Hydroxybutyrate	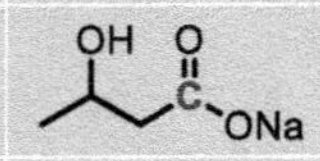	28	11.7	H_2O			[873]
Sodium [1-^{13}C]lactate	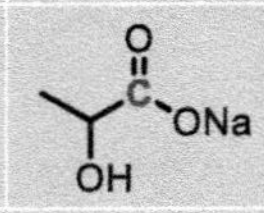	25	3.0	In vivo			[843]

Continued

TABLE A.3 ^{13}C T_1 **values of selected compounds.—cont'd**

Name	Chemical structure	Apparent T_1 (s)	B_0 (T)	Solvent	Temp. (°C)	pH	References
Sodium [1-^{13}C]lactate	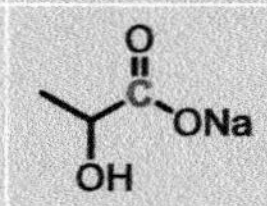	55	3.0	H_2O		7.0	[616]
		51	3.0	H_2O			[275]
		33	14.1	H_2O			[737,874]
		45	3.0	H_2O			
		39	3.0	H_2O		7.0	[875]
		48	3.0	H_2O			[743]
Sodium [1-^{13}C, 2-^{2}H]lactate	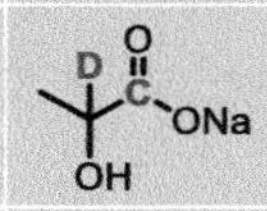	64	3.0	H_2O		7.0	[616]
Sodium [2-^{13}C]lactate	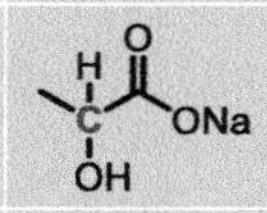	7	3.0	H_2O		7.0	[616]
Sodium [2-^{13}C,2-^{2}H]lactate	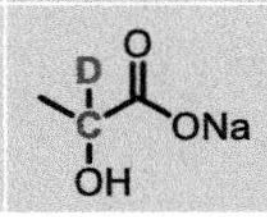	34	3.0	H_2O		7.0	[616]

[a]*Non HP.*

TABLE A.4 ^{15}N T_1 **values of selected compounds.**

Name	Chemical structure	Apparent T_1 (s)	B_0 (T)	Solvent	Temp. (°C)	pH	References
Sodium [^{15}N]nitrate	$Na[^{15}N]NO_3$	109	5.8	D_2O	34–44		[512]
		172			10–19		
		98		H_2O	34–44		
Sodium [^{15}N]nitrite	$Na[^{15}N]NO_2$	15	5.8	D_2O			[512]
		14		H_2O			
[^{15}N]Choline chloride		189	7.0	D_2O	25	8	[876]
		285	11.7	$H_2O:D_2O$ (9:1)	25	8	[513]
		120	11.7	Human blood	37		
		203	11.7	H_2O/OX063	37	8	
		232	14.1	H_2O	30		[390]
		215	9.4	H_2O	25		[207]
		200	9.4	D_2O	25		[593]
[^{15}N]Choline-d_9 chloride	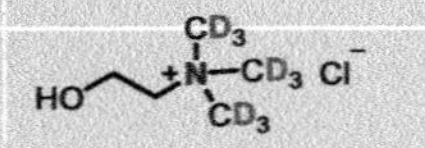	390	7.0	D_2O	25		[876]
		625	14.1	D_2O	30		[390]
[$^2H_{13}$]Choline	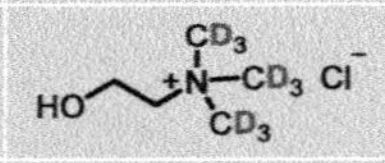	380	9.4	H_2O	25		[593]
[^{15}N-$^2H_{13}$]Choline	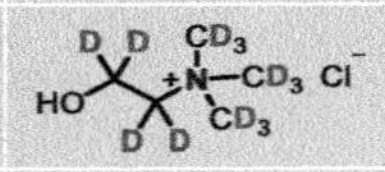	580	9.4	H_2O	25		[593]
α-Trideuteromethyl-glutamine	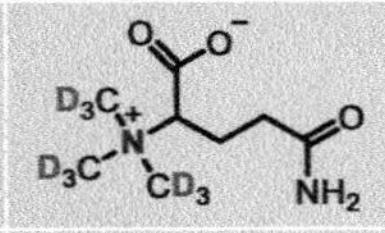	252	14.1	D_2O	37		[307]
α-Trideuteromethyl-glutamate	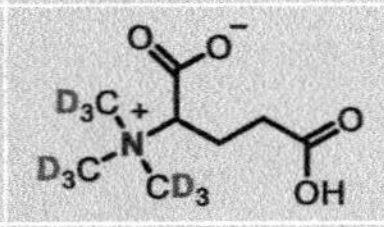	255	14.1	D_2O	37		[307]
ε-Trideuteromethyl-lysine	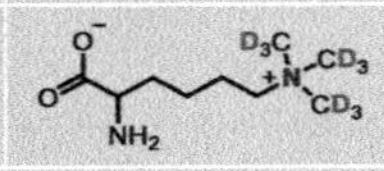	220	14.1	D_2O	37		[307]
L-[^{15}N]Carnitine-d_9	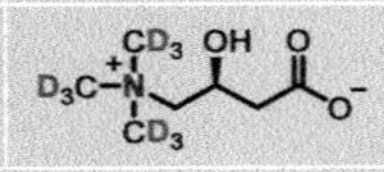	210	4.7	H_2O			[514]
		160		In vivo			

Continued

TABLE A.4 ^{15}N T_1 values of selected compounds.—cont'd

Name	Chemical structure	Apparent T_1 (s)	B_0 (T)	Solvent	Temp. (°C)	pH	References
3′-([3-^{15}N]Azido)-3′-deoxy- thymidine	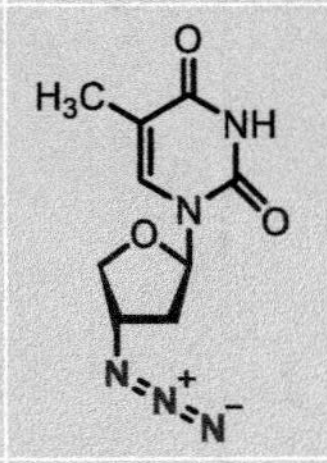	37	9.4	D_2O	25		[877]
		140	0.050				
([$^{15}N_3$]Azidoethyl)choline	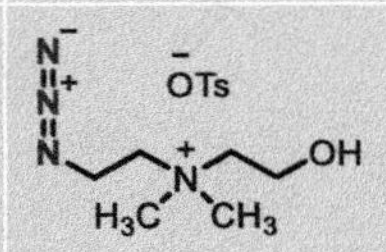	144 (α-^{15}N)	1.0	D_2O			[878]
		174 (β-^{15}N)					
		198 (γ-^{15}N)					
		22 (α-^{15}N)	16.4				
		6 (β-^{15}N)					
		15 (γ-^{15}N)					
2-[$^{15}N_3$]azido-2-deoxy-glucose	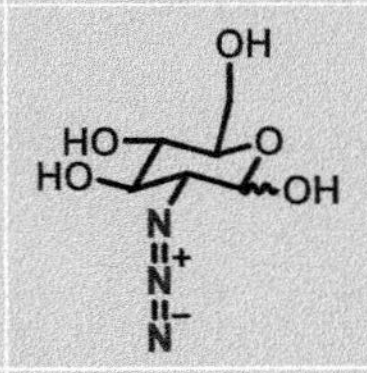	180 (α-^{15}N)	1.0	D_2O			[878]
		240 (β-^{15}N)					
		456 (γ-^{15}N)					
[$^{15}N_3$]Azidoacetyl-glucosamine	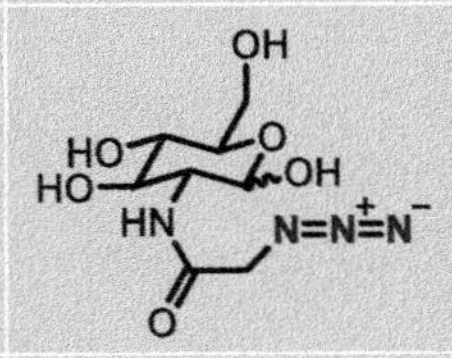	180 (α-^{15}N)	1.0	D_2O			[878]
		336 (β-^{15}N)					
		486 (γ-^{15}N)					
[$^{15}N_3$]Azidoethyl tyrosine	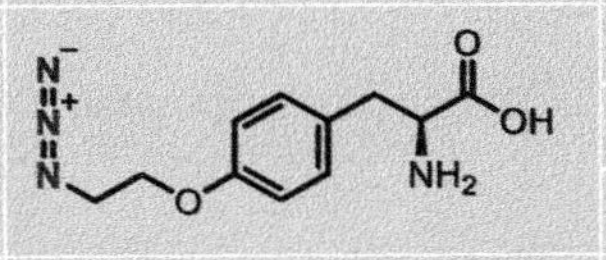	210 (α-^{15}N)	1.0	D_2O			[878]
		372 (β-^{15}N)					
		588 (γ-^{15}N)					
[$^{15}N_3$]Azidoalanine	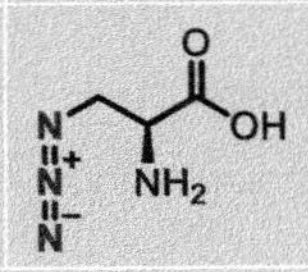	114 (α-^{15}N)	1.0	D_2O			[878]
		240 (β-^{15}N)					
		408 (γ-^{15}N)					

3′-([$^{15}N_3$]Azido)-3′-deoxy- thymidine	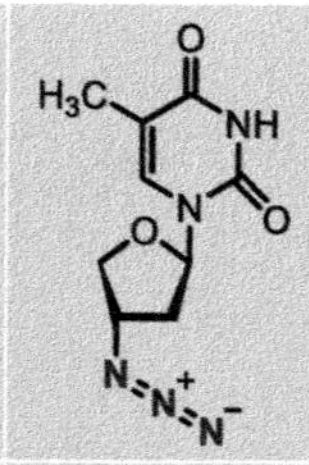	150 (α-^{15}N)	1.0	D_2O			[878]
		216 (β-^{15}N)					
		318 (γ-^{15}N)					
[^{15}N]Nitrobenzene	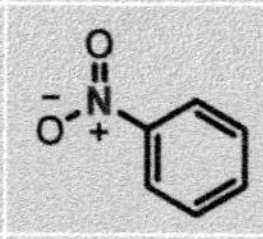	100	9.4	CH_3OH	25	7.8	[207]
		135	9.4	CH_3OH	25		[216]
[^{15}N]Pyridine	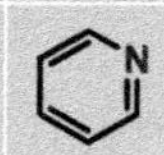	41	9.4	H_2O	23		[385]
		40	9.4	H_2O			[207]
		63	8.5	CD_3OD			[879]
		123	1.0				
Pyridine	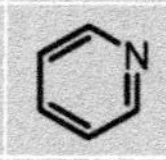	90 199	8.5 1.0	Neat			[879]
Pyridine		58	9.4	H_2O	25		[216]
Pyridine-d_5	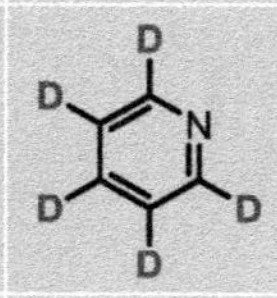	96 221	8.5 1.0	CD_3OD			[879]
[^{15}N]-2,6-Lutidine	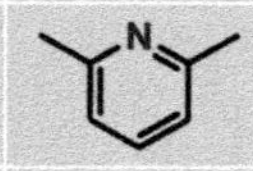	31	9.4	H_2O		9.5	[385]
[^{15}N]-2-Picoline	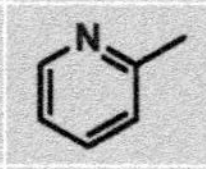	38	9.4	H_2O		8.5	[385]
[^{15}N]-2,4,6-Collidine	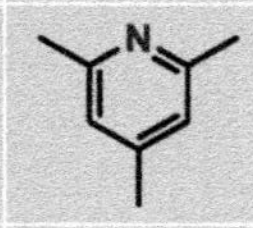	36	9.4	CH_3OH			[385]
[^{15}N]nicotinamide	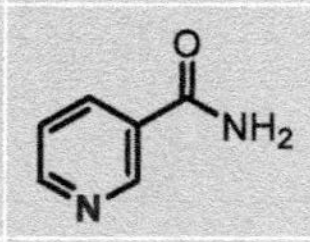	22	9.4	H_2O		5.9	[385]
		126	1.0	D_2O			[879]
1-Allyl-[^{15}N]pyridinium bromide-d_{10}	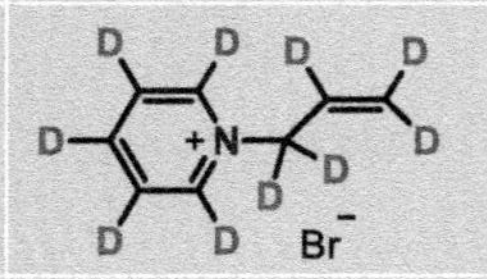	140	7	CH_3OD	47		[880]
		220	7	D_2O	80		

Continued

TABLE A.4 ^{15}N T_1 values of selected compounds.—cont'd

Name	Chemical structure	Apparent T_1 (s)	B_0 (T)	Solvent	Temp. (°C)	pH	References
1Propyl-[^{15}N]pyridinium bromide-d_{10}	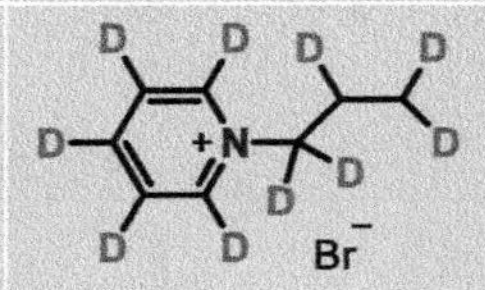	240	7	CH_3OD	47		[880]
		360	1	CH_3OD	47		
		570	0.1	CH_3OD	47		
		380	0.01	CH_3OD	47		
		120	7	D_2O	80		
		330	1	D_2O	80		
		500	0.1	D_2O	80		
		290	0.01	D_2O	80		
N-Allyl-N-methyl [^{15}N]aniline-d_{13}	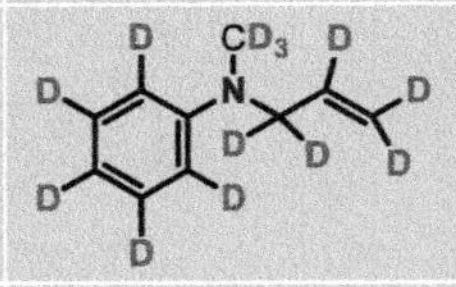	570	9.4	CH_3OH	25		[880]
N-Propyl-N-methyl [^{15}N]aniline-d_{13}	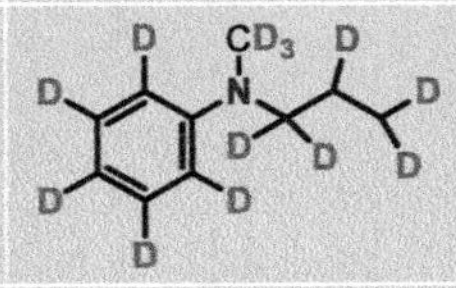	150 90	7 0.1	$CH_3OD:D_2O$ $CH_3OD:D_2O$	47 47		[880]
N,N-Dimethyl-N-propyl [^{15}N]benzenaminium-d_{16} iodide	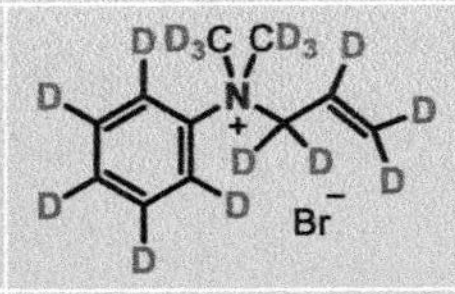	420	9.4	D_2O	25		[880]
[$^{15}N_2$]Imidazole	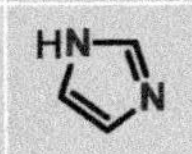	24	9.4	CD_3OD		12	[879]
		24	9.4	$CH_3OH:H_2O$ (1:1)			[386,879]
		86		$CH_3OH:H_2O$ (1:9)		12	
[^{15}N]Trimethylphenylammonium (TMPA) iodide	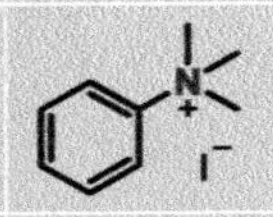	275 262	14.1 9.4	D_2O D_2O	30		[297,390]
[^{15}N]TMPA-d_9	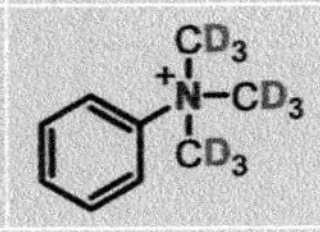	816 880	14.1 9.4	D_2O D_2O	30		[297,390]

[15N]TMPA-d_{14}	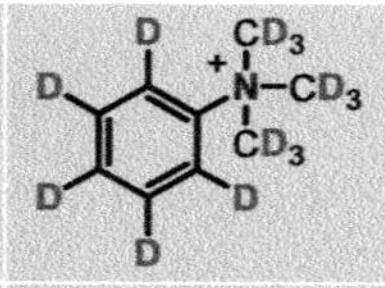	1128 1177	14.1 9.4	D_2O D_2O			[297,390]
4-(Bis(carboxymethyl)amino)-3-(carboxymethoxy)-N,N,N-trimethyl [15N]benzenaminium-d_9 iodide	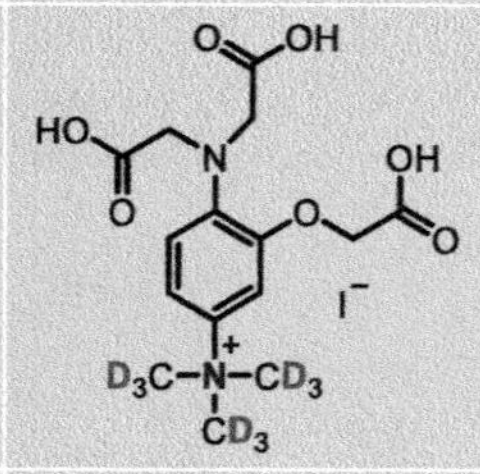	129 130 (Ca^{2+})	14.1	H_2O	30	7.4	[390]
4-Hydroxy-N,N,N-trimethyl [15N]benzenaminium-d_9 iodide	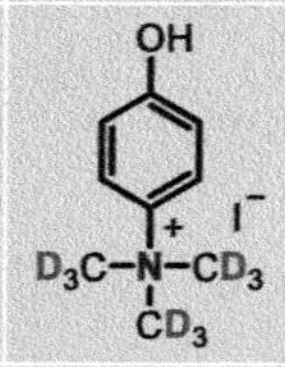	486	14.1	H_2O	30	7.4	[390]
4-Acetoxy-N,N,N-trimethyl [15N]benzenaminium-d_9 iodide	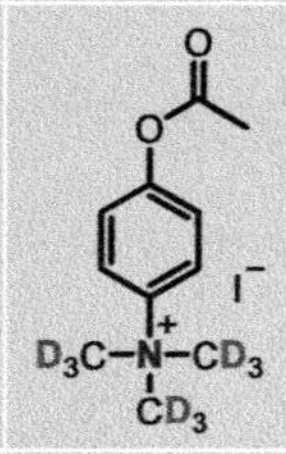	536	14.1	H_2O	30	7.4	[390]
		570	9.4	D_2O			[297]
		450		H_2O:D_2O (9:1)			
4-Acetoxy-N,N,N-trimethyl [15N]benzenaminium-d_{16} iodide	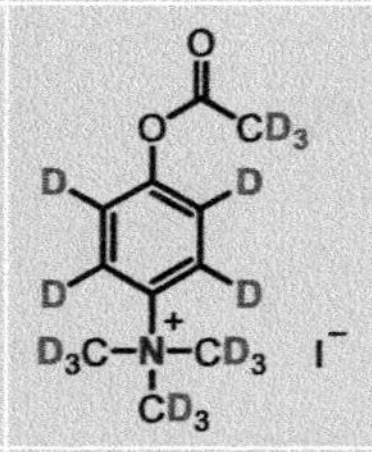	795	9.4	D_2O	30	7.4	[297]
		602	9.4	H_2O:D_2O (9:1)			
[15N]-Tris(2-pyridylmethyl)amine	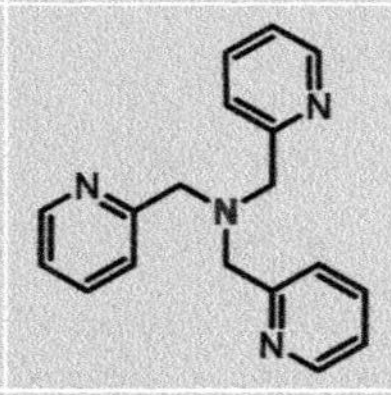	26 18 (Zn^{2+}-complex)	9.4	H_2O	25	6.8	[26]

Continued

TABLE A.4 ^{15}N T_1 values of selected compounds.—cont'd

Name	Chemical structure	Apparent T_1 (s)	B_0 (T)	Solvent	Temp. (°C)	pH	References
[^{15}N]-Tris(2-pyridylmethyl)amine-d_6	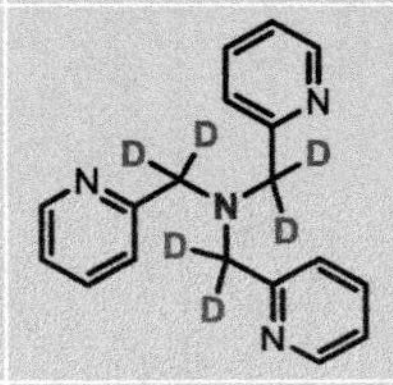	71 57 (Zn^{2+}-complex)	9.4	H_2O	25	6.8	[26]
4-(5-(2-(3-Methyl-[$^{15}N_2$]-*3H*-diazirin-3-yl)ethoxy)-3-(trifluoro-methyl)-1H-pyrazol-1-yl)benzenesulfonamide	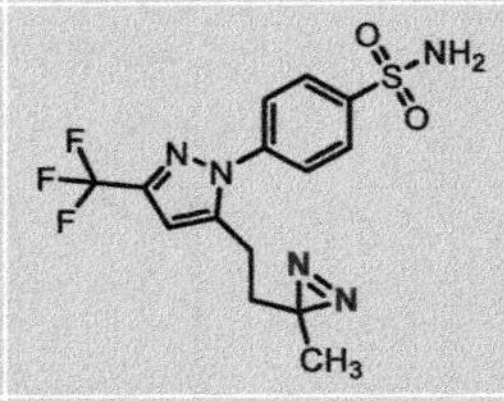	246	1.0	CD_3OD			[381]
(*S*)-2-Amino-4-(3-methyl-[$^{15}N_2$]-*3H*-diazirin-3-yl)butanoic acid	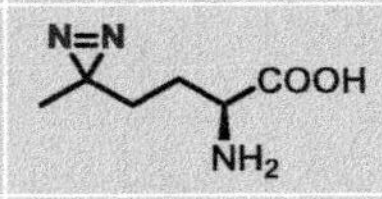	236	1.0	D_2O			[381]
N-3-(3-methyl-[$^{15}N_2$]-*3H*-diazirin-3-yl)propionyl-D-glucosamine	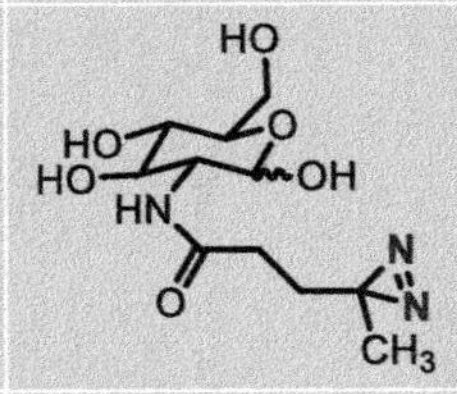	178	1.0	D_2O			[381]
N-3-(3-methyl-d_3-[$^{15}N_2$]-*3H*-diazirin-3-yl)propionyl-D-glucosamine	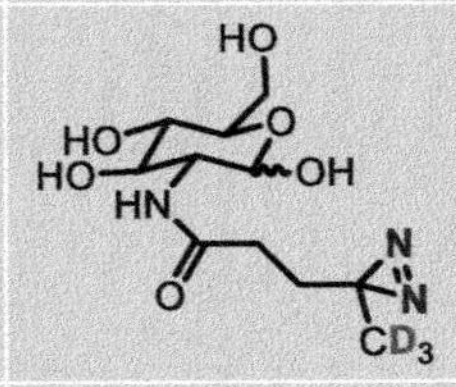	184	1.0	D_2O			[381]
^{15}N-Acetonitrile		123 224	1.0 8.5	CD_3OD			[879]
Acetonitrile	 Natural abundance	138	1.0	Neat			[879]
		132	8.5				
		110	9.4	MeOH	25		[216]

Acetonitrile-d_3	$N \equiv CD_3$ Natural abundance	121 142	1.0 8.5	CD_3OD			[879]
[^{15}N]Benzonitrile	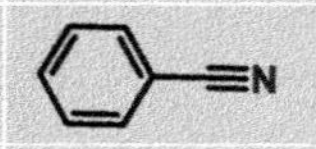	717 129	1 8.5	CD_3OD			[879]
Benzonitrile	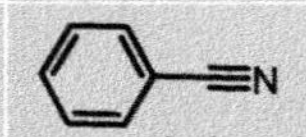 Natural abundance	630 173	1 8.5	Neat			[879]
α-Cyano-4-hydroxycinnamic acid (CHCA)	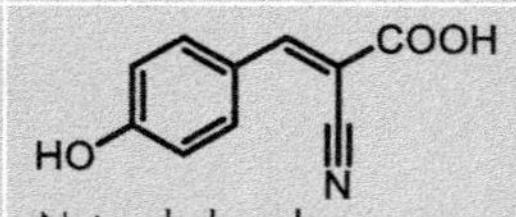 Natural abundance	72	1.0	D_2O			[879]
Metronidazole	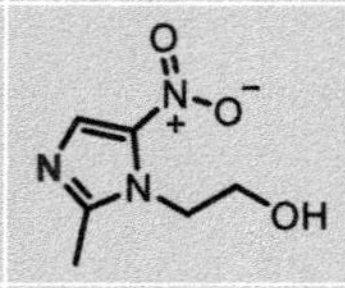	36	9.4	CD_3OD			[879]
Methyl [^{15}N]-N-benzylideneglycinate	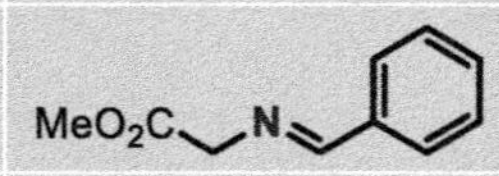	132	1	CD_3OD			[879]
Dichloropyridazine	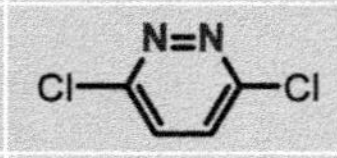	15	8.5	CD_3OD			[879]
Azobenzene	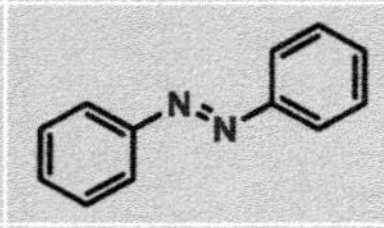	9.0	8.5	CD_3OD			[879]
2-Cyano-3-(3-(methyl-d_3)-[$^{15}N_2$]-3*H*-diazirin-3-yl)propanoic-2-*d* acid	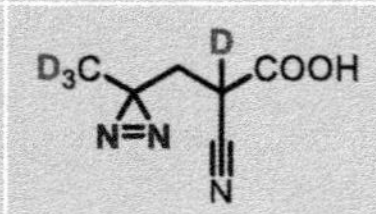	300	0.3 mT	CD_3OD			[879]
		10	8.5				
		261	1.0				
Pentaerythrityl tetraazide	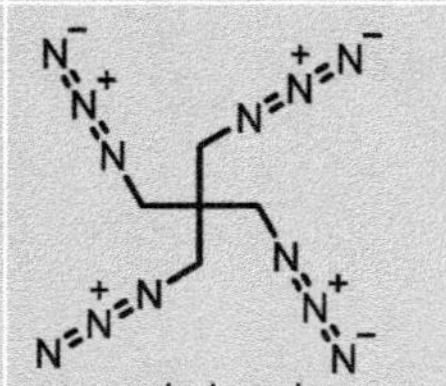 Natural abundance	122	9.4	CH_3OH	25		[215]
		147					
		64					

TABLE A.5 ^{31}P T_1 **values of selected compounds.**

Name	Chemical structure	Apparent T_1 (s)	B_0 (T)	Solvent	Temp. (°C	pH	References
Triphenylphosphine oxide (Ph_3PO)[a]	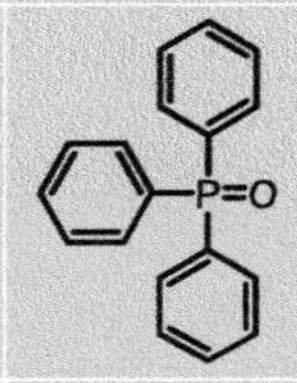	4 20 (1 M)	9.4 40.5 MHz	9:1 THF/H_2O $CDCl_3$	29		[272] [399]
Triphenylphosphine (PPh_3)[a]	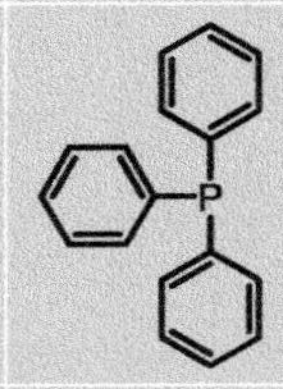	16 31 (0.7 M)	9.4 40.5 MHz	9:1 THF/H_2O $CDCl_3$	29		[272] [399]
Phosphoric acid[a]	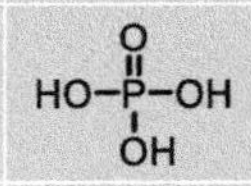	10 (0.2 M)	2.35	H_2O	31		[401]
Phosphoric acid-d_3[a]	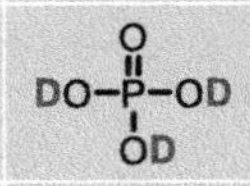	34 (0.2 M)	2.35	D_2O	31		[401]
Phosphoric acid[a]	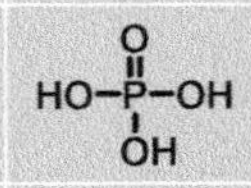	0.5 (14.1 M)	2.35	H_2O	31		[401]
Phosphoric acid-d_3[a]	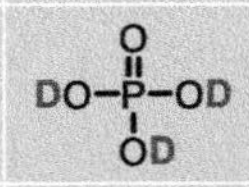	3.2 (14.1 M)	2.35	D_2O	31		[401]
Disodium hydrogen phosphate	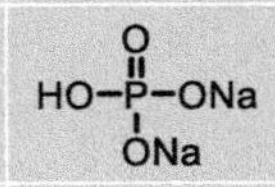	51 (0.10 M)	2.35	D_2O	25		[402]
Trimethyl phosphate	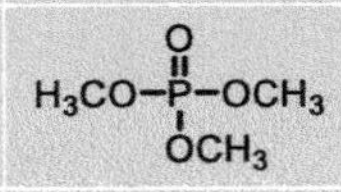	20	9.4	H_2O			[207]
Diethyl pyridin-4-ylphosphonate		**8**	**9.4**	**CD_3OD**			**[881]**
Di(ethyl-d_5) pyridin-4-ylphosphonate	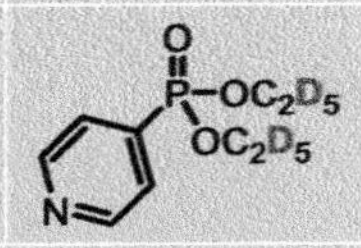	**8**	**9.4**	**CD_3OD**			**[881]**

Diethyl pyridin-3-ylphosphonate	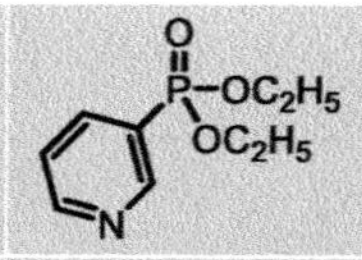	9	9.4	CD_3OD			[881]
Di(ethyl-d_5) pyridin-4-ylphosphonate	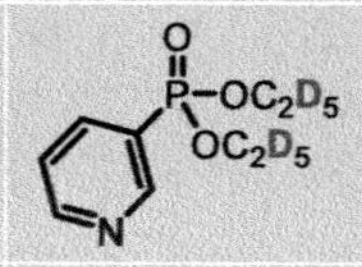	9	9.4	CD_3OD			[881]
Tetraethyl pyridine-3,5-diylbis(phosphonate)	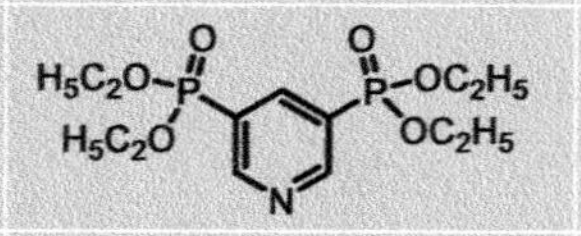	5	9.4	CD_3OD			[881]
Tetra(ethyl-d_5) pyridine-3,5-diylbis(phosphonate)	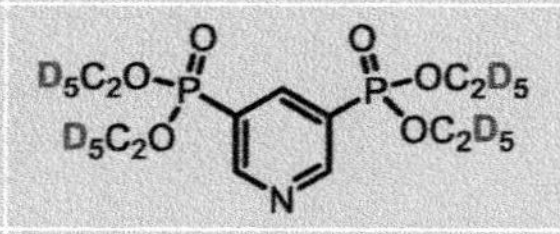	6 7 4	9.4	CD_3OD CD_3CD_2OD D_2O			[881]
Potassium dihydrogen phosphate	KH_2PO_4	16	5.8	D_2O		5.1	[400]
Potassium dihydrogen phosphate	KH_2PO_4	6	5.8	Saline and citrate-Tris buffer			[400]
Dipotassium hydrogen phosphate	K_2HPO_4	29	5.8	D_2O		8.1	[400]
Dipotassium hydrogen phosphate	K_2HPO_4	15.5	5.8	Pure H_2O			[400]
Dipotassium hydrogen phosphate	K_2HPO_4	13	5.8	Physiological saline			[400]
Phosphocreatine Phosphocreatine Phosphocreatine	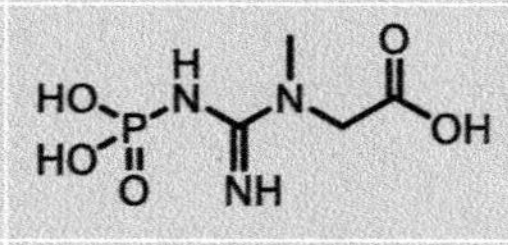	10 8 11	5.8 5.8 5.8	D_2O H_2O D_2O with phosphate buffer			[400] [400] [400]
ATP	α-, β, and γ-phosphate	1.2	5.8	H_2O		7.4	[400]
Phosphocreatine	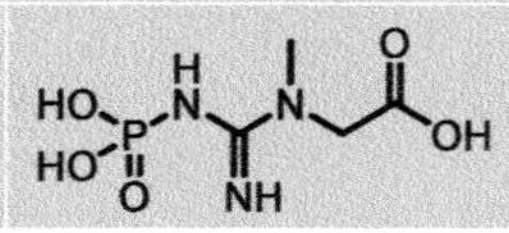	4.78 10.25	7.0 2.35 T	H_2O	32		[882]

[a]*Non HP.*

TABLE A.6 ^{89}Y T_1 **values of selected compounds.**

Name	Chemical structure	Apparent T_1 (s)	B_0 (T)	Solvent	Temp. (°C)	pH	References
Yttrium nitrate[a]	$Y(NO_3)_3$	270 (1 M)	3.52	H_2O	5	1.8	[305]
		240 (1 M)			15		
		180 (1 M)			25		
		240 (1 M)			35		
		130 (1 M)			65		
		63 (3 M)			25		
		85 (3 M)			35		
		75 (3 M)			65		
Yttrium nitrate[a]	$Y(NO_3)_3$	260	3.52	DMSO	35		[305]
		550		DMSO-d_6			
Yttrium chloride	YCl_3	620	14.1	H_2O			[406]
Yttrium triflate	$Y(OTf)_3$	460	9.4	H2O			[207]
Yttrium(III) 12-crown-4 complex[a]	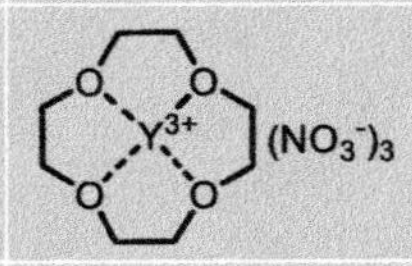	300	3.52	DMSO	35		[305]
		560			65		
		800		DMSO-d_6	35		
Yttrium(III) 15-crown-5 complex[a]	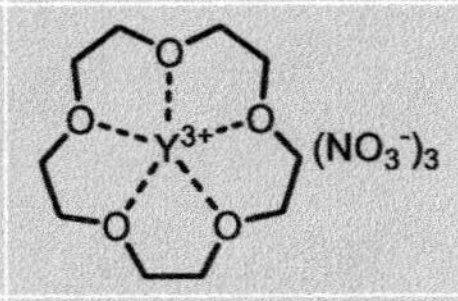	400	3.52	DMSO	35		[305]
Yttrium(III) 18-crown-6 complex[a]	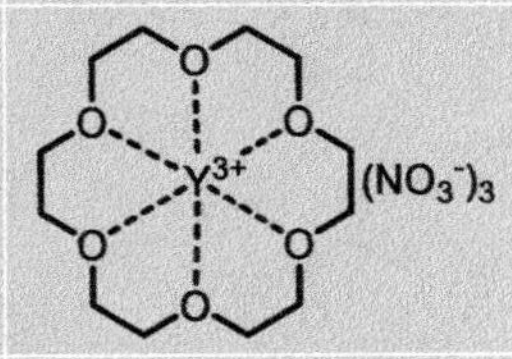	330	3.52	DMSO	35		[305]
$[Y(EGTA)(H_2O)]^-$	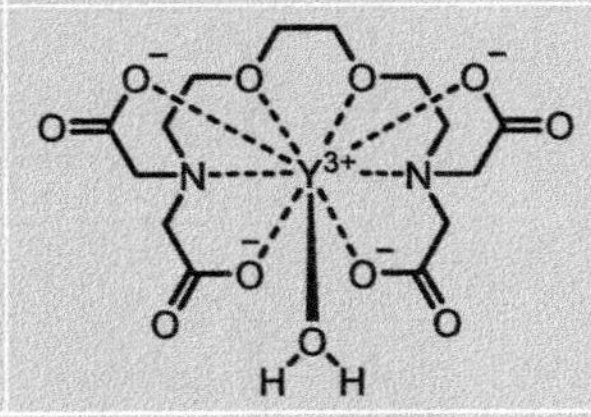	474	9.4	H_2O	25	7.4	[407]

$[Y(NTA)_2]^{3-}$	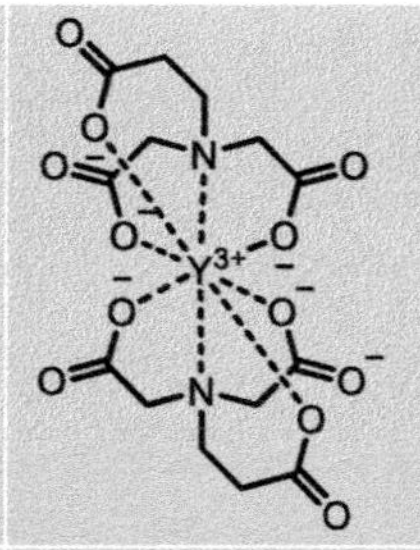	534	9.4	H_2O	25	7.4	[407]
$[Y(DTPA)(H_2O)]^{2-}$	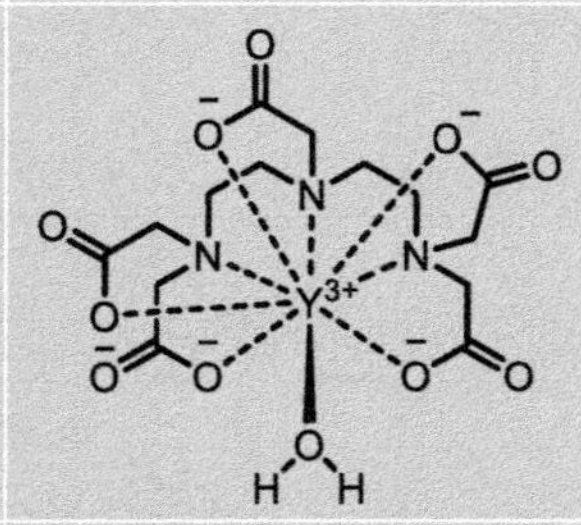	227	9.4	H_2O	25	7.4	[407]
		451	14.1				[406]
$[Y(DO3A)(H_2O)_2]$	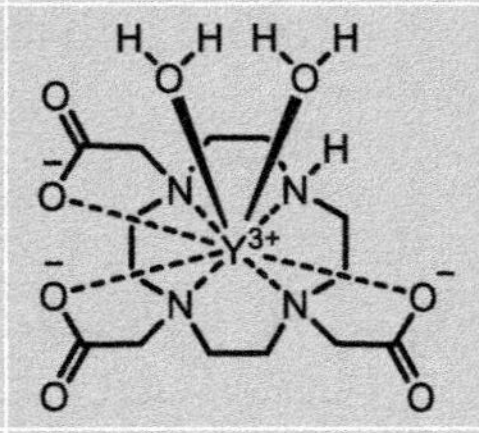	247	9.4	H_2O	25	7.4	[407]
$[Y(EDTA)(H_2O)_2]^-$	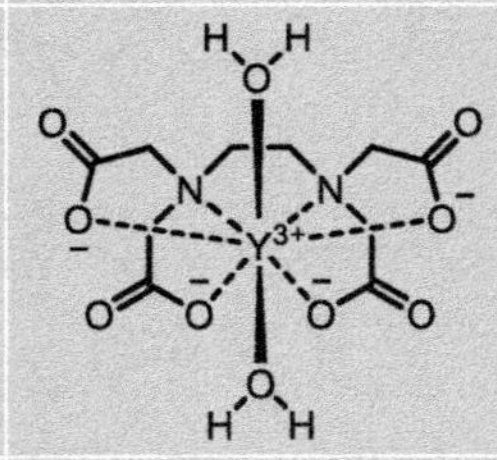	310	9.4	H_2O	25	7.4	[407]
$[Y(DOTA)(H_2O)]^-$	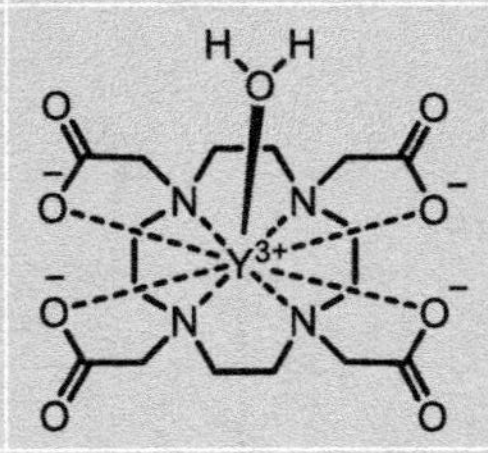	446	9.4	H_2O	25	7.4	[407]
		410	9.4				[115]
		499	14.1				[406]

Continued

TABLE A.6 ^{89}Y T_1 values of selected compounds.—cont'd

Name	Chemical structure	Apparent T_1 (s)	B_0 (T)	Solvent	Temp. (°C)	pH	References
$[Y(TTHA)]^{3-}$	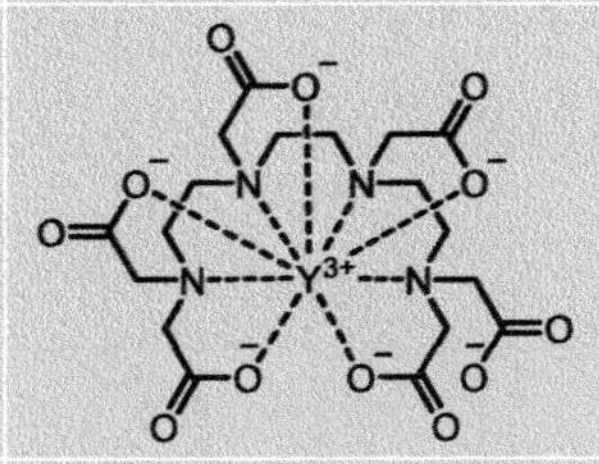	245 (96.9 ppm)	9.4	H_2O	25	7.4	[407]
		230 (103.2 ppm)					
$[Y(DOTP)]^{5-}$	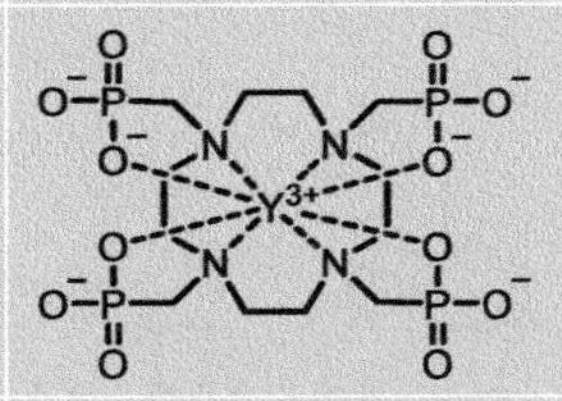	264	14.1	H_2O	25		[406]
		202	9.4			4	[405]
		123				7	
		57				9	
$Y(EDTMP)^{5-}$	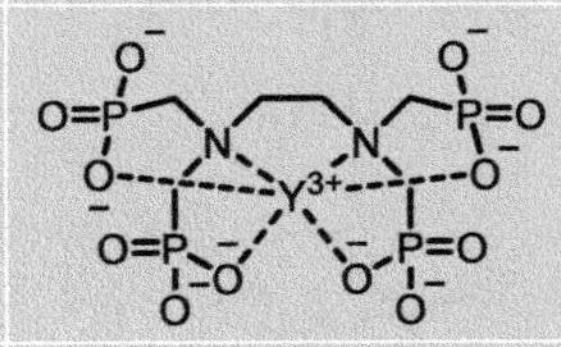	90	9.4	H_2O		7.01	[408]
		80				5.8	
		110				8.3	
Yttrium acetylacetonate $[Y(acac)]_3$	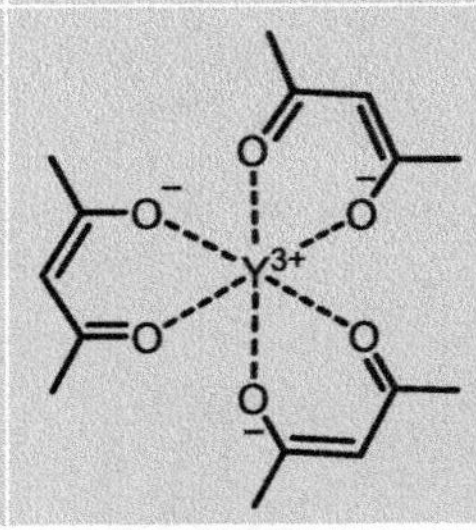	287	9.4	MeOH			[207]

[a]*Non HP.*

TABLE A.7 ^{107}Ag and ^{109}Ag T_1 values of selected compounds, natural abundance mixture of the two isotopes (51.8% ^{107}Ag and 48.2% ^{109}Ag).

Name	Chemical structure	Apparent T_1 (s)	B_0 (T)	Solvent	Temp. (°C)	pH	References
Silver triflate	AgOTf	370 (^{109}Ag)	9.4	H_2O	25		[410]
		424 (^{107}Ag)					
[Ag(cryptand 2.2.2)]OTf	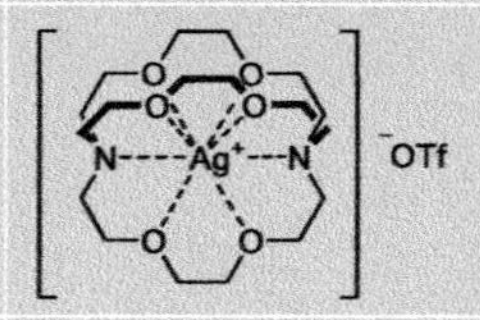	108 (^{109}Ag)	9.4	MeOH	25		[410]
		123 (^{107}Ag)					
[Ag(DOTETE)]OTf	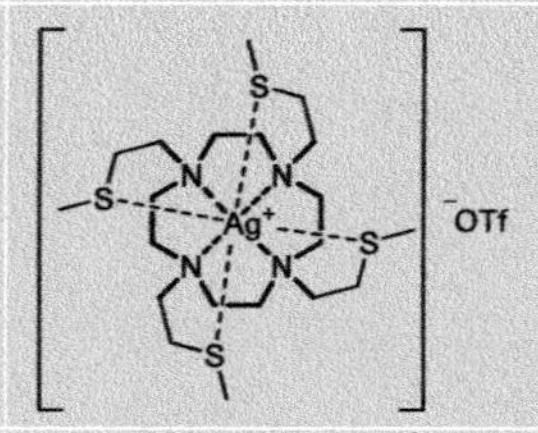	59 (^{109}Ag)	9.4	MeOH	25		[410]
		73 (^{107}Ag)					
[Ag([D_8]bipy)$_2$]OTf	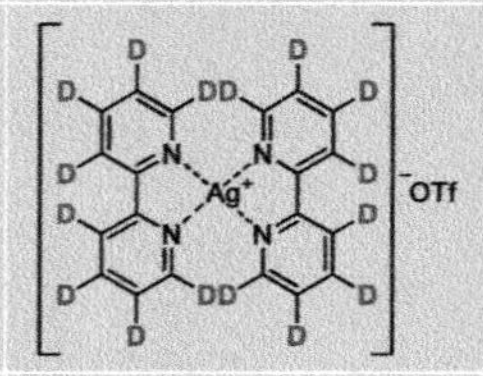	28 (^{109}Ag)	9.4	MeOH	25		[410]
[Ag(bipy)$_2$]OTf	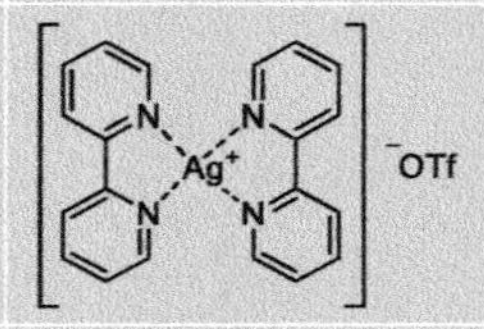	27 (^{109}Ag)	9.4	MeOH	25		[410]
[Ag(dppp)$_2$]OTf	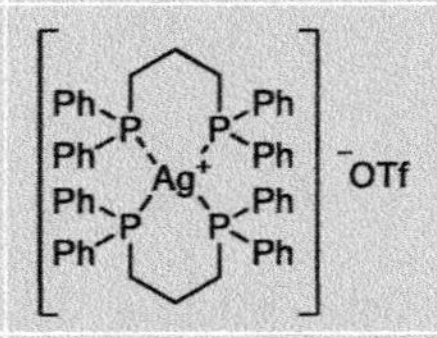	47 (^{109}Ag)	9.4	MeOH	25		[410]
		55 (^{107}Ag)					
Silver nitrate[a]	$AgNO_3$	971 (^{109}Ag)	2.114	H_2O			[883]
		1034 (^{109}Ag)		D_2O			
		1102 (^{109}Ag)		H_2O degassed			
		1186(^{109}Ag)		D_2O degassed			

[a]*Non HP.*

TABLE A.8 ^{129}Xe of T_1 values for several compounds.

Name	Chemical structure	Apparent T_1 (s)	B_0 (T)	Solvent	Temp. (°C)	pH	References
Xenon (gas phase)	^{129}Xe	276	8.0				[884]
Xenon inside Tedlar reservoir	^{129}Xe (natural abundance)	1440–5,040, depending on the surface-to-volume ratio in a single sealed Tedlar bag	2.0	Pure xenon			[411]
^{129}Xe in distilled water	^{129}Xe (natural abundance)	26	2.35		36		[413]
^{129}Xe in blood	^{129}Xe (natural abundance)	4.5	2.35		36		[413]
Xenon (gas phase) in Tedlar bag	^{129}Xe (natural abundance)	21,240 2280	3.0 1.5	38% Xe in N_2			[417]
Xenon in a glass cell, gas phase	^{129}Xe (natural abundance)	1500	1.4		−73		[885]
Xenon in a glass cell treated with $Gd(NO_3)_3$, gas phase	^{129}Xe (natural abundance)	240	1.4		−73		[885]
Xenon in a glass cell, liquid phase	^{129}Xe (natural abundance)	1506	1.4		−73		[885]
Xenon in a glass cell treated with $Gd(NO_3)_3$, liquid phase	^{129}Xe (natural abundance)	1314	1.4		−73		[885]

TABLE A.9 ^{133}Cs T_1 values of selected compounds.

Name	Chemical structure	Apparent T_1 (s)	B_0 (T)	Solvent	Temp. (°C)	pH	References
Cesium [1-^{13}C]acetate with Gd(III) (ProHance)		16 (^{133}Cs) 75 (^{13}C)	9.4	D_2O	45		[464]
Cesium [1-^{13}C]acetate without Gd(III) (ProHance)		16	9.4	D_2O	45		[464]
Cesium acetate with Gd(III) (ProHance)		14	9.4	H_2O	45		[464]
Cesium chloride	$^{133}CsCl$	9–14 (outside RBCs) 1–3 (inside RBCs)	9.4	In red blood cell suspension			[466]
Cesium hydride[a]	^{133}CsH	1000 (H) 300 (Cs)		Solid-state			[885]
Cesium fluoride[a]	^{133}CsF	463 (Cs) 50 (F)		Solid-state			[886]
Cesium halide[a]	^{133}Cs halide	13.3		H_2O			[887]

[a]*Non HP.*

References

[1] N. Zumbulyadis, Whence spin? Concepts Magn. Reson. Part A Bridg. Educ. Res. 3 (2) (1991) 89–107.

[2] R.K. Harris, E.D. Becker, S.M.C. De Menezes, R. Goodfellow, P. Granger, NMR nomenclature. Nuclear spin properties and conventions for chemical shifts (IUPAC Recommendations 2001), Pure Appl. Chem. 73 (11) (2001) 1795–1818.

[3] L. Ronconi, P.J. Sadler, Applications of heteronuclear NMR spectroscopy in biological and medicinal inorganic chemistry, Coord. Chem. Rev. 252 (21–22) (2008) 2239–2277.

[4] A. Abragam, The Principles of Nuclear Magnetism, Oxford University Press, 1961.

[5] C.P. Slichter, Principles of Magnetic Resonance, vol. 1, Springer Science & Business Media, 2013.

[6] R.R. Ernst, G. Bodenhausen, A. Wokaun, Principles of Nuclear Magnetic Resonance in One and Two Dimensions, 1987.

[7] B. Cowan, B. Cowan, Nuclear Magnetic Resonance and Relaxation, vol. 427, Cambridge University Press Cambridge, 1997.

[8] M.H. Levitt, P. Tekely, Spin Dynamics: Basics of Nuclear Magnetic Resonance, John Wiley & Sons, Wiley Online Library, Chichester, UK, 2001, p. 686.

[9] J.B. Lambert, E.P. Mazzola, C.D. Ridge, Nuclear Magnetic Resonance Spectroscopy: An Introduction to Principles, Applications, and Experimental Methods, John Wiley & Sons, 2019.

[10] J.W. Akitt, B.E. Mann, NMR and Chemistry: An Introduction to Modern NMR Spectroscopy, CRC Press, 2017.

[11] J.H. Lee, Y. Okuno, S. Cavagnero, Sensitivity enhancement in solution NMR: emerging ideas and new frontiers, J. Magn. Reson. 241 (2014) 18–31.

[12] R.D. Alvares, D.A. Szulc, H.-L.M. Cheng, A scale to measure MRI contrast agent sensitivity, Sci. Rep. 7 (1) (2017) 15493.

[13] D.I. Hoult, R. Richards, The signal-to-noise ratio of the nuclear magnetic resonance experiment, J. Magn. Reson. 24 (1) (1976) 71–85.

[14] H. Hill, R. Richards, Limits of measurement in magnetic resonance, J. Phys. E: Sci. Instr. 1 (10) (1968) 977.

[15] R. Chidambaram, Signal to noise ratio in nuclear magnetic resonance, Proc. Phys. Soc. 75 (1) (1960) 163.

[16] A. Webb, Increasing the sensitivity of magnetic resonance spectroscopy and imaging, Anal. Chem. 84 (1) (2012) 9–16.

[17] N. Spengler, P.T. While, M.V. Meissner, U. Wallrabe, J.G. Korvink, Magnetic Lenz lenses improve the limit-of-detection in nuclear magnetic resonance, PLoS One 12 (8) (2017) e0182779.

[18] T.L. Peck, R.L. Magin, P.C. Lauterbur, Design and analysis of microcoils for NMR microscopy, J. Magn. Reson. B 108 (2) (1995) 114–124.

[19] H. Kovacs, D. Moskau, M. Spraul, Cryogenically cooled probes—a leap in NMR technology, Prog. Nucl. Magn. Reson. Spectrosc. 2 (46) (2005) 131–155.

[20] L.G. Hanson, Is quantum mechanics necessary for understanding magnetic resonance? Concepts Magn. Reson. Part A Bridg. Educ. Res. 32 (5) (2008) 329–340.

[21] R. Freeman, Handbook of Nuclear Magnetic Resonance, 1987.

[22] P. Lepucki, A.P. Dioguardi, D. Karnaushenko, O.G. Schmidt, H.-J. Grafe, The normalized limit of detection in NMR spectroscopy, J. Magn. Reson. 332 (2021) 107077.

[23] M.E. Lacey, R. Subramanian, D.L. Olson, A.G. Webb, J.V. Sweedler, High-resolution NMR spectroscopy of sample volumes from 1 nL to 10 μL, Chem. Rev. 99 (10) (1999) 3133–3152.

[24] P.R. Jensen, S. Meier, J.H. Ardenkjær-Larsen, J.Ø. Duus, M. Karlsson, M.H. Lerche, Detection of low-populated reaction intermediates with hyperpolarized NMR, Chem. Commun. 34 (2009) 5168–5170.

[25] P.R. Jensen, F. Sannelli, L.T. Stauning, S. Meier, Enhanced 13 C NMR detects extended reaction networks in living cells, Chem. Commun. 57 (81) (2021) 10572–10575.

[26] E.H. Suh, J.M. Park, L. Lumata, A.D. Sherry, Z. Kovacs, Hyperpolarized 15 N-labeled, deuterated tris (2-pyridylmethyl) amine as an MRI sensor of freely available Zn 2+, Commun. Chem. 3 (1) (2020) 1–9.

[27] K. Hashi, S. Ohki, S. Matsumoto, G. Nishijima, A. Goto, K. Deguchi, K. Yamada, T. Noguchi, S. Sakai, M. Takahashi, Y. Yanagisawa, S. Iguchi, T. Yamazaki, H. Maeda, R. Tanaka, T. Nemoto, H. Suematsu, T. Miki, K. Saito, T. Shimizu, Achievement of 1020MHz NMR, J. Magn. Reson. 256 (2015) 30–33.

[28] H. Schwalbe, New 1.2 GHz NMR Spectrometers—New Horizons?, vol. 56, Wiley Online Library, 2017, pp. 10252–10253.

[29] R.R. Ernst, W.A. Anderson, Application of Fourier transform spectroscopy to magnetic resonance, Rev. Sci. Instrum. 37 (1) (1966) 93–102.

[30] O. Sørensen, G. Eich, M.H. Levitt, G. Bodenhausen, R. Ernst, Product operator formalism for the description of NMR pulse experiments, Prog. Nucl. Magn. Reson. Spectrosc. 16 (1984) 163–192.

[31] R.R. Ernst, Nuclear magnetic resonance Fourier transform spectroscopy (Nobel lecture), Angew. Chem. Int. Ed. Engl. 31 (7) (1992) 805–823.

[32] G. Bačić, A. Pavićević, F. Peyrot, In vivo evaluation of different alterations of redox status by studying pharmacokinetics of nitroxides using magnetic resonance techniques, Redox Biol. 8 (2016) 226–242.

[33] J.H. Ardenkjaer-Larsen, G.S. Boebinger, A. Comment, S. Duckett, A.S. Edison, F. Engelke, C. Griesinger, R.G. Griffin, C. Hilty, H. Maeda, Facing and overcoming sensitivity challenges in biomolecular NMR spectroscopy, Angew. Chem. Int. Ed. 54 (32) (2015) 9162–9185.

[34] D.G. Gadian, K.S. Panesar, A.J.P. Linde, A.J. Horsewill, W. Köckenberger, J.R. Owers-Bradley, Preparation of highly polarized nuclear spin systems using brute-force and low-field thermal mixing, Phys. Chem. Chem. Phys. 14 (16) (2012) 5397–5402.

[35] M.L. Hirsch, N. Kalechofsky, A. Belzer, M. Rosay, J.G. Kempf, Brute-force hyperpolarization for NMR and MRI, J. Am. Chem. Soc. 137 (26) (2015) 8428–8434.

[36] L.T. Kuhn, Ü. Akbey, Hyperpolarization Methods in NMR Spectroscopy, vol. 338, Springer, 2013.

[37] P. Nikolaou, B.M. Goodson, E.Y. Chekmenev, NMR hyperpolarization techniques for biomedicine, Chem. Eur J. 21 (8) (2015) 3156–3166.

[38] S. Glöggler, J. Colell, S. Appelt, Para-hydrogen perspectives in hyperpolarized NMR, J. Magn. Reson. 235 (2013) 130–142.

[39] M.E. Halse, Perspectives for hyperpolarisation in compact NMR, TrAC, Trends Anal. Chem. 83 (2016) 76–83.

[40] K.V. Kovtunov, E.V. Pokochueva, O.G. Salnikov, S.F. Cousin, D. Kurzbach, B. Vuichoud, S. Jannin, E.Y. Chekmenev, B.M. Goodson, D.A. Barskiy, I.V. Koptyug, Hyperpolarized NMR spectroscopy: d-DNP, PHIP, and SABRE techniques, Chem. Asian J. 13 (15) (2018) 1857–1871.

[41] Eills J., Budker D., Cavagnero S., Chekmenev E.Y., Elliott S.J., Jannin S., Lesage A., Matysik J.r., Meersmann T., Prisner T. Reimer J.A., Yang H., Koptyug I.V. Spin hyperpolarization in modern magnetic resonance. Chem. Rev. 2023;123(4):1417-1551.

[42] J.B. Hövener, A.N. Pravdivtsev, B. Kidd, C.R. Bowers, S. Glöggler, K.V. Kovtunov, M. Plaumann, R. Katz-Brull, K. Buckenmaier, A. Jerschow, F. Reineri, T. Theis, R.V. Shchepin, S. Wagner, P. Bhattacharya, N.M. Zacharias, E.Y. Chekmenev, Parahydrogen-based hyperpolarization for biomedicine, Angew. Chem. Int. Ed. 57 (35) (2018) 11140–11162.

[43] A. Abragam, M. Goldman, Principles of dynamic nuclear polarisation, Rep. Prog. Phys. 41 (3) (1978) 395.

[44] W.T. Wenckebach, Essentials of Dynamic Nuclear Polarization, Spindrift Publications, 2016.

[45] M.-T. Türke, I. Tkach, M. Reese, P. Höfer, M. Bennati, Optimization of dynamic nuclear polarization experiments in aqueous solution at 15 MHz/9.7 GHz: a comparative study with DNP at 140 MHz/94 GHz, Phys. Chem. Chem. Phys. 12 (22) (2010) 5893–5901.

[46] S. Jannin, A. Comment, J. Van der Klink, Dynamic nuclear polarization by thermal mixing under partial saturation, Appl. Magn. Reson. 43 (1) (2012) 59–68.

[47] T. Maly, G.T. Debelouchina, V.S. Bajaj, K.-N. Hu, C.-G. Joo, M.L. Mak–Jurkauskas, J.R. Sirigiri, P.C. Van Der Wel, J. Herzfeld, R.J. Temkin, R.G. Griffin, Dynamic nuclear polarization at high magnetic fields, J. Chem. Phys. 128 (5) (2008) 02B611.

[48] K.-N. Hu, Polarizing agents and mechanisms for high-field dynamic nuclear polarization of frozen dielectric solids, Solid State Nucl. Magn. Reson. 40 (2) (2011) 31–41.

[49] C. Jeffries, Polarization of nuclei by resonance saturation in paramagnetic crystals, Phys. Rev. 106 (1) (1957) 164.

[50] B. Provotorov, Theory of double magnetic resonance in solids, Phys. Rev. 128 (1) (1962) 75.

[51] M. Goldman, A. Landesman, Dynamic polarization by thermal mixing between two spin systems, Phys. Rev. 132 (2) (1963) 610.

[52] G. Liu, M. Levien, N. Karschin, G. Parigi, C. Luchinat, M. Bennati, One-thousand-fold enhancement of high field liquid nuclear magnetic resonance signals at room temperature, Nat. Chem. 9 (7) (2017) 676–680.

[53] A.S.L. Thankamony, J.J. Wittmann, M. Kaushik, B. Corzilius, Dynamic nuclear polarization for sensitivity enhancement in modern solid-state NMR, Prog. Nucl. Magn. Reson. Spectrosc. 102 (2017) 120–195.

[54] M. Kaushik, T. Bahrenberg, T.V. Can, M.A. Caporini, R. Silvers, J. Heiliger, A.A. Smith, H. Schwalbe, R.G. Griffin, B. Corzilius, Gd (III) and Mn (II) complexes for dynamic nuclear polarization: small molecular chelate polarizing agents and applications with site-directed spin labeling of proteins, Phys. Chem. Chem. Phys. 18 (39) (2016) 27205–27218.

[55] W. de Boer, Dynamic orientation of nuclei at low temperatures, J. Low Temp. Phys. 22 (1) (1976) 185–212.

[56] M.-T. Türke, G. Parigi, C. Luchinat, M. Bennati, Overhauser DNP with 15 N labelled Frémy's salt at 0.35 Tesla, Phys. Chem. Chem. Phys. 14 (2) (2012) 502–510.

[57] H. Jóhannesson, S. Macholl, J.H. Ardenkjaer-Larsen, Dynamic nuclear polarization of [1-13C] pyruvic acid at 4.6 tesla, J. Magn. Reson. 197 (2) (2009) 167–175.

[58] A. Comment, B. van den Brandt, K. Uffmann, F. Kurdzesau, S. Jannin, J. Konter, P. Hautle, W.T. Wenckebach, R. Gruetter, J. Van Der Klink, Design and performance of a DNP prepolarizer coupled to a rodent MRI scanner, Concepts Magn. Reson. Part B Magn. Reson. Eng. 31 (4) (2007) 255–269.

[59] S. Jannin, A. Comment, F. Kurdzesau, J. Konter, P. Hautle, B. Van Den Brandt, J. Van Der Klink, A 140 GHz prepolarizer for dissolution dynamic nuclear polarization, J. Chem. Phys. 128 (24) (2008) 241102.

[60] W.T. Wenckebach, Thermodynamics of dynamic nuclear polarization, Nucl. Instrum. Methods. Phys. Res. Sect. A 356 (1) (1995) 1–4.

[61] A.W. Overhauser, Polarization of nuclei in metals, Phys. Rev. 92 (2) (1953) 411.

[62] T.R. Carver, C.P. Slichter, Polarization of nuclear spins in metals, Phys. Rev. 92 (1) (1953) 212.

[63] C.P. Slichter, The discovery and renaissance of dynamic nuclear polarization, Rep. Prog. Phys. 77 (7) (2014) 072501.

[64] D. Crabb, W. Meyer, Solid polarized targets for nuclear and particle physics experiments, Annu. Rev. Nucl. Part Sci. 47 (1) (1997) 67–109.

[65] W. Meyer, G. Reicherz, Polarized H, D and 3He targets for particle physics experiments, Prog. Part. Nucl. Phys. 49 (2) (2002) 403–489.

[66] W. de Boer, M. Borghini, K. Morimoto, T. Niinikoski, F. Udo, Dynamic polarization of protons, deuterons, and carbon-13 nuclei: thermal contact between nuclear spins and an electron spin-spin interaction reservoir, J. Low Temp. Phys. 15 (3) (1974) 249–267.

[67] L.R. Becerra, G.J. Gerfen, R.J. Temkin, D.J. Singel, R.G. Griffin, Dynamic nuclear polarization with a cyclotron resonance maser at 5 T, Phys. Rev. Lett. 71 (21) (1993) 3561.

[68] D.A. Hall, D.C. Maus, G.J. Gerfen, S.J. Inati, L.R. Becerra, F.W. Dahlquist, R.G. Griffin, Polarization-enhanced NMR spectroscopy of biomolecules in frozen solution, Science 276 (5314) (1997) 930–932.

[69] Q.Z. Ni, E. Daviso, T.V. Can, E. Markhasin, S.K. Jawla, T.M. Swager, R.J. Temkin, J. Herzfeld, R.G. Griffin, High frequency dynamic nuclear polarization, Acc. Chem. Res. 46 (9) (2013) 1933–1941.

[70] I. Kaminker, Recent advances in magic angle spinning-dynamic nuclear polarization methodology, Isr. J. Chem. 59 (11–12) (2019) 990–1000.

[71] M. Rosay, M. Blank, F. Engelke, Instrumentation for solid-state dynamic nuclear polarization with magic angle spinning NMR, J. Magn. Reson. 264 (2016) 88–98.

[72] K.R. Thurber, W.-M. Yau, R. Tycko, Low-temperature dynamic nuclear polarization at 9.4 T with a 30 mW microwave source, J. Magn. Reson. 204 (2) (2010) 303–313.
[73] K.R. Thurber, A. Potapov, W.-M. Yau, R. Tycko, Solid state nuclear magnetic resonance with magic-angle spinning and dynamic nuclear polarization below 25 K, J. Magn. Reson. 226 (2013) 100–106.
[74] D. Lee, E. Bouleau, P. Saint-Bonnet, S. Hediger, G. De Paëpe, Ultra-low temperature MAS-DNP, J. Magn. Reson. 264 (2016) 116–124.
[75] Y. Matsuki, T. Idehara, J. Fukazawa, T. Fujiwara, Advanced instrumentation for DNP-enhanced MAS NMR for higher magnetic fields and lower temperatures, J. Magn. Reson. 264 (2016) 107–115.
[76] K. Jaudzems, A. Bertarello, S.R. Chaudhari, A. Pica, D. Cala-De Paepe, E. Barbet-Massin, A.J. Pell, I. Akopjana, S. Kotelovica, D. Gajan, O. Ouari, K. Tars, G. Pintacuda, A. Lesage, Dynamic nuclear polarization-enhanced biomolecular NMR spectroscopy at high magnetic field with fast magic-angle spinning, Angew. Chem. Int. Ed. 57 (25) (2018) 7458–7462.
[77] R. Rogawski, A.E. McDermott, New NMR tools for protein structure and function: spin tags for dynamic nuclear polarization solid state NMR, Arch. Biochem. Biophys. 628 (2017) 102–113.
[78] A. Barnes, G. De Paëpe, P. Van Der Wel, K.-N. Hu, C.-G. Joo, V. Bajaj, M. Mak-Jurkauskas, J. Sirigiri, J. Herzfeld, R. Temkin, R.G. Griffin, High-field dynamic nuclear polarization for solid and solution biological NMR, Appl. Magn. Reson. 34 (3) (2008) 237–263.
[79] S.Y. Liao, M. Lee, T. Wang, I.V. Sergeyev, M. Hong, Efficient DNP NMR of membrane proteins: sample preparation protocols, sensitivity, and radical location, J. Biomol. NMR 64 (3) (2016) 223–237.
[80] G.T. Debelouchina, M.J. Bayro, A.W. Fitzpatrick, V. Ladizhansky, M.T. Colvin, M.A. Caporini, C.P. Jaroniec, V.S. Bajaj, M. Rosay, C.E. MacPhee, M. Vendruscolo, W.E. Maas, C.M. Dobson, R.G. Griffin, Higher order amyloid fibril structure by MAS NMR and DNP spectroscopy, J. Am. Chem. Soc. 135 (51) (2013) 19237–19247.
[81] D. Gauto, O. Dakhlaoui, I. Marin-Montesinos, S. Hediger, G. De Paëpe, Targeted DNP for biomolecular solid-state NMR, Chem. Sci. 12 (18) (2021) 6223–6237.
[82] C.-G. Joo, K.-N. Hu, J.A. Bryant, R.G. Griffin, In situ temperature jump high-frequency dynamic nuclear polarization experiments: enhanced sensitivity in liquid-state NMR spectroscopy, J. Am. Chem. Soc. 128 (29) (2006) 9428–9432.
[83] C. Song, K.-N. Hu, C.-G. Joo, T.M. Swager, R.G. Griffin, TOTAPOL: a biradical polarizing agent for dynamic nuclear polarization experiments in aqueous media, J. Am. Chem. Soc. 128 (35) (2006) 11385–11390.
[84] J.H. Ardenkjaer-Larsen, B. Fridlund, A. Gram, G. Hansson, L. Hansson, M.H. Lerche, R. Servin, M. Thaning, K. Golman, Increase in signal-to-noise ratio of $>$ 10,000 times in liquid-state NMR, Proc. Natl. Acad. Sci. USA 100 (18) (2003) 10158–10163.
[85] J.H. Ardenkjaer-Larsen, H. Jóhannesson, J.S. Petersson, J. Wolber, Hyperpolarized molecules in solution, in: In Vivo NMR Imaging: Methods and Protocols, 2011, pp. 205–226.
[86] W. Köckenberger, Dissolution Dynamic Nuclear Polarization, eMagRes, 2014, pp. 161–170.
[87] F. Jähnig, G. Kwiatkowski, M. Ernst, Conceptual and instrumental progress in dissolution DNP, J. Magn. Reson. 264 (2016) 22–29.
[88] J. Wolber, F. Ellner, B. Fridlund, A. Gram, H. Johannesson, G. Hansson, L. Hansson, M.H. Lerche, S. Månsson, R. Servin, M. Thaning, K. Golman, J.H. Ardenkjær-Larsen, Generating highly polarized nuclear spins in solution using dynamic nuclear polarization, Nucl. Instrum. Methods. Phys. Res. Sect. A 526 (1–2) (2004) 173–181.
[89] L. Lumata, C. Yang, M. Ragavan, N. Carpenter, R.J. DeBerardinis, M.E. Merritt, Hyperpolarized (13)C magnetic resonance and its use in metabolic assessment of cultured cells and perfused organs, Methods Enzymol. 561 (2015) 73–106.
[90] S. Bowen, J.H. Ardenkjaer-Larsen, Enhanced performance large volume dissolution-DNP, J. Magn. Reson. 240 (2014) 90–94.
[91] J.H. Ardenkjaer-Larsen, A.M. Leach, N. Clarke, J. Urbahn, D. Anderson, T.W. Skloss, Dynamic nuclear polarization polarizer for sterile use intent, NMR Biomed. 24 (8) (2011) 927–932.
[92] A.C. Pinon, A. Capozzi, J.H. Ardenkjær-Larsen, Hyperpolarization via dissolution dynamic nuclear polarization: new technological and methodological advances, Magn. Reson. Mater. Phys. Biol. Med. 34 (1) (2021) 5–23.
[93] R.M. Malinowski, K.W. Lipsø, M.H. Lerche, J.H. Ardenkjær-Larsen, Dissolution Dynamic Nuclear Polarization capability study with fluid path, J. Magn. Reson. 272 (2016) 141–146.
[94] T.P. Lê, J.-N. Hyacinthe, A. Capozzi, How to improve the efficiency of a traditional dissolution dynamic nuclear polarization (dDNP) apparatus: design and performance of a fluid path compatible dDNP/LOD-ESR probe, J. Magn. Reson. 338 (2022) 107197.
[95] J.H. Ardenkjær-Larsen, S. Bowen, J.R. Petersen, O. Rybalko, M.S. Vinding, M. Ullisch, N.C. Nielsen, Cryogen-free dissolution dynamic nuclear polarization polarizer operating at 3.35 T, 6.70 T, and 10.1 T, Magn. Reson. Med. 81 (3) (2019) 2184–2194.
[96] A. Ferrari, J. Peters, M. Anikeeva, A. Pravdivtsev, F. Ellermann, K. Them, O. Will, E. Peschke, H. Yoshihara, O. Jansen, J.-B. Hövener, Performance and reproducibility of 13C and 15N hyperpolarization using a cryogen-free DNP polarizer, Sci. Rep. 12 (1) (2022) 11694.
[97] Q. Stern, S.F. Cousin, F. Mentink-Vigier, A.C. Pinon, S.J. Elliott, O. Cala, S. Jannin, Direct observation of hyperpolarization breaking through the spin diffusion barrier, Sci. Adv. 7 (18) (2021) eabf5735.
[98] M. Ceillier, O. Cala, T. El Daraï, S.F. Cousin, Q. Stern, S. Guibert, S.J. Elliott, A. Bornet, B. Vuichoud, J. Milani, C. Pages, D. Eshchenko, J.G. Kempf, C. Jose, S.A. Lambert, S. Jannin, An automated system for fast transfer and injection of hyperpolarized solutions, J. Magn. Reson. Open 8 (2021) 100017.
[99] H.A. Yoshihara, E. Can, M. Karlsson, M.H. Lerche, J. Schwitter, A. Comment, High-field dissolution dynamic nuclear polarization of [1-13 C] pyruvic acid, Phys. Chem. Chem. Phys. 18 (18) (2016) 12409–12413.
[100] A. Comment, B. Van Den Brandt, K. Uffmann, F. Kurdzesau, S. Jannin, J. Konter, P. Hautle, W.T. Wenckebach, R. Gruetter, J. Van Der Klink, Principles of operation of a DNP prepolarizer coupled to a rodent MRI scanner, Appl. Magn. Reson. 34 (3) (2008) 313–319.

[101] L.L. Lumata, R. Martin, A.K. Jindal, Z. Kovacs, M.S. Conradi, M.E. Merritt, Development and performance of a 129-GHz dynamic nuclear polarizer in an ultra-wide bore superconducting magnet, Magn. Reson. Mater. Phys. Biol. Med. 28 (2) (2015) 195–205.

[102] A. Kiswandhi, P. Niedbalski, C. Parish, S. Ferguson, D. Taylor, G. McDonald, L. Lumata, Construction and 13C hyperpolarization efficiency of a 180 GHz dissolution dynamic nuclear polarization system, Magn. Reson. Chem. 55 (9) (2017) 828–836.

[103] Y. Crémillieux, F. Goutailler, B. Montcel, D. Grand, G. Vermeulen, P.-E. Wolf, A super-wide bore DNP system for multiple sample polarization: cryogenic performance and polarization at low temperature, Appl. Magn. Reson. 43 (1) (2012) 167–180.

[104] A. Kiswandhi, P. Niedbalski, C. Parish, Q. Wang, L. Lumata, Assembly and performance of a 6.4 T cryogen-free dynamic nuclear polarization system, Magn. Reson. Chem. 55 (9) (2017) 846–852.

[105] M. Baudin, B. Vuichoud, A. Bornet, G. Bodenhausen, S. Jannin, A cryogen-consumption-free system for dynamic nuclear polarization at 9.4 T, J. Magn. Reson. 294 (2018) 115–121.

[106] T. Cheng, A.P. Gaunt, I. Marco-Rius, M. Gehrung, A.P. Chen, J.J. van der Klink, A. Comment, A multisample 7 T dynamic nuclear polarization polarizer for preclinical hyperpolarized MR, NMR Biomed. 33 (5) (2020) e4264.

[107] M. Batel, M. Krajewski, K. Weiss, O. With, A. Däpp, A. Hunkeler, M. Gimersky, K.P. Pruessmann, P. Boesiger, B.H. Meier, S. Kozerke, M. Ernst, A multi-sample 94 GHz dissolution dynamic-nuclear-polarization system, J. Magn. Reson. 214 (2012) 166–174.

[108] R. Balzan, L. Fernandes, A. Comment, L. Pidial, B. Tavitian, P.R. Vasos, Dissolution dynamic nuclear polarization instrumentation for real-time enzymatic reaction rate measurements by NMR, JoVE 108 (2016) e53548.

[109] K. Kouřil, H. Kouřilová, S. Bartram, M.H. Levitt, B. Meier, Scalable dissolution-dynamic nuclear polarization with rapid transfer of a polarized solid, Nat. Commun. 10 (1) (2019) 1–6.

[110] K. Kouřil, M. Gramberg, M. Jurkutat, H. Kouřilová, B. Meier, A cryogen-free, semi-automated apparatus for bullet-dynamic nuclear polarization with improved resolution, Magn. Reson. 2 (2) (2021) 815–825.

[111] J. Leggett, R. Hunter, J. Granwehr, R. Panek, A.J. Perez-Linde, A.J. Horsewill, J. McMaster, G. Smith, W. Köckenberger, A dedicated spectrometer for dissolution DNP NMR spectroscopy, Phys. Chem. Chem. Phys. 12 (22) (2010) 5883–5892.

[112] J.H. Ardenkjaer-Larsen, On the present and future of dissolution-DNP, J. Magn. Reson. 264 (2016) 3–12.

[113] A. Eldirdiri, A. Clemmensen, S. Bowen, A. Kjær, J.H. Ardenkjær-Larsen, Simultaneous imaging of hyperpolarized [1, 4-13C2] fumarate,[1-13C] pyruvate and 18F–FDG in a rat model of necrosis in a clinical PET/MR scanner, NMR Biomed. 30 (12) (2017) e3803.

[114] D.M. Wilson, K.R. Keshari, P.E. Larson, A.P. Chen, S. Hu, M. Van Criekinge, R. Bok, S.J. Nelson, J.M. Macdonald, D.B. Vigneron, J. Kurhanewicz, Multi-compound polarization by DNP allows simultaneous assessment of multiple enzymatic activities in vivo, J. Magn. Reson. 205 (1) (2010) 141–147.

[115] L. Lumata, A.K. Jindal, M.E. Merritt, C.R. Malloy, A.D. Sherry, Z. Kovacs, DNP by thermal mixing under optimized conditions yields> 60 000-fold enhancement of 89Y NMR signal, J. Am. Chem. Soc. 133 (22) (2011) 8673–8680.

[116] L. Lumata, M. Merritt, C. Malloy, A.D. Sherry, Z. Kovacs, Fast dissolution dynamic nuclear polarization NMR of 13C-enriched 89Y-DOTA complex: experimental and theoretical considerations, Appl. Magn. Reson. 43 (1) (2012) 69–79.

[117] J.L. Muñoz-Gómez, E. Monteagudo, V. Lloveras, T. Parella, J. Veciana, J. Vidal-Gancedo, Optimized polarization build-up times in dissolution DNP-NMR using a benzyl amino derivative of BDPA, RSC Adv. 6 (32) (2016) 27077–27082.

[118] A. Flori, G. Giovannetti, M.F. Santarelli, G.D. Aquaro, D. De Marchi, S. Burchielli, F. Frijia, V. Positano, L. Landini, L. Menichetti, Biomolecular imaging of 13C-butyrate with dissolution-DNP: polarization enhancement and formulation for in vivo studies, Spectrochim. Acta 199 (2018) 153–160.

[119] X. Ji, A. Bornet, B. Vuichoud, J. Milani, D. Gajan, A.J. Rossini, L. Emsley, G. Bodenhausen, S. Jannin, Transportable hyperpolarized metabolites, Nat. Commun. 8 (1) (2017) 1–7.

[120] T.-C. Ong, M.L. Mak-Jurkauskas, J.J. Walish, V.K. Michaelis, B.R. Corzilius, A.A. Smith, A.M. Clausen, J.C. Cheetham, T.M. Swager, R.G. Griffin, Solvent-free dynamic nuclear polarization of amorphous and crystalline ortho-terphenyl, J. Phys. Chem. B 117 (10) (2013) 3040–3046.

[121] R.E. Hurd, Y.F. Yen, A. Chen, J.H. Ardenkjaer-Larsen, Hyperpolarized 13C metabolic imaging using dissolution dynamic nuclear polarization, J. Magn. Reson. Imaging 36 (6) (2012) 1314–1328.

[122] S.J. Elliott, Q. Stern, M. Ceillier, T. El Daraï, S.F. Cousin, O. Cala, S. Jannin, Practical dissolution dynamic nuclear polarization, Prog. Nucl. Magn. Reson. Spectrosc. 126 (2021) 59–100.

[123] T. Cheng, M. Mishkovsky, J.A. Bastiaansen, O. Ouari, P. Hautle, P. Tordo, B. van den Brandt, A. Comment, Automated transfer and injection of hyperpolarized molecules with polarization measurement prior to in vivo NMR, NMR Biomed. 26 (11) (2013) 1582–1588.

[124] S. Reynolds, A. Bucur, M. Port, T. Alizadeh, S.M. Kazan, G.M. Tozer, M.N. Paley, A system for accurate and automated injection of hyperpolarized substrate with minimal dead time and scalable volumes over a large range, J. Magn. Reson. 239 (2014) 1–8.

[125] S. Bowen, C. Hilty, Time-resolved dynamic nuclear polarization enhanced NMR spectroscopy, Angew. Chem. Int. Ed. 47 (28) (2008) 5235–5237.

[126] E.M. Serrao, K.M. Brindle, Dynamic nuclear polarisation: the future of imaging in oncology? Porto Biomed. J. 2 (3) (2017) 71–75.

[127] S. Bowen, C. Hilty, Rapid sample injection for hyperpolarized NMR spectroscopy, Phys. Chem. Chem. Phys. 12 (22) (2010) 5766–5770.

[128] S. Katsikis, I. Marin-Montesinos, M. Pons, C. Ludwig, U.L. Günther, Improved stability and spectral quality in ex situ dissolution DNP using an improved transfer device, Appl. Magn. Reson. 46 (7) (2015) 723–729.

[129] G. Olsen, E. Markhasin, O. Szekely, C. Bretschneider, L. Frydman, Optimizing water hyperpolarization and dissolution for sensitivity-enhanced 2D biomolecular NMR, J. Magn. Reson. 264 (2016) 49–58.

[130] H.Y. Chen, C. Hilty, Implementation and characterization of flow injection in dissolution dynamic nuclear polarization NMR spectroscopy, ChemPhysChem 16 (12) (2015) 2646–2652.

[131] E.T. Peterson, J.W. Gordon, M.G. Erickson, S.B. Fain, I.J. Rowland, Dynamic nuclear polarization system output volume reduction using inert fluids, J. Magn. Reson. Imaging 33 (4) (2011) 1003–1008.

[132] T. Harris, C. Bretschneider, L. Frydman, Dissolution DNP NMR with solvent mixtures: substrate concentration and radical extraction, J. Magn. Reson. 211 (1) (2011) 96–100.

[133] J.Y. Lau, A.P. Chen, Y.P. Gu, C.H. Cunningham, A calibration-based approach to real-time in vivo monitoring of pyruvate C1 and C2 polarization using the JCC spectral asymmetry, NMR Biomed. 26 (10) (2013) 1233–1241.

[134] B. Vuichoud, J. Milani, Q. Chappuis, A. Bornet, G. Bodenhausen, S. Jannin, Measuring absolute spin polarization in dissolution-DNP by spin PolarimetrY magnetic resonance (SPY-MR), J. Magn. Reson. 260 (2015) 127–135.

[135] W. Meyer, J. Heckmann, C. Hess, E. Radtke, G. Reicherz, L. Triebwasser, L. Wang, Dynamic polarization of 13C nuclei in solid 13C labeled pyruvic acid, Nucl. Instrum. Methods. Phys. Res. Sect. A 631 (1) (2011) 1–5.

[136] L. Lumata, Z. Kovacs, A.D. Sherry, C. Malloy, S. Hill, J. Van Tol, L. Yu, L. Song, M.E. Merritt, Electron spin resonance studies of trityl OX063 at a concentration optimal for DNP, Phys. Chem. Chem. Phys. 15 (24) (2013) 9800–9807.

[137] W.T. Wenckebach, T. Swanenburg, N. Poulis, Thermodynamics of spin systems in paramagnetic crystals, Phys. Rep. 14 (5) (1974) 181–255.

[138] A. Bornet, S. Jannin, Optimizing dissolution dynamic nuclear polarization, J. Magn. Reson. 264 (2016) 13–21.

[139] A. Dey, B. Charrier, K. Lemaitre, V. Ribay, D. Eshchenko, M. Schnell, R. Melzi, Q. Stern, S.F. Cousin, J.G. Kempf, S. Jannin, J.N. Dumez, P. Giraudeau, Fine optimization of a dissolution dynamic nuclear polarization experimental setting for 13 C NMR of metabolic samples, Magn. Reson. 3 (2) (2022) 183–202.

[140] A. Bornet, J. Milani, B. Vuichoud, A.J.P. Linde, G. Bodenhausen, S. Jannin, Microwave frequency modulation to enhance dissolution dynamic nuclear polarization, Chem. Phys. Lett. 602 (2014) 63–67.

[141] K.W. Lipsø, S. Bowen, O. Rybalko, J.H. Ardenkjær-Larsen, Large dose hyperpolarized water with dissolution-DNP at high magnetic field, J. Magn. Reson. 274 (2017) 65–72.

[142] Y. Hovav, A. Feintuch, S. Vega, D. Goldfarb, Dynamic nuclear polarization using frequency modulation at 3.34 T, J. Magn. Reson. 238 (2014) 94–105.

[143] Cross-polarization principles, in: R. Kimmich (Ed.), NMR: Tomography, Diffusometry, Relaxometry, Springer, 1997, pp. 360–366.

[144] S. Hartmann, E. Hahn, Nuclear double resonance in the rotating frame, Phys. Rev. 128 (5) (1962) 2042.

[145] S. Jannin, A. Bornet, R. Melzi, G. Bodenhausen, High field dynamic nuclear polarization at 6.7 T: carbon-13 polarization above 70% within 20 min, Chem. Phys. Lett. 549 (2012) 99–102.

[146] A. Bornet, R. Melzi, A.J. Perez Linde, P. Hautle, B. van den Brandt, S. Jannin, G. Bodenhausen, Boosting dissolution dynamic nuclear polarization by cross polarization, J. Phys. Chem. Lett. 4 (1) (2013) 111–114.

[147] S. Jannin, A. Bornet, S. Colombo, G. Bodenhausen, Low-temperature cross polarization in view of enhancing dissolution dynamic nuclear polarization in NMR, Chem. Phys. Lett. 517 (4–6) (2011) 234–236.

[148] A. Bornet, R. Melzi, S. Jannin, G. Bodenhausen, Cross polarization for dissolution dynamic nuclear polarization experiments at readily accessible temperatures 1.2< T< 4.2 K, Appl. Magn. Reson. 43 (1) (2012) 107–117.

[149] M. Batel, M. Krajewski, A. Däpp, A. Hunkeler, B.H. Meier, S. Kozerke, M. Ernst, Dissolution dynamic nuclear polarization efficiency enhanced by Hartmann–Hahn cross polarization, Chem. Phys. Lett. 554 (2012) 72–76.

[150] M. Batel, A. Däpp, A. Hunkeler, B.H. Meier, S. Kozerke, M. Ernst, Cross-polarization for dissolution dynamic nuclear polarization, Phys. Chem. Chem. Phys. 16 (39) (2014) 21407–21416.

[151] A.J.P. Linde, A. Bornet, J. Milani, B. Vuichoud, R. Melzi, S. Jannin, G. Bodenhausen, Cross polarization from 1 H to quadrupolar 6 Li nuclei for dissolution DNP, Phys. Chem. Chem. Phys. 16 (45) (2014) 24813–24817.

[152] S.J. Elliott, O. Cala, Q. Stern, S.F. Cousin, D. Eshchenko, R. Melzi, J.G. Kempf, S. Jannin, Pulse sequence and sample formulation optimization for dipolar order mediated 1 H→ 13 C cross-polarization, Phys. Chem. Chem. Phys. 23 (15) (2021) 9457–9465.

[153] E. Chiavazza, E. Kubala, C.V. Gringeri, S. Düwel, M. Durst, R.F. Schulte, M.I. Menzel, Earth's magnetic field enabled scalar coupling relaxation of 13C nuclei bound to fast-relaxing quadrupolar 14N in amide groups, J. Magn. Reson. 227 (2013) 35–38.

[154] J. Milani, B. Vuichoud, A. Bornet, P. Miéville, R. Mottier, S. Jannin, G. Bodenhausen, A magnetic tunnel to shelter hyperpolarized fluids, Rev. Sci. Instrum. 86 (2) (2015) 024101.

[155] K. Halbach, Design of permanent multipole magnets with oriented rare earth cobalt material, Nucl. Instrum. Methods 169 (1) (1980) 1–10.

[156] T.W. Skloss, J.A. Murray, J.H. Ardenkjær-Larsen, M.K.C. Cherukuri, M. Bernardo, N. Devasahayam, Hyperpolarized Media Transport Vessel, 2015.

[157] J.W. Gordon, S.B. Fain, I.J. Rowland, Effect of lanthanide ions on dynamic nuclear polarization enhancement and liquid-state T1 relaxation, Magn. Reson. Med. 68 (6) (2012) 1949–1954.

[158] J.H. Ardenkjær-Larsen, S. Macholl, H. Johannesson, Dynamic nuclear polarization with trityls at 1.2 K, Appl. Magn. Reson. 34 (3) (2008) 509–522.

[159] E. Ravera, D. Shimon, A. Feintuch, D. Goldfarb, S. Vega, A. Flori, C. Luchinat, L. Menichetti, G. Parigi, The effect of Gd on trityl-based dynamic nuclear polarisation in solids, Phys. Chem. Chem. Phys. 17 (40) (2015) 26969–26978.

[160] A. Kiswandhi, P. Niedbalski, C. Parish, P. Kaur, A. Martins, L. Fidelino, C. Khemtong, L. Song, A.D. Sherry, L. Lumata, Impact of Ho 3+-doping on 13 C dynamic nuclear polarization using trityl OX063 free radical, Phys. Chem. Chem. Phys. 18 (31) (2016) 21351–21359.

[161] L. Lumata, M.E. Merritt, C.R. Malloy, A.D. Sherry, Z. Kovacs, Impact of Gd3+ on DNP of [1-13C] pyruvate doped with trityl OX063, BDPA, or 4-oxo-TEMPO, J. Phys. Chem. A 116 (21) (2012) 5129–5138.

[162] S.A. Walker, D.T. Edwards, T.A. Siaw, B.D. Armstrong, S. Han, Temperature dependence of high field 13 C dynamic nuclear polarization processes with trityl radicals below 35 Kelvin, Phys. Chem. Chem. Phys. 15 (36) (2013) 15106–15120.

[163] F. Jähnig, G. Kwiatkowski, A. Däpp, A. Hunkeler, B.H. Meier, S. Kozerke, M. Ernst, Dissolution DNP using trityl radicals at 7 T field, Phys. Chem. Chem. Phys. 19 (29) (2017) 19196–19204.

[164] X. Wang, J.E. McKay, B. Lama, J. van Tol, T. Li, K. Kirkpatrick, Z. Gan, S. Hill, J.R. Long, H.C. Dorn, Gadolinium based endohedral metallofullerene Gd 2@ C 79 N as a relaxation boosting agent for dissolution DNP at high fields, Chem. Commun. 54 (19) (2018) 2425–2428.

[165] A. Capozzi, S. Patel, W.T. Wenckebach, M. Karlsson, M.H. Lerche, J.H. Ardenkjær-Larsen, Gadolinium effect at high-magnetic-field DNP: 70% 13C polarization of [U-13C] glucose using trityl, J. Phys. Chem. Lett. 10 (12) (2019) 3420–3425.

[166] F. Kurdzesau, B. Van Den Brandt, A. Comment, P. Hautle, S. Jannin, J. Van Der Klink, J. Konter, Dynamic nuclear polarization of small labelled molecules in frozen water–alcohol solutions, J. Phys. D Appl. Phys. 41 (15) (2008) 155506.

[167] L. Lumata, M.E. Merritt, Z. Kovacs, Influence of deuteration in the glassing matrix on 13 C dynamic nuclear polarization, Phys. Chem. Chem. Phys. 15 (19) (2013) 7032–7035.

[168] T. Cheng, A. Capozzi, Y. Takado, R. Balzan, A. Comment, Over 35% liquid-state 13 C polarization obtained via dissolution dynamic nuclear polarization at 7 T and 1 K using ubiquitous nitroxyl radicals, Phys. Chem. Chem. Phys. 15 (48) (2013) 20819–20822.

[169] A. Kiswandhi, B. Lama, P. Niedbalski, M. Goderya, J. Long, L. Lumata, The effect of glassing solvent deuteration and Gd 3+ doping on 13 C DNP at 5 T, RSC Adv. 6 (45) (2016) 38855–38860.

[170] T. El Daraï, S. Jannin, Sample formulations for dissolution dynamic nuclear polarization, Chem. Phys. Rev. 2 (4) (2021) 041308.

[171] A. Flori, M. Liserani, S. Bowen, J.H. Ardenkjaer-Larsen, L. Menichetti, Dissolution dynamic nuclear polarization of non-self-glassing agents: spectroscopy and relaxation of hyperpolarized [1-13C] acetate, J. Phys. Chem. A 119 (10) (2015) 1885–1893.

[172] L. Lumata, Z. Kovacs, C. Malloy, A.D. Sherry, M. Merritt, The effect of 13C enrichment in the glassing matrix on dynamic nuclear polarization of [1-13C] pyruvate, Phys. Med. Biol. 56 (5) (2011) N85.

[173] J.R. Brender, S. Kishimoto, G.R. Eaton, S.S. Eaton, Y. Saida, J. Mitchell, M.C. Krishna, Trehalose as an alternative to glycerol as a glassing agent for in vivo DNP MRI, Magn. Reson. Med. 85 (1) (2021) 42–48.

[174] C. Ni, Y. Gong, X. Liu, C.Q. Sun, Z. Zhou, The anti-frozen attribute of sugar solutions, J. Mol. Liq. 247 (2017) 337–344.

[175] J.C. Fanning, The solubilities of the alkali metal salts and the precipitation of Cs+ from aqueous solution, Coord. Chem. Rev. 140 (1995) 27–36.

[176] F.A. Gallagher, M.I. Kettunen, S.E. Day, D.-E. Hu, J.H. Ardenkjaer-Larsen, P.R. Jensen, M. Karlsson, K. Golman, M.H. Lerche, K.M. Brindle, Magnetic resonance imaging of pH in vivo using hyperpolarized 13 C-labelled bicarbonate, Nature 453 (7197) (2008) 940–943.

[177] W.-Y. Wen, S. Saito, Activity coefficients and molal volumes of two tetraethanolammonium halides in aqueous solutions at 25, J. Phys. Chem. 69 (10) (1965) 3569–3574.

[178] A.P. Abbott, D. Boothby, G. Capper, D.L. Davies, R.K. Rasheed, Deep eutectic solvents formed between choline chloride and carboxylic acids: versatile alternatives to ionic liquids, J. Am. Chem. Soc. 126 (29) (2004) 9142–9147.

[179] A.P. Abbott, G. Capper, D.L. Davies, R.K. Rasheed, V. Tambyrajah, Novel solvent properties of choline chloride/urea mixtures, Chem. Commun 1 (2003) 70–71.

[180] E.L. Smith, A.P. Abbott, K.S. Ryder, Deep eutectic solvents (DESs) and their applications, Chem. Rev. 114 (21) (2014) 11060–11082.

[181] S. Bowen, J.H. Ardenkjaer-Larsen, Formulation and utilization of choline based samples for dissolution dynamic nuclear polarization, J. Magn. Reson. 236 (2013) 26–30.

[182] M. Karlsson, P.R. Jensen, J.Ø. Duus, S. Meier, M.H. Lerche, Development of dissolution DNP-MR substrates for metabolic research, Appl. Magn. Reson. 43 (1) (2012) 223–236.

[183] B. Lama, J.H. Collins, D. Downes, A.N. Smith, J.R. Long, Expeditious dissolution dynamic nuclear polarization without glassing agents, NMR Biomed. 29 (3) (2016) 226–231.

[184] A.J. Rossini, A. Zagdoun, M. Lelli, A. Lesage, C. Coperet, L. Emsley, Dynamic nuclear polarization surface enhanced NMR spectroscopy, Acc. Chem. Res. 46 (9) (2013) 1942–1951.

[185] B. Corzilius, High-field dynamic nuclear polarization, Annu. Rev. Phys. Chem. 71 (2020) 143–170.

[186] D. Banerjee, D. Shimon, A. Feintuch, S. Vega, D. Goldfarb, The interplay between the solid effect and the cross effect mechanisms in solid state 13C DNP at 95 GHz using trityl radicals, J. Magn. Reson. 230 (2013) 212–219.

[187] Y. Hovav, A. Feintuch, S. Vega, Theoretical aspects of dynamic nuclear polarization in the solid state–the cross effect, J. Magn. Reson. 214 (2012) 29–41.

[188] C. Farrar, D. Hall, G. Gerfen, S. Inati, R. Griffin, Mechanism of dynamic nuclear polarization in high magnetic fields, J. Chem. Phys. 114 (11) (2001) 4922–4933.

[189] T.V. Can, M.A. Caporini, F. Mentink-Vigier, B. Corzilius, J.J. Walish, M. Rosay, W.E. Maas, M. Baldus, S. Vega, T.M. Swager, R.G. Griffin, Overhauser effects in insulating solids, J. Chem. Phys. 141 (6) (2014) 064202.

[190] S. Pylaeva, P. Marx, G. Singh, T.D. Kühne, M. Roemelt, H. Elgabarty, Organic mixed-valence compounds and the overhauser effect in insulating solids, J. Phys. Chem. A 125 (3) (2021) 867–874.

[191] M. Serda, Y.-K. Wu, E.D. Barth, H.J. Halpern, V.H. Rawal, EPR imaging spin probe trityl radical OX063: a method for its isolation from animal effluent, redox chemistry of its quinone methide oxidation product, and in vivo application in a mouse, Chem. Res. Toxicol. 29 (12) (2016) 2153–2156.

[192] Y. Song, Y. Liu, W. Liu, F.A. Villamena, J.L. Zweier, Characterization of the binding of the Finland trityl radical with bovine serum albumin, RSC Adv. 4 (88) (2014) 47649–47656.

[193] P. Kuppusamy, P. Wang, M. Chzhan, J.L. Zweier, High resolution electron paramagnetic resonance imaging of biological samples with a single line paramagnetic label, Magn. Reson. Med. 37 (4) (1997) 479–483.

[194] J. Ardenkjaer-Larsen, I. Laursen, I. Leunbach, G. Ehnholm, L.-G. Wistrand, J. Petersson, K. Golman, EPR and DNP properties of certain novel single electron contrast agents intended for oximetric imaging, J. Magn. Reson. 133 (1) (1998) 1–12.

[195] L. Chen, L. Wu, X. Tan, A. Rockenbauer, Y. Song, Y. Liu, Synthesis and redox properties of water-soluble asymmetric trityl radicals, J. Org. Chem. 86 (12) (2021) 8351–8364.

[196] C. Decroos, Y. Li, G. Bertho, Y. Frapart, D. Mansuy, J.-L. Boucher, Oxidative and reductive metabolism of tris (p-carboxyltetrathiaaryl) methyl radicals by liver microsomes, Chem. Res. Toxicol. 22 (7) (2009) 1342–1350.

[197] U. Sanzhaeva, M. Poncelet, O. Tseytlin, M. Tseytlin, M. Gencheva, T.D. Eubank, V.V. Khramtsov, B. Driesschaert, Synthesis, characterization, and application of a highly hydrophilic triarylmethyl radical for biomedical EPR, J. Org. Chem. 85 (16) (2020) 10388–10398.

[198] M. Poncelet, J.L. Huffman, V.V. Khramtsov, I. Dhimitruka, B. Driesschaert, Synthesis of hydroxyethyl tetrathiatriarylmethyl radicals OX063 and OX071, RSC Adv. 9 (60) (2019) 35073–35076.

[199] I. Dhimitruka, M. Velayutham, A.A. Bobko, V.V. Khramtsov, F.A. Villamena, C.M. Hadad, J.L. Zweier, Large-scale synthesis of a persistent trityl radical for use in biomedical EPR applications and imaging, Bioorg. Med. Chem. Lett. 17 (24) (2007) 6801–6805.

[200] T.J. Reddy, T. Iwama, H.J. Halpern, V.H. Rawal, General synthesis of persistent trityl radicals for EPR imaging of biological systems, J. Org. Chem. 67 (14) (2002) 4635–4639.

[201] O.Y. Rogozhnikova, V.G. Vasiliev, T.I. Troitskaya, D.V. Trukhin, T.V. Mikhalina, H.J. Halpern, V.M. Tormyshev, Generation of trityl radicals by nucleophilic quenching of tris (2, 3, 5, 6-tetrathiaaryl) methyl cations and practical and convenient large-scale synthesis of persistent tris (4-carboxy-2, 3, 5, 6-tetrathiaaryl) methyl radical, Eur. J. Org. Chem. 2013 (16) (2013) 3347–3355.

[202] I. Marin-Montesinos, J.C. Paniagua, M. Vilaseca, A. Urtizberea, F. Luis, M. Feliz, F. Lin, S. Van Doorslaer, M. Pons, Self-assembled trityl radical capsules—implications for dynamic nuclear polarization, Phys. Chem. Chem. Phys. 17 (8) (2015) 5785–5794.

[203] I. Ratera, J. Vidal-Gancedo, D. Maspoch, S.T. Bromley, N. Crivillers, M. Mas-Torrent, Perspectives for polychlorinated trityl radicals, J. Mater. Chem. C 9 (33) (2021) 10610–10623.

[204] C. Gabellieri, V. Mugnaini, J.C. Paniagua, N. Roques, M. Oliveros, M. Feliz, J. Veciana, M. Pons, Dynamic nuclear polarization with polychlorotriphenylmethyl radicals: supramolecular polarization-transfer effects, Angew. Chem. 122 (19) (2010) 3432–3434.

[205] J.C. Paniagua, V. Mugnaini, C. Gabellieri, M. Feliz, N. Roques, J. Veciana, M. Pons, Polychlorinated trityl radicals for dynamic nuclear polarization: the role of chlorine nuclei, Phys. Chem. Chem. Phys. 12 (22) (2010) 5824–5829.

[206] P. Demay-Drouhard, H.V. Ching, C. Decroos, R. Guillot, Y. Li, L.C. Tabares, C. Policar, H.C. Bertrand, S. Un, Understanding the g-tensors of perchlorotriphenylmethyl and Finland-type trityl radicals, Phys. Chem. Chem. Phys. 22 (36) (2020) 20792–20800.

[207] L. Lumata, S.J. Ratnakar, A. Jindal, M. Merritt, A. Comment, C. Malloy, A.D. Sherry, Z. Kovacs, BDPA: an efficient polarizing agent for fast dissolution dynamic nuclear polarization NMR spectroscopy, Chem. Eur J. 17 (39) (2011) 10825–10827.

[208] R.B. Khattri, A.A. Sirusi, E.H. Suh, Z. Kovacs, M.E. Merritt, The influence of Ho 3+ doping on 13 C DNP in the presence of BDPA, Phys. Chem. Chem. Phys. 21 (34) (2019) 18629–18635.

[209] E.L. Dane, T.M. Swager, Synthesis of a water-soluble 1, 3-bis (diphenylene)-2-phenylallyl radical, J. Org. Chem. 75 (10) (2010) 3533–3536.

[210] O. Haze, B.r. Corzilius, A.A. Smith, R.G. Griffin, T.M. Swager, Water-soluble narrow-line radicals for dynamic nuclear polarization, J. Am. Chem. Soc. 134 (35) (2012) 14287–14290.

[211] S. Mandal, S.T. Sigurdsson, Water-soluble BDPA radicals with improved persistence, Chem. Commun. 56 (86) (2020) 13121–13124.

[212] J.L. Muñoz-Gómez, E. Monteagudo, V. Lloveras, T. Parella, J. Veciana, J. Vidal-Gancedo, A benzyl alcohol derivative of the BDPA radical for fast dissolution dynamic nuclear polarization NMR spectroscopy, Org. Biomol. Chem. 13 (9) (2015) 2689–2693.

[213] C. Parish, P. Niedbalski, Q. Wang, F. Khashami, Z. Hayati, M. Liu, L. Song, L. Lumata, Effects of glassing matrix deuteration on the relaxation properties of hyperpolarized 13C spins and free radical electrons at cryogenic temperatures, J. Chem. Phys. 150 (23) (2019) 234307.

[214] P. Niedbalski, C. Parish, Q. Wang, A. Kiswandhi, L. Lumata, 13C dynamic nuclear polarization using derivatives of TEMPO free radical, Appl. Magn. Reson. 48 (9) (2017) 933–942.

[215] L. Lumata, M. Merritt, C. Khemtong, S.J. Ratnakar, J. Van Tol, L. Yu, L. Song, Z. Kovacs, The efficiency of DPPH as a polarising agent for DNP-NMR spectroscopy, RSC Adv. 2 (33) (2012) 12812–12817.

[216] L.L. Lumata, M.E. Merritt, C.R. Malloy, A.D. Sherry, J. Van Tol, L. Song, Z. Kovacs, Dissolution DNP-NMR spectroscopy using galvinoxyl as a polarizing agent, J. Magn. Reson. 227 (2013) 14–19.

[217] J. Heckmann, W. Meyer, E. Radtke, G. Reicherz, S. Goertz, Electron spin resonance and its implication on the maximum nuclear polarization of deuterated solid target materials, Phys. Rev. B 74 (13) (2006) 134418.

[218] J.-L. Muñoz-Gómez, I. Marín-Montesinos, V. Lloveras, M. Pons, J. Vidal-Gancedo, J. Veciana, Novel PTM–TEMPO biradical for fast dissolution dynamic nuclear polarization, Org. Lett. 16 (20) (2014) 5402–5405.

[219] L.F. Pinto, I. Marín-Montesinos, V. Lloveras, J.L. Muñoz-Gómez, M. Pons, J. Veciana, J. Vidal-Gancedo, NMR signal enhancement of >50000 times in fast dissolution dynamic nuclear polarization, Chem. Commun. 53 (26) (2017) 3757–3760.

[220] H. Van Kesteren, W.T. Wenckebach, J. Schmidt, N. Poulis, Dynamic nuclear polarization of proton spins via photoexcited triplet states: the system phenanthrene in fluorene, Chem. Phys. Lett. 89 (1) (1982) 67–70.

[221] K. Takeda, Triplet State Dynamic Nuclear Polarization: Basics, Concepts, Methods, VDM Publishing, 2009.

[222] A. Van Strien, J. Schmidt, An EPR study of the triplet state of pentacene by electron spin-echo techniques and laser flash excitation, Chem. Phys. Lett. 70 (3) (1980) 513–517.

[223] T. Eichhorn, M. Haag, B. Van Den Brandt, P. Hautle, W.T. Wenckebach, High proton spin polarization with DNP using the triplet state of pentacene-d14, Chem. Phys. Lett. 555 (2013) 296–299.

[224] Y. Quan, B. van den Brandt, J. Kohlbrecher, W.T. Wenckebach, P. Hautle, A transportable neutron spin filter, Nucl. Instrum. Methods. Phys. Res. Sect. A 921 (2019) 22–26.

[225] A. Henstra, T.-S. Lin, J. Schmidt, W.T. Wenckebach, High dynamic nuclear polarization at room temperature, Chem. Phys. Lett. 165 (1) (1990) 6–10.

[226] M. Iinuma, Y. Takahashi, I. Shaké, M. Oda, A. Masaike, T. Yabuzaki, H. Shimizu, Proton polarization with p-terphenyl crystal by integrated solid effect on photoexcited triplet state, J. Magn. Reson. 175 (2) (2005) 235–241.

[227] K. Tateishi, M. Negoro, S. Nishida, A. Kagawa, Y. Morita, M. Kitagawa, Room temperature hyperpolarization of nuclear spins in bulk, Proc. Natl. Acad. Sci. 111 (21) (2014) 7527–7530.

[228] T.R. Eichhorn, M. Haag, B. van den Brandt, P. Hautle, W.T. Wenckebach, S. Jannin, J. Van der Klink, A. Comment, An apparatus for pulsed ESR and DNP experiments using optically excited triplet states down to liquid helium temperatures, J. Magn. Reson. 234 (2013) 58–66.

[229] T.R. Eichhorn, B.V.D. Brandt, P. Hautle, A. Henstra, W.T. Wenckebach, Dynamic nuclear polarisation via the integrated solid effect II: experiments on naphthalene-h 8 doped with pentacene-d 14, Mol. Phys. 112 (13) (2014) 1773–1782.

[230] T.R. Eichhorn, A.J. Parker, F. Josten, C. Müller, J. Scheuer, J.M. Steiner, M. Gierse, J. Handwerker, M. Keim, S. Lucas, Hyperpolarized solution-state NMR spectroscopy with optically polarized crystals, J. Am. Chem. Soc. 144 (6) (2022) 2511–2519.

[231] G.A. Morris, R. Freeman, Enhancement of nuclear magnetic resonance signals by polarization transfer, J. Am. Chem. Soc. 101 (3) (1979) 760–762.

[232] J. Waugh, C. Wang, L. Huber, R. Vold, Multiple-pulse NMR experiments, J. Chem. Phys. 48 (2) (1968) 662–670.

[233] P. Mieville, P. Ahuja, R. Sarkar, S. Jannin, P.R. Vasos, S. Gerber-Lemaire, M. Mishkovsky, A. Comment, R. Gruetter, O. Ouari, P. Tordo, G. Bodenhausen, Scavenging free radicals to preserve enhancement and extend relaxation times in NMR using dynamic nuclear polarization, Angew. Chem. Int. Ed. Engl. 49 (35) (2010) 6182–6185.

[234] D. Gajan, M. Schwarzwälder, M.P. Conley, W.R. Grüning, A.J. Rossini, A. Zagdoun, M. Lelli, M. Yulikov, G. Jeschke, C. Sauvée, O. Ouari, P. Tordo, L. Veyre, A. Lesage, C. Thieuleux, L. Emsley, C. Copéret, Solid-phase polarization matrixes for dynamic nuclear polarization from homogeneously distributed radicals in mesostructured hybrid silica materials, J. Am. Chem. Soc. 135 (41) (2013) 15459–15466.

[235] D. Gajan, A. Bornet, B. Vuichoud, J. Milani, R. Melzi, H.A. Van Kalkeren, L. Veyre, C. Thieuleux, M.P. Conley, W.R. Grüning, M. Schwarzwälde, A. Lesage, C. Copéret, G. Bodenhausen, L. Emsley, S. Jannin, Hybrid polarizing solids for pure hyperpolarized liquids through dissolution dynamic nuclear polarization, Proc. Natl. Acad. Sci. 111 (41) (2014) 14693–14697.

[236] D. Baudouin, H.A. Van Kalkeren, A. Bornet, B. Vuichoud, L. Veyre, M. Cavaillès, M. Schwarzwälder, W.-C. Liao, D. Gajan, G. Bodenhausen, L. Emsley, A. Lesage, S. Jannin, C. Copéret, C. Thieuleu, Cubic three-dimensional hybrid silica solids for nuclear hyperpolarization, Chem. Sci. 7 (11) (2016) 6846–6850.

[237] D.L. Silverio, H.A. Van Kalkeren, T.C. Ong, M. Baudin, M. Yulikov, L. Veyre, P. Berruyer, S. Chaudhari, D. Gajan, D. Baudouin, M. Cavaillès, B. Vuichoud, A. Bornet, G. Jeschke, G. Bodenhausen, A. Lesage, L. Emsley, S. Jannin, C. Thieuleux, C. Copéret, Tailored polarizing hybrid solids with nitroxide radicals localized in mesostructured silica walls, Helv. Chim. Acta 100 (6) (2017) e1700101.

[238] M. Cavaillès, A. Bornet, X. Jaurand, B. Vuichoud, D. Baudouin, M. Baudin, L. Veyre, G. Bodenhausen, J.N. Dumez, S. Jannin, C. Copéret, C. Thieuleux, Tailored microstructured hyperpolarizing matrices for optimal magnetic resonance imaging, Angew. Chem. 130 (25) (2018) 7575–7579.

[239] T. El Daraï, S.F. Cousin, Q. Stern, M. Ceillier, J. Kempf, D. Eshchenko, R. Melzi, M. Schnell, L. Gremillard, A. Bornet, A. Bornet, J. Milani, B. Vuichoud, O. Cala, D. Montarnal, S. Jannin, Porous functionalized polymers enable generating and transporting hyperpolarized mixtures of metabolites, Nat. Commun. 12 (1) (2021) 1–9.

[240] T. Cheng, M. Mishkovsky, M.J. Junk, K. Münnemann, A. Comment, Producing radical-free hyperpolarized perfusion agents for in vivo magnetic resonance using spin-labeled thermoresponsive hydrogel, Macromol. Rapid Commun. 37 (13) (2016) 1074–1078.

[241] B.C. Dollmann, M.J. Junk, M. Drechsler, H.W. Spiess, D. Hinderberger, K. Münnemann, Thermoresponsive, spin-labeled hydrogels as separable DNP polarizing agents, Phys. Chem. Chem. Phys. 12 (22) (2010) 5879–5882.

[242] T.R. Eichhorn, Y. Takado, N. Salameh, A. Capozzi, T. Cheng, J.-N. Hyacinthe, M. Mishkovsky, C. Roussel, A. Comment, Hyperpolarization without persistent radicals for in vivo real-time metabolic imaging, Proc. Natl. Acad. Sci. 110 (45) (2013) 18064–18069.

[243] I. Marco-Rius, T. Cheng, A.P. Gaunt, S. Patel, F. Kreis, A. Capozzi, A.J. Wright, K.M. Brindle, O. Ouari, A. Comment, Photogenerated radical in phenylglyoxylic acid for in vivo hyperpolarized 13C MR with photosensitive metabolic substrates, J. Am. Chem. Soc. 140 (43) (2018) 14455–14463.

[244] A. Capozzi, S. Patel, C.P. Gunnarsson, I. Marco-Rius, A. Comment, M. Karlsson, M.H. Lerche, O. Ouari, J.H. Ardenkjær-Larsen, Efficient hyperpolarization of U-13C-glucose using narrow-line UV-generated labile free radicals, Angew. Chem. Int. Ed. 58 (5) (2019) 1334–1339.

[245] A.P. Gaunt, J.S. Lewis, F. Hesse, T. Cheng, I. Marco-Rius, K.M. Brindle, A. Comment, Labile photo-induced free radical in α-ketoglutaric acid: a universal endogenous polarizing agent for in vivo hyperpolarized 13C magnetic resonance, Angew. Chem. 134 (2) (2022) e202112982.

[246] P.A. Leermakers, G.F. Vesley, The photochemistry of α-keto acids and α-keto esters. I. Photolysis of pyruvic acid and benzoylformic acid, J. Am. Chem. Soc. 85 (23) (1963) 3776–3779.

[247] D.S. Kendall, P.A. Leermakers, Photoreduction of pyruvic acid by isopropyl alcohol and t-butyl alcohol. A kinetic Study1, J. Am. Chem. Soc. 88 (12) (1966) 2766–2768.

[248] A. Capozzi, J.-N.L. Hyacinthe, T. Cheng, T.R. Eichhorn, G. Boero, C. Roussel, J.J. van der Klink, A. Comment, Photoinduced nonpersistent radicals as polarizing agents for X-nuclei dissolution dynamic nuclear polarization, J. Phys. Chem. C 119 (39) (2015) 22632–22639.

[249] J.S. Lewis, A.P. Gaunt, A. Comment, Photochemistry of pyruvic acid is governed by photo-induced intermolecular electron transfer through hydrogen bonds, Chem. Sci. 13 (40) (2022) 11849–11855.

[250] E.C. Griffith, B.K. Carpenter, R.K. Shoemaker, V. Vaida, Photochemistry of aqueous pyruvic acid, Proc. Natl. Acad. Sci. 110 (29) (2013) 11714–11719.

[251] M.I. Guzman, A.J. Eugene, Aqueous photochemistry of 2-oxocarboxylic acids: evidence, mechanisms, and atmospheric impact, Molecules 26 (17) (2021) 5278.

[252] S. Patel, A.C. Pinon, M.H. Lerche, M. Karlsson, A. Capozzi, J.H. Ardenkjær-Larsen, UV-irradiated 2-keto-(1-13C) isocaproic acid for high-performance 13C hyperpolarized MR, J. Phys. Chem. C 124 (43) (2020) 23859–23866.

[253] C.C. Zanella, A. Capozzi, H.A. Yoshihara, A. Radaelli, A.L. Mackowiak, L.P. Arn, R. Gruetter, J.A. Bastiaansen, Radical-free hyperpolarized MRI using endogenously occurring pyruvate analogues and UV-induced nonpersistent radicals, NMR Biomed. 34 (11) (2021) e4584.

[254] J.A. Bastiaansen, H.A. Yoshihara, A. Capozzi, J. Schwitter, R. Gruetter, M.E. Merritt, A. Comment, Probing cardiac metabolism by hyperpolarized 13 C MR using an exclusively endogenous substrate mixture and photo-induced nonpersistent radicals, Magn. Reson. Med. 79 (5) (2018) 2451–2459.

[255] Y. Rao, S.T. Gammon, M.N. Sutton, N.M. Zacharias, P. Bhattacharya, D. Piwnica-Worms, Excess exogenous pyruvate inhibits lactate dehydrogenase activity in live cells in an MCT1-dependent manner, J. Biol. Chem. (1) (2021) 297.

[256] I. Katz, A. Blank, Dynamic nuclear polarization in solid samples by electrical-discharge-induced radicals, J. Magn. Reson. 261 (2015) 95–100.

[257] I. Katz, A. Feintuch, R. Carmieli, A. Blank, Proton polarization enhancement of up to 150 with dynamic nuclear polarization of plasma-treated glucose powder, Solid State Nucl. Magn. Reson. 100 (2019) 26–35.

[258] W. Blumberg, Nuclear spin-lattice relaxation caused by paramagnetic impurities, Phys. Rev. 119 (1) (1960) 79.

[259] M. Jurkutat, H. Kourilová, D. Peat, K. Kouril, A.S. Khan, A.J. Horsewill, J.F. MacDonald, J. Owers-Bradley, B. Meier, Radical-induced low-field 1H relaxation in solid pyruvic acid doped with trityl-OX063, J. Phys. Chem. Lett. 13 (44) (2022) 10370–10376.

[260] A. Capozzi, T. Cheng, G. Boero, C. Roussel, A. Comment, Thermal annihilation of photo-induced radicals following dynamic nuclear polarization to produce transportable frozen hyperpolarized 13C-substrates, Nat. Commun. 8 (1) (2017) 1–7.

[261] A. Capozzi, M. Karlsson, J.R. Petersen, M.H. Lerche, J.H. Ardenkjaer-Larsen, Liquid-state 13C polarization of 30% through photoinduced nonpersistent radicals, J. Phys. Chem. C 122 (13) (2018) 7432–7443.

[262] A. Capozzi, J. Kilund, M. Karlsson, S. Patel, A.C. Pinon, F. Vibert, O. Ouari, M.H. Lerche, J.H. Ardenkjær-Larsen, Metabolic contrast agents produced from transported solid 13C-glucose hyperpolarized via dynamic nuclear polarization, Commun. Chem. 4 (1) (2021) 1–11.

[263] A. Capozzi, M. Karlsson, E.S.S. Hansen, J.D. Sanchez, C. Laustsen, M.H. Lerche, J. Van Der Klink, J.H. Ardenkjær-Larsen, A 320 km hyperpolarization journey: performing [U-13C, d7] glucose DNP in Copenhagen and hyperpolarized 13C-MR in Aarhus, in: Ann. Meeting Proc., 2022.

[264] N. Bloembergen, E.M. Purcell, R.V. Pound, Relaxation effects in nuclear magnetic resonance absorption, Phys. Rev. 73 (7) (1948) 679.

[265] A. Gupta, T. Stait-Gardner, W.S. Price, Is it time to forgo the use of the terms "spin–lattice" and "spin–spin" relaxation in NMR and MRI? J. Phys. Chem. Lett. 12 (27) (2021) 6305–6312.

[266] J.P. Ridgway, Cardiovascular magnetic resonance physics for clinicians: part I, J. Cardiovasc. Magn. Reson. 12 (1) (2010) 1–28.

[267] R. Vold, J. Waugh, M.P. Klein, D. Phelps, Measurement of spin relaxation in complex systems, J. Chem. Phys. 48 (8) (1968) 3831–3832.

[268] R. Freeman, H. Hill, High-resolution studies of nuclear spin–lattice relaxation, J. Chem. Phys. 51 (7) (1969) 3140–3141.

[269] M. Huisjen, J.S. Hyde, Saturation recovery measurements of electron spin-lattice relaxation times of free radicals in solution, J. Chem. Phys. 60 (4) (1974) 1682–1683.

[270] P.W. Percival, J.S. Hyde, Saturation-recovery measurements of the spin-lattice relaxation times of some nitroxides in solution, J. Magn. Reson. 23 (2) (1976) 249–257.

[271] R. Freeman, H. Hill, Fourier transform study of NMR spin–lattice relaxation by "progressive saturation", J. Chem. Phys. 54 (8) (1971) 3367–3377.

[272] R. Wei, C.L. Dickson, D.A. Uhrín, G.C. Lloyd-Jones, Rapid estimation of T 1 for quantitative NMR, J. Org. Chem. 86 (2021) 9023–9029.

[273] P.B. Kingsley, Methods of measuring spin-lattice (T1) relaxation times: an annotated bibliography, Concepts Magn. Reson. Part A Bridg. Educ. Res. 11 (4) (1999) 243–276.

[274] V. Mitrović, E. Sigmund, W.P. Halperin, Progressive saturation NMR relaxation, Phys. Rev. B 64 (2) (2001) 024520.

[275] A.P. Chen, J. Tropp, R.E. Hurd, M. Van Criekinge, L.G. Carvajal, D. Xu, J. Kurhanewicz, D.B. Vigneron, In vivo hyperpolarized 13C MR spectroscopic imaging with 1H decoupling, J. Magn. Reson. 197 (1) (2009) 100–106.

[276] B.R. Patyal, J.-H. Gao, R.F. Williams, J. Roby, B. Saam, B.A. Rockwell, R.J. Thomas, D.J. Stolarski, P.T. Fox, Longitudinal relaxation and diffusion measurements using magnetic resonance signals from laser-hyperpolarized129Xe nuclei, J. Magn. Reson. 126 (1) (1997) 58–65.

[277] A.P. Chen, R.E. Hurd, Y.P. Gu, D.M. Wilson, C.H. Cunningham, 13C MR reporter probe system using dynamic nuclear polarization, NMR Biomed. 24 (5) (2011) 514–520.

[278] R.Y. Dong, NMR Relaxation Rates, 2017.

[279] R.T. Boeré, R.G. Kidd, Rotational correlation times in nuclear magnetic relaxation, in: Annual Reports on NMR Spectroscopy, vol. 13, Elsevier, 1983, pp. 319–385.

[280] T. James, Conformational flexibility in nucleic acids investigated by NMR relaxation, Bull. Magn. Reson. 4 (1982) 119.

[281] A. Gupta, T. Stait-Gardner, M.J. Moghaddam, W.S. Price, Dipolar relaxation revisited: a complete derivation for the two spin case, Concepts Magn. Reson. Part A Bridg. Educ. Res. 44 (2) (2015) 74–113.

[282] N. Bloembergen, E.M. Purcell, R.V. Pound, Nuclear magnetic relaxation, Nature 160 (4066) (1947) 475–476.

[283] T. Peters Jr., All About Albumin: Biochemistry, Genetics, and Medical Applications, Academic Press, 1995.

[284] C.K. Larive, G.A. Barding Jr., M.M. Dinges, NMR spectroscopy for metabolomics and metabolic profiling, Anal. Chem. 87 (1) (2015) 133–146.

[285] G. Sudlow, D. Birkett, D. Wade, The characterization of two specific drug binding sites on human serum albumin, Mol. Pharmacol. 11 (6) (1975) 824–832.

[286] L.V. Elst, S. Laurent, H.M. Bintoma, R.N. Muller, Albumin-bound MRI contrast agents: the dilemma of the rotational correlation time, Magn. Reson. Mater. Phys. Biol. Med. 12 (2) (2001) 135–140.

[287] N.M. Zacharias, A. Ornelas, J. Lee, J. Hu, J.S. Davis, N. Uddin, S. Pudakalakatti, D.G. Menter, J.A. Karam, C.G. Wood, Real-time interrogation of Aspirin reactivity, biochemistry, and biodistribution by hyperpolarized magnetic resonance spectroscopy, Angew. Chem. 131 (13) (2019) 4223–4227.

[288] I. Matias, A. Ceballos, F. Gonzalez-Velasco, J. Cachaza, Thermodynamics of the binding of salicylate to human serum albumin: evidence of non-competition with imidazole, J. Pharm. Pharmacol. 41 (2) (1989) 123–125.

[289] R.A. Copeland, Reversible inhibitors. Enzymes: A Practical Introduction to Structure, Mechanism, and Data Analysis 8, 2000, pp. 266–304.

[290] E.H. Suh, Z. Kovacs, Competitive displacement restores the hyperpolarized 15N NMR signal in blood plasma, J. Phys. Chem. Au 3 (2022) 167–171.

[291] G.H. Mudge, N.I. Desbiens, G. Stibitz, Binding of iophenoxate and iopanoate to human serum albumin, Drug Metab. Dispos. 6 (4) (1978) 432–439.

[292] P. Caravan, N.J. Cloutier, M.T. Greenfield, S.A. McDermid, S.U. Dunham, J.W. Bulte, J.C. Amedio, R.J. Looby, R.M. Supkowski, W.D. Horrocks, T.J. McMurry, R.B. Lauffer, The interaction of MS-325 with human serum albumin and its effect on proton relaxation rates, J. Am. Chem. Soc. 124 (12) (2002) 3152–3162.

[293] A.F. Martins, V. Clavijo Jordan, F. Bochner, S. Chirayil, N. Paranawithana, S. Zhang, S.-T. Lo, X. Wen, P. Zhao, M. Neeman, Imaging insulin secretion from mouse pancreas by MRI is improved by use of a zinc-responsive MRI sensor with lower affinity for Zn2+ ions, J. Am. Chem. Soc. 140 (50) (2018) 17456–17464.

[294] C. Burt, Phosphorus NMR in Biology, CRC Press, 2018.

[295] G.C. Levy, Carbon-13-spin-lattice relaxation studies and their application to organic chemical problems, Acc. Chem. Res. 6 (5) (1973) 161–169.

[296] E.D. Becker, R. Shoup, 13C NMR spectroscopy: relaxation times of 13C and methods for sensitivity enhancement, Pure Appl. Chem. 32 (1–4) (1972) 51–66.

[297] H. Nonaka, M. Hirano, Y. Imakura, Y. Takakusagi, K. Ichikawa, S. Sando, Design of a 15 N molecular unit to achieve long retention of hyperpolarized spin state, Sci. Rep. 7 (1) (2017) 1–6.

[298] Y. Kondo, H. Nonaka, Y. Takakusagi, S. Sando, Design of nuclear magnetic resonance molecular probes for hyperpolarized bioimaging, Angew. Chem. Int. Ed. 60 (27) (2021) 14779–14799.

[299] R. Gil, A. Navarro-Vázquez, Application of the Nuclear Overhauser Effect to the Structural Elucidation of Natural Products, 2016.

[300] E. Breitmaier, K.H. Spohn, S. Berger, 13C spin-lattice relaxation times and the mobility of organic molecules in solution, Angew. Chem. Int. Ed. Engl. 14 (3) (1975) 144–159.

[301] M. Goldman, Formal theory of spin–lattice relaxation, J. Magn. Reson. 149 (2) (2001) 160–187.

[302] G. Morris, Spin Dynamics: Basics of Nuclear Magnetic Resonance, Malcolm H. Levitt. Wiley Chichester, 2008.

[303] A. Allerhand, D. Doddrell, R. Komoroski, Natural abundance carbon-13 partially relaxed Fourier transform nuclear magnetic resonance spectra of complex molecules, J. Chem. Phys. 55 (1) (1971) 189–198.

[304] R.S. Norton, A. Allerhand, Effect of carbon-13-nitrogen dipolar interactions on spin-lattice relaxation times and intensities of nonprotonated carbon resonances, J. Am. Chem. Soc. 98 (4) (1976) 1007–1014.

[305] G.C. Levy, P.L. Rinaldi, J.T. Bailey, Yttrium-89 NMR. A possible spin relaxation probe for studying metal ion interactions with organic ligands, J. Magn. Reson. 40 (1) (1980) 167–173.

[306] Y. Imakura, H. Nonaka, Y. Takakusagi, K. Ichikawa, N.R. Maptue, A.M. Funk, C. Khemtong, S. Sando, Rational design of [13C, D14] *tert*-butylbenzene as a scaffold structure for designing long-lived hyperpolarized 13C probes, Chem. Asian J. 13 (3) (2018) 280–283.

[307] E. Chiavazza, A. Viale, M. Karlsson, S. Aime, 15N-permethylated amino acids as efficient probes for MRI-DNP applications, Contrast Media Mol. Imaging 8 (5) (2013) 417–421.

[308] J. Blicharski, Nuclear magnetic relaxation by anisotropy of the chemical shift, Z. Naturforsch. 27 (10) (1972) 1456–1458.

[309] G.H. Penner, G.M. Bernard, R.E. Wasylishen, A. Barrett, R.D. Curtis, A solid-state nitrogen-15 NMR and ab initio study of nitrobenzenes, J. Org. Chem. 68 (11) (2003) 4258–4264.

[310] T.D. Alger, W.D. Hamill Jr., R.J. Pugmire, D.M. Grant, G.D. Silcox, M. Solum, Carbon-13 spin-lattice relaxation in condensed aromatic compounds, J. Phys. Chem. 84 (6) (1980) 632–636.

[311] B. Borah, R.G. Bryant, NMR relaxation dispersion in an aqueous nitroxide system, J. Chem. Phys. 75 (7) (1981) 3297–3300.

[312] P. Miéville, S. Jannin, G. Bodenhausen, Relaxometry of insensitive nuclei: optimizing dissolution dynamic nuclear polarization, J. Magn. Reson. 210 (1) (2011) 137–140.
[313] N. Chattergoon, F. Martínez-Santiesteban, W. Handler, J.H. Ardenkjær-Larsen, T. Scholl, Field dependence of T1 for hyperpolarized [1-13C] pyruvate, Contrast Media Mol. Imaging 8 (1) (2013) 57–62.
[314] D.T. Peat, M.L. Hirsch, D.G. Gadian, A.J. Horsewill, J.R. Owers-Bradley, J.G. Kempf, Low-field thermal mixing in [1-13 C] pyruvic acid for brute-force hyperpolarization, Phys. Chem. Chem. Phys. 18 (28) (2016) 19173–19182.
[315] S. Elliott, C. Bengs, L. Brown, J. Hill-Cousins, D. O'Leary, G. Pileio, M. Levitt, Nuclear singlet relaxation by scalar relaxation of the second kind in the slow-fluctuation regime, J. Chem. Phys. 150 (6) (2019) 064315.
[316] A. Gryff-Keller, D. Kubica, Scalar relaxation of the second kind: a potential source of information on the dynamics of molecular movements. 1. investigation of solution reorientation of N-methylpyridone and 1, 3-dimethyluracil using measurements of longitudinal relaxation rates in the rotating frame, J. Phys. Chem. A 116 (39) (2012) 9632–9638.
[317] D. Kubica, A. Wodynski, A. Kraska-Dziadecka, A. Gryff-Keller, Scalar relaxation of the second kind. A potential source of information on the dynamics of molecular movements. 3. A 13C nuclear spin relaxation study of CBrX3 (X= Cl, CH3, Br) molecules, J. Phys. Chem. A 118 (16) (2014) 2995–3003.
[318] P.J. Rayner, M.J. Burns, A.M. Olaru, P. Norcott, M. Fekete, G.G. Green, L.A. Highton, R.E. Mewis, S.B. Duckett, Delivering strong 1H nuclear hyperpolarization levels and long magnetic lifetimes through signal amplification by reversible exchange, Proc. Natl. Acad. Sci. 114 (16) (2017) E3188–E3194.
[319] P. Norcott, M.J. Burns, P.J. Rayner, R.E. Mewis, S.B. Duckett, Using 2H labelling to improve the NMR detectability of pyridine and its derivatives by SABRE, Magn. Reson. Chem. 56 (7) (2018) 663–671.
[320] J. Hindman, A. Svirmickas, M. Wood, Relaxation processes in water. A study of the proton spin-lattice relaxation time, J. Chem. Phys. 59 (3) (1973) 1517–1522.
[321] K. Krynicki, Proton spin-lattice relaxation in pure water between 0°C and 100°C, Physica 32 (1) (1966) 167–178.
[322] K. Wigh Lipsø, E.S.S. Hansen, R.S. Tougaard, C. Laustsen, J.H. Ardenkjær-Larsen, Renal MR angiography and perfusion in the pig using hyperpolarized water, Magn. Reson. Med. 78 (3) (2017) 1131–1135.
[323] M.D. Lingwood, T.A. Siaw, N. Sailasuta, B.D. Ross, P. Bhattacharya, S. Han, Continuous flow Overhauser dynamic nuclear polarization of water in the fringe field of a clinical magnetic resonance imaging system for authentic image contrast, J. Magn. Reson. 205 (2) (2010) 247–254.
[324] E. Vahala, M. Ylihautala, G. Ehnholm, N. Etelä, I. Young, K. Golman, I. Leunbach, A study of the use of overhauser enhancement to assist with needle and catheter placement during interventional MRI, J. Magn. Reson. 157 (2002) 298–303.
[325] J.G. Krummenacker, V.P. Denysenkov, M. Terekhov, L.M. Schreiber, T.F. Prisner, DNP in MRI: an in-bore approach at 1.5 T, J. Magn. Reson. 215 (2012) 94–99.
[326] P. Neugebauer, J.G. Krummenacker, V.P. Denysenkov, G. Parigi, C. Luchinat, T.F. Prisner, Liquid state DNP of water at 9.2 T: an experimental access to saturation, Phys. Chem. Chem. Phys. 15 (16) (2013) 6049–6056.
[327] N. Abhyankar, V. Szalai, Challenges and advances in the application of dynamic nuclear polarization to liquid-state NMR spectroscopy, J. Phys. Chem. B 125 (20) (2021) 5171–5190.
[328] K.H. Hausser, D. Stehlik, in: A. Magn, O. Reson (Eds.), Dynamic Nuclear Polarization in Liquids, vol. 3, Elsevier, 1968, pp. 79–139.
[329] D.J. Lurie, D.M. Bussell, L.H. Bell, J.R. Mallard, Proton-electron double magnetic resonance imaging of free radical solutions, J. Magn. Reson. 76 (2) (1988) 366–370.
[330] D.J. Lurie, Progress toward whole-body proton—electron double-resonance imaging of free radicals, Magn. Reson. Mater. Phys. Biol. Med. 2 (3) (1994) 267–271.
[331] D.J. Lurie, M.A. Foster, D. Yeung, J.M. Hutchison, Design, construction and use of a large-sample field-cycled PEDRI imager, Phys. Med. Biol. 43 (7) (1998) 1877.
[332] D.J. Lurie, G.R. Davies, M.A. Foster, J.M. Hutchison, Field-cycled PEDRI imaging of free radicals with detection at 450 mT, Magn. Reson. Imaging 23 (2) (2005) 175–181.
[333] W. Youngdee, G. Planinsic, D.J. Lurie, Optimization of field-cycled PEDRI for in vivo imaging of free radicals, Phys. Med. Biol. 46 (10) (2001) 2531.
[334] M.D. Lingwood, T.A. Siaw, N. Sailasuta, O.A. Abulseoud, H.R. Chan, B.D. Ross, P. Bhattacharya, S. Han, Hyperpolarized water as an MR imaging contrast agent: feasibility of in vivo imaging in a rat model, Radiology 265 (2) (2012) 418–425.
[335] A.C. Pinon, A. Capozzi, J.H. Ardenkjær-Larsen, Hyperpolarized water through dissolution dynamic nuclear polarization with UV-generated radicals, Commun. Chem. 3 (1) (2020) 1–9.
[336] T. Harris, O. Szekely, L. Frydman, On the potential of hyperpolarized water in biomolecular NMR studies, J. Phys. Chem. B 118 (12) (2014) 3281–3290.
[337] B. Vuichoud, A. Bornet, F. De Nanteuil, J. Milani, E. Canet, X. Ji, P. Miéville, E. Weber, D. Kurzbach, A.G. Flamm, R. Konrat, A.D. Gossert, S. Jannin, G. Bodenhausen, Filterable agents for hyperpolarization of water, metabolites, and proteins, Chem. Eur J. 22 (41) (2016) 14696–14700.
[338] J.H. Ardenkjaer-Larsen, C. Laustsen, S. Bowen, R. Rizi, Hyperpolarized H2O MR angiography, Magn. Reson. Med. 71 (1) (2014) 50–56.
[339] J. Kim, M. Liu, C. Hilty, Modeling of polarization transfer kinetics in protein hydration using hyperpolarized water, J. Phys. Chem. B 121 (27) (2017) 6492–6498.
[340] O. Szekely, G.L. Olsen, I.C. Felli, L. Frydman, High-resolution 2D NMR of disordered proteins enhanced by hyperpolarized water, Anal. Chem. 90 (10) (2018) 6169–6177.

[341] P. Kadeřávek, F. Ferrage, G. Bodenhausen, D. Kurzbach, High-resolution NMR of folded proteins in hyperpolarized physiological solvents, Chem. Eur J. 24 (51) (2018) 13418–13423.

[342] J. Kim, R. Mandal, C. Hilty, 2D NMR spectroscopy of refolding RNase Sa using polarization transfer from hyperpolarized water, J. Magn. Reson. 326 (2021) 106942.

[343] J. Kim, R. Mandal, C. Hilty, Observation of fast two-dimensional NMR spectra during protein folding using polarization transfer from hyperpolarized water, J. Phys. Chem. Lett. 10 (18) (2019) 5463–5467.

[344] Q. Chappuis, J. Milani, B. Vuichoud, A. Bornet, A.D. Gossert, G. Bodenhausen, S. Jannin, Hyperpolarized water to study protein–ligand interactions, J. Phys. Chem. Lett. 6 (9) (2015) 1674–1678.

[345] Y. Wang, M. Ragavan, C. Hilty, Site specific polarization transfer from a hyperpolarized ligand of dihydrofolate reductase, J. Biomol. NMR 65 (1) (2016) 41–48.

[346] J. Kim, R. Mandal, C. Hilty, Characterization of membrane protein-lipid interactions in unfolded OmpX with enhanced time resolution by hyperpolarized NMR, Chembiochem 21 (19) (2020) 2861–2867.

[347] R. Balzan, M. Mishkovsky, Y. Simonenko, R.B. van Heeswijk, R. Gruetter, U. Eliav, G. Navon, A. Comment, Hyperpolarized 6Li as a probe for hemoglobin oxygenation level, Contrast Media Mol. Imaging 11 (1) (2016) 41–46.

[348] R.B. van Heeswijk, K. Uffmann, A. Comment, F. Kurdzesau, C. Perazzolo, C. Cudalbu, S. Jannin, J.A. Konter, P. Hautle, B. van den Brandt, G. Navon, J.J. van der Klink, R. Gruetter, Hyperpolarized lithium-6 as a sensor of nanomolar contrast agents, Magn. Reson. Med.: J. Int. Soc. Magn. Reson. Med. 61 (6) (2009) 1489–1493.

[349] C.R. Malloy, E. Maher, I. Marin-Valencia, B. Mickey, R.J. DeBerardinis, A. Dean Sherry, N. Lutz, J. Sweedler, R. Wevers, Carbon-13 nuclear magnetic resonance for analysis of metabolic pathways, in: Methodologies for Metabolomics, 2012, pp. 415–445.

[350] F.M.H. Jeffrey, A. Rajagopal, C.R. Malloy, A.D. Sherry, C-NMR: a simple yet comprehensive method for analysis of intermediary metabolism, Trends Biochem. Sci. 16 (1991) 5–10.

[351] N.R. Pace, The universal nature of biochemistry, Proc. Natl. Acad. Sci. USA 98 (3) (2001) 805–808.

[352] V.K. Michaelis, A.A. Smith, B. Corzilius, O. Haze, T.M. Swager, R.G. Griffin, High-field 13C dynamic nuclear polarization with a radical mixture, J. Am. Chem. Soc. 135 (8) (2013) 2935–2938.

[353] A. Zagdoun, G. Casano, O. Ouari, M. Schwarzwalder, A.J. Rossini, F. Aussenac, M. Yulikov, G. Jeschke, C. Coperet, A. Lesage, P. Tordo, L. Emsley, Large molecular weight nitroxide biradicals providing efficient dynamic nuclear polarization at temperatures up to 200 K, J. Am. Chem. Soc. 135 (34) (2013) 12790–12797.

[354] F. Reineri, A. Viale, W. Dastrù, R. Gobetto, S. Aime, How to design 13C para-hydrogen-induced polarization experiments for MRI applications, Contrast Media Mol. Imaging 6 (2) (2011) 77–84.

[355] S. Kim, F. Martinez-Santiesteban, T.J. Scholl, Measuring the spin-lattice relaxation magnetic field dependence of hyperpolarized [1-13C] pyruvate, JoVE (151) (2019) e59399.

[356] F.A. Anet, D.J. O'leary, The shielding tensor part II: understanding its strange effects on relaxation, Concepts Magn. Reson. Part A Bridg. Educ. Res. 4 (1) (1992) 35–52.

[357] C. Hundshammer, M. Grashei, A. Greiner, S.J. Glaser, F. Schilling, pH dependence of T1 for 13C-labelled small molecules commonly used for hyperpolarized magnetic resonance imaging, ChemPhysChem 20 (6) (2019) 798–802.

[358] H. Allouche-Arnon, M.H. Lerche, M. Karlsson, R.E. Lenkinski, R. Katz-Brull, Deuteration of a molecular probe for DNP hyperpolarization–a new approach and validation for choline chloride, Contrast Media Mol. Imaging 6 (6) (2011) 499–506.

[359] S.S. Tee, V. DiGialleonardo, R. Eskandari, S. Jeong, K.L. Granlund, V. Miloushev, A.J. Poot, S. Truong, J.A. Alvarez, H.N. Aldeborgh, Sampling hyperpolarized molecules utilizing a 1 Tesla permanent magnetic field, Sci. Rep. 6 (1) (2016) 1–6.

[360] S.E. Day, M.I. Kettunen, F.A. Gallagher, D.-E. Hu, M. Lerche, J. Wolber, K. Golman, J.H. Ardenkjaer-Larsen, K.M. Brindle, Detecting tumor response to treatment using hyperpolarized 13C magnetic resonance imaging and spectroscopy, Nat. Med. 13 (11) (2007) 1382–1387.

[361] C.G. Moreland, F. Carroll, $^{13}C{-}^{13}C$ dipolar interactions as a relaxation mechanism, J. Magn. Reson. 15 (3) (1974) 596–599.

[362] A.P. Chen, R.E. Hurd, M.A. Schroeder, A.Z. Lau, Y.p. Gu, W.W. Lam, J. Barry, J. Tropp, C.H. Cunningham, Simultaneous investigation of cardiac pyruvate dehydrogenase flux, Krebs cycle metabolism and pH, using hyperpolarized [1, 2-13C2] pyruvate in vivo, NMR Biomed. 25 (2) (2012) 305–311.

[363] J. Singh, E.H. Suh, G. Sharma, J. Chen, E.P. Hackett, X. Wen, A.D. Sherry, C. Khemtong, C.R. Malloy, J.M. Park, Z. Kovacs, 13C-Labeled diethyl ketoglutarate derivatives as hyperpolarized probes of 2-ketoglutarate dehydrogenase activity, Anal. Sens. 1 (4) (2021) 156–160.

[364] J. Kim, Y. Kim, Q.S. Luu, J. Kim, C. Qi, C. Hilty, Y. Lee, Indirect detection of intermediate in decarboxylation reaction of phenylglyoxylic acid by hyperpolarized 13 C NMR, Chem. Commun. 56 (95) (2020) 15000–15003.

[365] H.Y. Chen, M. Ragavan, C. Hilty, Protein folding studied by dissolution dynamic nuclear polarization, Angew. Chem. 125 (35) (2013) 9362–9365.

[366] M. Ragavan, L.I. Iconaru, C.G. Park, R.W. Kriwacki, C. Hilty, Real-time analysis of folding upon binding of a disordered protein by using dissolution DNP NMR spectroscopy, Angew. Chem. Int. Ed. 56 (25) (2017) 7070–7073.

[367] S. Katsikis, I. Marin-Montesinos, C. Ludwig, U.L. Günther, Detecting acetylated amino acids in blood serum using hyperpolarized 13C-1H-2D-NMR, J. Magn. Reson. 305 (2019) 175–179.

[368] J. Katz, H.G. Wood, The use of glucose-C14 for the evaluation of the pathways of glucose metabolism, J. Biol. Chem. 235 (8) (1960) 2165–2177.

[369] D.S. Fredrickson, R.S. Gordon, The metabolism of albumin bound C 14-labeled unesterified fatty acids in normal human subjects, J. Clin. Investig. 37 (11) (1958) 1504–1515.

[370] J.A. Muntz, The pattern of radioactivity in glucose and glucose 6-phosphate following intraportal injection of either fructose-1-14C or fructose-6-14C in rats, J. Biol. Chem. 243 (10) (1968) 2788–2792.

[371] C. Jang, L. Chen, J.D. Rabinowitz, Metabolomics and isotope tracing, Cell 173 (4) (2018) 822–837.

[372] M.R. Antoniewicz, A guide to 13C metabolic flux analysis for the cancer biologist, Exp. Mol. Med. 50 (4) (2018) 1–13.

[373] C. Des Rosiers, S. Lloyd, B. Comte, J.C. Chatham, A critical perspective of the use of 13C-isotopomer analysis by GCMS and NMR as applied to cardiac metabolism, Metab. Eng. 6 (1) (2004) 44–58.

[374] T.B. Rodrigues, S. Cerdán, 13C MRS: an outstanding tool for metabolic studies, Concepts Magn. Reson. Part A Bridg. Educ. Res. 27 (1) (2005) 1–16.

[375] I.-Y. Kim, S.-H. Suh, I.-K. Lee, R.R. Wolfe, Applications of stable, nonradioactive isotope tracers in in vivo human metabolic research, Exp. Mol. Med. 48 (1) (2016) e203-e203.

[376] J. Kurhanewicz, D.B. Vigneron, J.H. Ardenkjaer-Larsen, J.A. Bankson, K. Brindle, C.H. Cunningham, F.A. Gallagher, K.R. Keshari, A. Kjaer, C. Laustsen, Hyperpolarized 13C MRI: path to clinical translation in oncology, Neoplasia 21 (1) (2019) 1–16.

[377] M. Vaeggemose, R. F Schulte, C. Laustsen, Comprehensive literature review of hyperpolarized carbon-13 MRI: the road to clinical application, Metabolites 11 (4) (2021) 219.

[378] R.A. de Graaf, G.F. Mason, A.B. Patel, K.L. Behar, D.L. Rothman, In vivo 1H-[13C]-NMR spectroscopy of cerebral metabolism, NMR Biomed. 16 (6–7) (2003) 339–357.

[379] J. Valette, B. Tiret, F. Boumezbeur, Experimental strategies for in vivo 13C NMR spectroscopy, Anal. Biochem. 529 (2017) 216–228.

[380] K. Kurmi, M.C. Haigis, Nitrogen metabolism in cancer and immunity, Trends Cell Biol. 30 (5) (2020) 408–424.

[381] H. Park, G. Zhang, J. Bae, T. Theis, W.S. Warren, Q. Wang, Application of 15N2-diazirines as a versatile platform for hyperpolarization of biological molecules by d-DNP, Bioconjug. Chem. 31 (3) (2020) 537–541.

[382] M.J. Cowley, R.W. Adams, K.D. Atkinson, M.C. Cockett, S.B. Duckett, G.G. Green, J.A. Lohman, R. Kerssebaum, D. Kilgour, R.E. Mewis, Iridium N-heterocyclic carbene complexes as efficient catalysts for magnetization transfer from para-hydrogen, J. Am. Chem. Soc. 133 (16) (2011) 6134–6137.

[383] R.W. Adams, J.A. Aguilar, K.D. Atkinson, M.J. Cowley, P.I. Elliott, S.B. Duckett, G.G. Green, I.G. Khazal, J. López-Serrano, D.C. Williamson, Reversible interactions with para-hydrogen enhance NMR sensitivity by polarization transfer, Science 323 (5922) (2009) 1708–1711.

[384] N.V. Chukanov, R.V. Shchepin, S.M. Joshi, M.S. Kabir, O.G. Salnikov, A. Svyatova, I.V. Koptyug, J.G. Gelovani, E.Y. Chekmenev, Synthetic approaches for 15N-labeled hyperpolarized heterocyclic molecular imaging agents for 15N NMR signal amplification by reversible exchange in microtesla magnetic fields, Chem. Eur J. 27 (38) (2021) 9727–9736.

[385] W. Jiang, L. Lumata, W. Chen, S. Zhang, Z. Kovacs, A.D. Sherry, C. Khemtong, Hyperpolarized 15 N-pyridine derivatives as pH-sensitive MRI agents, Sci. Rep. 5 (1) (2015) 1–6.

[386] R.V. Shchepin, D.A. Barskiy, A.M. Coffey, T. Theis, F. Shi, W.S. Warren, B.M. Goodson, E.Y. Chekmenev, 15N hyperpolarization of imidazole-15N2 for magnetic resonance pH sensing via SABRE-SHEATH, ACS Sens. 1 (6) (2016) 640–644.

[387] C. Cudalbu, A. Comment, F. Kurdzesau, R.B. van Heeswijk, K. Uffmann, S. Jannin, V. Denisov, D. Kirik, R. Gruetter, Feasibility of in vivo 15N MRS detection of hyperpolarized 15N labeled choline in rats, Phys. Chem. Chem. Phys. 12 (22) (2010) 5818–5823.

[388] T. Saluvere, E. Lippmaa, Spin-lattice relaxation of 15N nuclei, Chem. Phys. Lett. 7 (5) (1970) 545–548.

[389] W. Litchman, M. Alei Jr., 15N spin lattice relaxation in liquid 15NH3 and 15ND3, J. Chem. Phys. 56 (12) (1972) 5818–5822.

[390] H. Nonaka, R. Hata, T. Doura, T. Nishihara, K. Kumagai, M. Akakabe, M. Tsuda, K. Ichikawa, S. Sando, A platform for designing hyperpolarized magnetic resonance chemical probes, Nat. Commun. 4 (1) (2013) 1–7.

[391] M. Lu, R. Ishima, T. Polenova, A.M. Gronenborn, 19F NMR relaxation studies of fluorosubstituted tryptophans, J. Biomol. NMR 73 (8) (2019) 401–409.

[392] Y. Lee, H. Zeng, S. Ruedisser, A.D. Gossert, C. Hilty, Nuclear magnetic resonance of hyperpolarized fluorine for characterization of protein–ligand interactions, J. Am. Chem. Soc. 134 (42) (2012) 17448–17451.

[393] Y. Kim, C. Hilty, Affinity screening using competitive binding with fluorine-19 hyperpolarized ligands, Angew. Chem. Int. Ed. 54 (16) (2015) 4941–4944.

[394] S. Reynolds, H. Patel, Monitoring the solid-state polarization of 13C, 15N, 2H, 29Si and 31P, Appl. Magn. Reson. 34 (3) (2008) 495–508.

[395] A. Dementyev, D. Cory, C. Ramanathan, Dynamic nuclear polarization in silicon microparticles, Phys. Rev. Lett. 100 (12) (2008) 127601.

[396] M. Lin, V. Breukels, T.W. Scheenen, J.M. Paulusse, Dynamic nuclear polarization of silicon carbide micro-and nanoparticles, ACS Appl. Mater. Interfaces 13 (26) (2021) 30835–30843.

[397] J.W. Aptekar, M.C. Cassidy, A.C. Johnson, R.A. Barton, M. Lee, A.C. Ogier, C. Vo, M.N. Anahtar, Y. Ren, S.N. Bhatia, C. Ramanathan, D.G. Cory, A.L. Hill, R.W. Mair, M.S. Rosen, R.L. Walsworth, C.M. Marcus, Silicon nanoparticles as hyperpolarized magnetic resonance imaging agents, ACS Nano 3 (12) (2009) 4003–4008.

[398] M.C. Cassidy, H.R. Chan, B.D. Ross, P.K. Bhattacharya, C.M. Marcus, In vivo magnetic resonance imaging of hyperpolarized silicon particles, Nat. Nanotechnol. 8 (5) (2013) 363–368.

[399] N. Koole, A. De Koning, M. De Bie, 31P relaxation in some organophosphorus compounds, J. Magn. Reson. 25 (2) (1977) 375–378.

[400] A. Nardi-Schreiber, A. Gamliel, T. Harris, G. Sapir, J. Sosna, J.M. Gomori, R. Katz-Brull, Biochemical phosphates observed using hyperpolarized 31 P in physiological aqueous solutions, Nat. Commun. 8 (1) (2017) 1–7.

[401] W.E. Morgan, J.R. Van Wazer, Phosphorus-31 spin-lattice relaxation. I. Aqueous orthophosphoric acid and some related systems, J. Am. Chem. Soc. 97 (22) (1975) 6347–6352.

[402] D.C. McCain, J.L. Markley, Phosphorus-31 spin-lattice relaxation in aqueous orthophosphate solutions, J. Am. Chem. Soc. 102 (17) (1980) 5559–5565.

[403] E.H. Suh, J. Ratnakar, J. Singh, Z. Kovacs, Dissolution dynamic nuclear polarization of the 77Se nucleus, Anal. Sens. (2022) e202200086.

[404] M. Pérez-Malo, G. Szabó, E. Eppard, A. Vagner, E. Brucher, I. Tóth, A. Maiocchi, E.H. Suh, Z. Kovács, Z. Baranyai, Improved efficacy of synthesizing* MIII-labeled DOTA complexes in binary mixtures of water and organic solvents. A combined radio-and physicochemical study, Inorg. Chem. 57 (10) (2018) 6107–6117.

[405] A.K. Jindal, M.E. Merritt, E.H. Suh, C.R. Malloy, A.D. Sherry, Z. Kovács, Hyperpolarized 89Y complexes as pH sensitive NMR probes, J. Am. Chem. Soc. 132 (6) (2010) 1784–1785.

[406] M.E. Merritt, C. Harrison, Z. Kovacs, P. Kshirsagar, C.R. Malloy, A.D. Sherry, Hyperpolarized 89Y offers the potential of direct imaging of metal ions in biological systems by magnetic resonance, J. Am. Chem. Soc. 129 (43) (2007) 12942–12943.

[407] Y. Xing, A.K. Jindal, M. Regueiro-Figueroa, M. Le Fur, N. Kervarec, P. Zhao, Z. Kovacs, L. Valencia, P. Pérez-Lourido, R. Tripier, D. Esteban-Gómez, C. Platas-Iglesias, A.D. Sherry, On the relationship between NMR chemical shifts of thermally polarized and hyperpolarized 89Y complexes and their solution structures, Chem. Eur J. 22 (46) (2016) 16657.

[408] S. Jannin, L. Helm, G. Bodenhausen, Kinetics of yttrium– ligand complexation monitored using hyperpolarized 89Y as a model for gadolinium in contrast agents, J. Am. Chem. Soc. 132 (14) (2010) 5006–5007.

[409] K. Zangger, Silver and gold NMR, Met. Base. Drugs 6 (4–5) (1999) 239–245.

[410] L. Lumata, M.E. Merritt, Z. Hashami, S.J. Ratnakar, Z. Kovacs, Production and NMR characterization of hyperpolarized 107,109 Ag complexes, Angew. Chem. 124 (2) (2012) 540–542.

[411] H.E. Möller, Z.I. Cleveland, B. Driehuys, Relaxation of hyperpolarized 129Xe in a deflating polymer bag, J. Magn. Reson. 212 (1) (2011) 109–115.

[412] T. Asanuma, T. Hiraga, M. Hattori, Long-distance delivery of laser-polarized 129Xe using a capillary tube, J. Phys. Commun. 4 (10) (2020) 101001.

[413] M.S. Albert, V.D. Schepkin, T.F. Budinger, Measurement of 129Xe T1 in blood to explore the feasibility of hyperpolarized 129Xe MRI, J. Comput. Assist. Tomogr. 19 (6) (1995) 975–978.

[414] J. Wolber, A. Cherubini, A.S. Dzik-Jurasz, M.O. Leach, A. Bifone, Spin-lattice relaxation of laser-polarized xenon in human blood, Proc. Natl. Acad. Sci. 96 (7) (1999) 3664–3669.

[415] T.G. Walker, W. Happer, Spin-exchange optical pumping of noble-gas nuclei, Rev. Mod. Phys. 69 (2) (1997) 629.

[416] J.R. Birchall, M.R.H. Chowdhury, P. Nikolaou, Y.A. Chekmenev, A. Shcherbakov, M.J. Barlow, B.M. Goodson, E.Y. Chekmenev, Pilot quality-assurance study of a third-generation batch-mode clinical-scale automated xenon-129 hyperpolarizer, Molecules 27 (4) (2022) 1327.

[417] P. Nikolaou, A.M. Coffey, L.L. Walkup, B.M. Gust, N. Whiting, H. Newton, S. Barcus, I. Muradyan, M. Dabaghyan, G.D. Moroz, M.S. Rosen, S. Patz, M.J. Barlow, E.Y. Chekmenev, B.M. Goodson, Near-unity nuclear polarization with an open-source 129Xe hyperpolarizer for NMR and MRI, Proc. Natl. Acad. Sci. 110 (35) (2013) 14150–14155.

[418] A.L. Zook, B.B. Adhyaru, C.R. Bowers, High capacity production of> 65% spin polarized xenon-129 for NMR spectroscopy and imaging, J. Magn. Reson. 159 (2) (2002) 175–182.

[419] I. Ruset, S. Ketel, F.W. Hersman, Optical pumping system design for large production of hyperpolarized Xe 129, Phys. Rev. Lett. 96 (5) (2006) 053002.

[420] C. Witte, M. Kunth, J. Döpfert, F. Rossella, L. Schröder, Hyperpolarized xenon for NMR and MRI applications, JoVE 67 (2012) e4268.

[421] J.-N. Hyacinthe, A. Capozzi, A. Comment, Beyond spin exchange optical pumping: hyperpolarization of 129Xe via sublimation dynamic nuclear polarization, in: Hyperpolarized Xenon-129 Magnetic Resonance, 2015, pp. 442–452.

[422] G.L. Pollack, R.P. Kennan, J.F. Himm, P.W. Carr, Solubility of xenon in 45 organic solvents including cycloalkanes, acids, and alkanals: experiment and theory, J. Chem. Phys. 90 (11) (1989) 6569–6579.

[423] A. Capozzi, C. Roussel, A. Comment, J.-N. Hyacinthe, Optimal glass-forming solvent brings sublimation dynamic nuclear polarization to 129Xe hyperpolarization biomedical imaging standards, J. Phys. Chem. C 119 (9) (2015) 5020–5025.

[424] M. Pourfathi, J. Clapp, S.J. Kadlecek, C.D. Keenan, R.K. Ghosh, N.N. Kuzma, R.R. Rizi, Low-temperature dynamic nuclear polarization of gases in Frozen mixtures, Magn. Reson. Med. 76 (3) (2016) 1007–1014.

[425] N. Kuzma, B. Patton, K. Raman, W. Happer, Fast nuclear spin relaxation in hyperpolarized solid 129 Xe, Phys. Rev. Lett. 88 (14) (2002) 147602.

[426] B. Vuichoud, E. Canet, J. Milani, A.l. Bornet, D. Baudouin, L. Veyre, D. Gajan, L. Emsley, A. Lesage, C. Copéret, Hyperpolarization of frozen hydrocarbon gases by dynamic nuclear polarization at 1.2 K, J. Phys. Chem. Lett. 7 (16) (2016) 3235–3239.

[427] M. Albert, G. Cates, B. Driehuys, W. Happer, B. Saam, C. Springer, A. Wishnia, Biological magnetic resonance imaging using laser-polarized 129Xe, Nature 370 (6486) (1994) 199–201.

[428] J.P. Mugler III, B. Driehuys, J.R. Brookeman, G.D. Cates, S.S. Berr, R.G. Bryant, T.M. Daniel, E.E. De Lange, J.H. Downs, C.J. Erickson, W. Happer, D.P. Hinton, N.F. Kassel, T. Maier, C.D. Phillips, B.T. Saam, K.L. Sauer, M.E. Wagshul, MR imaging and spectroscopy using hyperpolarized 129Xe gas: preliminary human results, Magn. Reson. Med. 37 (6) (1997) 809–815.

[429] H.E. Möller, X.J. Chen, B. Saam, K.D. Hagspiel, G.A. Johnson, T.A. Altes, E.E. de Lange, H.U. Kauczor, MRI of the lungs using hyperpolarized noble gases, Magn. Reson. Med.: J. Int. Soc. Magn. Reson. Med. 47 (6) (2002) 1029–1051.

[430] E.E. de Lange, J.P. Mugler III, J.R. Brookeman, J. Knight-Scott, J.D. Truwit, C.D. Teates, T.M. Daniel, P.L. Bogorad, G.D. Cates, Lung air spaces: MR imaging evaluation with hyperpolarized 3He gas, Radiology 210 (3) (1999) 851–857.

[431] H. Middleton, R.D. Black, B. Saam, G.D. Cates, G.P. Cofer, R. Guenther, W. Happer, L.W. Hedlund, G.A. Johnson, K. Juvan, MR imaging with hyperpolarized 3He gas, Magn. Reson. Med. 33 (2) (1995) 271–275.

[432] M. Albert, D. Balamore, Development of hyperpolarized noble gas MRI, Nucl. Instrum. Methods. Phys. Res. Sect. A 402 (2–3) (1998) 441–453.

[433] L.L. Walkup, J.C. Woods, Translational applications of hyperpolarized 3He and 129Xe, NMR Biomed. 27 (12) (2014) 1429–1438.

[434] M. Rao, N.J. Stewart, G. Norquay, P.D. Griffiths, J.M. Wild, High resolution spectroscopy and chemical shift imaging of hyperpolarized 129Xe dissolved in the human brain in vivo at 1.5 tesla, Magn. Reson. Med. 75 (6) (2016) 2227–2234.

[435] J. Kershaw, K. Nakamura, Y. Kondoh, A. Wakai, N. Suzuki, I. Kanno, Confirming the existence of five peaks in 129Xe rat head spectra, Magn. Reson. Med.: J. Int. Soc. Magn. Reson. Med. 57 (4) (2007) 791–797.

[436] Y. Wang, I.J. Dmochowski, An expanded palette of xenon-129 NMR biosensors, Acc. Chem. Res. 49 (10) (2016) 2179–2187.

[437] K.K. Palaniappan, M.B. Francis, A. Pines, D.E. Wemmer, Molecular sensing using hyperpolarized xenon NMR spectroscopy, Isr. J. Chem. 54 (1-2) (2014) 104–112.

[438] M.M. Spence, E.J. Ruiz, S.M. Rubin, T.J. Lowery, N. Winssinger, P.G. Schultz, D.E. Wemmer, A. Pines, Development of a functionalized xenon biosensor, J. Am. Chem. Soc. 126 (46) (2004) 15287–15294.

[439] C. Hilty, S. Bowen, Applications of dynamic nuclear polarization to the study of reactions and reagents in organic and biomolecular chemistry, Org. Biomol. Chem. 8 (15) (2010) 3361–3365.

[440] N. Kotera, N. Tassali, E. Léonce, C. Boutin, P. Berthault, T. Brotin, J.P. Dutasta, L. Delacour, T. Traoré, D.A. Buisson, P. Berthault, T. Brotin, J.-P. Dutasta, L. Delacour, T. Traoré, D.-A. Buisson, F. Taran, S. Coudert, B. Rousseau, A sensitive zinc-activated 129Xe MRI probe, Angew. Chem. Int. Ed. 51 (17) (2012) 4100–4103.

[441] J. Jayapaul, L. Schröder, Molecular sensing with host systems for hyperpolarized 129Xe, Molecules 25 (20) (2020) 4627.

[442] D.R. Jacobson, N.S. Khan, R. Collé, R. Fitzgerald, L. Laureano-Pérez, Y. Bai, I.J. Dmochowski, Measurement of radon and xenon binding to a cryptophane molecular host, Proc. Natl. Acad. Sci. 108 (27) (2011) 10969–10973.

[443] T. Brotin, J.-P. Dutasta, Cryptophanes and their complexes present and future, Chem. Rev. 109 (1) (2009) 88–130.

[444] K. Bartik, M. Luhmer, J.-P. Dutasta, A. Collet, J. Reisse, 129Xe and 1H NMR study of the reversible trapping of xenon by cryptophane-A in organic solution, J. Am. Chem. Soc. 120 (4) (1998) 784–791.

[445] S.-I. Han, S. Garcia, T.J. Lowery, E.J. Ruiz, J.A. Seeley, L. Chavez, D.S. King, D.E. Wemmer, A. Pines, NMR-based biosensing with optimized delivery of polarized 129Xe to solutions, Anal. Chem. 77 (13) (2005) 4008–4012.

[446] L. Schröder, T.J. Lowery, C. Hilty, D.E. Wemmer, A. Pines, Molecular imaging using a targeted magnetic resonance hyperpolarized biosensor, Science 314 (5798) (2006) 446–449.

[447] L. Schröder, Xenon for NMR biosensing—inert but alert, Phys. Med. 29 (1) (2013) 3–16.

[448] S.D. Zemerov, I.J. Dmochowski, Cryptophane—xenon complexes for 129 Xe MRI applications, RSC Adv. 11 (13) (2021) 7693–7703.

[449] M.M. Spence, S.M. Rubin, I.E. Dimitrov, E.J. Ruiz, D.E. Wemmer, A. Pines, S.Q. Yao, F. Tian, P.G. Schultz, Functionalized xenon as a biosensor, Proc. Natl. Acad. Sci. 98 (19) (2001) 10654–10657.

[450] A.E. Truxal, L. Cao, L. Isaacs, D.E. Wemmer, A. Pines, Directly functionalized cucurbit [7] uril as a biosensor for the selective detection of protein interactions by 129Xe HyperCEST NMR, Chem. Eur J. 25 (24) (2019) 6108–6112.

[451] Y. Wang, I.J. Dmochowski, Cucurbit [6] uril is an ultrasensitive 129 Xe NMR contrast agent, Chem. Commun. 51 (43) (2015) 8982–8985.

[452] M.G. Shapiro, R.M. Ramirez, L.J. Sperling, G. Sun, J. Sun, A. Pines, D.V. Schaffer, V.S. Bajaj, Genetically encoded reporters for hyperpolarized xenon magnetic resonance imaging, Nat. Chem. 6 (7) (2014) 629–634.

[453] K. Du, S.D. Zemerov, S. Hurtado Parra, J.M. Kikkawa, I.J. Dmochowski, Paramagnetic organocobalt capsule revealing xenon host—guest chemistry, Inorg. Chem. 59 (19) (2020) 13831–13844.

[454] J. Jayapaul, L. Schröder, Nanoparticle-based contrast agents for 129Xe HyperCEST NMR and MRI applications, Contrast Media Mol. Imaging 2019 (2019).

[455] L. Delacour, N. Kotera, T. Traoré, S. Garcia-Argote, C. Puente, F. Leteurtre, E. Gravel, N. Tassali, C. Boutin, E. Léonce, Y. Boulard, P. Berthault, B. Rousseau, "Clickable" hydrosoluble PEGylated cryptophane as a universal platform for 129Xe magnetic resonance imaging biosensors, Chem. Eur J. 19 (19) (2013) 6089–6093.

[456] H.A. Fogarty, P. Berthault, T. Brotin, G. Huber, H. Desvaux, J.-P. Dutasta, A cryptophane core optimized for xenon encapsulation, J. Am. Chem. Soc. 129 (34) (2007) 10332–10333.

[457] Y. Bai, P.A. Hill, I.J. Dmochowski, Utilizing a water-soluble cryptophane with fast xenon exchange rates for picomolar sensitivity NMR measurements, Anal. Chem. 84 (22) (2012) 9935–9941.

[458] J.A. Finbloom, C.C. Slack, C.J. Bruns, K. Jeong, D.E. Wemmer, A. Pines, M.B. Francis, Rotaxane-mediated suppression and activation of cucurbit [6] uril for molecular detection by 129 Xe hyperCEST NMR, Chem. Commun. 52 (15) (2016) 3119–3122.

[459] K.I. Assaf, W.M. Nau, Cucurbiturils: from synthesis to high-affinity binding and catalysis, Chem. Soc. Rev. 44 (2) (2015) 394–418.

[460] C.C. Slack, J.A. Finbloom, K. Jeong, C.J. Bruns, D.E. Wemmer, A. Pines, M.B. Francis, Rotaxane probes for protease detection by 129 Xe hyperCEST NMR, Chem. Commun. 53 (6) (2017) 1076–1079.

[461] S.H. Klass, A.E. Truxal, T.A. Fiala, J. Kelly, D. Nguyen, J.A. Finbloom, D.E. Wemmer, A. Pines, M.B. Francis, Rotaxane probes for the detection of hydrogen peroxide by 129Xe hyperCEST NMR spectroscopy, Angew. Chem. Int. Ed. 58 (29) (2019) 9948–9953.

[462] J. Goodman, J.J. Neil, J.J. Ackerman, Biomedical applications of 133Cs NMR, NMR Biomed. 18 (2) (2005) 125–134.

[463] F.W. Wehrli, Spin-lattice relaxation of some less common weakly quadrupolar nuclei, J. Magn. Reson. 30 (2) (1978) 193–209.

[464] M. Karlsson, J.H. Ardenkjær-Larsen, M.H. Lerche, Hyperpolarized 133 Cs is a sensitive probe for real-time monitoring of biophysical environments, Chem. Commun. 53 (49) (2017) 6625–6628.

[465] L. Wittenkeller, D. Mota de Freitas, C.F. Geraldes, A.J. Tome, Physical basis for the resolution of intra-and extracellular cesium-133 NMR resonances in cesium (+) loaded human erythrocyte suspensions in the presence and absence of shift reagents, Inorg. Chem. 31 (7) (1992) 1135–1144.

[466] P.W. Kuchel, M. Karlsson, M.H. Lerche, D. Shishmarev, J.H. Ardenkjaer-Larsen, Rapid zero-trans kinetics of Cs+ exchange in human erythrocytes quantified by dissolution hyperpolarized 133 Cs+ NMR spectroscopy, Sci. Rep. 9 (1) (2019) 1–12.

[467] J. Singh, E.H. Suh, G. Sharma, C. Khemtong, A.D. Sherry, Z. Kovacs, Probing carbohydrate metabolism using hyperpolarized (13) C-labeled molecules, NMR Biomed. 32 (10) (2019) e4018.

[468] K.R. Keshari, D.M. Wilson, Chemistry and biochemistry of 13 C hyperpolarized magnetic resonance using dynamic nuclear polarization, Chem. Soc. Rev. 43 (5) (2014) 1627–1659.

[469] J.G. Skinner, L. Menichetti, A. Flori, A. Dost, A.B. Schmidt, M. Plaumann, F.A. Gallagher, J.-B. Hövener, Metabolic and molecular imaging with hyperpolarised tracers, Mol. Imaging Biol. 20 (2018) 902–918.

[470] N.J. Stewart, T. Sato, N. Takeda, H. Hirata, S. Matsumoto, Hyperpolarized 13C magnetic resonance imaging as a tool for imaging tissue redox state, oxidative stress, inflammation, and cellular metabolism, Antioxid. Redox Signal. 36 (1–3) (2022) 81–94.

[471] K. Golman, R. in 't Zandt, M. Thaning, Real-time metabolic imaging, Proc. Natl. Acad. Sci. 103 (30) (2006) 11270–11275.

[472] K. Golman, J.S. Petersson, Metabolic imaging and other applications of hyperpolarized 13C1, Acad. Radiol. 13 (8) (2006) 932–942.

[473] A. Apps, J. Lau, M. Peterzan, S. Neubauer, D. Tyler, O. Rider, Hyperpolarised magnetic resonance for in vivo real-time metabolic imaging, Heart 104 (18) (2018) 1484–1491.

[474] M.A. Schroeder, K. Clarke, S. Neubauer, D.J. Tyler, Hyperpolarized magnetic resonance: a novel technique for the in vivo assessment of cardiovascular disease, Circulation 124 (14) (2011) 1580–1594.

[475] S. Månsson, E. Johansson, P. Magnusson, C.-M. Chai, G. Hansson, J.S. Petersson, F. Ståhlberg, K. Golman, 13 C imaging—a new diagnostic platform, Eur. Radiol. 16 (2006) 57–67.

[476] F. Gallagher, M. Kettunen, K. Brindle, Biomedical applications of hyperpolarized 13C magnetic resonance imaging, Prog. Nucl. Magn. Reson. Spectrosc. 4 (55) (2009) 285–295.

[477] K.M. Brindle, Imaging metabolism with hyperpolarized 13C-labeled cell substrates, J. Am. Chem. Soc. 137 (20) (2015) 6418–6427.

[478] M. Mishkovsky, A. Comment, Hyperpolarized MRS: new tool to study real-time brain function and metabolism, Anal. Biochem. 529 (2017) 270–277.

[479] A. Comment, M.E. Merritt, Hyperpolarized magnetic resonance as a sensitive detector of metabolic function, Biochemistry 53 (47) (2014) 7333–7357.

[480] K.M. Brindle, S.E. Bohndiek, F.A. Gallagher, M.I. Kettunen, Tumor imaging using hyperpolarized 13C magnetic resonance spectroscopy, Magn. Reson. Med. 66 (2) (2011) 505–519.

[481] A. Cho, J.Y. Lau, B.J. Geraghty, C.H. Cunningham, K.R. Keshari, Noninvasive interrogation of cancer metabolism with hyperpolarized 13C MRI, J. Nucl. Med. 58 (8) (2017) 1201–1206.

[482] K. Golman, M. Lerche, R. Pehrson, J.H. Ardenkjaer-Larsen, Metabolic imaging by hyperpolarized 13C magnetic resonance imaging for in vivo tumor diagnosis, Cancer Res. 66 (22) (2006) 10855–10860.

[483] O. Warburg, F. Wind, E. Negelein, The metabolism of tumors in the body, J. Gen. Physiol. 8 (6) (1927) 519–530.

[484] S. Josan, K. Billingsley, J. Orduna, J.M. Park, R. Luong, L. Yu, R. Hurd, A. Pfefferbaum, D. Spielman, D. Mayer, Assessing inflammatory liver injury in an acute CCl4 model using dynamic 3D metabolic imaging of hyperpolarized [1-(13)C]pyruvate, NMR Biomed. 28 (12) (2015) 1671–1677.

[485] A.Z. Lau, A.P. Chen, N.R. Ghugre, V. Ramanan, W.W. Lam, K.A. Connelly, G.A. Wright, C.H. Cunningham, Rapid multislice imaging of hyperpolarized 13C pyruvate and bicarbonate in the heart, Magn. Reson. Med. 64 (5) (2010) 1323–1331.

[486] H.A. Yoshihara, HP agents and biochemical interactions, Adv. Magn. Reson. Technol. Appl. 3 (2021) 93–127.

[487] C. Hundshammer, M. Braeuer, C.A. Müller, A.E. Hansen, M. Schillmaier, S. Düwel, B. Feuerecker, S.J. Glaser, A. Haase, W. Weichert, Y. Boulard, P. Berthault, B. Rousseau, Simultaneous characterization of tumor cellularity and the Warburg effect with PET, MRI and hyperpolarized 13C-MRSI, Theranostics 8 (17) (2018) 4765.

[488] J. Ma, C.R. Malloy, S. Pena, C.E. Harrison, J. Ratnakar, V.G. Zaha, J.M. Park, Dual-phase imaging of cardiac metabolism using hyperpolarized pyruvate, Magn. Reson. Med. 87 (1) (2022) 302–311.

[489] Q. Wang, C. Parish, P. Niedbalski, J. Ratnakar, Z. Kovacs, L. Lumata, Hyperpolarized 89Y-EDTMP complex as a chemical shift-based NMR sensor for pH at the physiological range, J. Magn. Reson. 320 (2020) 106837.

[490] R. Hata, H. Nonaka, Y. Takakusagi, K. Ichikawa, S. Sando, Design of a hyperpolarized (15)N NMR probe that induces a large chemical-shift change upon binding of calcium ions, Chem. Commun. (Camb) 51 (61) (2015) 12290–12292.

[491] S. Duwel, C. Hundshammer, M. Gersch, B. Feuerecker, K. Steiger, A. Buck, A. Walch, A. Haase, S.J. Glaser, M. Schwaiger, F. Schilling, Imaging of pH in vivo using hyperpolarized (13)C-labelled zymonic acid, Nat. Commun. 8 (2017) 15126.

[492] R.R. Flavell, C. von Morze, J.E. Blecha, D.E. Korenchan, M. Van Criekinge, R. Sriram, J.W. Gordon, H.Y. Chen, S. Subramaniam, R.A. Bok, Z.J. Wang, D.B. Vigneron, P.E. Larson, J. Kurhanewicz, D.M. Wilson, Application of Good's buffers to pH imaging using hyperpolarized (13)C MRI, Chem. Commun. (Camb) 51 (74) (2015) 14119–14122.

[493] D.E. Korenchan, C. Taglang, C. von Morze, J.E. Blecha, J.W. Gordon, R. Sriram, P.E.Z. Larson, D.B. Vigneron, H.F. VanBrocklin, J. Kurhanewicz, D.M. Wilson, R.R. Flavell, Dicarboxylic acids as pH sensors for hyperpolarized (13)C magnetic resonance spectroscopic imaging, Analyst 142 (9) (2017) 1429–1433.

[494] T. Nishihara, Y. Kameyama, H. Nonaka, Y. Takakusagi, F. Hyodo, K. Ichikawa, S. Sando, A strategy to design hyperpolarized 13C magnetic resonance probes using [1-13C] α-amino acid as a scaffold structure, Chem. Asian J. 12 (9) (2017) 949–953.

[495] A. Mishra, G. Pariani, T. Oerther, M. Schwaiger, G.G. Westmeyer, Hyperpolarized multi-metal 13C-sensors for magnetic resonance imaging, Anal. Chem. 88 (22) (2016) 10790–10794.

[496] K. W Lipso, P. Magnusson, J. Henrik Ardenkjaer-Larsen, Hyperpolarized 13C MR angiography, Curr. Pharm. Des. 22 (1) (2016) 90–95.

[497] V.A. Levin, Relationship of octanol/water partition coefficient and molecular weight to rat brain capillary permeability, J. Med. Chem. 23 (6) (1980) 682–684.

[498] K. Golman, J.H. Ardenkjaer-Larsen, J. Svensson, O. Axelsson, G. Hansson, L. Hansson, H. Johannesson, I. Leunbach, S. Mansson, J.S. Petersson, G. Pettersson, R. Servin, L.G. Wistrand, 13C-angiography, Acad. Radiol. 9 (Suppl. 2) (2002) S507–S510.

[499] E. Johansson, S. Månsson, R. Wirestam, J. Svensson, J. Petersson, K. Golman, F. Ståhlberg, Cerebral perfusion assessment by bolus tracking using hyperpolarized 13C, Magn. Reson. Med.: J. Int. Soc. Magn. Reson. Med. 51 (3) (2004) 464–472.

[500] J. Svensson, S. Mansson, E. Johansson, J.S. Petersson, L.E. Olsson, Hyperpolarized 13C MR angiography using trueFISP, Magn. Reson. Med. 50 (2) (2003) 256–262.

[501] S.J. Lee, I. Park, J.F. Talbott, J. Gordon, Investigating the feasibility of in vivo perfusion imaging methods for spinal cord using hyperpolarized [(13)C]t-Butanol and [(13)C,(15)N2]Urea, Mol. Imaging Biol. 24 (2021) 371–376.

[502] K. Golman, J.H. Ardenkjaer-Larsen, J.S. Petersson, S. Mansson, I. Leunbach, Molecular imaging with endogenous substances, Proc. Natl. Acad. Sci. USA 100 (18) (2003) 10435–10439.

[503] G.D. Reed, C. von Morze, R. Bok, B.L. Koelsch, M. Van Criekinge, K.J. Smith, S. Hong, P.E. Larson, J. Kurhanewicz, D.B. Vigneron, High resolution (13)C MRI with hyperpolarized urea: in vivo T(2) mapping and (15)N labeling effects, IEEE Trans. Med. Imaging 33 (2) (2014) 362–371.

[504] A.K. Grant, E. Vinogradov, X. Wang, R.E. Lenkinski, D.C. Alsop, Perfusion imaging with a freely diffusible hyperpolarized contrast agent, Magn. Reson. Med. 66 (3) (2011) 746–755.

[505] C. von Morze, P.E. Larson, S. Hu, K. Keshari, D.M. Wilson, J.H. Ardenkjaer-Larsen, A. Goga, R. Bok, J. Kurhanewicz, D.B. Vigneron, Imaging of blood flow using hyperpolarized [(13)C]urea in preclinical cancer models, J. Magn. Reson. Imaging 33 (3) (2011) 692–697.

[506] C. Laustsen, T. Stokholm Norlinger, D. Christoffer Hansen, H. Qi, P. Mose Nielsen, L. Bonde Bertelsen, J. Henrik Ardenkjaer-Larsen, H. Stodkilde Jorgensen, Hyperpolarized 13C urea relaxation mechanism reveals renal changes in diabetic nephropathy, Magn. Reson. Med. 75 (2) (2016) 515–518.

[507] E. Milshteyn, G.D. Reed, J.W. Gordon, C. von Morze, P. Cao, S. Tang, A.P. Leynes, P.E.Z. Larson, D.B. Vigneron, Simultaneous T1 and T2 mapping of hyperpolarized (13)C compounds using the bSSFP sequence, J. Magn. Reson. 312 (2020) 106691.

[508] P.M. Nielsen, E.S. Szocska Hansen, T.S. Norlinger, R. Norregaard, L. Bonde Bertelsen, H. Stodkilde Jorgensen, C. Laustsen, Renal ischemia and reperfusion assessment with three-dimensional hyperpolarized (13) C,(15) N2-urea, Magn. Reson. Med. 76 (5) (2016) 1524–1530.

[509] G.D. Reed, C. von Morze, A.S. Verkman, B.L. Koelsch, M.M. Chaumeil, M. Lustig, S.M. Ronen, R.A. Bok, J.M. Sands, P.E. Larson, Z.J. Wang, J.H. Larsen, J. Kurhanewicz, D.B. Vigneron, Imaging renal urea handling in rats at millimeter resolution using hyperpolarized magnetic resonance relaxometry, Tomography 2 (2) (2016) 125–135.

[510] M. Durst, E. Chiavazza, A. Haase, S. Aime, M. Schwaiger, R.F. Schulte, alpha-trideuteromethyl[15N]glutamine: a long-lived hyperpolarized perfusion marker, Magn. Reson. Med. 76 (6) (2016) 1900–1904.

[511] K.W. Lipsø, E.S.S. Hansen, R.S. Tougaard, C. Laustsen, J.H. Ardenkjær-Larsen, Dynamic coronary MR angiography in a pig model with hyperpolarized water, Magn. Reson. Med. 80 (3) (2018) 1165–1169.

[512] A. Gamliel, S. Uppala, G. Sapir, T. Harris, A. Nardi-Schreiber, D. Shaul, J. Sosna, J.M. Gomori, R. Katz-Brull, Hyperpolarized [(15)N]nitrate as a potential long lived hyperpolarized contrast agent for MRI, J. Magn. Reson. 299 (2019) 188–195.

[513] C. Gabellieri, S. Reynolds, A. Lavie, G.S. Payne, M.O. Leach, T.R. Eykyn, Therapeutic target metabolism observed using hyperpolarized 15N choline, J. Am. Chem. Soc. 130 (14) (2008) 4598–4599.

[514] C. von Morze, J.A. Engelbach, G.D. Reed, A.P. Chen, J.D. Quirk, T. Blazey, R. Mahar, C.R. Malloy, J.R. Garbow, M.E. Merritt, (15) N-carnitine, a novel endogenous hyperpolarized MRI probe with long signal lifetime, Magn. Reson. Med. 85 (4) (2021) 1814–1820.

[515] X. Li, G. Chen, B. Yang, Urea transporter physiology studied in knockout mice, Front. Physiol. 3 (2012) 217.

[516] J. Kim, D. Jo, S.H. Yang, C.G. Joo, N. Whiting, S. Pudakalakatti, H. Seo, H.Y. Son, S.J. Min, P. Bhattacharya, Y.M. Huh, J.H. Shim, Y. Lee, (29) Si isotope-enriched silicon nanoparticles for an efficient hyperpolarized magnetic resonance imaging probe, ACS Appl. Mater. Interfaces 13 (48) (2021) 56923–56930.

[517] E.R. McCarney, B.D. Armstrong, M.D. Lingwood, S. Han, Hyperpolarized water as an authentic magnetic resonance imaging contrast agent, Proc. Natl. Acad. Sci. 104 (6) (2007) 1754–1759.

[518] R.E. Lenkinski, Science to practice: can hyperpolarized water be used to enhance MR angiography and flow measurement? Radiology 265 (2) (2012) 325–326.

[519] M.R. Rao, N.J. Stewart, P.D. Griffiths, G. Norquay, J.M. Wild, Imaging human brain perfusion with inhaled hyperpolarized 129Xe MR imaging, Radiology 286 (2) (2018) 659–665.

[520] Y. Shepelytskyi, F.T. Hane, V. Grynko, T. Li, A. Hassan, M.S. Albert, Hyperpolarized 129Xe time-of-flight MR imaging of perfusion and brain function, Diagnostics 10 (9) (2020) 630.

[521] J.E. Roos, H.P. McAdams, S.S. Kaushik, B. Driehuys, Hyperpolarized gas MR imaging: technique and applications, Magn. Reson. Imaging Clin. 23 (2) (2015) 217–229.

[522] M. Goyen, Gadofosveset-enhanced magnetic resonance angiography, Vasc. Health Risk Manag. 4 (1) (2008) 1.

[523] H. Qin, S. Tang, A.M. Riselli, R.A. Bok, R. Delos Santos, M. van Criekinge, J.W. Gordon, R. Aggarwal, R. Chen, G. Goddard, Clinical translation of hyperpolarized 13C pyruvate and urea MRI for simultaneous metabolic and perfusion imaging, Magn. Reson. Med. 87 (1) (2022) 138–149.

[524] H. Qin, V. Zhang, R.A. Bok, R.D. Santos, J.A. Cunha, I.-C. Hsu, J.D. Santos, J.E. Lee, S. Sukumar, P.E. Larson, Simultaneous metabolic and perfusion imaging using hyperpolarized 13C MRI can evaluate early and dose-dependent response to radiation therapy in a prostate cancer mouse model, Int. J. Radiat. Oncol. Biol. Phys. 107 (5) (2020) 887–896.

[525] G. Zhang, C. Hilty, Applications of dissolution dynamic nuclear polarization in chemistry and biochemistry, Magn. Reson. Chem. 56 (7) (2018) 566–582.

[526] M.M. Chaumeil, C. Najac, S.M. Ronen, Studies of metabolism using 13C MRS of hyperpolarized probes, Methods Enzymol. 561 (2015) 1–71.

[527] N. Salvi, Applications of hyperpolarisation and NMR long-lived states in drug screening, in: Annual Reports on NMR Spectroscopy vol. 96, Elsevier, 2019, pp. 1–33.

[528] O. Szekely, G.L. Olsen, M. Novakovic, R. Rosenzweig, L. Frydman, Assessing site-specific enhancements imparted by hyperpolarized water in folded and unfolded proteins by 2D HMQC NMR, J. Am. Chem. Soc. 142 (20) (2020) 9267–9284.

[529] T.H. Witney, M.I. Kettunen, K.M. Brindle, Kinetic modeling of hyperpolarized 13C label exchange between pyruvate and lactate in tumor cells, J. Biol. Chem. 286 (28) (2011) 24572–24580.

[530] M.F. Penet, T. Shah, F. Wildes, B. Krishnamachary, S.K. Bharti, J. Pacheco-Torres, D. Artemov, Z.M. Bhujwalla, MRI and MRS of intact perfused cancer cell metabolism, invasion, and stromal cell interactions, NMR Biomed. 32 (10) (2019) e4053.

[531] F. Cruz, S. Cerdán, Cells studied by NMR, in: J.C. Lindon (Ed.), Encyclopedia of Spectroscopy and Spectrometry, second ed., Academic Press, Oxford, 1999, pp. 212–220.

[532] M. Neeman, H. Degani, Metabolic studies of estrogen-and tamoxifen-treated human breast cancer cells by nuclear magnetic resonance spectroscopy, Cancer Res. 49 (3) (1989) 589–594.

[533] R.M. Tostões, S.B. Leite, J.P. Miranda, M. Sousa, D.I. Wang, M.J. Carrondo, P.M. Alves, Perfusion of 3D encapsulated hepatocytes—a synergistic effect enhancing long-term functionality in bioreactors, Biotechnol. Bioeng. 108 (1) (2011) 41–49.

[534] S. Ronen, E. Rushkin, H. Degani, Lipid metabolism in T47D human breast cancer cells: 31P and13C-NMR studies of choline and ethanolamine uptake, Biochim. Biophys. Acta, Mol. Cell. Res. 1095 (1) (1991) 5–16.

[535] C.S. Ward, H.S. Venkatesh, M.M. Chaumeil, A.H. Brandes, M. VanCriekinge, H. Dafni, S. Sukumar, S.J. Nelson, D.B. Vigneron, J. Kurhanewicz, Noninvasive detection of target modulation following phosphatidylinositol 3-kinase inhibition using hyperpolarized 13C magnetic resonance spectroscopy, Cancer Res. 70 (4) (2010) 1296–1305.

[536] M.M. Chaumeil, P.E. Larson, S.M. Woods, L. Cai, P. Eriksson, A.E. Robinson, J.M. Lupo, D.B. Vigneron, S.J. Nelson, R.O. Pieper, Hyperpolarized [1-13C] glutamate: a metabolic imaging biomarker of IDH1 mutational status in glioma, Cancer Res. 74 (16) (2014) 4247–4257.

[537] S.P. Wolfe, E. Hsu, L.M. Reid, J.M. Macdonald, A novel multi-coaxial hollow fiber bioreactor for adherent cell types. Part 1: hydrodynamic studies, Biotechnol. Bioeng. 77 (1) (2002) 83–90.

[538] K.M. Brindle, Investigating the performance of intensive mammalian cell bioreactor systems using magnetic resonance imaging and spectroscopy, Biotechnol. Genet. Eng. Rev. 15 (1) (1998) 499–520.

[539] H.D. Ritchie, J.D. Hardcastle, J. Hermon-Taylor, A.D. Maclean, Isolated organ perfusion for physiologic studies, Am. J. Dig. Dis. 16 (1) (1971) 81–87.

[540] D.J. Hearse, F.J. Sutherland, Experimental models for the study of cardiovascular function and disease, Pharmacol. Res. 41 (6) (2000) 597–603.

[541] C. Khemtong, N.R. Carpenter, L.L. Lumata, M.E. Merritt, K.X. Moreno, Z. Kovacs, C.R. Malloy, A.D. Sherry, Hyperpolarized 13C NMR detects rapid drug-induced changes in cardiac metabolism, Magn. Reson. Med. 74 (2) (2015) 312–319.

[542] J. Ackerman, P. Bore, D. Gadian, T. Grove, G.K. Radda, Nmr studies of metabolism in perfused organs, Philos. Trans. R. Soc. B: Biol. Sci. 289 (1037) (1980) 425–436.

[543] S. Ad, N. Rl, P. Rm, Metabolic studies of pyruvate-and lactate-perfused Guinea pig hearts by 13C NMR, J. Biol. Chem. 260 (1985) 9272–9279.

[544] M.E. Merritt, C. Harrison, C. Storey, F.M. Jeffrey, A.D. Sherry, C.R. Malloy, Hyperpolarized 13C allows a direct measure of flux through a single enzyme-catalyzed step by NMR, Proc. Natl. Acad. Sci. 104 (50) (2007) 19773–19777.

[545] W. Chen, G. Sharma, W. Jiang, N.R. Maptue, C.R. Malloy, A.D. Sherry, C. Khemtong, Metabolism of hyperpolarized 13C-acetoacetate to β-hydroxybutyrate detects real-time mitochondrial redox state and dysfunction in heart tissue, NMR Biomed. 32 (6) (2019) e4091.

[546] G. Sharma, X. Wen, N.R. Maptue, T. Hever, C.R. Malloy, A.D. Sherry, C. Khemtong, Co-polarized [1-13C]pyruvate and [1,3-13C2]Acetoacetate provide a simultaneous view of cytosolic and mitochondrial redox in a single experiment, ACS Sens. 6 (2021) 3967–3977.

[547] M.A. Schroeder, L.E. Cochlin, L.C. Heather, K. Clarke, G.K. Radda, D.J. Tyler, In vivo assessment of pyruvate dehydrogenase flux in the heart using hyperpolarized carbon-13 magnetic resonance, Proc. Natl. Acad. Sci. 105 (2008) 12051–12056.

[548] G. Sharma, C.Y. Wu, R.M. Wynn, W. Gui, C.R. Malloy, A.D. Sherry, D.T. Chuang, C. Khemtong, Real-time hyperpolarized (13)C magnetic resonance detects increased pyruvate oxidation in pyruvate dehydrogenase kinase 2/4-double knockout mouse livers, Sci. Rep. 9 (1) (2019) 16480.

[549] K.X. Moreno, C.L. Moore, S.C. Burgess, A.D. Sherry, C.R. Malloy, M.E. Merritt, Production of hyperpolarized (13)CO2 from [1-(13)C]pyruvate in perfused liver does reflect total anaplerosis but is not a reliable biomarker of glucose production, Metabolomics 11 (5) (2015) 1144–1156.

[550] M.E. Merritt, C. Harrison, A.D. Sherry, C.R. Malloy, S.C. Burgess, Flux through hepatic pyruvate carboxylase and phosphoenolpyruvate carboxykinase detected by hyperpolarized 13C magnetic resonance, Proc. Natl. Acad. Sci. USA 108 (47) (2011) 19084–19089.

[551] C.O. Mariager, E.S.S. Hansen, S.K. Bech, A. Munk, U. Kjaergaard, M.D. Lyhne, K. Soberg, P.F. Nielsen, S. Ringgaard, C. Laustsen, Graft assessment of the ex vivo perfused porcine kidney using hyperpolarized [1-(13) C]pyruvate, Magn. Reson. Med. 84 (5) (2020) 2645–2655.

[552] S.C. Kolwicz Jr., R. Tian, Assessment of cardiac function and energetics in isolated mouse hearts using 31P NMR spectroscopy, JoVE (42) (2010) e2069.

[553] A. Ferrigno, P. Richelmi, M. Vairetti, Troubleshooting and improving the mouse and rat isolated perfused liver preparation, J. Pharmacol. Toxicol. Methods 67 (2) (2013) 107–114.

[554] M. Watanabe, T. Okada, Langendorff perfusion method as an ex vivo model to evaluate heart function in rats. Experimental Models of Cardiovascular Diseases: Methods and Protocols, 2018, pp. 107–116.

[555] R.M. Bell, M.M. Mocanu, D.M. Yellon, Retrograde heart perfusion: the Langendorff technique of isolated heart perfusion, J. Mol. Cell. Cardiol. 50 (6) (2011) 940–950.

[556] L.E. Bailey, S.D. Ong, Krebs-Henseleit solution as a physiological buffer in perfused and superfused preparations, J. Pharmacol. Methods 1 (2) (1978) 171–175.

[557] S. Siddiqui, S. Kadlecek, M. Pourfathi, Y. Xin, W. Mannherz, H. Hamedani, N. Drachman, K. Ruppert, J. Clapp, R. Rizi, The use of hyperpolarized carbon-13 magnetic resonance for molecular imaging, Adv. Drug Deliv. Rev. 113 (2017) 3–23.

[558] J. Bankson, P.E.Z. Larson, Analysis and visualization of hyperpolarized 13C MR data, in: P.E.Z. Larson (Ed.), Hyperpolarized Carbon-13 Magnetic Resonance Imaging and Spectroscopy, vol. 3, Academic Press, 2021, pp. 129–155.

[559] J.C. Crane, J.W. Gordon, H.Y. Chen, A.W. Autry, Y. Li, M.P. Olson, J. Kurhanewicz, D.B. Vigneron, P.E.Z. Larson, D. Xu, Hyperpolarized C-13 MRI data acquisition and analysis in prostate and brain at University of California, San Francisco, NMR Biomed. 34 (5) (2021).

[560] R.E. Hurd, D. Spielman, S. Josan, Y.F. Yen, A. Pfefferbaum, D. Mayer, Exchange-linked dissolution agents in dissolution-DNP (13)C metabolic imaging, Magn. Reson. Med. 70 (4) (2013) 936–942.

[561] T. Xu, D. Mayer, M. Gu, Y.F. Yen, S. Josan, J. Tropp, A. Pfefferbaum, R. Hurd, D. Spielman, Quantification of in vivo metabolic kinetics of hyperpolarized pyruvate in rat kidneys using dynamic 13C MRSI, NMR Biomed. 24 (8) (2011) 997–1005.

[562] M.L. Zierhut, Y.-F. Yen, A.P. Chen, R. Bok, M.J. Albers, V. Zhang, J. Tropp, I. Park, D.B. Vigneron, J. Kurhanewicz, Kinetic modeling of hyperpolarized 13C1-pyruvate metabolism in normal rats and TRAMP mice, J. Magn. Reson. 202 (1) (2010) 85–92.

[563] C. Harrison, C. Yang, A. Jindal, R.J. DeBerardinis, M. Hooshyar, M. Merritt, A. Dean Sherry, C.R. Malloy, Comparison of kinetic models for analysis of pyruvate-to-lactate exchange by hyperpolarized 13C NMR, NMR Biomed. 25 (11) (2012) 1286–1294.

[564] C.L. Swisher, P.E. Larson, K. Kruttwig, A.B. Kerr, S. Hu, R.A. Bok, A. Goga, J.M. Pauly, S.J. Nelson, J. Kurhanewicz, Quantitative measurement of cancer metabolism using stimulated echo hyperpolarized carbon-13 MRS, Magn. Reson. Med. 71 (1) (2014) 1–11.

[565] F. Ahamed, M. Van Criekinge, Z.J. Wang, J. Kurhanewicz, P. Larson, R. Sriram, Modeling hyperpolarized lactate signal dynamics in cells, patient-derived tissue slice cultures and murine models, NMR Biomed. 34 (3) (2021) e4467.

[566] J. Bankson, P.E. Larson, Analysis and visualization of hyperpolarized 13C MR data, Adv. Magn. Reson. Technol. Appl. 3 (2021) 129–155.

[567] Y. Zhuo, C.D. Cordeiro, S.K. Hekmatyar, R. Docampo, J.H. Prestegard, Dynamic nuclear polarization facilitates monitoring of pyruvate metabolism in Trypanosoma brucei, J. Biol. Chem. 292 (44) (2017) 18161–18168.

[568] C. Yang, C. Harrison, E.S. Jin, D.T. Chuang, A.D. Sherry, C.R. Malloy, M.E. Merritt, R.J. DeBerardinis, Simultaneous steady-state and dynamic 13C NMR can differentiate alternative routes of pyruvate metabolism in living cancer cells, J. Biol. Chem. 289 (9) (2014) 6212–6224.

[569] D. Mammoli, J. Gordon, A. Autry, P.E. Larson, Y. Li, H.-Y. Chen, B. Chung, P. Shin, M. Van Criekinge, L. Carvajal, Kinetic modeling of hyperpolarized carbon-13 pyruvate metabolism in the human brain, IEEE Trans. Med. Imaging 39 (2) (2019) 320–327.

[570] S. Kazan, S. Reynolds, A. Kennerley, E. Wholey, J. Bluff, J. Berwick, V. Cunningham, M. Paley, G. Tozer, Kinetic modeling of hyperpolarized 13C pyruvate metabolism in tumors using a measured arterial input function, Magn. Reson. Med. 70 (4) (2013) 943–953.

[571] C.R. Malloy, A.D. Sherry, F. Jeffrey, Analysis of tricarboxylic acid cycle of the heart using 13C isotope isomers, Am. J. Physiol. Heart Circ. 259 (3) (1990) H987–H995.

[572] A.D. Sherry, C.R. Malloy, Cardiac metabolism, in: NMR Physiol. Biomed., Elsevier, 1994, pp. 439–449.

[573] B. Bequette, N. Sunny, S. El-Kadi, S. Owens, Application of stable isotopes and mass isotopomer distribution analysis to the study of intermediary metabolism of nutrients, Anim. Sci. J. 84 (Suppl. l_13) (2006) E50–E59.

[574] M. Ruiz, R. Gélinas, F. Vaillant, B. Lauzier, C. Des Rosiers, Metabolic tracing using stable isotope-labeled substrates and mass spectrometry in the perfused mouse heart, Methods Enzymol. 561 (2015) 107–147.

[575] J.R. Alger, A. Minhajuddin, A. Dean Sherry, C.R. Malloy, Analysis of steady-state carbon tracer experiments using akaike information criteria, Metabolomics 17 (7) (2021) 61.

[576] M. Rahim, M. Ragavan, S. Deja, M.E. Merritt, S.C. Burgess, J.D. Young, INCA 2.0: a tool for integrated, dynamic modeling of NMR- and MS-based isotopomer measurements and rigorous metabolic flux analysis, Metab. Eng. 69 (2022) 275–285.

[577] H.C. Urey, F.G. Brickwedde, G.M. Murphy, A hydrogen isotope of mass 2, Phys. Rev. 39 (1) (1932) 164.

[578] S. Sadhukhan, Y. Han, G.-F. Zhang, H. Brunengraber, G.P. Tochtrop, Using isotopic tools to dissect and quantitate parallel metabolic pathways, J. Am. Chem. Soc. 132 (18) (2010) 6309–6311.

[579] E. Ciccimaro, I.A. Blair, Stable-isotope dilution LC–MS for quantitative biomarker analysis, Bioanalysis 2 (2) (2010) 311–341.

[580] X. Song, J. Li, M. Mofidfar, R.N. Zare, Distinguishing between isobaric ions using microdroplet hydrogen–deuterium exchange mass spectrometry, Metabolites 11 (11) (2021) 728.

[581] D. Shemin, An illustration of the use of isotopes: the biosynthesis of porphyrins, Bioessays 10 (1) (1989) 30–35.

[582] G.P. Tochtrop, G.T. DeKoster, D.P. Cistola, D.F. Covey, Synthesis of [3, 4-13C2]-enriched bile salts as NMR probes of Protein– ligand interactions, J. Org. Chem. 67 (19) (2002) 6764–6771.

[583] H. Park, Q. Wang, State-of-the-art accounts of hyperpolarized 15 N-labeled molecular imaging probes for magnetic resonance spectroscopy and imaging, Chem. Sci. 13 (25) (2022) 7378–7391.

[584] W. Fernelius, T. Coyle, W. Powell, Nomenclature of inorganic chemistry. 12. Isotopically modified compounds, Pure Appl. Chem. 51 (1979) 1981–1994.

[585] K.E. Wilzbach, Tritium-labeling by exposure of organic compounds to tritium GAS1, J. Am. Chem. Soc. 79 (4) (1957), 1013-1013.

[586] M. Shree, K. Masakapalli, S., Intracellular fate of universally labelled 13C isotopic tracers of glucose and xylose in central metabolic pathways of Xanthomonas oryzae, Metabolites 8 (4) (2018) 66.

[587] L. Bluck, D.A. Volmer, The role of naturally occurring stable isotopes in mass spectrometry, part I: the theory, Spectroscopy 23 (10) (2009) 36.

[588] P. Muller, Glossary of terms used in physical organic chemistry (IUPAC Recommendations 1994), Pure Appl. Chem. 66 (5) (1994) 1077–1184.

[589] H. Allouche-Arnon, A. Gamliel, C.M. Barzilay, R. Nalbandian, J.M. Gomori, M. Karlsson, M.H. Lerche, R. Katz-Brull, A hyperpolarized choline molecular probe for monitoring acetylcholine synthesis, Contrast Media Mol. Imaging 6 (3) (2011) 139–147.

[590] T. Maltseva, A. Földesi, J. Chattopadhyaya, T1 and T2 relaxations of the 13C nuclei of deuterium-labeled nucleosides, Magn. Reson. Chem. 36 (4) (1998) 227–239.

[591] T.B. Rodrigues, E.M. Serrao, B.W. Kennedy, D.-E. Hu, M.I. Kettunen, K.M. Brindle, Magnetic resonance imaging of tumor glycolysis using hyperpolarized 13 C-labeled glucose, Nat. Med. 20 (1) (2014) 93–97.

[592] T. Doura, R. Hata, H. Nonaka, K. Ichikawa, S. Sando, Design of a 13C magnetic resonance probe using a deuterated methoxy group as a long-lived hyperpolarization unit, Angew. Chem. 124 (40) (2012) 10261–10264.

[593] K. Kumagai, K. Kawashima, M. Akakabe, M. Tsuda, T. Abe, M. Tsuda, Synthesis and hyperpolarized 15N NMR studies of 15N-choline-d13, Tetrahedron 69 (19) (2013) 3896–3900.

[594] A. Blomquist, R. Cedergren, B.F. Hiscock, S.L. Tripp, D.N. Harpp, Synthesis of highly deuterated amino acids, Proc. Natl. Acad. Sci. USA 55 (3) (1966) 453.

[595] J. Atzrodt, V. Derdau, T. Fey, J. Zimmermann, The renaissance of H/D exchange, Angew. Chem. Int. Ed. 46 (41) (2007) 7744–7765.

[596] J. Atzrodt, V. Derdau, W.J. Kerr, M. Reid, Deuterium-and tritium-labelled compounds: applications in the life sciences, Angew. Chem. Int. Ed. 57 (7) (2018) 1758–1784.

[597] A. Michelotti, M. Roche, 40 Years of hydrogen–deuterium exchange adjacent to heteroatoms: a survey, Synthesis 51 (06) (2019) 1319–1328.

[598] P. Grocholska, R. Bąchor, Trends in the hydrogen– deuterium exchange at the carbon centers. Preparation of internal standards for quantitative analysis by LC-MS, Molecules 26 (10) (2021) 2989.

[599] C. Schmidt, First deuterated drug approved, Nat. Biotechnol. 35 (6) (2017) 493–495.

[600] S. Cargnin, M. Serafini, T. Pirali, A primer of deuterium in drug design, Future Sci. 11 (2019) 2039–2042.

[601] G.S. Timmins, Deuterated drugs: where are we now? Expert Opin. Ther. Pat. 24 (10) (2014) 1067–1075.

[602] M. Zhan, T. Zhang, H. Huang, Y. Xie, Y. Chen, A simple method for α-position deuterated carbonyl compounds with pyrrolidine as catalyst, J. Label. Compd. Radiopharm. 57 (8) (2014) 533–539.

[603] T. Maegawa, Y. Fujiwara, Y. Inagaki, Y. Monguchi, H. Sajiki, A convenient and effective method for the regioselective deuteration of alcohols, Adv. Synth. Catal. 350 (14-15) (2008) 2215–2218.

[604] Y. Sawama, Y. Monguchi, H. Sajiki, Efficient H–D exchange reactions using heterogeneous platinum-group metal on carbon–H2–D2O system, Synlett 23 (07) (2012) 959–972.

[605] A. Barb, S. Hekmatyar, J. Glushka, J. Prestegard, Probing alanine transaminase catalysis with hyperpolarized 13CD3-pyruvate, J. Magn. Reson. 228 (2013) 59–65.

[606] R.V. Shchepin, A.M. Coffey, K.W. Waddell, E.Y. Chekmenev, Parahydrogen Induced polarization of 1-13C-phospholactate-d 2 for Biomedical imaging with> 30,000,000-fold NMR signal enhancement in water, Anal. Chem. 86 (12) (2014) 5601–5605.

[607] B. Vuichoud, J. Milani, A. Bornet, R. Melzi, S. Jannin, G. Bodenhausen, Hyperpolarization of deuterated metabolites via remote cross-polarization and dissolution dynamic nuclear polarization, J. Phys. Chem. B 118 (5) (2014) 1411–1415.

[608] A. Copper, A. Redfield, Proton magnetic resonance studies of alpha-keto acids, J. Biol. Chem. 250 (2) (1975) 527–532.

[609] A. Michelotti, F. Rodrigues, M. Roche, Development and scale-up of stereoretentive α-deuteration of amines, Org. Process Res. Dev. 21 (11) (2017) 1741–1744.

[610] C. Taglang, L.M. Martínez-Prieto, I. del Rosal, L. Maron, R. Poteau, K. Philippot, B. Chaudret, S. Perato, A. Sam Lone, C. Puente, Enantiospecific C-H activation using ruthenium nanocatalysts, Angew. Chem. 127 (36) (2015) 10620–10623.

[611] B. Chatterjee, C. Gunanathan, Ruthenium catalyzed selective α-and α, β-deuteration of alcohols using D2O, Org. Lett. 17 (19) (2015) 4794–4797.

[612] B. Chatterjee, V. Krishnakumar, C. Gunanathan, Selective α-deuteration of amines and amino acids using D2O, Org. Lett. 18 (22) (2016) 5892–5895.

[613] Y. Sawama, Y. Yabe, H. Iwata, Y. Fujiwara, Y. Monguchi, H. Sajiki, Stereo-and regioselective direct multi-deuterium-labeling methods for sugars, Chem. Eur J. 18 (51) (2012) 16436–16442.

[614] G. Pieters, C. Taglang, E. Bonnefille, T. Gutmann, C. Puente, J.C. Berthet, C. Dugave, B. Chaudret, B. Rousseau, Regioselective and stereospecific deuteration of bioactive aza compounds by the use of ruthenium nanoparticles, Angew. Chem. Int. Ed. 53 (1) (2014) 230–234.

[615] L. Neubert, D. Michalik, S. Bähn, S. Imm, H. Neumann, J. Atzrodt, V. Derdau, W. Holla, M. Beller, Ruthenium-catalyzed selective α, β-deuteration of bioactive amines, J. Am. Chem. Soc. 134 (29) (2012) 12239–12244.
[616] C. Taglang, D.E. Korenchan, C. Von Morze, J. Yu, C. Najac, S. Wang, J.E. Blecha, S. Subramaniam, R. Bok, H.F. VanBrocklin, Late-stage deuteration of 13 C-enriched substrates for T 1 prolongation in hyperpolarized 13 C MRI, Chem. Commun. 54 (41) (2018) 5233–5236.
[617] K.B. Wiberg, The deuterium isotope effect, Chem. Rev. 55 (4) (1955) 713–743.
[618] A.M. Funk, B.L. Anderson, X. Wen, T. Hever, C. Khemtong, Z. Kovacs, A.D. Sherry, C.R. Malloy, The rate of lactate production from glucose in hearts is not altered by per-deuteration of glucose, J. Magn. Reson. 284 (2017) 86–93.
[619] M.H. O'Leary, Multiple isotope effects on enzyme-catalyzed reactions, Annu. Rev. Biochem. 58 (1) (1989) 377–401.
[620] J. Lehn, J. Kintzinger, Nitrogen-14 nuclear quadrupole effects, in: Nitrogen NMR, Springer, 1973, pp. 79–161.
[621] A. Cho, R. Eskandari, V.Z. Miloushev, K.R. Keshari, A non-synthetic approach to extending the lifetime of hyperpolarized molecules using D2O solvation, J. Magn. Reson. 295 (2018) 57–62.
[622] R. Eskandari, N. Kim, A. Mamakhanyan, M. Saoi, G. Zhang, M. Berishaj, K.L. Granlund, A.J. Poot, J. Cross, C.B. Thompson, Hyperpolarized [5-13C, 4, 4-2H2, 5-15N]-L-glutamine provides a means of annotating in vivo metabolic utilization of glutamine, Proc. Natl. Acad. Sci. 119 (19) (2022) e2120595119.
[623] A. Cho, R. Eskandari, K.L. Granlund, K.R. Keshari, Hyperpolarized [6-13C, 15N3]-arginine as a probe for in vivo arginase activity, ACS Chem. Biol. 14 (4) (2019) 665–673.
[624] N. Miura, C. Mushti, D. Sail, J.E. AbuSalim, K. Yamamoto, J.R. Brender, T. Seki, D.I. AbuSalim, S. Matsumoto, K.A. Camphausen, Synthesis of [1-13C-5-12C]-alpha-ketoglutarate enables noninvasive detection of 2-hydroxyglutarate, NMR Biomed. 34 (11) (2021) e4588.
[625] H.-L. Li, Y.-L. Ju, L.-J. Li, D.-G. Xu, Separation of isotope 13C using high-performance structured packing, Chem. Eng. Process 49 (3) (2010) 255–261.
[626] W. Zhang, S. Zhao, A.S. Serianni, Labeling monosaccharides with stable isotopes, Methods Enzymol. 565 (2015) 423–458.
[627] D. Ott, V. Kerr, T. Sanchez, T. Whaley, Syntheses with stable isotopes: sodium cyanide-13C, methane-13C, methane-13C-d4, and methane-d4, J. Label. Compd. Radiopharm. 17 (2) (1980) 255–262.
[628] M. Mishkovsky, B. Anderson, M. Karlsson, M.H. Lerche, A.D. Sherry, R. Gruetter, Z. Kovacs, A. Comment, Measuring glucose cerebral metabolism in the healthy mouse using hyperpolarized 13C magnetic resonance, Sci. Rep. 7 (1) (2017) 1–8.
[629] H. Allouche-Arnon, A. Gamliel, J. Sosna, J.M. Gomori, R. Katz-Brull, In vitro visualization of betaine aldehyde synthesis and oxidation using hyperpolarized magnetic resonance spectroscopy, Chem. Commun. 49 (63) (2013) 7076–7078.
[630] H. Allouche-Arnon, Y. Hovav, L. Friesen-Waldner, J. Sosna, J. Moshe Gomori, S. Vega, R. Katz-Brull, Quantification of rate constants for successive enzymatic reactions with DNP hyperpolarized MR, NMR Biomed. 27 (6) (2014) 656–662.
[631] L.J. Friesen-Waldner, T.P. Wade, K. Thind, A.P. Chen, J.M. Gomori, J. Sosna, C.A. McKenzie, R. Katz-Brull, Hyperpolarized choline as an MR imaging molecular probe: feasibility of in vivo imaging in a rat model, J. Magn. Reson. Imaging 41 (4) (2015) 917–923.
[632] P. Dzien, M.I. Kettunen, I. Marco-Rius, E.M. Serrao, T.B. Rodrigues, T.J. Larkin, K.N. Timm, K.M. Brindle, 13C magnetic resonance spectroscopic imaging of hyperpolarized [1-13C, U-2H5] ethanol oxidation can be used to assess aldehyde dehydrogenase activity in vivo, Magn. Reson. Med. 73 (5) (2015) 1733–1740.
[633] L. Jouhi, V. Koljonen, T. Böhling, C. Haglund, J. Hagström, The expression of Toll-like receptors 2, 4, 5, 7 and 9 in Merkel cell carcinoma, Anticancer Res. 35 (4) (2015) 1843–1849.
[634] K. Saito, D. Sail, K. Yamamoto, S. Matsumoto, B. Blackman, S. Kishimoto, J.R. Brender, R.E. Swenson, J.B. Mitchell, M.C. Krishna, Synthesis and evaluation of 13C-labeled 5-5-dimethyl-1-pyrroline-N-oxide aimed at in vivo detection of reactive oxygen species using hyperpolarized 13C-MRI, Free Radic. Biol. Med. 131 (2019) 18–26.
[635] D. Chen, Y. Shan, J. Li, J. You, X. Sun, G. Qiu, External Reductant-free palladium-catalyzed reductive insertion of Isocyanide: synthesis of Polysubstituted Pyrroles and its applications as a cysteine probe, Org. Lett. 21 (11) (2019) 4044–4048.
[636] K. Datta, S.-C. Liu, S.R. Lynch, Z. Chen, R. Hurd, J. Rao, D.M. Spielman, [13C]-tert-butanol-2-β-D-galactose: A Potential New Hyperpolarized Imaging Agent for in Vivo Imaging of Senescent Cells.
[637] D.H. Marchbank, R.G. Kerr, Semisynthesis of fuscoside B analogues and eunicosides, and analysis of anti-inflammatory activity, Tetrahedron 67 (17) (2011) 3053–3061.
[638] S. Borthakur, S. Baruah, B. Sarma, S. Gogoi, Pd (II)-catalyzed synthesis of alkylidene phthalides via a decarbonylative annulation reaction, Org. Lett. 21 (8) (2019) 2768–2771.
[639] M. Perkins, Spin trapping, Adv. Phys. Org. Chem. 17 (1980) 1–64.
[640] K.X. Moreno, S. Satapati, R.J. DeBerardinis, S.C. Burgess, C.R. Malloy, M.E. Merritt, Real-time detection of hepatic gluconeogenic and glycogenolytic states using hyperpolarized [2-13C] dihydroxyacetone, J. Biol. Chem. 289 (52) (2014) 35859–35867.
[641] J. Szammer, E. Simon-Trompler, J. Máté, M. Abermann, I. Klebovich, Synthesis and stability of [1, 3-14C2]-glyceryl trinitrate, J. Label. Compd. Radiopharm. 34 (11) (1994) 1111–1115.
[642] S. Sumino, A. Fusano, T. Fukuyama, I. Ryu, Carbonylation reactions of alkyl iodides through the interplay of carbon radicals and Pd catalysts, Acc. Chem. Res. 47 (5) (2014) 1563–1574.
[643] A. Brennführer, H. Neumann, M. Beller, Palladium-catalyzed carbonylation reactions of aryl halides and related compounds, Angew. Chem. Int. Ed. 48 (23) (2009) 4114–4133.
[644] M.-Y. She, D.-W. Xiao, B. Yin, Z. Yang, P. Liu, J.-L. Li, Z. Shi, An efficiently cobalt-catalyzed carbonylative approach to phenylacetic acid derivatives, Tetrahedron 69 (35) (2013) 7264–7268.

[645] L. Shi, B.P. Tu, Acetyl-CoA and the regulation of metabolism: mechanisms and consequences, Curr. Opin. Cell Biol. 33 (2015) 125–131.

[646] U. Koellisch, C.V. Gringeri, G. Rancan, E.V. Farell, M.I. Menzel, A. Haase, M. Schwaiger, R.F. Schulte, Metabolic imaging of hyperpolarized [1-13C] acetate and [1-13C] acetylcarnitine-investigation of the influence of dobutamine induced stress, Magn. Reson. Med. 74 (4) (2015) 1011–1018.

[647] J.A. Bastiaansen, T. Cheng, H. Lei, R. Gruetter, A. Comment, Direct noninvasive estimation of myocardial tricarboxylic acid cycle flux in vivo using hyperpolarized 13C magnetic resonance, J. Mol. Cell. Cardiol. 87 (2015) 129–137.

[648] M. Mishkovsky, A. Comment, R. Gruetter, In vivo detection of brain Krebs cycle intermediate by hyperpolarized magnetic resonance, J. Cereb. Blood Flow Metab 32 (12) (2012) 2108–2113.

[649] J.A. Bastiaansen, T. Cheng, M. Mishkovsky, J.M. Duarte, A. Comment, R. Gruetter, In vivo enzymatic activity of acetylCoA synthetase in skeletal muscle revealed by 13C turnover from hyperpolarized [1-13C] acetate to [1-13C] acetylcarnitine, Biochim. Biophys. Acta. Gen. Subj. 1830 (8) (2013) 4171–4178.

[650] P.R. Jensen, T. Peitersen, M. Karlsson, A. Gisselsson, G. Hansson, S. Meier, M.H. Lerche, Tissue-specific short chain fatty acid metabolism and slow metabolic recovery after ischemia from hyperpolarized NMR in vivo, J. Biol. Chem. 284 (52) (2009) 36077–36082.

[651] A. Flori, M. Liserani, F. Frijia, G. Giovannetti, V. Lionetti, V. Casieri, V. Positano, G.D. Aquaro, F.A. Recchia, M.F. Santarelli, Real-time cardiac metabolism assessed with hyperpolarized [1-13C] acetate in a large-animal model, Contrast Media Mol. Imaging 10 (3) (2015) 194–202.

[652] J. Steinhauser, P. Wespi, G. Kwiatkowski, S. Kozerke, Production of highly polarized [1-13C] acetate by rapid decarboxylation of [2-13C] pyruvate—application to hyperpolarized cardiac spectroscopy and imaging, Magn. Reson. Med. 82 (3) (2019) 1140–1149.

[653] D.R. Ball, B. Rowlands, M.S. Dodd, L. Le Page, V. Ball, C.A. Carr, K. Clarke, D.J. Tyler, Hyperpolarized butyrate: a metabolic probe of short chain fatty acid metabolism in the heart, Magn. Reson. Med. 71 (5) (2014) 1663–1669.

[654] J.A. Bastiaansen, M.E. Merritt, A. Comment, Measuring changes in substrate utilization in the myocardium in response to fasting using hyperpolarized [1-13C] butyrate and [1-13C] pyruvate, Sci. Rep. 6 (1) (2016) 25573.

[655] H. Yoshihara, J.A. Bastiaansen, M. Karlsson, M.H. Lerche, A. Comment, J. Schwitter, Myocardial fatty acid metabolism probed with hyperpolarized [1-13C] octanoate, J. Cardiovasc. Magn. Reson. 17 (1) (2015) 1–2.

[656] F.A. Gallagher, M.I. Kettunen, D.-E. Hu, P.R. Jensen, M. Karlsson, A. Gisselsson, S.K. Nelson, T.H. Witney, S.E. Bohndiek, G. Hansson, Production of hyperpolarized [1, 4-13C2] malate from [1, 4-13C2] fumarate is a marker of cell necrosis and treatment response in tumors, Proc. Natl. Acad. Sci. 106 (47) (2009) 19801–19806.

[657] N.M. Zacharias, H.R. Chan, N. Sailasuta, B.D. Ross, P. Bhattacharya, Real-time molecular imaging of tricarboxylic acid cycle metabolism in vivo by hyperpolarized 1-13C diethyl succinate, J. Am. Chem. Soc. 134 (2) (2012) 934–943.

[658] K.L. Billingsley, S. Josan, J.M. Park, S.S. Tee, E. Spielman-Sun, R. Hurd, D. Mayer, D. Spielman, Hyperpolarized [1, 4-13C]-diethylsuccinate: a potential DNP substrate for in vivo metabolic imaging, NMR Biomed. 27 (3) (2014) 356–362.

[659] W. Sakami, W.E. Evans, S. Gurin, The synthesis of organic compounds labelled with isotopic Carbon1, J. Am. Chem. Soc. 69 (5) (1947) 1110–1112.

[660] J.S. Lee, N. Lifson, Studies on the conversion of acetate, lactate and malonate to succinate in the intact rat, J. Biol. Chem. 193 (1) (1951) 253–263.

[661] J. Micklefield, K.J. Harris, S. Groeger, U. Mocek, H. Hilbi, P. Dimroth, H.G. Floss, Stereochemical course of malonate decarboxylation in Malonomonas rubra, J. Am. Chem. Soc. 117 (3) (1995) 1153–1154.

[662] P.M. Jordan, J.B. Spencer, D.L. Corina, The synthesis of chiral malonates and the determination of their chirality using fatty acid synthase and mass spectrometry, J. Chem. Soc., Chem. Commun. 12 (1986) 911–913.

[663] J.M. Park, S. Josan, T. Grafendorfer, Y.F. Yen, R.E. Hurd, D.M. Spielman, D. Mayer, Measuring mitochondrial metabolism in rat brain in vivo using MR Spectroscopy of hyperpolarized [2-13C] pyruvate, NMR Biomed. 26 (10) (2013) 1197–1203.

[664] M.A. Schroeder, H.J. Atherton, D.R. Ball, M.A. Cole, L.C. Heather, J.L. Griffin, K. Clarke, G.K. Radda, D.J. Tyler, Real-time assessment of Krebs cycle metabolism using hyperpolarized 13C magnetic resonance spectroscopy, Faseb. J. 23 (8) (2009) 2529.

[665] M.A. Schroeder, H.J. Atherton, M.S. Dodd, P. Lee, L.E. Cochlin, G.K. Radda, K. Clarke, D.J. Tyler, The cycling of acetyl-coenzyme A through acetylcarnitine buffers cardiac substrate supply: a hyperpolarized 13C magnetic resonance study, Circ. Cardiovasc. Imaging 5 (2) (2012) 201–209.

[666] B.T. Chung, H.-Y. Chen, J. Gordon, D. Mammoli, R. Sriram, A.W. Autry, L.M. Le Page, M.M. Chaumeil, P. Shin, J. Slater, First hyperpolarized [2-13C] pyruvate MR studies of human brain metabolism, J. Magn. Reson. 309 (2019) 106617.

[667] A.P. Chen, A.Z. Lau, Y.p. Gu, M.A. Schroeder, J. Barry, C.H. Cunningham, Probing the cardiac malate—aspartate shuttle non-invasively using hyperpolarized [1, 2-13C2] pyruvate, NMR Biomed. 31 (1) (2018) e3845.

[668] F. Winkler, K. Kühnl, R. Medina, R.S. Kaske, H. Schmidt, Principles and results of stable isotope labelling of L-α-Aminoacids by combined chemical and enzymatic methods, Isot. Environ. Health Stud. 31 (2) (1995) 161–190.

[669] U. Hollstein, D.L. Mock, R.R. Sibbitt, U. Roisch, F. Lingens, Synthesis of Shikimic-6-13C acid, J. Label. Compd. Radiopharm. 17 (2) (1980) 289–296.

[670] P. Roth, A. Hädener, C. Tamm, Further studies on the biosynthesis of tabtoxin (wildfire toxin): incorporation of [2, 3-13C2] pyruvate into the β-lactam moiety, Helv. Chim. Acta 73 (2) (1990) 476–482.

[671] N.M. Kelly, B.C. O'Neill, J. Probert, G. Reid, R. Stephen, T. Wang, C.L. Willis, P. Winton, Chemo-enzymatic syntheses of isotopically labelled L-amino acids, Tetrahedron Lett. 35 (35) (1994) 6533–6536.

[672] X. Creary, M. Keller, J.P. Dinnocenzo, Steric and conformational effects in the solvolysis of ring-fused tertiary cyclopropyl derivatives, J. Org. Chem. 43 (20) (1978) 3874–3878.

[673] X. Creary, Reaction of organometallic reagents with ethyl trifluoroacetate and diethyl oxalate. Formation of trifluoromethyl ketones and. alpha.-keto esters via stable tetrahedral adducts, J. Org. Chem. 52 (22) (1987) 5026–5030.

[674] E.J. Corey, D. Seebach, Synthesis of 1, n-dicarbonyl derivates using carbanions from 1, 3-dithianes, Angew. Chem. Int. Ed. Engl. 4 (12) (1965) 1077–1078.

[675] D. Seebach, E.J. Corey, Generation and synthetic applications of 2-lithio-1, 3-dithianes, J. Org. Chem. 40 (2) (1975) 231–237.

[676] J. Thirkettle, J. Baldwin, J. Griffin, C. Schofield, The origin of the β-lactam carbons of clavulanic acid, Chem. Commun. 11 (1997) 1025–1026.

[677] H. Anker, Synthesis of carbonyl-labeled pyruvic acid, J. Biol. Chem. 176 (3) (1948) 1333–1335.

[678] S. Margolis, B. Coxon, Identification and quantitation of the impurities in sodium pyruvate, Anal. Chem. 58 (12) (1986) 2504–2510.

[679] T. Harris, A. Gamliel, J. Sosna, J.M. Gomori, R. Katz-Brull, Impurities of [1-13C] pyruvic acid and a method to minimize their signals for hyperpolarized pyruvate metabolism studies, Appl. Magn. Reson. 49 (10) (2018) 1085–1098.

[680] R. Von Korff, [68] Purity and stability of pyruvate and α-ketoglutarate, Methods Enzymol. 13 (1969) 519–523.

[681] R. Von Korff, Pyruvate-C14, purity and stability, Anal. Biochem. 8 (2) (1964) 171–178.

[682] H. Goldfine, The formation of γ-hydroxy-γ-methylglutamic acid from a common impurity in pyruvic acid, Biochim. Biophys. Acta 40 (1960) 557–559.

[683] J. Bloomer, M. Gross, F. Kappler, G. Pandey, Identity of "zymonic acid" with a pyruvate derivative, J. Chem. Soc. D Chem. Commun. 16 (1970), 1030a-1030a.

[684] A.C. Rios, P.P. Bera, J.A. Moreno, G. Cooper, Pyruvate aldol condensation product: a metabolite that escaped synthetic preparation for over a century, ACS Omega 5 (25) (2020) 15063–15068.

[685] G. Buldain, C.D.L. Santos, B. Frydman, Carbon-13 nuclear magnetic resonance spectra of the hydrate, keto and enol forms of oxalacetic acid, Magn. Reson. Chem. 23 (6) (1985) 478–481.

[686] F.H. Stodola, O.L. Shotwell, L.B. Lockwood, Zymonic acid, a new metabolic product of some yeasts grown in aerated culture. I. Structure Studies2, J. Am. Chem. Soc. 74 (21) (1952) 5415–5418.

[687] C.M. Montgomery, J.L. Webb, Detection of a new inhibitor of the tricarboxylic acid cycle, Science 120 (3125) (1954) 843–844.

[688] C.M. Montgomery, J.L. Webb, Metabolic studies on heart mitochondria: II. The inhibitory action of parapyruvate on the tricarboxylic acid cycle, J. Biol. Chem. 221 (1) (1956) 359–368.

[689] I. Lee, N.-C. Yang, Using taguchi method to determine the Optimum conditions for synthesizing parapyruvate, Molecules 27 (6) (2022) 1870.

[690] C. Hundshammer, S. Düwel, S.S. Köcher, M. Gersch, B. Feuerecker, C. Scheurer, A. Haase, S.J. Glaser, M. Schwaiger, F. Schilling, Deuteration of hyperpolarized 13C-labeled zymonic acid enables sensitivity-enhanced dynamic MRI of pH, ChemPhysChem 18 (18) (2017) 2422–2425.

[691] M. Grashei, P. Wodtke, J.G. Skinner, S. Sühnel, N. Setzer, T. Metzler, S. Gulde, M. Park, D. Witt, H. Mohr, C. Hundshammer, N. Strittmatter, N.S. Pellegata, K. Steiger, F. Schilling, Simultaneous magnetic resonance imaging of pH, perfusion and renal filtration using hyperpolarized ^{13}C-labelled Z-OMPD, Nat. Commun. 14 (1) (2023) 5060.

[692] R.E. Hurd, Y.F. Yen, D. Mayer, A. Chen, D. Wilson, S. Kohler, R. Bok, D. Vigneron, J. Kurhanewicz, J. Tropp, Metabolic imaging in the anesthetized rat brain using hyperpolarized [1-13C] pyruvate and [1-13C] ethyl pyruvate, Magn. Reson. Med. 63 (5) (2010) 1137–1143.

[693] Y. Lee, N. Zacharias, D. Piwnica-Worms, P.K. Bhattacharya, Chemical reaction-induced multi-molecular polarization (CRIMP), Chem. Commun. 50 (86) (2014) 13030–13033.

[694] L.F. Huergo, R. Dixon, The emergence of 2-oxoglutarate as a master regulator metabolite, Microbiol. Mol. Biol. Rev. 79 (4) (2015) 419–435.

[695] L. Dang, D.W. White, S. Gross, B.D. Bennett, M.A. Bittinger, E.M. Driggers, V.R. Fantin, H.G. Jang, S. Jin, M.C. Keenan, Cancer-associated IDH1 mutations produce 2-hydroxyglutarate, Nature 462 (7274) (2009) 739–744.

[696] X. Du, H. Hu, The roles of 2-Hydroxyglutarate, Front. Cell Dev. Biol. 9 (2021) 486.

[697] C. Houillier, X. Wang, G. Kaloshi, K. Mokhtari, R. Guillevin, J. Laffaire, S. Paris, B. Boisselier, A. Idbaih, F. Laigle-Donadey, IDH1 or IDH2 mutations predict longer survival and response to temozolomide in low-grade gliomas, Neurology 75 (17) (2010) 1560–1566.

[698] M.M. Chaumeil, P.E. Larson, H.A. Yoshihara, O.M. Danforth, D.B. Vigneron, S.J. Nelson, R.O. Pieper, J.J. Phillips, S.M. Ronen, Non-invasive in vivo assessment of IDH1 mutational status in glioma, Nat. Commun. 4 (1) (2013) 1–12.

[699] T.S. Viswanathan, R.E. Johnson, H.F. Fisher, alpha.-Ketoglutaric acid: solution structure and the active form for reductive amination by bovine liver glutamate dehydrogenase, Biochemistry 21 (2) (1982) 339–345.

[700] T.T. Zengeya, R.A. Kulkarni, J.L. Meier, Modular synthesis of cell-permeating 2-ketoglutarate esters, Org. Lett. 17 (10) (2015) 2326–2329.

[701] J. Cappon, J. Baart, G. van der Walk, J. Raap, J. Lugtenburg, Chemo-enzymatic synthesis of specifically stable-isotope labelledL-glutamic acid and 2-oxoglutaric acid, Recl. Trav. Chim. Pays-Bas 110 (5) (1991) 158–166.

[702] E.M.M. Bottorff, α-Oxoglutaric acid, Org. Synth. 44 (1964) 67–69.

[703] S.J. Parker, J. Encarnación-Rosado, K.E. Hollinshead, D.M. Hollinshead, L.J. Ash, J.A. Rossi, E.Y. Lin, A.S. Sohn, M.R. Philips, D.R. Jones, Spontaneous hydrolysis and spurious metabolic properties of α-ketoglutarate esters, Nat. Commun. 12 (1) (2021) 1–15.

[704] E.H. Suh, E.P. Hackett, R.M. Wynn, D.T. Chuang, B. Zhang, W. Luo, A.D. Sherry, J.M. Park, In vivo assessment of increased oxidation of branched-chain amino acids in glioblastoma, Sci. Rep. 9 (1) (2019) 1–9.

[705] R.J. Lichtenecker, N. Coudevylle, R. Konrat, W. Schmid, Selective isotope labelling of leucine residues by using α-ketoacid precursor compounds, Chembiochem 14 (7) (2013) 818–821.

[706] C. Von Morze, R.A. Bok, M.A. Ohliger, Z. Zhu, D.B. Vigneron, J. Kurhanewicz, Hyperpolarized [13C] ketobutyrate, a molecular analog of pyruvate with modified specificity for LDH isoforms, Magn. Reson. Med. 75 (5) (2016) 1894–1900.

[707] J.D. Gross, V.M. Gelev, G. Wagner, A sensitive and robust method for obtaining intermolecular NOEs between side chains in large protein complexes, J. Biomol. NMR 25 (2003) 235–242.

[708] A.R. Lippert, K.R. Keshari, J. Kurhanewicz, C.J. Chang, A hydrogen peroxide-responsive hyperpolarized 13C MRI contrast agent, J. Am. Chem. Soc. 133 (11) (2011) 3776–3779.

[709] H.H. Wasserman, W.-B. Ho, (Cyanomethylene) phosphoranes as novel carbonyl 1, 1-dipole synthons: an efficient synthesis of. alpha.-keto acids, esters, and amides, J. Org. Chem. 59 (16) (1994) 4364–4366.

[710] P. Nshimiyimana, L. Liu, G. Du, Engineering of L-amino acid deaminases for the production of α-keto acids from L-amino acids, Bioengineered 10 (1) (2019) 43–51.

[711] C. Najac, M. Radoul, L.M. Le Page, G. Batsios, E. Subramani, P. Viswanath, A.M. Gillespie, S.M. Ronen, In vivo investigation of hyperpolarized [1, 3-13C2] acetoacetate as a metabolic probe in normal brain and in glioma, Sci. Rep. 9 (1) (2019) 1–10.

[712] D. Abdurrachim, C.C. Woo, X.Q. Teo, W.X. Chan, G.K. Radda, P.T.H. Lee, A new hyperpolarized 13C ketone body probe reveals an increase in acetoacetate utilization in the diabetic rat heart, Sci. Rep. 9 (1) (2019) 1–13.

[713] P.R. Jensen, S.C. Serra, L. Miragoli, M. Karlsson, C. Cabella, L. Poggi, L. Venturi, F. Tedoldi, M.H. Lerche, Hyperpolarized [1, 3-13C2] ethyl acetoacetate is a novel diagnostic metabolic marker of liver cancer, Int. J. Cancer 136 (4) (2015) E117–E126.

[714] K.A. Reynolds, D. O'Hagan, D. Gani, J.A. Robinson, Butyrate metabolism in streptomycetes. Characterization of an intramolecular vicinal interchange rearrangement linking isobutyrate and butyrate in Streptomyces cinnamonensis, J. Chem. Soc., Perkin Trans. 1 (12) (1988) 3195–3207.

[715] G. D'alessandro, G. Sleiter, The synthesis of pyrrole derivatives labelled with 13C in selected positions: improved procedures, J. Label. Compd. Radiopharm. 17 (6) (1980) 813–824.

[716] P. Hütter, K.-P. Zeller, Eine einfache Methode zur Herstellung 13C-markierter Ethyl-acetoacetate, Synthesis 1985 (03) (1985) 334–335.

[717] D.S. Breslow, E. Baumgarten, C.R. Hauser, A new synthesis of β-keto esters of the type RCOCH2COOC2H51, 2, J. Am. Chem. Soc. 66 (8) (1944) 1286–1288.

[718] W.G. Dauben, H.L. Bradlow, The synthesis of C14-labeled "Squalene", J. Am. Chem. Soc. 74 (20) (1952) 5204–5206.

[719] H. Dahn, H. Hauth, Über die Synthese von Acetessigester-[1-14C] und α, β-Diketobuttersäureester-[1-14C]. 12. Mitteilung über Reduktone und 1, 2, 3-Tricarbonylverbindungen, Helv. Chim. Acta 42 (4) (1959) 1214–1224.

[720] A.F. Creemers, J. Lugtenburg, The preparation of all-trans uniformly 13C-labeled retinal via a modular total organic synthetic strategy. Emerging central contribution of organic synthesis toward the structure and function study with atomic resolution in protein research, J. Am. Chem. Soc. 124 (22) (2002) 6324–6334.

[721] W. Qu, B. Hu, J.W. Babich, N. Waterhouse, M. Dooley, S. Ponnala, J. Urgiles, A general 11C-labeling approach enabled by fluoride-mediated desilylation of organosilanes, Nat. Commun. 11 (1) (2020) 1–10.

[722] R. Hay, M. Bond, Kinetics of the decarboxylation of acetoacetic acid, Aust. J. Chem. 20 (9) (1967) 1823–1828.

[723] J.P. Guthrie, F. Jordan, Amine-catalyzed decarboxylation of acetoacetic acid. The rate constant for decarboxylation of a. beta.-imino acid, J. Am. Chem. Soc. 94 (26) (1972) 9136–9141.

[724] I. Fritzsche, P. Bührdel, R. Melcher, H. Böhme, Stability of ketone bodies in serum in dependence on storage time and storage temperature, Clin. Lab. 47 (7–8) (2001) 399–403.

[725] P.J. Vlachostergios, K.G. Oikonomou, E. Gibilaro, G. Apergis, Elevated lactic acid is a negative prognostic factor in metastatic lung cancer, Cancer Biomark. 15 (6) (2015) 725–734.

[726] H.-W. Sim, R. Nejad, W. Zhang, F. Nassiri, W. Mason, K.D. Aldape, G. Zadeh, E.X. Chen, Tissue 2-hydroxyglutarate as a biomarker for Isocitrate Dehydrogenase mutations in gliomas, Clin. Cancer Res. 25 (11) (2019) 3366–3373.

[727] M. Sheikh-Ali, B.S. Karon, A. Basu, Y.C. Kudva, L.A. Muller, J. Xu, W.F. Schwenk, J.M. Miles, Can serum β-hydroxybutyrate be used to diagnose diabetic ketoacidosis? Diabetes Care 31 (4) (2008) 643–647.

[728] M. Winitz, L. Bloch-Frankenthal, N. Izumiya, S.M. Birnbaum, C.G. Baker, J.P. Greenstein, Studies on diastereoisomeric α-amino acids and corresponding α-hydroxy acids. VII. Influence of β-configuration on enzymic susceptibility, J. Am. Chem. Soc. 78 (11) (1956) 2423–2430.

[729] N. Stuhr-Hansen, S. Padrah, K. Strømgaard, Facile synthesis of α-hydroxy carboxylic acids from the corresponding α-amino acids, Tetrahedron Lett. 55 (30) (2014) 4149–4151.

[730] H. Kunz, H.-G. Lerchen, Stereocontrolled synthesis of d-α-hydroxy carboxylic acid from l-amino acids, Tetrahedron Lett. 28 (17) (1987) 1873–1876.

[731] T.M. Phillips, G.W. Kosicki, D.E. Schmidt Jr., Stereoselective reduction of pyruvate by sodium borohydride catalyzed by pyruvate kinase, Biochim. Biophys. Acta Enzymol. 293 (1) (1973) 125–133.

[732] S. Noël, E. Caronia, H. Bricout, I.C. Tichá, S. Menuel, A. Ponchel, S. Tilloy, A. Galia, É. Monflier, J. Jindřich, Asymmetric hydrogenation of ethyl pyruvate over aqueous dispersed Pt nanoparticles stabilized by a cinchonidine-functionalized β-cyclodextrin, Catal. Commun. 150 (2021) 106272.

[733] Z. Zheng, B. Sheng, C. Gao, H. Zhang, T. Qin, C. Ma, P. Xu, Highly stereoselective biosynthesis of (R)-α-hydroxy carboxylic acids through rationally re-designed mutation of d-lactate dehydrogenase, Sci. Rep. 3 (1) (2013) 1–6.

[734] F.Q. Nuttall, A. Ngo, M.C. Gannon, Regulation of hepatic glucose production and the role of gluconeogenesis in humans: is the rate of gluconeogenesis constant? Diabetes/Metab. Res. Rev. 24 (6) (2008) 438–458.

[735] G.A. Brooks, Mammalian fuel utilization during sustained exercise, Comp. Biochem. Physiol. B Biochem. Mol. Biol. 120 (1) (1998) 89–107.

[736] R.S.D.O. Cruz, R.A. de Aguiar, T. Turnes, R. Penteado Dos Santos, M. Fernandes Mendes de Oliveira, F. Caputo, Intracellular shuttle: the lactate aerobic metabolism, Sci. World J. 2012 (2012).

[737] A.P. Chen, J. Kurhanewicz, R. Bok, D. Xu, D. Joun, V. Zhang, S.J. Nelson, R.E. Hurd, D.B. Vigneron, Feasibility of using hyperpolarized [1-13C] lactate as a substrate for in vivo metabolic 13C MRSI studies, Magn. Reson. Imaging 26 (6) (2008) 721–726.

[738] J. Chen, E.P. Hackett, Z. Kovacs, C.R. Malloy, J.M. Park, Assessment of hepatic pyruvate carboxylase activity using hyperpolarized [1-13C]-l-lactate, Magn. Reson. Med. 85 (3) (2021) 1175–1182.

[739] J.M. Park, S. Josan, R.E. Hurd, J. Graham, P.J. Havel, D. Bendahan, D. Mayer, Y. Chung, D.M. Spielman, T. Jue, Hyperpolarized NMR study of the impact of pyruvate dehydrogenase kinase inhibition on the pyruvate dehydrogenase and TCA flux in type 2 diabetic rat muscle, Pflug. Arch. Eur. J. Physiol. 473 (11) (2021) 1761–1773.

[740] J.-N. Hyacinthe, L. Buscemi, T.P. Lê, M. Lepore, L. Hirt, M. Mishkovsky, Evaluating the potential of hyperpolarised [1-13C] L-lactate as a neuroprotectant metabolic biosensor for stroke, Sci. Rep. 10 (1) (2020) 1–11.

[741] A.Z. Lau, A.P. Chen, C.H. Cunningham, Cardiac metabolic imaging using hyperpolarized [1-13C] lactate as a substrate, NMR Biomed. 34 (7) (2021) e4532.

[742] J.N. Starr, G. Westhoff, Lactic acid, Ullmann's Encycl. Ind. Chem. (2000) 1–8.

[743] A.P. Chen, J.Y. Lau, R.D. Alvares, C.H. Cunningham, Using [1-13C] lactic acid for hyperpolarized 13C MR cardiac studies, Magn. Reson. Med. 73 (6) (2015) 2087–2093.

[744] H. Borsook, H.M. Huffman, Y.-P. Liu, The preparation of crystalline lactic acid, J. Biol. Chem. 102 (2) (1933) 449–460.

[745] H. Krebs, L(+)-Lactic acid, Biochem. Prep. 8 (1961) 75–79.

[746] E.N. Jacobsen, I. Marko, W.S. Mungall, G. Schroeder, K.B. Sharpless, Asymmetric dihydroxylation via ligand-accelerated catalysis, J. Am. Chem. Soc. 110 (6) (1988) 1968–1970.

[747] J.M. Park, M. Wu, K. Datta, S.-C. Liu, A. Castillo, H. Lough, D.M. Spielman, K.L. Billingsley, Hyperpolarized sodium [1-13C]-glycerate as a probe for assessing glycolysis in vivo, J. Am. Chem. Soc. 139 (19) (2017) 6629–6634.

[748] K. Pichumani, T. Mashimo, H.-M. Baek, J. Ratnakar, B. Mickey, R.J. DeBerardinis, E.A. Maher, R.M. Bachoo, C.R. Malloy, Z. Kovacs, Conditions for 13C NMR detection of 2-hydroxyglutarate in tissue extracts from isocitrate dehydrogenase-mutated gliomas, Anal. Biochem. 481 (2015) 4–6.

[749] S.J. Gelman, N.G. Mahieu, K. Cho, E.M. Llufrio, T.A. Wencewicz, G.J. Patti, Evidence that 2-hydroxyglutarate is not readily metabolized in colorectal carcinoma cells, Cancer Metab. 3 (1) (2015) 1–10.

[750] G.K. Schroeder, W.H. Johnson Jr., J.P. Huddleston, H. Serrano, K.A. Johnson, C.P. Whitman, Reaction of cis-3-chloroacrylic acid dehalogenase with an allene substrate, 2, 3-butadienoate: hydration via an enamine, J. Am. Chem. Soc. 134 (1) (2012) 293–304.

[751] J.J. Lacharity, A.K. Mailyan, K.Y. Chen, A. Zakarian, Concise synthesis of (+)-[13C4]-Anatoxin-a by dynamic kinetic resolution of a cyclic Iminium ion, Angew. Chem. Int. Ed. 59 (28) (2020) 11364–11368.

[752] I. Ndiege, J. Staunton, Acyl-meldrums acid in regiospecific synthesis of isotopically labeled compounds for polyketide biosynthetic-studies, Bull. Chem. Soc. Ethiop. 9 (1) (1995) 43–49.

[753] P. Waser, M. Rueping, D. Seebach, E. Duchardt, H. Schwalbe, On the solution structure of PHB: preparation and NMR analysis of isotopically labeled oligo [(R)-3-hydroxybutanoic acids](OHBs), Helv. Chim. Acta 84 (6) (2001) 1821–1845.

[754] J. Genêt, V. Ratovelomanana-Vidal, M.C. De Andrade, X. Pfister, P. Guerreiro, J. Lenoir, Practical asymmetric hydrogenation of β-keto esters at atmospheric pressure using chiral Ru (II) catalysts, Tetrahedron Lett. 36 (27) (1995) 4801–4804.

[755] M. Beylot, B. Beaufrere, S. Normand, J. Riou, R. Cohen, R. Momex, Determination of human ketone body kinetics using stable-isotope labelled tracers, Diabetologia 29 (2) (1986) 90–96.

[756] B. Passingham, R. Barton, An ensymic method for the preparation of millimole quantities of d (−)-β-hydroxybutyrate, Anal. Biochem. 65 (1–2) (1975) 418–421.

[757] R.K. Ghosh, S.J. Kadlecek, M. Pourfathi, R.R. Rizi, Efficient production of hyperpolarized bicarbonate by chemical reaction on a DNP precursor to measure pH, Magn. Reson. Med. 74 (5) (2015) 1406–1413.

[758] D. Korenchan, R. Flavell, C. Baligand, R. Sriram, K. Neumann, S. Sukumar, H. VanBrocklin, D. Vigneron, D. Wilson, J. Kurhanewicz, Dynamic nuclear polarization of biocompatible 13 C-enriched carbonates for in vivo pH imaging, Chem. Commun. 52 (14) (2016) 3030–3033.

[759] A. Gordillo, G.C. Lloyd-Jones, Pd-Catalyzed reaction of allyl carbonate with polyols: the role of CO2 in transesterification versus etherification of glycerol, Chem. Eur J. 18 (9) (2012) 2660–2665.

[760] N. Maptue, W. Jiang, C. Harrison, A.M. Funk, G. Sharma, C.R. Malloy, D. Sherry, C. Khemtong, Esterase-catalyzed production of hyperpolarized 13C-enriched carbon dioxide in tissues for measuring pH, ACS Sens. 3 (11) (2018) 2232–2236.

[761] V.H. Kollman, J.L. Hanners, R.E. London, E.G. Adame, T.E. Walker, Photosynthetic preparation and characterization of 13C-labeled carbohydrates in Agmenellum quadruplicatum, Carbohydr. Res. 73 (1) (1979) 193–202.

[762] E. Fischer, Reduction von säuren der Zuckergruppe, Ber. Dtsch. Chem. Ges. 22 (2) (1889) 2204–2205.

[763] A. Serianni, H. Nunez, M. Hayes, R. Barker, Chemical synthesis of monosaccharides enriched with carbon isotopes, Methods Enzymol. 89 (1982) 64–73.

[764] J.C. Sowden, H.O. Fischer, The condensation of nitromethane with D-and L-arabinose: preparation of L-glucose and L-mannose, J. Am. Chem. Soc. 69 (8) (1947) 1963–1965.

[765] M.L. Hayes, N.J. Pennings, A.S. Serianni, R. Barker, Epimerization of aldoses by molybdate involving a novel rearrangement of the carbon skeleton, J. Am. Chem. Soc. 104 (24) (1982) 6764–6769.

[766] M.D. Bednarski, E.S. Simon, N. Bischofberger, W.D. Fessner, M.J. Kim, W. Lees, T. Saito, H. Waldmann, G.M. Whitesides, Rabbit muscle aldolase as a catalyst in organic synthesis, J. Am. Chem. Soc. 111 (2) (1989) 627–635.

[767] A. Serianni, E. Cadman, J. Pierce, M. Hayes, R. Barker, Enzymic synthesis of13C-enriched aldoses, ketoses, and their phosphate esters, Methods Enzymol. 89 (1982) 83–92.
[768] A.S. Serianni, T. Vuorinen, P.B. Bondo, Stable isotopically-enriched D-glucose: strategies to introduce carbon, hydrogen and oxygen isotopes at various sites, J. Carbohydr. Chem. 9 (1990).
[769] H. Allouche-Arnon, T. Wade, L.F. Waldner, V.N. Miller, J.M. Gomori, R. Katz-Brull, C.A. McKenzie, In vivo magnetic resonance imaging of glucose—initial experience, Contrast Media Mol. Imaging 8 (1) (2013) 72–82.
[770] E. Fischer, E. Fischer, Ueber die Glucoside der Alkohole, Springer, 1909.
[771] M. Jaseja, A.S. Perlin, P. Dais, Two-dimensional NMR spectral study of the tautomeric equilibria of D-fructose and related compounds, Magn. Reson. Chem. 28 (4) (1990) 283–289.
[772] K.R. Keshari, D.M. Wilson, A.P. Chen, R. Bok, P.E. Larson, S. Hu, M.V. Criekinge, J.M. Macdonald, D.B. Vigneron, J. Kurhanewicz, Hyperpolarized [2-13C]-fructose: a hemiketal DNP substrate for in vivo metabolic imaging, J. Am. Chem. Soc. 131 (48) (2009) 17591–17596.
[773] S.S. Tee, N. Kim, Q. Cullen, R. Eskandari, A. Mamakhanyan, R.M. Srouji, R. Chirayil, S. Jeong, M. Shakiba, E.R. Kastenhuber, Ketohexokinase-mediated fructose metabolism is lost in hepatocellular carcinoma and can be leveraged for metabolic imaging, Sci. Adv. 8 (14) (2022) eabm7985.
[774] K.X. Moreno, C.E. Harrison, M.E. Merritt, Z. Kovacs, C.R. Malloy, A.D. Sherry, Hyperpolarized δ-[1-13C] gluconolactone as a probe of the pentose phosphate pathway, NMR Biomed. 30 (6) (2017) e3713.
[775] G. Batsios, C. Taglang, P. Cao, A.M. Gillespie, C. Najac, E. Subramani, D.M. Wilson, R.R. Flavell, P.E. Larson, S.M. Ronen, Imaging 6-phosphogluconolactonase activity in brain tumors in Vivo using hyperpolarized δ-[1-13C] gluconolactone, Front. Oncol. 11 (2021) 1194.
[776] R.J. Ferrier, Unsaturated carbohydrates. Part 21. A carbocyclic ring closure of a hex-5-enopyranoside derivative, J. Chem. Soc., Perkin Trans. 1 (1979) 1455–1458.
[777] M. Moure, Y. Zhuo, G. Boons, J.H. Prestegard, Perdeuterated and 13 C-enriched myo-inositol for DNP assisted monitoring of enzymatic phosphorylation by inositol-3-kinase, Chem. Commun. 53 (92) (2017) 12398–12401.
[778] S.E. Bohndiek, M.I. Kettunen, D.-e. Hu, B.W. Kennedy, J. Boren, F.A. Gallagher, K.M. Brindle, Hyperpolarized [1-13C]-ascorbic and dehydroascorbic acid: vitamin C as a probe for imaging redox status in vivo, J. Am. Chem. Soc. 133 (30) (2011) 11795–11801.
[779] K.R. Keshari, V. Sai, Z.J. Wang, H.F. VanBrocklin, J. Kurhanewicz, D.M. Wilson, Hyperpolarized [1-13C] dehydroascorbate MR spectroscopy in a murine model of prostate cancer: comparison with 18F-FDG PET, J. Nucl. Med. 54 (6) (2013) 922–928.
[780] K.N. Drew, T.J. Church, B. Basu, T. Vuorinen, A.S. Serianni, l-(1-13C)-and (2-13C) ascorbic acid: synthesis and NMR characterization, Carbohydr. Res. 284 (1) (1996) 135–143.
[781] K.R. Keshari, J. Kurhanewicz, R. Bok, P.E. Larson, D.B. Vigneron, D.M. Wilson, Hyperpolarized 13C dehydroascorbate as an endogenous redox sensor for in vivo metabolic imaging, Proc. Natl. Acad. Sci. 108 (46) (2011) 18606–18611.
[782] E.K. Koliou, P.V. Ioannou, Preparation of dehydro-L-ascorbic acid dimer by air oxidation of L-ascorbic acid in the presence of catalytic amounts of copper (II) acetate and pyridine, Carbohydr. Res. 340 (2) (2005) 315–318.
[783] P.M. Reddy, A.V. Prasad, C. Reddy, V. Ravinder, Synthesis of new macrocyclic rhodium (III) compounds and their utility as catalysts for the oxidation of ascorbic acid, Transit. Met. Chem. 33 (2) (2008) 251–258.
[784] J. Hvoslef, B. Pedersen, The structure of dehydroascorbic acid in solution, Acta Chem. Scand. B 33 (1979) 503.
[785] B.M. Tolbert, J.B. Ward, Dehydroascorbic Acid, ACS Publications, 1982.
[786] P. Pastore, T. Rizzetto, O. Curcuruto, M.D. Cin, A. Zaramella, D. Marton, Characterization of dehydroascorbic acid solutions by liquid chromatography/mass spectrometry, Rapid Commun. Mass Spectrom. 15 (22) (2001) 2051–2057.
[787] N. Miyake, T. Kurata, Possible formation of dehydro-L-ascorbic acid from 2, 3-diketo-L-gulonic acid in an aqueous solution, Biosci. Biotechnol. Biochem. 62 (7) (1998) 1419–1421.
[788] K.N. Timm, D.-E. Hu, M. Williams, A.J. Wright, M.I. Kettunen, B.W. Kennedy, T.J. Larkin, P. Dzien, I. Marco-Rius, S.E. Bohndiek, Assessing oxidative stress in tumors by measuring the rate of hyperpolarized [1-13C] dehydroascorbic acid reduction using 13C magnetic resonance spectroscopy, J. Biol. Chem. 292 (5) (2017) 1737–1748.
[789] K. Yamamoto, A. Opina, D. Sail, B. Blackman, K. Saito, J.R. Brender, R.M. Malinowski, T. Seki, N. Oshima, D.R. Crooks, Real-Time insight into in vivo redox status utilizing hyperpolarized [1-13C] N-acetyl cysteine, Sci. Rep. 11 (1) (2021) 1–9.
[790] S.D. Yelamanchi, S. Jayaram, J.K. Thomas, S. Gundimeda, A.A. Khan, A. Singhal, T. Keshava Prasad, A. Pandey, B. Somani, H. Gowda, A pathway map of glutamate metabolism, J. Cell. Commun. Signal. 10 (1) (2016) 69–75.
[791] B.J. Altman, Z.E. Stine, C.V. Dang, From Krebs to clinic: glutamine metabolism to cancer therapy, Nat. Rev. Cancer 16 (10) (2016) 619–634.
[792] P.R. Jensen, M. Karlsson, S. Meier, J.Ø. Duus, M.H. Lerche, Hyperpolarized amino acids for in vivo assays of transaminase activity, Chem. Eur J. 15 (39) (2009) 10010–10012.
[793] S. Hu, M. Zhu, H.A. Yoshihara, D.M. Wilson, K.R. Keshari, P. Shin, G. Reed, C. von Morze, R. Bok, P.E. Larson, In vivo measurement of normal rat intracellular pyruvate and lactate levels after injection of hyperpolarized [1-13C] alanine, Magn. Reson. Imaging 29 (8) (2011) 1035–1040.
[794] J.M. Park, C. Khemtong, S.C. Liu, R.E. Hurd, D.M. Spielman, In vivo assessment of intracellular redox state in rat liver using hyperpolarized [1-13C] Alanine, Magn. Reson. Med. 77 (5) (2017) 1741–1748.
[795] A. Radaelli, R. Gruetter, H.A. Yoshihara, In vivo detection of d-amino acid oxidase with hyperpolarized d-[1-13C] alanine, NMR Biomed. 33 (7) (2020) e4303.
[796] L. Mazuel, R.F. Schulte, A. Cladière, C. Spéziale, M. Lagrée, M. Leremboure, B. Jean, F. Durif, C. Chassain, Intracerebral synthesis of glutamine from hyperpolarized glutamate, Magn. Reson. Med. 78 (4) (2017) 1296–1305.

[797] L. Salamanca-Cardona, H. Shah, A.J. Poot, F.M. Correa, V. Di Gialleonardo, H. Lui, V.Z. Miloushev, K.L. Granlund, S.S. Tee, J.R. Cross, In vivo imaging of glutamine metabolism to the oncometabolite 2-hydroxyglutarate in IDH1/2 mutant tumors, Cell Metab. 26 (6) (2017) 830–841.

[798] Z. Sha, A. Peth, A.L. Goldberg, Keeping proteasomes under control–a role for phosphorylation in the nucleus, Proc. Natl. Acad. Sci. USA 108 (46) (2011) 18573–18574, https://doi.org/10.1073/pnas.1115315108.

[799] F.A. Gallagher, M.I. Kettunen, S.E. Day, M. Lerche, K.M. Brindle, 13C MR spectroscopy measurements of glutaminase activity in human hepatocellular carcinoma cells using hyperpolarized 13C-labeled glutamine, Magn. Reson. Med.: J. Int. Soc. Magn. Reson. Med. 60 (2) (2008) 253–257.

[800] C. Cabella, M. Karlsson, C. Canape, G. Catanzaro, S.C. Serra, L. Miragoli, L. Poggi, F. Uggeri, L. Venturi, P.R. Jensen, In vivo and in vitro liver cancer metabolism observed with hyperpolarized [5-13C] glutamine, J. Magn. Reson. 232 (2013) 45–52.

[801] C. Canapè, G. Catanzaro, E. Terreno, M. Karlsson, M.H. Lerche, P.R. Jensen, Probing treatment response of glutaminolytic prostate cancer cells to natural drugs with hyperpolarized [5-13C] glutamine, Magn. Reson. Med. 73 (6) (2015) 2296–2305.

[802] C. Najac, M.M. Chaumeil, G. Kohanbash, C. Guglielmetti, J.W. Gordon, H. Okada, S.M. Ronen, Detection of inflammatory cell function using 13C magnetic resonance spectroscopy of hyperpolarized [6-13C]-arginine, Sci. Rep. 6 (1) (2016) 1–10.

[803] C. Hundshammer, S. Düwel, D. Ruseckas, G. Topping, P. Dzien, C. Müller, B. Feuerecker, J.B. Hövener, A. Haase, M. Schwaiger, Hyperpolarized amino acid derivatives as multivalent magnetic resonance pH sensor molecules, Sensors 18 (2) (2018) 600.

[804] S. Wang, D.E. Korenchan, P.M. Perez, C. Taglang, T.R. Hayes, R. Sriram, R. Bok, A.S. Hong, Y. Wu, H. Li, Amino acid-derived sensors for specific Zn2+ detection using hyperpolarized 13C magnetic resonance spectroscopy, Chem. Eur J. 25 (51) (2019) 11842–11846.

[805] J. Chen, E.P. Hackett, J. Singh, Z. Kovács, J.M. Park, Simultaneous assessment of intracellular and extracellular pH using hyperpolarized [1-13C] alanine ethyl ester, Anal. Chem. 92 (17) (2020) 11681–11686.

[806] N.M. Kelly, A. Sutherland, C.L. Willis, Syntheses of amino acids incorporating stable isotopes, Nat. Prod. Rep. 14 (3) (1997) 205–219.

[807] R. Fitzi, D. Seebach, Resolution and use in α-amino acid synthesis of imidazolidinone glycine derivatives, Tetrahedron 44 (17) (1988) 5277–5292.

[808] Y.-C. Luo, H.-H. Zhang, Y. Wang, P.-F. Xu, Synthesis of α-amino acids based on chiral tricycloiminolactone derived from natural (+)-camphor, Acc. Chem. Res. 43 (10) (2010) 1317–1330.

[809] D.O. Rees, N. Bushby, J.R. Harding, C. Song, C.L. Willis, Synthesis of isotopically labelled amino acids, J. Label. Compd. Radiopharm. 50 (5-6) (2007) 399–401.

[810] H. Josien, A. Martin, G. Chassaing, Asymmetric synthesis of L-diphenylalanine and L-9-fluorenylglycine via room temperature alkylations of a sultam-derived glycine imine, Tetrahedron Lett. 32 (45) (1991) 6547–6550.

[811] R.M. Williams, M.N. Im, Asymmetric synthesis of monosubstituted and. alpha.,. alpha.-disubstituted. alpha.-amino acids via diastereoselective glycine enolate alkylations, J. Am. Chem. Soc. 113 (24) (1991) 9276–9286.

[812] T.K. Chakraborty, A. Ghosh, Synthesis of chiral α-amino acids, Tetrahedron Lett. 43 (52) (2002) 9691–9693.

[813] D.A. Evans, T.C. Britton, J.A. Ellman, R.L. Dorow, The asymmetric synthesis of. alpha.-amino acids. Electrophilic azidation of chiral imide enolates, a practical approach to the synthesis of (R)-and (S)-. alpha.-azido carboxylic acids, J. Am. Chem. Soc. 112 (10) (1990) 4011–4030.

[814] L. Lecointe, V. Rolland, L. Pappalardo, M. Roumestant, P. Viallefont, J. Martine, Diastereoselective synthesis of non-proteinogenic α-amino acids, J. Pept. Res. 55 (4) (2000) 300–307.

[815] M.J. O'Donnell, W.D. Bennett, S. Wu, The stereoselective synthesis of. alpha.-amino acids by phase-transfer catalysis, J. Am. Chem. Soc. 111 (6) (1989) 2353–2355.

[816] M.J. O'donnell, The enantioselective synthesis of α-amino acids by phase-transfer catalysis with achiral Schiff base esters, Acc. Chem. Res. 37 (8) (2004) 506–517.

[817] B. Lygo, B.I. Andrews, Asymmetric phase-transfer catalysis utilizing chiral quaternary ammonium salts: asymmetric alkylation of glycine imines, Acc. Chem. Res. 37 (8) (2004) 518–525.

[818] W. He, Q. Wang, Q. Wang, B. Zhang, X. Sun, S. Zhang, Synthesis of novel chiral phase-transfer catalysts and their application to asymmetric synthesis of α-amino acid derivatives, Synlett 2009 (08) (2009) 1311–1314.

[819] H.G. Park, B.S. Jeong, M.S. Yoo, J.H. Lee, M.K. Park, Y.J. Lee, M.J. Kim, S.S. Jew, Highly enantioselective and practical cinchona-derived phase-transfer catalysts for the synthesis of α-amino acids, Angew. Chem. Int. Ed. 41 (16) (2002) 3036–3038.

[820] M.J. O'Donnell, F. Delgado, E. Domınguez, J. de Blas, W.L. Scott, Enantioselective solution-and solid-phase synthesis of glutamic acid derivatives via Michael addition reactions, Tetrahedron Asymmetry 12 (6) (2001) 821–828.

[821] A.H. Siebum, W.S. Woo, J. Raap, J. Lugtenburg, Access to any site-directed isotopomer of methionine, selenomethionine, cysteine, and selenocysteine–use of simple, efficient modular synthetic reaction schemes for isotope incorporation, Eur. J. Org. Chem. 2004 (13) (2004) 2905–2913.

[822] A.H. Siebum, R.K. Tsang, R.V.D. Steen, J. Raap, J. Lugtenburg, Synthesis of (ϵ-13C-, ϵ-15N)-Enriched l-lysine– establishing schemes for the preparation of all possible 13C and 15N isotopomers of l-lysine, l-ornithine, and l-proline, Eur. J. Org. Chem. 2004 (21) (2004) 4391–4396.

[823] A.H. Siebum, W.S. Woo, J. Lugtenburg, Preparation and characterization of [5-13C]-(2S, 4R)-leucine and [4-13C]-(2S, 3S)-Valine–establishing synthetic schemes to prepare any site-directed isotopomer of L-leucine, L-Isoleucine and L-Valine, Eur. J. Org. Chem. 2003 (23) (2003) 4664–4678.

[824] L. Duhamel, P. Duhamel, J.-C. Plaquevent, Enantioselective protonations: fundamental insights and new concepts, Tetrahedron: Asymmetry 15 (23) (2004) 3653–3691.

[825] M. Calmès, C. Glot, J. Martinez, Investigation into the enantioselective protonation of enolate Schiff bases with (R)-pantolactone, Tetrahedron Asymmetry 12 (1) (2001) 49–52.

[826] K. Futatsugi, A. Yanagisawa, H. Yamamoto, Asymmetric protonation of lithium enolates of α-amino acid derivatives with α-amino acid-based chiral Brønsted acids, Chem. Commun. (5) (2003) 566–567.

[827] W. Qu, Z. Zha, B.P. Lieberman, A. Mancuso, M. Stetz, R. Rizzi, K. Ploessl, D. Wise, C. Thompson, H.F. Kung, Facile synthesis [5-13C-4-2H2]-L-glutamine for hyperpolarized MRS imaging of cancer cell metabolism, Acad. Radiol. 18 (8) (2011) 932–939.

[828] R. Appel, Tertiary phosphane/tetrachloromethane, a versatile reagent for chlorination, dehydration, and P—N linkage, Angew. Chem. Int. Ed. Engl. 14 (12) (1975) 801–811.

[829] D. Kang, J. Lee, H.Y. Lee, Anhydrous hydration of nitriles to amides: p-carbomethoxybenzamide, Org. Synth. 89 (2003) 66–72.

[830] N.E. Good, G.D. Winget, W. Winter, T.N. Connolly, S. Izawa, R.M. Singh, Hydrogen ion buffers for biological research, Biochemistry 5 (2) (1966) 467–477.

[831] C. Morze, M. Criekinge, Application of good's buffers to pH imaging using hyperpolarized 13 C MRI, Chem. Commun. 51 (74) (2015) 14119–14122.

[832] J.A. Mótyán, F. Tóth, J. Tőzsér, Research applications of proteolytic enzymes in molecular biology, Biomolecules 3 (4) (2013) 923–942.

[833] A. Radaelli, D. Ortiz, A. Michelotti, M. Roche, R. Hata, S. Sando, O. Bonny, R. Gruetter, H.A. Yoshihara, Hyperpolarized (1-13C) alaninamide is a multifunctional in vivo sensor of aminopeptidase N activity, pH, and CO2, ACS Sens. 7 (10) (2022) 2987–2994.

[834] Y. Saito, H. Yatabe, I. Tamura, Y. Kondo, R. Ishida, T. Seki, K. Hiraga, A. Eguchi, Y. Takakusagi, K. Saito, Structure-guided design enables development of a hyperpolarized molecular probe for the detection of aminopeptidase N activity in vivo, Sci. Adv. 8 (13) (2022) eabj2667.

[835] D. Hedley, L. Ogilvie, C. Springer, Carboxypeptidase G2-based gene-directed enzyme–prodrug therapy: a new weapon in the GDEPT armoury, Nat. Rev. Cancer 7 (11) (2007) 870–879.

[836] L.C. Davies, F. Friedlos, D. Hedley, J. Martin, L.M. Ogilvie, I.J. Scanlon, C.J. Springer, Novel fluorinated prodrugs for activation by carboxypeptidase G2 showing good in vivo antitumor activity in gene-directed enzyme prodrug therapy, J. Med. Chem. 48 (16) (2005) 5321–5328.

[837] L. F Tietze, K. Schmuck, Prodrugs for targeted tumor therapies: recent developments in ADEPT, GDEPT and PMT, Curr. Pharm. Des. 17 (32) (2011) 3527–3547.

[838] T. Nishihara, H.A. Yoshihara, H. Nonaka, Y. Takakusagi, F. Hyodo, K. Ichikawa, E. Can, J.A. Bastiaansen, Y. Takado, A. Comment, Direct monitoring of γ-glutamyl transpeptidase activity in vivo using a hyperpolarized 13C-labeled molecular probe, Angew. Chem. Int. Ed. 55 (36) (2016) 10626–10629.

[839] Y. Kondo, Y. Saito, A.E. Elhelaly, F. Hyodo, T. Nishihara, M. Itoda, H. Nonaka, M. Matsuo, S. Sando, Evaluation of enzymatic and magnetic properties of γ-glutamyl-[1-13 C] glycine and its deuteration toward longer retention of the hyperpolarized state, RSC Adv. 11 (59) (2021) 37011–37018.

[840] J.R. Lakowicz, B. Maliwal, H. Cherek, A. Balter, Rotational freedom of tryptophan residues in proteins and peptides, Biochemistry 22 (8) (1983) 1741–1752.

[841] K. Uma, H. Balaram, S. Raghothama, P. Balaram, Simultaneous observation of positive and negative nuclear overhauser effects in oligopeptides due to segmental motion, Biochem. Biophys. Res. Commun. 151 (1) (1988) 153–157.

[842] Y. Kondo, Y. Saito, T. Seki, Y. Takakusagi, J. Morimoto, H. Nonaka, K. Miyanishi, W. Mizukami, M. Negoro, A.E. Elhelaly, Structure-based relaxation analysis reveals C-terminal [1-13C] glycine-d2 in peptides has long spin-lattice relaxation time that is applicable to in vivo hyperpolarized magnetic resonance studies, ChemRxiv (2022), https://doi.org/10.26434/chemrxiv-2022-tfkk1.

[843] S.J. Nelson, J. Kurhanewicz, D.B. Vigneron, P.E. Larson, A.L. Harzstark, M. Ferrone, M. Van Criekinge, J.W. Chang, R. Bok, I. Park, Metabolic imaging of patients with prostate cancer using hyperpolarized [1-13C] pyruvate, Sci. Transl. Med. 5 (198) (2013), 198ra108-198ra108.

[844] M. Fiedorowicz, M. Wieteska, K. Rylewicz, B. Kossowski, E. Piątkowska-Janko, A.M. Czarnecka, B. Toczylowska, P. Bogorodzki, Hyperpolarized 13C tracers: technical advancements and perspectives for clinical applications, Biocybern. Biomed. Eng. 41 (4) (2021) 1466–1485.

[845] Z.J. Wang, M.A. Ohliger, P.E. Larson, J.W. Gordon, R.A. Bok, J. Slater, J.E. Villanueva-Meyer, C.P. Hess, J. Kurhanewicz, D.B. Vigneron, Hyperpolarized 13C MRI: state of the art and future directions, Radiology 291 (2) (2019) 273–284.

[846] A. Zhuravlev, V. Berdinskii, A. Buchachenko, Generation of high-frequency current by the products of a photochemical reaction, JETP Lett. 28 (3) (1978) 140–142.

[847] S. Lehmkuhl, S. Fleischer, L. Lohmann, M.S. Rosen, E.Y. Chekmenev, A. Adams, T. Theis, S. Appelt, RASER MRI: magnetic resonance images formed spontaneously exploiting cooperative nonlinear interaction, Sci. Adv. 8 (28) (2022) eabp8483.

[848] W.S. Warren, E. Jenista, R.T. Branca, X. Chen, Increasing hyperpolarized spin lifetimes through true singlet eigenstates, Science 323 (5922) (2009) 1711–1714.

[849] Y. Feng, T. Theis, X. Liang, Q. Wang, P. Zhou, W.S. Warren, Storage of hydrogen spin polarization in long-lived 13C2 singlet order and implications for hyperpolarized magnetic resonance imaging, J. Am. Chem. Soc. 135 (26) (2013) 9632–9635.

[850] G. Stevanato, J.T. Hill-Cousins, P. Håkansson, S.S. Roy, L.J. Brown, R.C. Brown, G. Pileio, M.H. Levitt, A nuclear singlet lifetime of more than one hour in room-temperature solution, Angew. Chem. 127 (12) (2015) 3811–3814.

[851] G. Pileio, J.T. Hill-Cousins, S. Mitchell, I. Kuprov, L.J. Brown, R.C. Brown, M.H. Levitt, Long-lived nuclear singlet order in near-equivalent 13C spin pairs, J. Am. Chem. Soc. 134 (42) (2012) 17494–17497.

[852] M.H. Levitt, Singlet and other states with extended lifetimes, eMagRes (2007), https://doi.org/10.1002/9780470034590.emrstm1036.

[853] G. Pileio, S. Bowen, C. Laustsen, M.C. Tayler, J.T. Hill-Cousins, L.J. Brown, R.C. Brown, J.H. Ardenkjaer-Larsen, M.H. Levitt, Recycling and imaging of nuclear singlet hyperpolarization, J. Am. Chem. Soc. 135 (13) (2013) 5084–5088.

[854] J.-N. Dumez, Perspective on long-lived nuclear spin states, Mol. Phys. 118 (6) (2020) e1644382.

[855] A. Moysiadi, F. Giustiniano, A.M. Hall, T.A. Cartlidge, L.J. Brown, G. Pileio, Nuclear spin relaxation of longitudinal and singlet order in liquid-co2 solutions, Front. Chem. 9 (2021) 668044.

[856] M. Carravetta, M.H. Levitt, Long-lived nuclear spin states in high-field solution NMR, J. Am. Chem. Soc. 126 (20) (2004) 6228–6229.

[857] F.W. Wehrli, Temperature-dependent spin-lattice relaxation of 6 Li in aqueous lithium chloride, J. Magn. Reson. 23 (3) (1976) 527–532.

[858] M. Mohammadi, S. Benders, A. Jerschow, Nuclear magnetic resonance spin-lattice relaxation of lithium ions in aqueous solution by NMR and molecular dynamics, J. Chem. Phys. 153 (18) (2020) 184502.

[859] Y. Jamin, C. Gabellieri, L. Smyth, S. Reynolds, S.P. Robinson, C.J. Springer, M.O. Leach, G.S. Payne, T.R. Eykyn, Hyperpolarized 13C magnetic resonance detection of carboxypeptidase G2 activity, Magn. Reson. Med.: J. Int. Soc. Magn. Reson. Med. 62 (5) (2009) 1300–1304.

[860] D.M. Wilson, R.E. Hurd, K. Keshari, M. Van Criekinge, A.P. Chen, S.J. Nelson, D.B. Vigneron, J. Kurhanewicz, Generation of hyperpolarized substrates by secondary labeling with [1, 1-13C] acetic anhydride, Proc. Natl. Acad. Sci. 106 (14) (2009) 5503–5507.

[861] B.L. Koelsch, K.R. Keshari, T.H. Peeters, P.E. Larson, D.M. Wilson, J. Kurhanewicz, Diffusion MR of hyperpolarized 13 C molecules in solution, Analyst 138 (4) (2013) 1011–1014.

[862] S. Colombo Serra, M. Karlsson, G.B. Giovenzana, C. Cavallotti, F. Tedoldi, S. Aime, Hyperpolarized 13C-labelled anhydrides as DNP precursors of metabolic MRI agents, Contrast Media Mol. Imaging 7 (5) (2012) 469–477.

[863] S. Meier, M. Karlsson, P.R. Jensen, M.H. Lerche, J.Ø. Duus, Metabolic pathway visualization in living yeast by DNP-NMR, Mol. Biosyst. 7 (10) (2011) 2834–2836.

[864] K.R. Keshari, J. Kurhanewicz, J.M. Macdonald, D.M. Wilson, Generating contrast in hyperpolarized 13 C MRI using ligand–receptor interactions, Analyst 137 (15) (2012) 3427–3429.

[865] Y. Lee, G.S. Heo, H. Zeng, K.L. Wooley, C. Hilty, Detection of living anionic species in polymerization reactions using hyperpolarized NMR, J. Am. Chem. Soc. 135 (12) (2013) 4636–4639.

[866] G.C. Levy, J.D. Cargioli, F. Anet, Carbon-13 spin-lattice relaxation in benzene and substituted aromatic compounds, J. Am. Chem. Soc. 95 (5) (1973) 1527–1535.

[867] S. Jonas, Development and optimization of techniques for paramagnetic and hyperpolarized contrast media, Acta Radiol. 44 (Suppl. 429) (2003) 1–30.

[868] C. Laustsen, G. Pileio, M.C. Tayler, L.J. Brown, R.C. Brown, M.H. Levitt, J.H. Ardenkjaer-Larsen, Hyperpolarized singlet NMR on a small animal imaging system, Magn. Reson. Med. 68 (4) (2012) 1262–1265.

[869] J.E. AbuSalim, K. Yamamoto, N. Miura, B. Blackman, J.R. Brender, C. Mushti, T. Seki, K.A. Camphausen, R.E. Swenson, M.C. Krishna, Simple esterification of [1-13C]-Alpha-Ketoglutarate enhances membrane permeability and allows for noninvasive tracing of glutamate and glutamine production, ACS Chem. Biol. 16 (11) (2021) 2144–2150.

[870] K. Golman, J.S. Petersson, P. Magnusson, E. Johansson, P. Åkeson, C.M. Chai, G. Hansson, S. Månsson, Cardiac metabolism measured non-invasively by hyperpolarized 13C MRI, Magn. Reson. Med.: J. Int. Soc. Magn. Reson. Med. 59 (5) (2008) 1005–1013.

[871] Y. Feng, R.M. Davis, W.S. Warren, Accessing long-lived nuclear singlet states between chemically equivalent spins without breaking symmetry, Nat. Phys. 8 (11) (2012) 831–837.

[872] M. Karlsson, P.R. Jensen, R. in't Zandt, A. Gisselsson, G. Hansson, J.Ø. Duus, S. Meier, M.H. Lerche, Imaging of branched chain amino acid metabolism in tumors with hyperpolarized 13C ketoisocaproate, Int. J. Cancer 127 (3) (2010) 729–736.

[873] J.J. Miller, D.R. Ball, A.Z. Lau, D.J. Tyler, Hyperpolarized ketone body metabolism in the rat heart, NMR Biomed. 31 (6) (2018) e3912.

[874] C. von Morze, G.Y. Chang, P.E. Larson, H. Shang, P.K. Allu, R.A. Bok, J.C. Crane, M.P. Olson, C.T. Tan, I. Marco-Rius, Detection of localized changes in the metabolism of hyperpolarized gluconeogenic precursors 13C-lactate and 13C-pyruvate in kidney and liver, Magn. Reson. Med. 77 (4) (2017) 1429–1437.

[875] D. Mayer, Y.F. Yen, S. Josan, J.M. Park, A. Pfefferbaum, R.E. Hurd, D.M. Spielman, Application of hyperpolarized [1-13C] lactate for the in vivo investigation of cardiac metabolism, NMR Biomed. 25 (10) (2012) 1119–1124.

[876] R. Sarkar, A. Comment, P.R. Vasos, S. Jannin, R. Gruetter, G. Bodenhausen, H. Hall, D. Kirik, V.P. Denisov, Proton NMR of 15N-choline metabolites enhanced by dynamic nuclear polarization, J. Am. Chem. Soc. 131 (44) (2009) 16014–16015.

[877] R.V. Shchepin, E.Y. Chekmenev, Toward hyperpolarized molecular imaging of HIV: synthesis and longitudinal relaxation properties of 15N-Azidothymidine, J. Label. Compd. Radiopharm. 57 (10) (2014) 621–624.

[878] J. Bae, G. Zhang, H. Park, W.S. Warren, Q. Wang, 15 N-Azides as practical and effective tags for developing long-lived hyperpolarized agents, Chem. Sci. 12 (42) (2021) 14309–14315.

[879] J.F. Colell, A.W. Logan, Z. Zhou, R.V. Shchepin, D.A. Barskiy, G.X. Ortiz Jr., Q. Wang, S.J. Malcolmson, E.Y. Chekmenev, W.S. Warren, Generalizing, extending, and maximizing nitrogen-15 hyperpolarization induced by parahydrogen in reversible exchange, J. Phys. Chem. C 121 (12) (2017) 6626–6634.

[880] A.P. Jagtap, L. Kaltschnee, S. Glöggler, Hyperpolarization of 15 N-pyridinium and 15 N-aniline derivatives by using parahydrogen: new opportunities to store nuclear spin polarization in aqueous media, Chem. Sci. 10 (37) (2019) 8577–8582.

[881] M.J. Burns, P.J. Rayner, G.G. Green, L.A. Highton, R.E. Mewis, S.B. Duckett, Improving the hyperpolarization of 31P nuclei by synthetic design, J. Phys. Chem. B 119 (15) (2015) 5020–5027.

[882] R. Gaspar Jr., W. Brey, E. Andrew, Phosphorus-31 magnetic relaxation of phosphocreatine in solution, Chem. Phys. Lett. 184 (1–3) (1991) 17–20.

[883] H. Pfister, A. Schwenk, D. Zeller, Longitudinal and transverse 109Ag relaxation of the silver ion in aqueous solution, J. Magn. Reson. 68 (1) (1986) 138−145.
[884] B. Anger, G. Schrank, A. Schoeck, K. Butler, M. Solum, R. Pugmire, B. Saam, Gas-phase spin relaxation of Xe 129, Phys. Rev. A 78 (4) (2008) 043406.
[885] K.L. Sauer, R. Fitzgerald, W. Happer, Laser-polarized liquid xenon, Chem. Phys. Lett. 277 (1−3) (1997) 153−158.
[886] D. Hutchins, S. Day, Pulsed nuclear magnetic resonance in cesium fluoride, Phys. Rev. 180 (2) (1969) 432.
[887] H. Hertz, M. Holz, G. Keller, H. Versmold, C. Yoon, Nuclear magnetic relaxation of alkali metal ions in aqueous solutions, Ber. Bunsenges. Phys. Chem. 78 (5) (1974) 493−509.

Chapter 2

Optimization of hyperpolarized NMR signals in dissolution dynamic nuclear polarization

Asiye Asaadzade[1,a], Fatemeh Khashami[1,2,a], Wirya Feizi[1,3,a], Daniel Anable[1], Cody Larsen[1], Qing Wang[1,4] and Lloyd Lumata[1,5,6,*]

[1]*Department of Physics, University of Texas at Dallas, Richardson, TX, United States;* [2]*Advanced Imaging Research Center, University of Texas Southwestern Medical Center, Dallas, TX, United States;* [3]*McLean Imaging Center, McLean Hospital, Harvard Medical School, Belmont, MA, United States;* [4]*Department of Imaging Physics, University of Texas MD Anderson Cancer Center, Houston, TX, United States;* [5]*Department of Neuroscience, University of Texas at Dallas, Richardson, TX, United States;* [6]*Department of Bioengineering, University of Texas at Dallas, Richardson, TX, United States*

**Corresponding author: E-mail: lloyd.lumata@utdallas.edu*

List of abbreviations

BDPA 1,3-bisdiphenylene-2-phenylallyl
DMSO Dimethyl sulfoxide
DNP Dynamic nuclear polarization
DOTA 1,4,7,10-tetraazacyclododecane-1,4,7,10-tetraacetic acid
DPPH 2,2-diphenylpicrylhydrazyl
ESR Electron spin resonance
ESSI Electron spin-spin interaction
MRI Magnetic resonance imaging
NMR Nuclear magnetic resonance
NOTA 1,4,7-triazacyclononane-1,4,7-triacetic acid
NZS Nuclear Zeeman system
P_{dnp} DNP-enhanced polarization
P_{th} Thermal polarization
SPION Superparamagnetic iron oxide nanoparticles
T_1 Spin-lattice relaxation time
T_{1e} Electron spin-lattice relaxation time
TEMPO 2,2,6,6-tetramethyl-1-piperidinyloxy
T_L Lattice temperature
T_S Spin temperature
γ Gyromagnetic ratio
ε DNP enhancement

2.1 Introduction

Due to its high specificity and non-destructive nature of measurement, NMR spectroscopy is a widely used analytical technique for the characterization and structural elucidation of numerous samples and materials in academic, medical, and industrial research [1–3]. Moreover, when magnetic field gradients are used, NMR spectroscopy can non-invasively

[a] These authors contributed equally.

The Chemistry of Hyperpolarized Magnetic Resonance Probes. https://doi.org/10.1016/B978-0-323-91842-8.00002-1

produce images of soft tissues or objects via its related technique magnetic resonance imaging (MRI) [3,4]. However, in spite of its numerous practical applications in science and medicine, NMR spectroscopy of samples with low spin count and nuclei with relatively low gyromagnetic ratios γ such as ^{13}C ($\gamma = 10.7$ MHz/T) can be challenging and time-consuming with substantial signal averaging needed in measurement [1–3]. This is because the strength of the NMR signal is proportional to the number of nuclear spins and a parameter known as polarization, which is dictated by Boltzmann spin statistics [1–3].

For an ensemble of spin $I = 1/2$ nuclei, the thermal equilibrium polarization P_{th} is given by:

$$P_{th} = \tanh(\gamma B / 2k_B T) \times 100\%$$

where B is the magnetic field, T is the temperature, and k_B is the Boltzmann constant [1,2]. At ambient NMR conditions, the Boltzmann-dictated thermal polarization of these nuclei could be very miniscule on the order of only a few parts per million, resulting in an NMR spectrum with a poor signal-to-noise ratio. Meanwhile, electron spins ($\gamma_e = 28{,}020$ MHz/T) have a much larger magnetic moment that dwarfs the magnetism of any nuclear spin [5,6]. The most sensitive NMR-active stable nuclear isotope is ^{1}H, or proton, which has $\gamma = 42.58$ MHz/T. Electrons have about three to four orders of magnitude larger polarization than the nuclear spins—the main reason why the related magnetic resonance technique electron spin resonance (ESR) is much more sensitive than NMR. At cryogenic temperatures close to 1 K and relatively high magnetic fields (e.g., 5 T), the electron spin polarization is almost 100%, whereas nuclei such as ^{13}C spins have P_{th} equal to a very small fraction of a percent (0.09%), as depicted in Fig. 2.1. The idea behind hyperpolarization is to create a non-Boltzmann nuclear spin distribution by transferring the high electron thermal polarization to the weakly polarized nuclear spins; in this way, the nuclear thermal polarization P_{th} is enhanced by a factor of ε, and thus the DNP-enhanced polarization is calculated as $P_{dnp} = \varepsilon \times P_{th}$, where ε is the NMR signal enhancement of the hyperpolarized signal measured relative to a reference thermal NMR signal [5–9].

Hyperpolarization via dissolution dynamic nuclear polarization (DNP) provides a way to harness the high electron spin alignment for NMR signal amplification by first transferring it via microwave irradiation to the nuclear spins of a frozen solid sample at cryogenic temperature and high magnetic field, followed by the production of hyperpolarized liquid at room temperature after rapid dissolution of the highly polarized frozen sample pellet [8–10]. The DNP technique has been around since the 1950s and was used primarily in particle physics and nuclear scattering experiments [5–7]. In 2003, Ardenkjaer-Larsen et al. [8] introduced the dissolution method, in which the highly polarized sample in cryogenic

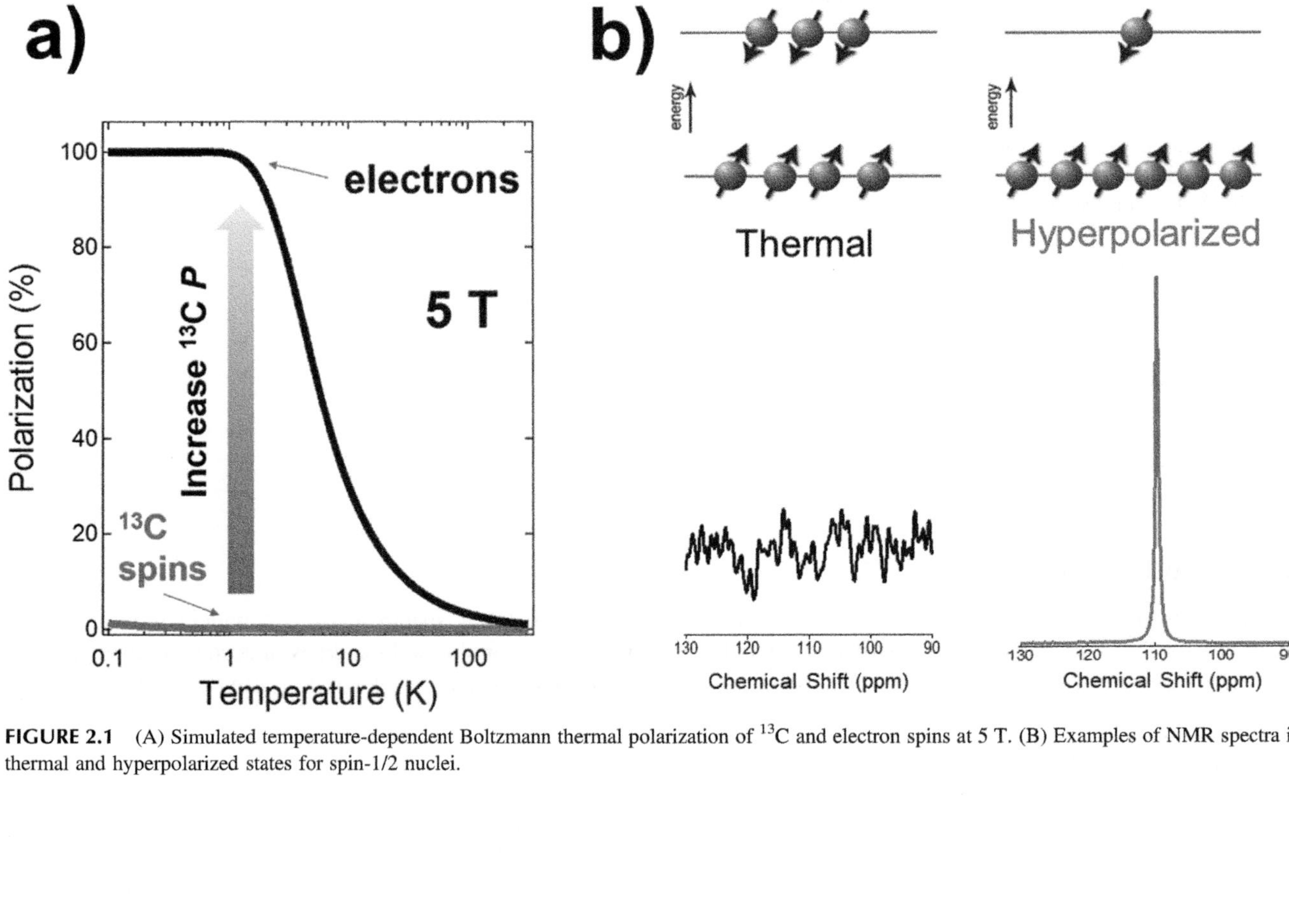

FIGURE 2.1 (A) Simulated temperature-dependent Boltzmann thermal polarization of ^{13}C and electron spins at 5 T. (B) Examples of NMR spectra in thermal and hyperpolarized states for spin-1/2 nuclei.

conditions is rapidly dissolved into a hyperpolarized liquid at a physiologically tolerable temperature, which opened a new door for real-time in vivo biochemical or metabolic NMR and MRI. The versatility of the dissolution DNP technique lies in its ability to hyperpolarize any NMR-active nuclei, although its practical utility in liquid-state NMR or MRI applications is tied up to the length of the hyperpolarization lifetime or the spin-lattice relaxation time T_1 of the hyperpolarized nuclei [8–11]. The dissolution DNP technology allows ultrasensitive detection of liquid-state NMR signals of low-γ nuclei at low concentrations and in ambient conditions with just one scan (Fig. 2.1). The DNP technique exploits the high Boltzmann thermal spin polarization of electrons due to their high gamma and transfers it to insensitive nuclear spins such as ^{13}C spins via microwave irradiation of target samples at cryogenic temperatures close to 1 K and high magnetic field [5–10]. Under cryogenic conditions, the polarization transfer from electrons to nuclear spins is done via slightly off-resonance microwave irradiation of the target sample. Once the sample is hyperpolarized, a dissolution device is inserted into the DNP hyperpolarizer, which injects pressurized superheated solvent (e.g., 4 mL of water) into the sample cup [8–11]. This protocol rapidly dissolves the frozen polarized samples and ejects the diluted liquid out of the polarizer. The hyperpolarized liquid is then rapidly transferred to an NMR tube for detection in a nearby high-resolution NMR spectrometer or injected into living subjects in a nearby MRI scanner [11–18]. Provided that the nuclei of interest have sufficiently long spin-lattice relaxation time T_1, the enhanced nuclear polarization from cryogenic temperature is mostly preserved in the dissolution liquid at room temperature [8–11]. The unprecedented increase in NMR signal by several thousand-fold in the liquid state has allowed high-sensitivity NMR spectroscopic chemical characterization and in vivo hyperpolarized MRI applications, in particular the use of hyperpolarized ^{13}C biomolecules in real-time metabolic NMR and MRI of cancer and other diseases [11–18]. Since the success of these hyperpolarized magnetic resonance experiments is dependent upon improved NMR sensitivity, it is thus paramount to optimize the DNP samples and the instrumental conditions to attain the highest NMR and MRI signal enhancements via dissolution DNP.

2.2 Optimization method in dissolution DNP: sample preparation

There are typically three main components of a sample for DNP: (1) Polarizing agent; (2) target nuclei or sample; and (3) glassing solvent [11]. The polarizing agent refers to the source of free electrons from which high-spin polarization is transferred to the nuclear spins. The target nuclei are the specific nuclear spins that are being polarized; these could also refer in general to the small molecules or macromolecules containing the target nuclei. Glassing solvent or matrix refers to solvent mixtures such as glycerol/water, DMSO/water, or ethanol/DMSO that are used to homogeneously distribute the polarizing agent and the target nuclei or compound by forming a glassy matrix at cryogenic temperature. In some cases, there are some DNP samples that are self-glassing, such as neat ^{13}C-labeled pyruvic acid. Certain chemical additives, such as paramagnetic complexes are also included as trace dopants in DNP samples in certain cases to further enhance the hyperpolarized NMR signals. Since the DNP process involves the interaction of the electrons and nuclear spins in the frozen glassing matrix at cryogenic conditions, the nature of the DNP sample composition would affect the maximum achievable DNP signal enhancements of the target nuclear spins.

2.2.1 Choice of free radicals for DNP

Central to the NMR signal amplification process of DNP are the polarizing agents, or the sources of free electrons, typically provided by stable, spin-1/2 organic free radicals (see Fig. 2.2) [19,20]. Among the organic free radicals successfully tested for dissolution DNP are BDPA (1,3-bisdiphenylene-2-phenylallyl), trityl OX063 (tris{8-carboxyl-2,2,6,6-benzo(1,2-d:4,5-d)-bis(1,3)dithiole-4-yl}methyl sodium salt), galvinoxyl ((2,6-di-tert-butyl-α-(3,5-di-tert-butyl-4-oxo-2,5-cyclohexadien-1-ylidene)-p-tolyloxy), 2,2-diphenylpicrylhydrazyl (DPPH), and 4-oxo-TEMPO (4-oxo-2,2,6,6-tetramethyl-1-piperidinyloxy) [19–23]. There are also variants of these free radicals, such as water-soluble BDPA [24] and nitroxide variants with different isotopic enrichments [25], that were successfully used in hyperpolarized magnetic resonance experiments. In addition, nitrogen-vacancy centers in nanodiamonds [26] and ultraviolet (UV) light-induced non-persistent free radicals were also utilized as polarizing agents in the dissolution of DNP [27–29]. It should be noted that the use of these UV-induced free radicals in DNP is advantageous since the free radicals disappear upon dissolution, pre-empting the need for radical removal or filtration in *in vivo* hyperpolarized MRI experiments [27–29].

Regardless of the source, these free electrons are thermally polarized close to 100% polarization at DNP instrumental conditions with a magnetic field >3 T and a low temperature close to 1 K. Although these free electron sources are theoretically polarized to the same degree under the same instrumental conditions, the ESR properties of these DNP polarizing agents, such as the ESR lineshape and electron relaxation, can greatly affect the maximum achievable DNP enhancements of the target nuclear spins [19,20]. For instance, the ESR lineshape function determines the breadth and

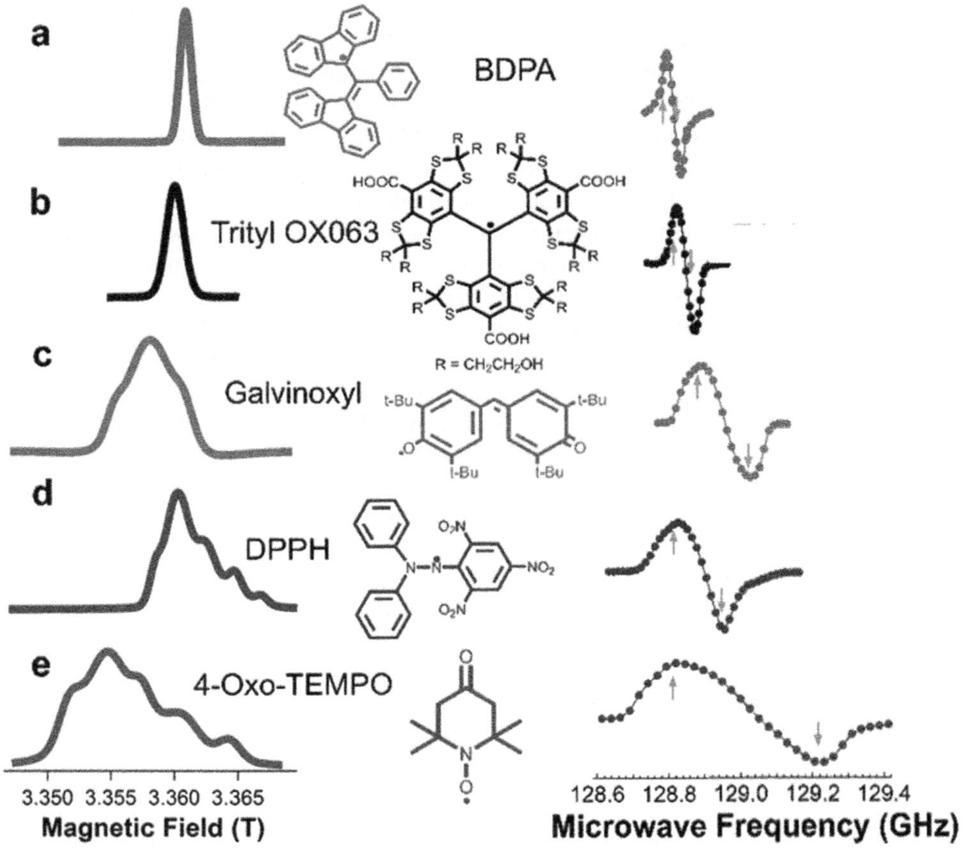

FIGURE 2.2 ESR spectra and 13C microwave DNP sweep profiles of various organic free radicals used as polarizing agents in dissolution DNP: (A) BDPA; (B) trityl OX063; (C) galvinoxyl; (D) DPPH; and (E) 4-oxo-TEMPO. *Figure adapted from Ref. [19] with permission from the American Chemical Society.*

shape of the so-called microwave DNP spectrum—a plot of nuclear polarization as a function of microwave irradiation frequency near the expected ESR frequency of the polarizing agent [30–32]. As can be seen in the middle column of Fig. 2.2, the separation distance between the optimum irradiation frequencies P(+) and P(−), denoted by up and down arrows in microwave DNP spectra, scales with the ESR linewidth D of the free radical. The varying ESR linewidths of the free radicals shown in Fig. 2.2 (BDPA, trityl OX063, galvinoxyl, DPPH, and 4-oxo-TEMPO) with ESR D from 60 to 465 MHz at 5% base present an exciting experimental opportunity on DNP wherein the ESR width D can be "chemically tuned" by using different free radicals.

The ESR lineshapes of the free radicals have important implications as to which DNP mechanism the polarization transfer from the electrons to the nuclear spins will proceed [5–7,30–32]. If the ESR width D of the polarizing agent is greater than or comparable to the nuclear Larmor frequency ω_n, then thermal mixing is the expected predominant mechanism for DNP. Thermal mixing, also known as the equal spin temperature model, is a thermodynamic mechanism of DNP in which a thermal link is established between the electron dipolar system or electron spin-spin interaction reservoir (ESSI) and the nuclear Zeeman system (NZS) due to their comparable energies [31,32]. Microwave irradiation of the DNP sample dynamically "cools" the ESSI reservoir, thereby lowering its spin temperature T_s relative to the actual temperature of the sample, which is denoted by the lattice temperature T_L. Due to the thermal link between the two thermodynamic systems, the ESSI reservoir in turn cools down the NZS, leading to the same low spin temperature for both systems, translating to enhanced nuclear polarization [5,31,32]. Meanwhile, when $D << \omega_n$, the solid-effect DNP mechanism is expected to be dominant, in which the optimum polarization peaks should appear at $\omega_e \pm \omega_n$, where ω_e is the ESR resonant frequency of the polarizing agent [5,6]. In solid effect, microwave irradiation of the sample at either of these optimum microwave frequencies causes a forbidden transition in a dipolar-coupled electron-nucleus system. In this scenario, an electron-nucleus system can be excited to a metastable forbidden state. Now, the electron relaxation time T_{1e} of free radicals is typically short on the order of μs to a fraction of a second even at cryogenic temperatures, which is several orders of magnitude faster than the T_{1n} nuclear relaxation time (typically >10,000 s), allowing for many thousands of electrons to flip to thermal equilibrium before the nuclei relax back. Through this process, the electron can chaperone

another nuclear spin to the metastable state, and eventually more nuclear spins will populate a particular Zeeman energy state, leading to a hyperpolarized state or enhanced nuclear polarization [5,33]. Another DNP mechanism is the cross effect, which is dominant when the polarizing agent is a radical with an inhomogeneously broadened ESR D larger than ω_n [5]. In this three-spin model, a nucleus is coupled to two electrons whose resonant frequencies are separated by ω_n and microwave irradiation causes partial saturation of the ESR spectrum of the polarizing agent [5,34]. Due to the numerous factors involved, it is likely that a combination of these DNP mechanisms is active during microwave irradiation of the DNP sample at high magnetic fields and low temperatures.

In light of these theoretical DNP mechanisms, it is thus expected that the use of free radicals with different ESR properties would yield different values of DNP enhancements. Previous experimental DNP studies [20,31] have shown that narrow ESR D radicals are more favorable polarizing agents for hyperpolarization of low-γ nuclei, such as ^{13}C spins [20,31]. In particular, the narrow ESR width-free radicals BDPA and trityl OX063 yielded are beneficial for ^{13}C DNP because they translate to a lower specific heat capacity of the electron spin system [12]. Microwave irradiation can dynamically lower the spin temperature of the electron dipolar system to a level that aligns with the spin temperature of the NZS, thereby leading to heightened nuclear polarization. In other words, the narrower ESR linewidths of trityl OX063 and BDPA are well-matched with the Larmor frequencies of low-γ nuclei like ^{13}C spins under the context of thermal mixing. This energy matching facilitates an efficient transfer of polarization from trityl OX063 electrons to the ^{13}C nuclei. Using the narrow ESR width trityl OX063 for ^{1}H DNP, however, yielded lower polarization since the trityl OX063 ESSI and the ^{1}H NZS thermal reservoirs are essentially decoupled due to energy mismatch; in this case, a well-resolved solid effect is the expected predominant mechanism, which may require higher microwave power than what is typically used in dissolution DNP to be more efficient [35]. Conversely, the larger ESR linewidths exhibited by free radicals such as galvinoxyl, DPPH, and 4-oxo-TEMPO resulted in relatively moderate levels of ^{13}C polarization. This suggests that these free radicals with wider ESR linewidths are better suited for DNP involving high-γ nuclear spins, whose nuclear Zeeman energies suitably match the energy of the electron dipolar system of the free radical polarizing agent [20,31,35,36]. Overall, it can be said that the choice of the free radical polarizing agent has a significant impact on the maximum achievable DNP enhancements, depending upon which nuclei are targeted for hyperpolarization.

2.2.2 Isotopic labeling of the glassing matrix

The vitreous glassing matrix (e.g., glycerol/water, ethanol/water, etc.) ensures that the molecules containing the target nuclei as well as the polarizing agents are homogeneously dispersed throughout the entire DNP sample, with a distribution of one free electron for every 100−1000 nuclear spins in a typical DNP sample. Since the glassing matrix is a crucial component of the DNP sample, it does warrant studies to ascertain its specific effects on DNP efficiency. In 2008, Kurdzesau et al. [36] reported the beneficial effect of deuterium (^{2}H) enrichment of the glassing matrices on ^{13}C DNP samples doped with TEMPO-free radicals polarized at 3.5 T and 1.2 K. In particular, glassing matrix deuteration led to a significant increase in the maximum polarization values, while the polarization buildup time was shortened.

In another study, Lumata et al. [37] indicated that one has to be cautious when employing glassing matrix deuteration for ^{13}C DNP as it may not always lead to increases in DNP signals depending upon the sample composition; in fact, the opposite effect may be observed in which ^{2}H enrichment of the glassing matrix led to significant decreases in the hyperpolarized signals in certain cases, as depicted in Fig. 2.3. Experimental data revealed that replacement of ^{1}H spins ($\gamma = 42.58$ MHz/T) by ^{2}H spins ($\gamma = 6.54$ MHz/T) in the glassing matrix further enhances ^{13}C ($\gamma = 10.7$ MHz/T) DNP by a factor of two to three for ^{13}C samples doped with large ESR linewidth free radicals such as TEMPO, DPPH, and galvinoxyl [37]. Meanwhile, the ^{13}C polarization substantially dropped to 30%−50% of its original value when glassing matrix deuteration is employed in DNP samples doped with narrow ESR widths such as trityl OX063 and BDPA [37]. The dramatic DNP response to glassing solvent deuteration can be explained in the context of thermal mixing mechanisms [36,37]. When large ESR-width free radicals such as TEMPO are used, all nuclei (^{1}H, ^{2}H, and ^{13}C) are in principle thermally linked to the electron dipolar system during DNP since the ESR D herein would be greater than or comparable to the Larmor frequencies of the aforementioned nuclear spins. In the thermodynamic DNP model, the heat capacity of a nuclear spin system is $C \propto N\gamma^2$, where N refers to the number of nuclear spins [32]. Since $\gamma(^{2}H)<\gamma(^{1}H)$, replacement of ^{1}H spins by ^{2}H spins in the glassing matrix will lead to a lower overall NZS heat load, thus an easier system for the ESSI reservoir to cool via a thermal mixing mechanism, leading to higher ^{13}C polarization as observed experimentally for samples with large ESR width radicals [36,37]. In the case of DNP with narrow ESR width radicals such as trityl OX063, the ^{1}H NZS is thermally decoupled from the free radical ESSI reservoir since the ESR D is less than the ^{1}H Larmor frequency. Thus, ^{2}H enrichment in the glassing matrix in this DNP sample would actually increase the overall NZS heat load since ^{2}H NZS now joins ^{13}C NZS in the common thermal link to the ESSI reservoir. This large overall heat load will

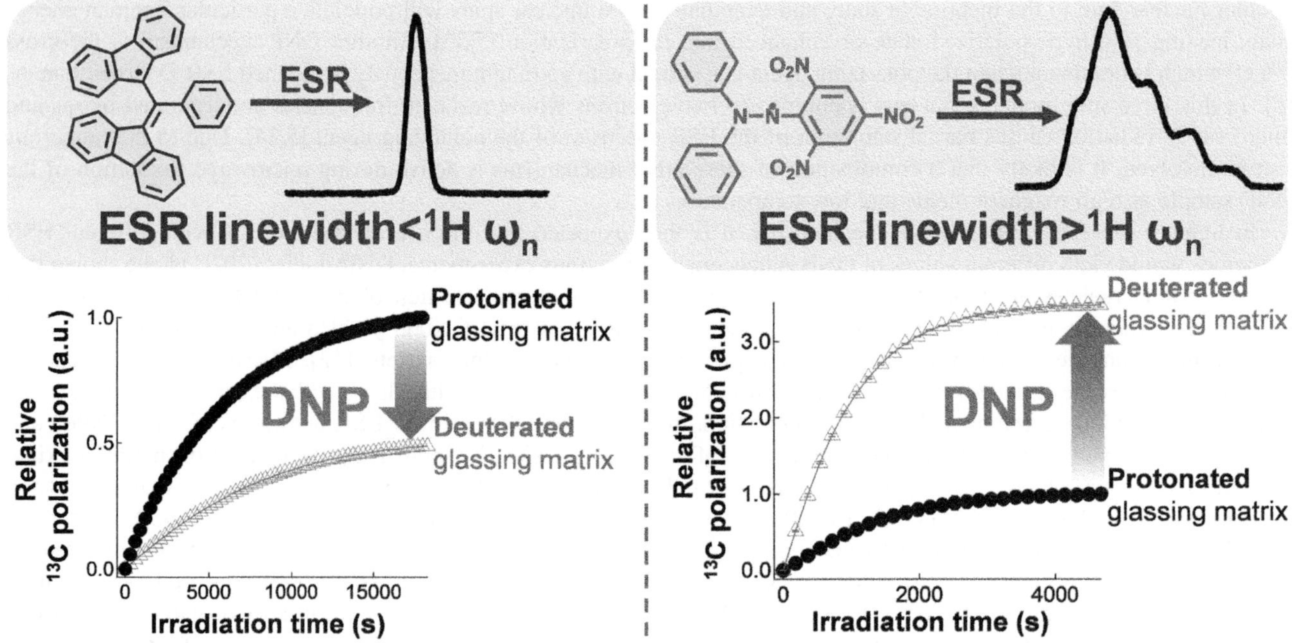

FIGURE 2.3 Schematic showing the effect of glassing matrix deuteration on ^{13}C DNP performed at 3.35 T and 1.4 K: ^{2}H-enriched solvents can be beneficial for ^{13}C DNP using large ESR width-free radicals such as DPPH, while they can lead to a decrease of ^{13}C DNP signal when narrow ESR width-free radicals such as BDPA are used. *Figure reprinted from Ref. [37] with permission from the Royal Society of Chemistry.*

lead to lower ^{13}C polarization as the ESSI reservoir is now also responsible for the dynamic cooling of the ^{2}H spins [36,37]. Thus, ^{2}H enrichment in the glassing matrix is only beneficial for ^{13}C DNP when it is paired with the right free radical polarizing agent (see Fig. 2.3).

A related investigation by Parish et al. [38] reported that the reduction in ^{13}C DNP signal due to glassing matrix deuteration in ^{13}C samples doped with small ESR width free radicals, such as trityl OX063, is accompanied by an increase in the solid-state ^{13}C T_1 relaxation time by as much as a factor of 2. Meanwhile, the solid-state ^{13}C T_1 values of samples doped with large ESR-width TEMPO-free radicals did not show any significant changes upon deuteration of the glassing matrix. These experimental data imply that changes in the relative ^{13}C DNP signals with glassing matrix deuteration appear to be unrelated to the changes that occur in the nuclear T_1 relaxation behavior. Thus, these results indicate that the ^{13}C DNP effects seen with glassing matrix deuteration are predominantly due to the thermodynamic interaction between the electron dipolar system and the ^{1}H, ^{2}H, and ^{13}C nuclear Zeeman reservoirs.

On another note, ^{13}C DNP work [39] at a higher magnetic field (5 T) shows that glassing matrix deuteration only improved the ^{13}C DNP signals of TEMPO-doped samples by about 40%, in contrast to the 100%−200% increase relative to the control DNP signal seen at 3.35 T. These results suggest that the beneficial effect of glassing matrix deuteration on ^{13}C DNP with large ESR-width free radicals appears to decrease at magnetic fields greater than 3.35 T.

In addition to deuteration, the effect of ^{13}C enrichment in the DMSO:water glassing matrix on ^{13}C DNP was also investigated in a prior research study [40]. The main finding of that research report was that the inclusion of the ^{13}C-labeled DMSO led to a faster solid-state ^{13}C polarization buildup rate for the ^{13}C DNP sample; however, the ^{13}C DNP signal of [1-^{13}C] pyruvate was largely unaffected, as confirmed by ^{13}C NMR enhancement measurements in the liquid state after dissolution. The extra benefit of a faster ^{13}C DNP buildup rate, which can significantly cut the required microwave irradiation time, is ascribed to increased $^{13}C-^{13}C$ spin diffusion in the sample [40,41].

2.2.3 Isotopic labeling of the molecules

Strategic selection of ^{13}C isotopic labeling in specific sites of biomolecules is a crucial aspect to consider in hyperpolarized ^{13}C magnetic resonance experiments, considering the specific biochemical pathway to target for study and the length of the ^{13}C hyperpolarization lifetime that is dictated by the ^{13}C T_1 relaxation of the ^{13}C isotopic label. There is a question, however, reminiscent of a familiar line in a real estate market, as to whether the location of ^{13}C isotopic labeling matters in pursuit of achieving high ^{13}C DNP-enhanced polarizations.

In a research study reported by Niedbalski et al. [42], it was found that the maximum ^{13}C DNP signals of various ^{13}C isotopic permutations of acetate were correlated with their solid-state ^{13}C longitudinal relaxation times when trityl OX063 was used as the polarizing agent. In particular, the DNP results reveal that the carboxyl ^{13}C spins have about twice the ^{13}C DNP signal of the methyl ^{13}C spins. Furthermore, deuteration of the methyl ^{13}C group, while proven to be beneficial in the liquid state in terms of prolonging the hyperpolarized state, resulted in a slight reduction of the solid-state ^{13}C DNP signals and shorter T_1 relaxation times, as shown in Fig. 2.4. The overall results revealed that isotopic location matters in ^{13}C DNP. Although further investigations are needed to confirm this, the observed ^{13}C DNP behavior of various $^{13}C/^{2}H$ acetate isotopomers is currently ascribed to the potential interplay of the nuclear relaxation leakage factor and the heteronuclear cross-talk between hyperpolarized ^{13}C and ^{2}H spins.

A comparative investigation of the ^{13}C DNP and relaxation behavior of single and double ^{13}C-labeled acetate isotopomers was also reported, this time using the wide ESR width 4-oxo-TEMPO free radical at 3.35 T and 1.4 K [43]. The study is interesting from the experimental and theoretical points of view because there are a total of six ^{13}C acetate isotopomers as a collective testbed for the effect of the intramolecular environment, and the use of 4-oxo-TEMPO allows simultaneous polarization of ^{13}C, ^{2}H, ^{1}H, and other NMR-active nuclei in the sample. With the DNP and sample glassing conditions kept equal, the results support that the ^{13}C DNP efficiency is highly dependent upon the intramolecular environment surrounding the ^{13}C spin of interest. In particular, ^{13}C labeling in a molecular site with a relatively lower intramolecular dipolar interaction to ^{13}C, such as the single ^{13}C-labeled [1-^{13}C], [1-^{13}C, $^{2}H_3$], and [2-^{13}C, $^{2}H_3$] acetate variants, appears to favor higher DNP signals compared to the other acetate isotopomers. For the double ^{13}C-labeled [1,2-$^{13}C_2$] and [1,2-$^{13}C_2$, $^{2}H_3$] acetate isotopomers, the $^{13}C-^{13}C$ dipolar interaction appears to dominate, leading to a lower ^{13}C DNP level comparable with that of [2-^{13}C] acetate. There is no significant difference between the solid-phase ^{13}C T_1 relaxation values of the acetate isotopomers, indicating that the paramagnetic relaxation contribution is dominant due to the relatively high concentration of free radicals in the DNP samples [43]. Overall, the location of the ^{13}C isotopic labeling in molecules of interest appears to have a significant impact on the DNP signal, depending upon the type of free radical polarizing agents used in DNP samples.

2.2.4 Inclusion of lanthanides in the DNP sample

The inclusion of certain paramagnetic chemical additives in DNP samples may also result in an improvement of the hyperpolarized NMR signals. A 2012 research report by Gordon et al. [44] showed that addition of trace amounts of gadolinium and holmium salts in trityl OX063 radical-doped ^{13}C DNP samples resulted in 50%–60% relative DNP signal improvements at 3.35 T and 1.4 K using the HyperSense polarizer. Since dissolution DNP is often used for biological applications, the toxic lanthanide is usually chelated into a ligand in the form of complexes such as Gd-DOTA, Gd-HP-DO3A, and Gd-EDTA, among others [45]. Friesen-Waldner et al. [46] studied the effect of specific chelates on the dissolution of DNP. The study showed that open-chain complexes like Gd-BOPTA, Gd-DTPA, and Gd-DTPA-BMA all significantly reduced the ^{13}C T_1 time of the substrate in the liquid form by over 50% in comparison to the 25% reduction at maximum caused by macro-cyclic complexes like Gd-DOTA and Gd-Hp-DO3A. The drastic reduction in liquid-state ^{13}C hyperpolarization lifetime in samples with Gd-complexes with open-chain ligands, compared to closed-chain macrocyclics such as the Gd-DOTA complex, is ascribed to the probable release of free gadolinium ions into the liquid [46].

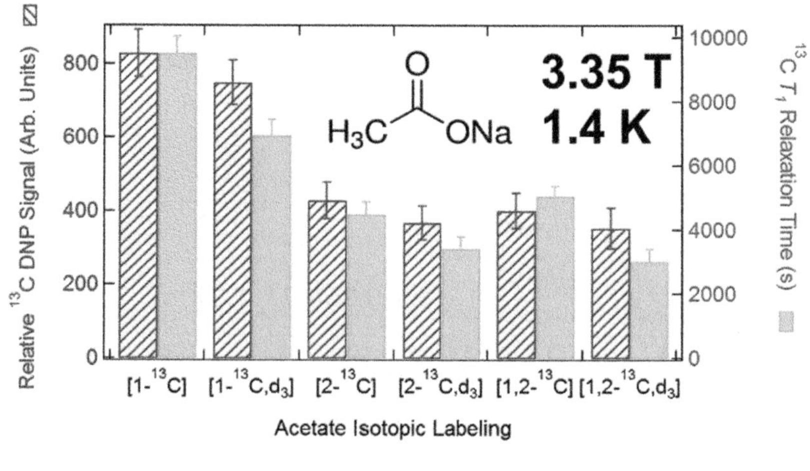

FIGURE 2.4 Influence of ^{13}C isotopic location on DNP: Bar plot showing the correlation of relative ^{13}C DNP signals and solid-state ^{13}C T_1 values of different isotopomers of ^{13}C acetate at 3.35 T and 1.4 K. *Figure reprinted from Ref. [42] with permission from the American Chemical Society.*

In addition, a trimeric Gd-complex (3-Gd), which consisted of 3 Gd-DOTA complexes chemically tethered together, was also tested for dissolution of ^{13}C DNP [47]. The study reported that doping the ^{13}C DNP sample with 3-Gd improved the DNP-enhanced solid-state ^{13}C polarization by a factor of 3.4, which is comparable to the enhancements obtained in DNP samples using Gd-DOTA, but it only requires about 0.4 mM concentration instead of the optimal 2 mM concentration for the monomeric Gd complexes [47].

In another research work [49], it was shown that with careful optimization, Dy^{3+} and Tb^{3+} complexes were also effective lanthanide agents that could double or triple the ^{13}C DNP-enhanced nuclear polarization relative to the signals of DNP samples without Gd^{3+} complexes under similar conditions. The best levels of Tb-DOTA and Dy-DOTA doping for optimal results in dissolution DNP were identified as 0.25 and 1.5 mM, respectively. At these optimal concentrations, the introduction of Dy^{3+} and Tb^{3+} as dopants led to notable enhancements in the ^{13}C solid-state DNP signals. These enhancements were approximately 1.8 times and 2.5 times higher for Dy^{3+} and Tb^{3+} doping, respectively, compared to the ^{13}C DNP signal of an undoped reference sample. Furthermore, a report by Kiswandhi et al. [48] indicated that the addition of 2 mM Ho-DOTA in ^{13}C DNP samples doped with trityl OX063 resulted in an improvement of the ^{13}C DNP signals by 2.7 times compared to undoped samples, as depicted in Fig. 2.5. Furthermore, the presence of Ho-DOTA in the hyperpolarized liquid has a minimal impact on liquid-state ^{13}C T_1, unlike Gd^{3+}-doping, which significantly diminishes the ^{13}C T_1. It is thus apparent that Ho^{3+} is a less effective relaxation agent compared to Gd^{3+}; thus, Ho-DOTA is advantageous over Gd-DOTA in terms of preserving the hyperpolarized state after the dissolution process.

In addition to increasing the ^{13}C DNP signals, another notable effect of lanthanide doping in ^{13}C DNP samples is the narrowing of the microwave DNP spectrum, in which the locations of the positive and negative ^{13}C maximum polarization peaks in the ^{13}C microwave DNP sweep are shifted toward each other upon lanthanide doping. Although it was initially reported that the W-band ESR spectra of trityl-doped DNP samples are largely unaffected by lanthanide doping at 4–10 K, Capozzi et al. [50] reported that the ESR spectra of the trityl OX063 free radical in ^{13}C DNP samples were slightly narrowed upon Gd^{3+} doping by doing ESR measurements at the actual DNP cryogenic temperature close to 1 K. An impressive 70% ^{13}C polarization was performed for trityl-doped [U–^{13}C] glucose with Gd-DOTA in DNP at 6.7 T and 1.1 K [50].

The narrowing of the ESR spectra of the free radical at the DNP operating temperature in the presence of lanthanide dopants could explain the shift observed in the ^{13}C DNP polarization peaks. This is generally in consonance with the Borghini model of DNP [30], which predicts the microwave DNP spectrum, including the locations of the optimum microwave frequencies for irradiation in DNP. Concomitant with the increase in ^{13}C DNP enhancement with lanthanide doping, the electron T_{1e} of trityl OX063 free radical in DNP samples was prominently reduced at cryogenic temperatures in the presence of Gd^{3+}, Ho^{3+}, Dy^{3+}, and Tb^{3+} dopants. The significant gain in ^{13}C DNP signal with lanthanide doping is ascribed to a thermal mixing model where shortened electron T_1 of the free radicals leads to dominant electron cooling in DNP, thus resulting in lower spin temperature in the electron dipolar system and consequently higher nuclear polarization [5,6,9].

It is important to note that so far, lanthanide doping has been quite effective in further increasing the hyperpolarized signals of DNP samples doped with the narrow ESR linewidth trityl OX063. A previous work by Lumata et al. [51]

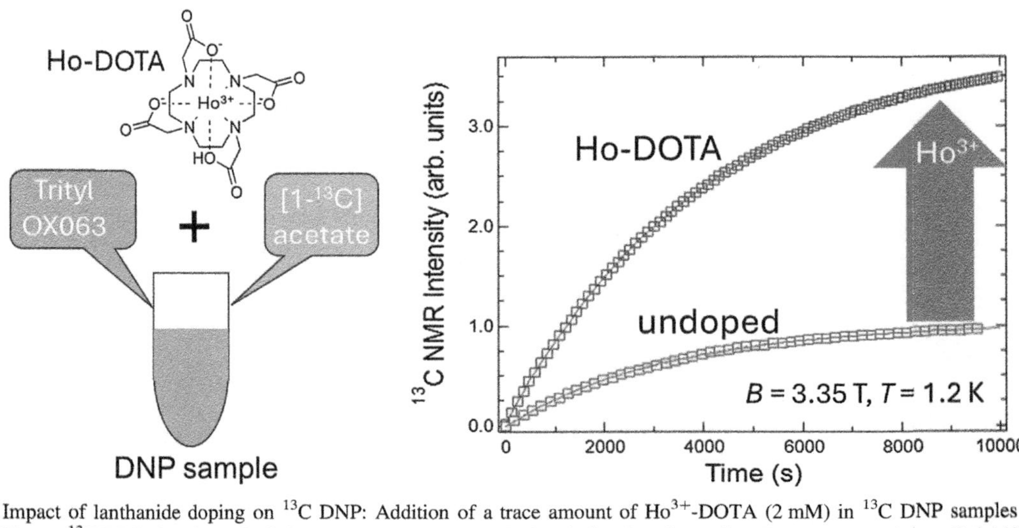

FIGURE 2.5 Impact of lanthanide doping on ^{13}C DNP: Addition of a trace amount of Ho^{3+}-DOTA (2 mM) in ^{13}C DNP samples can significantly improve the solid-state ^{13}C DNP signals by 2–3 times compared to the DNP signal of undoped samples. *Figure reprinted from Ref. [48] with permission from the American Chemical Society.*

showed that Gd-DOTA doping only yielded either minimal or no improvement in DNP signals in samples doped with the wide ESR linewidth TEMPO and the water-insoluble BDPA free radicals. A research work by Kattri et al. [52] revealed that Ho-DOTA doping on ^{13}C DNP samples doped with BDPA led to an improvement of the solid-state ^{13}C DNP signal at 3.35 T, though not on par with those achieved with the trityl OX063 radical. There is also a question as to the effect of doing DNP with lanthanide doping at higher magnetic fields than 3.35 T. A study by Kiswandhi et al. [39] showed the relative ^{13}C DNP improvement with Gd^{3+} doping observed at 3.35 T appeared to be less dramatic when performed at 5 T. In another research report by Sirusi et al. [53], it was shown that Ho^{3+} doping not only enhances the ^{13}C polarization but also decreases the polarization buildup time of trityl-doped DNP samples at 5 T and 1.2 K. It was also suggested that both the cross effect and thermal mixing mechanisms are co-existent for trityl radical at 5 T and 1.2 K [53]. It is thus apparent that the benefit of lanthanide doping in DNP varies with the type of free radical polarizing agent used and also with the strength of the operating magnetic field for DNP.

2.2.5 Transition metal doping in DNP

In addition to using lanthanide additives as DNP signal enhancers, transition metals can also be added to DNP samples to further boost the DNP signals. Transition metals are a group of elements located in the central block of the periodic table, specifically in the *d*-block, characterized by their partially filled *d* orbitals, which give rise to their unique chemical and physical properties.

In a research study reported by Niedbalski et al. [54], transition metal complexes *R*-NOTA (*R*=Mn^{2+}, Cu^{2+}, and Co^{2+}) were tested as ^{13}C DNP enhancers for ^{13}C DNP samples using trityl OX063 as the free radical polarizing agent at 3.35 T and 1.2 K. The results of that study showed that doping the ^{13}C DNP samples with 3 mM Mn^{2+}-NOTA further boosted the solid-state ^{13}C DNP signal by a factor of nearly 3, as displayed in Fig. 2.6. On the other hand, the other transition metal complexes Cu^{2+}-NOTA and Co^{2+}-NOTA complexes did not show any significant impact or improvement

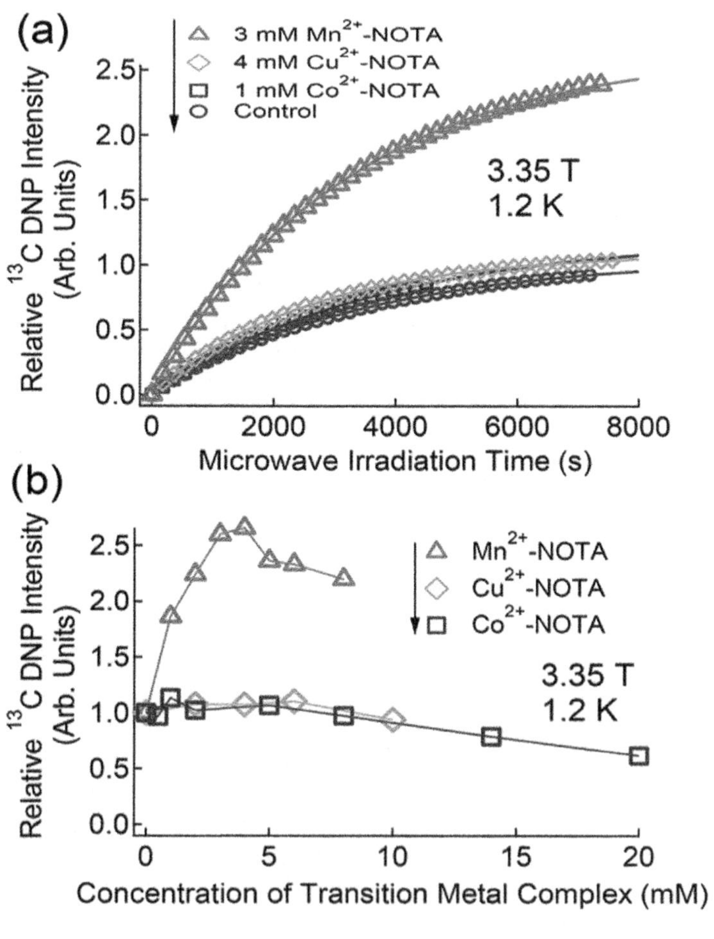

FIGURE 2.6 Impact of transition metal doping on ^{13}C DNP: (A) Increase in solid-state ^{13}C DNP signal was detected for samples doped with Mn^{2+}-NOTA, but not for samples with Cu^{2+}-NOTA and Co^{2+}-NOTA. (B) Relative ^{13}C DNP signal as a function of transition metal concentration. *Figure adapted from Ref. [54] with permission from the American Chemical Society.*

on solid-state ^{13}C DNP signals (see Fig. 2.6). EPR measurements demonstrate that there is a reduction in the electron T_1 relaxation time of Mn^{+2}-doped materials; however, this reduction was not present in Cu^{+2} and Co^{+2}-doped samples [54]. These data imply that the reduction of the T_{1e} relaxation time for electrons by a paramagnetic additive is necessary for the enhancement of solid-state DNP signals. Continually increasing the dopant concentration does not necessarily guarantee further enhancement of DNP signaling, as shown in Fig. 2.6, and as such, optimal dopant concentrations need to be investigated for each distinct dopant used. For example, Mn^{+2} had optimal signal enhancement at a concentration of 3–4 mM with a factor enhancement of approximately 2.5, whereas increasing the dopant amount to 5–10 mM led to a decrease in enhancement of around 2.2 times the original undoped signal. Lower concentrations similarly varied between a relative intensity of 1.0 and 2.2 for the undoped signal. Similar work has been performed for Co^{+2} and Cu^{+2} with maximum enhancement factors obtained at around 5 mM with a relative intensity of around 1.1 [54]. These experimental results suggest that not all paramagnetic transition metal dopants are beneficial to DNP. Furthermore, these data also imply that the electron T_1 reduction of the free radical polarizing agent by a paramagnetic dopant appears to be tied up to the observed improvement in ^{13}C DNP signals.

Another DNP study [55] reported the utilization of Feraheme (ferumoxytol), a superparamagnetic iron oxide nanoparticle (SPION) MRI contrast agent, as a dopant for boosting the hyperpolarized ^{13}C NMR signals. This water-soluble SPION consists of an iron oxide core with approximately 10 nm in diameter functionalized with polyglucose sorbitol carboxymethyl ether. Experimental data from this study indicate that the addition of 11 mM elemental iron concentration from Feraheme into trityl OX063-doped 3 M [1-^{13}C] acetate DNP samples led to a substantial enhancement of the ^{13}C DNP signal, nearly tripling its strength [55]. Alongside this significant boost in DNP signal, a narrowing of the ^{13}C microwave DNP spectra in samples doped with SPION was also observed, reminiscent of a similar experimental trend observed in ^{13}C DNP samples doped with lanthanide complexes [44–53]. Additionally, due to the relatively large size of these SPIONS, Feraheme can be conveniently filtered out from the hyperpolarized solution using straightforward techniques such as a syringe filter [54]. This characteristic of the FDA-approved Feraheme offers a significant edge over other paramagnetic DNP enhancers like lanthanide complexes in terms of readiness for translational or clinical use of these dopants in hyperpolarization.

2.2.6 Dissolution solvents

So far, the DNP sample optimization methods previously described were generally intended for achieving the maximum solid-state DNP signals at cryogenic temperature, which would translate to high liquid-state NMR signal enhancements upon dissolution. Since the hyperpolarized signal decays faster at room temperature after dissolution, it is also important to mitigate the signal loss by using appropriate solvents to prolong the preservation of the hyperpolarized state. One experimentally simple way to extend the lifetime of the hyperpolarized signal in the liquid state after rapid dissolution is the use of D_2O as the dissolution liquid [11,55]. This is especially effective for hyperpolarized liquids with ^{13}C compounds with labile protons exchanging with solvating deuterons due to the fact that deuterons have less dipolar relaxation than protons, leading to increased ^{13}C T_1 and higher NMR signal enhancements in the hyperpolarized solution [55]. Another reported technique is the inclusion of radical scavengers such as vitamin C or sodium ascorbate in the dissolution solvent, which can reduce TEMPO-free radicals and quench the paramagnetic relaxation from the radicals [56]. This simple method of adding vitamin C in the dissolution solvent led to significant gains in liquid-state T_1 of the target nuclei in TEMPO-doped samples and thus higher liquid-state NMR signal enhancements.

2.3 Optimization method in dissolution DNP: Instrumental advances

Although DNP technology has been used in nuclear and particle physics experiments since the 1960s, it was not until 2003, when Ardenkjaer-Larsen et al. first reported on the use of the dissolution method in DNP, that the NMR-enhancing power of this technology, which optimally occurs at cryogenic temperatures, was harnessed for biomedical applications at physiologically tolerable temperatures [8]. Ever since then, research on this field has exploded from in vitro to in vivo hyperpolarized NMR and MRI experiments of various pathologies to its first report of in-human hyperpolarized ^{13}C diagnostic MRI application [18], and even now, the practical uses of this technology are still counting on the pre-clinical and clinical fronts. In this section, a brief survey of the major instrumental advances in dissolution DNP will be discussed. In general, a typical dissolution DNP instrument or hyperpolarizer, whether homebuilt or commercial (see Fig. 2.7), consists of: (i) A superconducting magnet; (ii) cryostat; (iii) DNP probe; (iv) NMR circuit; (v) microwave system; (vi) vacuum or pumping system; and (vii) dissolution system. These components are assembled to subject the DNP sample to a high magnetic field and low-temperature environment where the electron thermal polarization is close

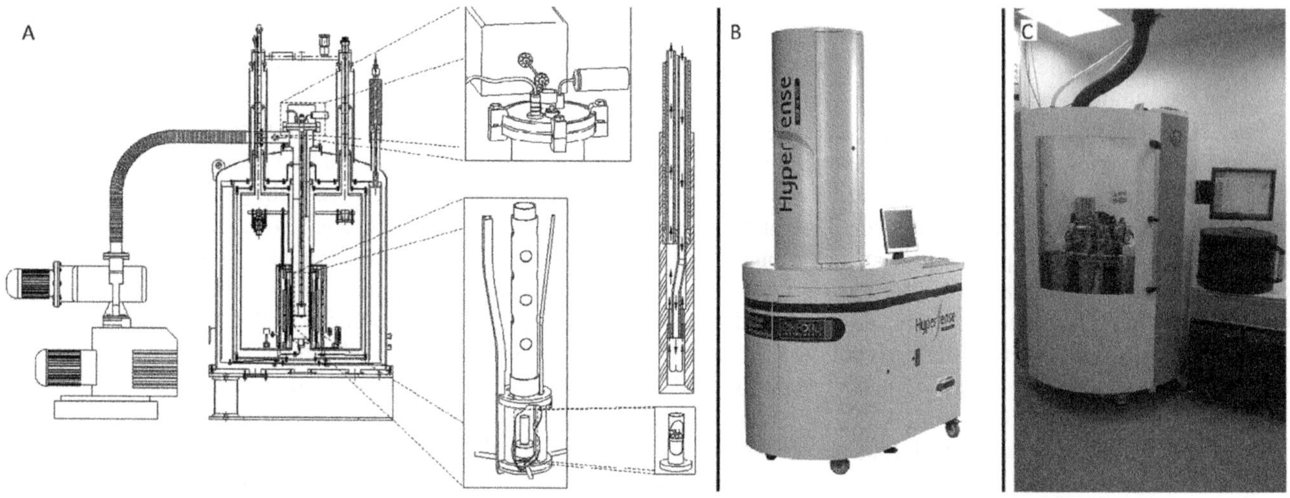

FIGURE 2.7 (A) Schematic drawing of the DNP polarizer and DNP probe of the first dissolution DNP design by Jan Henrik Ardenkjaer-Larsen et al. [8]. (B) Hypersense, Oxford Instruments. (C) SPINlab, GE Healthcare. *Figure adapted from Ref. [10] with permission from Elsevier.*

to 100%. A typical dissolution DNP system also has a variable-frequency microwave system to be able to detect the optimum microwave irradiation frequencies for specific samples, and it is also equipped with an NMR detection system to detect and monitor the DNP signals. Finally, once the irradiated samples are hyperpolarized in the cryogenic state, the dissolution system will rapidly dissolve the frozen hyperpolarized samples to produce hyperpolarized liquids at near room temperature. Other accessories in the dissolution DNP instrument include temperature and vacuum pressure sensors, a sample cup, specialized gas and liquid tubing, a liquid Helium level sensor and meter, pressurized He gas tanks, valves, pneumatics, and many others, depending on the complexity of the design or automation of the DNP experimental operation. Below are some of the instrumental advances that led to an improvement in the dissolution DNP technology's operation and efficiency.

2.3.1 High B/T conditions

A number of DNP instrumentation reports [20,57–62] have indicated that ^{13}C DNP done at higher magnetic field B, that is, using the W-band field ($B = 3.35$ T) of the first dissolution DNP system [8] as a base, generally led to higher solid-state ^{13}C DNP signals at temperatures close to 1 K. In particular, ^{13}C DNP experiments performed at ~5 T [5,57,58] typically yielded around twice the ^{13}C polarization values relative to previously reported W-band ^{13}C DNP results, and these numbers are even higher for ^{13}C DNP experiments done at ~7 T [59–62]. Using a DNP system with a variable magnetic field capability, a base temperature of 1.3 K in the sample space of the cryostat, and equipped with microwave sources at multiple frequencies, Ardenkjaer-Larsen et al. reported that there is a strong field dependence of the ^{13}C DNP signal in which the observed liquid-state ^{13}C polarization jumped from 27% to around 70% when the field was increased from 3.35 to 6.7 T, respectively [62]. This aforementioned DNP system is useful for making relative comparisons of ^{13}C DNP performance at different fields since the measurements were done in the same hyperpolarizer, meaning the same instrumental conditions except, of course, for the field. Similarly, Yoshihara et al. [61] reported >60% liquid-state ^{13}C polarization achieved for [1-^{13}C]pyruvic acid doped with trityl radical after the DNP was performed at 7 T and 1 K. Anderkjaer-Larsen et al. [62] further noticed that ^{13}C DNP done at 10.1 T yielded only slight polarization gain compared to the DNP results at 6.7 T, though it is possible that further adjustments are needed in sample preparation and instrumental tuning to optimize ^{13}C DNP at 10.1 T. A notable experimental observation when performing DNP at higher fields is the slower polarization buildup curves, which can be overcome by using higher free radical concentrations [61,62]. In addition to a higher magnetic field, another aspect to consider to achieve higher DNP signals is lowering the sample space temperature T in the cryostat. The current standard of ~1 K base operating temperature in dissolution DNP systems is due to the limitations of typical roots blower vacuum systems used in dissolution DNP instruments. It is conceivable that more powerful roots blower vacuum pumps could be employed to further decrease the base temperature below 1 K, which is expected to have significant DNP signal gains, albeit at the cost of higher liquid helium consumption during DNP operation. In general, a high B/T ratio is sought in dissolution DNP instrumental conditions to attain high DNP signals.

2.3.2 Cryogen-free DNP instrumentation

One significant obstacle hindering the broad adoption of dissolution DNP technology is the substantial expense associated with the consumption of liquid helium throughout the process of DNP operation. This issue is further compounded by the perennial liquid helium shortage in the United States and elsewhere. One way to offset or at least minimize this high expense in cryogens is to employ helium recovery or recycling systems. It should be noted that this would be a major facility undertaking that requires significant investment by a group or an institution, and thus, such liquid helium recycling infrastructure is not so widespread either. Alternatively, a number of research groups have reported on the use of cryogen-free DNP systems [60–64] to offset the high cost of liquid helium consumption in DNP experiments. For instance, Kiswandhi et al. [60] reported on the construction of a 180-GHz/6.4 T DNP system based on a cryogen-free and variable-field superconducting magnet (Cryomagnetics Inc., Oak Ridge, TN) that can be ramped up to 9 T. This DNP system is equipped with Cryomech cryocoolers (Cryomech Inc., Syracuse, NY) that can cool the DNP sample down to 1.8 K via the Joule–Thomson effect and only require a tank of high-purity helium gas every 4–6 months. ^{13}C DNP experiments on this system yielded solid-state ^{13}C polarization levels of 58 ± 3% and 18 ± 2% for [1-^{13}C] acetate samples doped with trityl OX063 and 4-oxo-TEMPO, respectively [60]. Meanwhile, Ardenkjaer-Larsen et al. [62] reported on the use of a 10.1 T field-sweepable and cryogen-free superconducting magnet (Cryogenic Ltd., UK), which can reach a base sample temperature of 1.3 K for DNP. This cryogen-free DNP system yielded a relatively high ^{13}C polarization of 70% at 6.7 T for ^{13}C DNP samples doped with trityl OX063 [62]. The relative stability and reliability of this DNP system formed the basis for Ardenkjaer-Larsen's spinoff company, which manufactures the commercial 7 T cryogen-free SpinAligner hyperpolarizer (Polarize ApS, Denmark).

2.3.3 Cross-polarization DNP

Large ESR-width free radicals such as TEMPO are known to be efficient polarizing agents for high-gamma nuclei such as ^{1}H spins, in which thermal mixing is the expected predominant DNP mechanism even at high B/T conditions of 6.7 T and 1 K. The idea behind cross-polarization-DNP [65–67] is to take advantage of the large DNP-enhanced polarization of high-gamma spins and transfer it to further boost the DNP signals of low-gamma spins such as ^{13}C spins using at least a double-resonance NMR coil and a specialized set of NMR pulses during DNP at cryogenic temperature. Jannin et al. [65] reported achieving high solid-state ^{13}C polarization of above 70% within a relatively fast time frame of 20 min after using an adiabatic half-passage cross-polarization-DNP technique in TEMPO-doped ^{13}C DNP samples at 6.7 T and 1.2 K. It should be noted that direct ^{13}C DNP with TEMPO under these conditions only yielded around 36% ^{13}C polarization with a buildup time that was four times slower [65]. In another study, Bornet et al. [66] showed that the high solid-phase ^{13}C polarization reported in the previous work [65] can be experimentally translated to a high liquid-state ^{13}C polarization after dissolution. To achieve this feat, certain modifications are required, such as employing a new DNP insert and a multiple-tuned saddle coil with dimensions suited for the dissolution process. After cross-polarization DNP and dissolution, a substantial liquid-state ^{13}C polarization of around 40% was recorded versus ~20% ^{13}C polarization achieved by direct polarization. These experimental results indicate that the extra DNP boost afforded by cross-polarization DNP was preserved after dissolution. Although it requires some specialized NMR hardware and pulses, the main advantages of cross-polarization DNP include the use of cheap and readily available TEMPO-free radicals plus the shorter polarization buildup times that go with high polarization levels.

2.3.4 Microwave frequency modulation

Conventional dissolution DNP employs monochromatic microwaves in which the sample is irradiated at only one frequency during DNP at cryogenic conditions. Bornet et al. [68] showed that ^{13}C DNP can be improved by using frequency-modulated microwave irradiation at 6.7 T and 1.2 K. The wide range of frequencies, e.g., on the order of ~100 MHz, irradiated using this technique takes the place of spectral spin diffusion to cover for partial saturation and is particularly useful in DNP using radicals with inhomogeneous broadening, in which the cross-effect DNP mechanism is dominant. It should be noted that the microwave frequency modulation is applicable to both direct polarization and cross-polarization DNP. The gain in ^{13}C DNP signal using this technique becomes more prominent with a lower radical concentration, in which conventionally only a few electrons will participate in DNP, leading to low polarization and long build-up times. With microwave frequency modulation, more electrons are forced to participate in the DNP process, which leads to a faster build-up rate and high polarizations, along with less paramagnetic broadening [68]. Furthermore, the decrease in radical concentration required for DNP is advantageous as it will lead to a longer preservation time of hyperpolarized magnetization after dissolution.

2.3.5 Magnetic tunnel

During the journey of the dissolution liquid from the polarizer to a nearby NMR spectrometer or MRI scanner, there is typically a low-field region in the middle that can be regarded as a "death valley" for hyperpolarized spins in which the hyperpolarized signal could die by rapid relaxation (see Fig. 2.8). To protect these hyperpolarized spins from perishing in the death valley, Milani et al. [69] devised a modular magnetic tunnel consisting of an array of permanent magnets to provide a lifesaving magnetic field to the hyperpolarized liquid during the transfer. Careful planning and magnetic field mapping are required for the construction of the magnetic tunnel, as the low or zero field spots in the valley vary in different laboratories due to different settings, shielding, configurations, and polarities of the polarizer and NMR or MRI magnets. The magnetic tunnel is made of an assembly of neodymium boron magnets secured by aluminum structures for support, wherein the tunnel center experiences a magnetic field of >0.9 T. To transport the dissolution liquid, Teflon tubing runs through the center of the tunnel. During testing, improvement factors between 1 and 25-fold were observed for ^{1}H and ^{13}C DNP signals compared to their values without the magnetic tunnel [69]. Overall, the use of a magnetic tunnel in this case is a mitigating solution to the longstanding but often overlooked issue of polarization loss due to low field relaxation during dissolution liquid transport.

2.3.6 Multisample capability

A typical dissolution DNP system is normally designed to hyperpolarize only one DNP sample at a time, which takes about 1−2 h of microwave irradiation before dissolution, and there will be some downtime of several minutes to allow for cleaning and maintaining the temperature and pressure conditions of the polarizer before another sample is inserted. To maximize experimental productivity and minimize the downtime in changing samples, dissolution DNP systems with multisample capability were constructed by various research groups [70−72]. Batel et al. [70] utilized a wide-bore (89 mm inner bore diameter) superconducting magnet to fit in a revolver-style sample changer in the cryostat and accommodate up to six samples simultaneously to reduce downtime in loading and unloading samples. The DNP conditions were set at 3.35 T/94 GHz and 1.3 K. In this configuration, microwave irradiation as well as dissolution are done on one sample at a time. ^{13}C DNP of trityl OX063-doped samples under these conditions yielded ~45% and ~13% ^{13}C polarization in the solid-state and liquid-state, respectively. On the other hand, Cremillieux et al. [71] used a super wide-bore (150 mm inner bore diameter) superconducting magnet to design a dissolution DNP system that allows up to three samples to be loaded

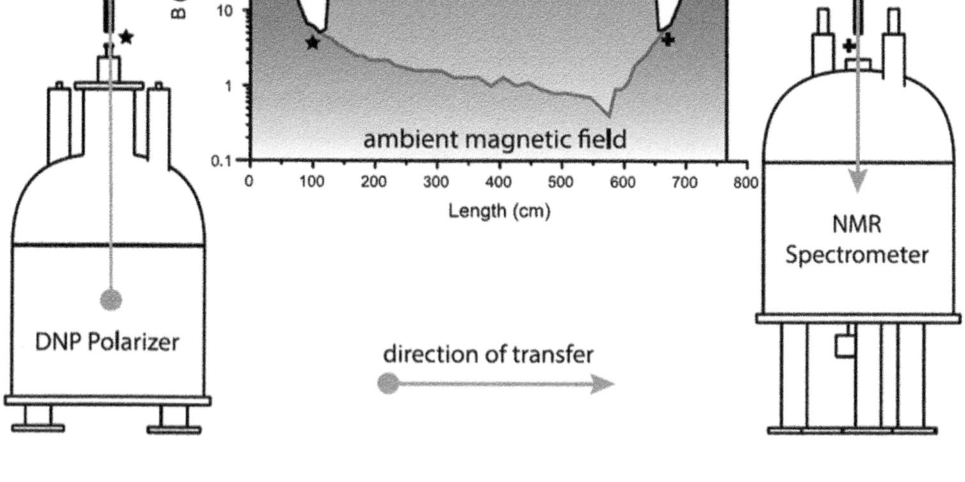

FIGURE 2.8 Schematic of the magnetic tunnel to protect the hyperpolarized spins from perishing via rapid relaxation in the low-field "death valley" between the polarizer and NMR magnets. *Figure adapted from Ref. [69] with permission from the American Institute of Physics.*

for DNP at 3.35 T and ~1.2 K. Meanwhile, Cheng et al. [72] constructed a zero boil-off 7 T dissolution DNP system that can hyperpolarize two DNP samples at the same time, but dissolution is done individually. Using this system, ^{13}C polarizations of ~55% and ~36% were obtained in the solid-state and liquid-state, respectively [72]. In summary, the incorporation of the multisample capability in dissolution DNP systems could potentially enable greater research productivity and minimal experimental downtime in DNP operation.

2.3.7 Fast transfer of dissolution liquids

There is a need for speed in dissolution liquid transport to limit nuclear spin relaxation losses outside the polarizer and/or the NMR/MRI magnet. Along with speed, the stability of the transferred liquid is also an important requirement since turbulence or the presence of bubbles in the transferred fluid can negatively impact the magnetic resonance measurements. Bowen et al. [73] addressed these challenges by repurposing an HPLC valve to initially load the hyperpolarized liquid after dissolution from the polarizer, then the liquid is rapidly transferred into a tube for NMR detection using a high-pressure nitrogen gas. This design is considered to be particularly useful for hyperpolarized ^{1}H detection since the dissolution liquid can be transferred for NMR detection in a second or less. Ceillier et al. [74] assembled a home-built transfer and injection system that features a 5 m/s speed and sub-Hz spectral resolution upon arrival at the detection spot. This transfer and injection system can accommodate fast transfer of dissolution liquid for inter-magnet distances of up to 10 m. Meanwhile, Cheng et al. [75] also designed an automated fast transfer system with a transport time of around 3 s and the added capability of polarization measurement prior to in vivo hyperpolarized NMR or MRI. In summary, the invention of the said fast transfer devices is certainly a welcome instrumental improvement to the status quo, considering the fact that the commercial HyperSense polarizer and a number of homebuilt hyperpolarizers take about 8–10 s to complete the dissolution liquid transfer.

2.4 Conclusion

In the ideal world for dissolution DNP, 100% liquid-state nuclear polarization is detected upon arrival of the dissolution liquid at the NMR scanner or MRI scanner from the polarizer in no time. In the real world, however, our pursuit of perfection of the dissolution DNP technology is beset with intrinsic natural limitations that we ought to accept while we strive our best to improve the things that are within our control in the context of DNP. As captured in this chapter, there is no doubt that, for those things within our control, a tremendous amount of DNP sample optimization and instrumental advances have been made in the field since the invention of dissolution DNP by Ardenkjaer-Larsen and co-workers in 2003. While liquid-state ^{13}C polarization as high as over 70% has been achieved via dissolution DNP, there remains a current challenge to sensitize the other insensitive nuclei, such as ^{15}N, ^{89}Y, $^{107,109}Ag$, ^{103}Rh, and other NMR-active spins in the periodic table, which are expected to require further fine-tuning than what we have learned from ^{13}C DNP.

Acknowledgment

The authors would like to thank these funding agencies for financial support of this work: the Welch Foundation grant number AT-2111-20220331 and the U.S. Department of Defense (DoD) Congressionally Directed Medical Research Program (CDMRP) grant numbers W81XWH-21-1-0176, W81XWH-22-1-0105, W81XWH-19-1-0741, HT9425-23-1-0062, and W81XWH-22-1-0003.

References

[1] C.P. Slichter, Principles of Magnetic Resonance, third ed., Springer-Verlag, Berlin, 1990.
[2] A. Abragam, The Principles of Nuclear Magnetism, Clarendon Press, Oxford, 1961.
[3] M.H. Levitt, Spin Dynamics: Basics of Nuclear Magnetic Resonance, second ed., John Wiley and Sons Ltd, West Sussex, 2008.
[4] R.H. Hashemi, W.G. Bradley, C.J. Lisanti, MRI: The Basics, third ed., Lippincot, Williams, and Wilkins, Philadelphia, PA, 2010.
[5] A. Abragam, M. Goldman, Principles of dynamic nuclear polarisation, Rep. Prog. Phys. 41 (1978) 395–467.
[6] D.G. Crabb, W. Meyer, Solid polarized targets for nuclear and particle physics experiments, Annu. Rev. Nucl. Part Sci. 47 (1997) 67–109.
[7] W. de Boer, Dynamic orientation of nuclei at low temperatures, J. Low Temp. Phys. 22 (1976) 185–212.
[8] J.H. Ardenkjær-Larsen, B. Fridlund, A. Gram, G. Hansson, L. Hansson, M.H. Lerche, et al., Increase in signal-to-noise ratio of >10,000 times in liquid-state NMR, Proc. Natl. Acad. Sci. U.S.A. 100 (2003) 10158–10163.
[9] L. Lumata, A.K. Jindal, M.E. Merritt, C.R. Malloy, A.D. Sherry, Z. Kovacs, DNP by thermal mixing under optimized conditions yields >60 000-fold enhancement of 89Y NMR signal, J. Am. Chem. Soc. 133 (2011) 8673–8680.
[10] J.H. Ardenkjaer-Larsen, On the present and future of dissolution-DNP, J. Magn. Reson. 264 (2016) 3–12.

[11] L. Lumata, C. Yang, M. Ragavan, N. Carpenter, R.J. DeBerardinis, M.E. Merritt, Hyperpolarized 13C magnetic resonance and its use in metabolic assessment of cultured cells and perfused organs, Methods Enzymol. (2015) 73–106.

[12] C. Yang, B. Ko, C.T. Hensley, L. Jiang, A.T. Wasti, J. Kim, et al., Glutamine oxidation maintains the TCA cycle and cell survival during impaired mitochondrial pyruvate transport, Mol. Cell 56 (2014) 414–424.

[13] F.A. Gallagher, M.I. Kettunen, K.M. Brindle, Biomedical applications of hyperpolarized 13C magnetic resonance imaging, Prog. Nucl. Magn. Reson. Spectrosc. 55 (2009) 285–295.

[14] K.M. Brindle, S.E. Bohndiek, F.A. Gallagher, M.I. Kettunen, Tumor imaging using hyperpolarized13C magnetic resonance spectroscopy, Magn. Reson. Med. 66 (2011) 505–519.

[15] K. Golman, J.S. Petersson, Metabolic imaging and other applications of hyperpolarized 13C, Acad. Radiol. 13 (2006) 932–942.

[16] J. Kurhanewicz, D.B. Vigneron, K. Brindle, E.Y. Chekmenev, A. Comment, C.H. Cunningham, et al., Analysis of cancer metabolism by imaging hyperpolarized nuclei: prospects for translation to clinical research, Neoplasia 13 (2011) 81–97.

[17] J. Kurhanewicz, D.B. Vigneron, J.H. Ardenkjaer-Larsen, J.A. Bankson, K. Brindle, C.H. Cunningham, et al., Hyperpolarized 13C MRI: path to clinical translation in oncology, Neoplasia 21 (2019) 1–16.

[18] S.J. Nelson, J. Kurhanewicz, D.B. Vigneron, P.E. Larson, A.L. Harzstark, M. Ferrone, et al., Metabolic imaging of patients with prostate cancer using hyperpolarized [1-13c] pyruvate, Sci. Transl. Med. 5 (2013).

[19] P. Niedbalski, A. Kiswandhi, C. Parish, Q. Wang, F. Khashami, L. Lumata, NMR spectroscopy unchained: attaining the highest signal enhancements in dissolution Dynamic Nuclear Polarization, J. Phys. Chem. Lett. 9 (2018) 5481–5489.

[20] L.L. Lumata, R. Martin, A.K. Jindal, Z. Kovacs, M.S. Conradi, M.E. Merritt, Development and performance of a 129-GHz dynamic nuclear polarizer in an ultra-wide bore superconducting magnet, Magn. Reson. Mater. Phys. Biol. Med. 28 (2014) 195–205.

[21] L.L. Lumata, M.E. Merritt, C.R. Malloy, A.D. Sherry, J. van Tol, L. Song, et al., Dissolution DNP-NMR spectroscopy using galvinoxyl as a polarizing agent, J. Magn. Reson. 227 (2013) 14–19.

[22] L. Lumata, S.J. Ratnakar, A. Jindal, M. Merritt, A. Comment, C. Malloy, et al., BDPA: an efficient polarizing agent for fast dissolution dynamic nuclear polarization NMR spectroscopy, Chem. Eur J. 17 (2011) 10825–10827.

[23] L. Lumata, M. Merritt, C. Khemtong, S.J. Ratnakar, J. van Tol, L. Yu, et al., The efficiency of DPPH as a polarising agent for DNP-NMR spectroscopy, RSC Adv. 2 (2012) 12812.

[24] A. Radaelli, H.A.I. Yoshihara, H. Nonaka, S. Sando, J.H. Ardenkjaer-Larsen, R. Gruetter, A. Capozzi, 13C dynamic nuclear polarization using SA-BDPA at 6.7 T and 1.1 K: coexistence of pure thermal mixing and well-resolved solid effect, J. Phys. Chem. Lett. 11 (2020) 6873–6879.

[25] P. Niedbalski, C. Parish, A. Kiswandhi, L. Lumata, 13C dynamic nuclear polarization using isotopically enriched 4-oxo-tempo free radicals, Magn. Reson. Chem. 54 (2016) 962–967.

[26] P. Dutta, G.V. Martinez, R.J. Gillies, Nanodiamond as a new hyperpolarizing agent and its 13C, MRS. J. Phys. Chem. Lett. 5 (3) (2014) 597–600.

[27] A.C. Pinon, A. Capozzi, J.H. Ardenkjær-Larsen, Hyperpolarized water through dissolution dynamic nuclear polarization with UV-generated radicals, Commun. Chem. 3 (3) (2020) 57.

[28] A. Capozzi, et al., Efficient hyperpolarization of U-(13) C-glucose using narrow-line UV-generated labile free radicals, Angew. Chem. 58 (2019) 1334–1339.

[29] T.R. Eichhorn, et al., Hyperpolarization without persistent radicals for in vivo real-time metabolic imaging, Proc. Natl. Acad. Sci. U.S.A. 110 (2013) 18064–18069.

[30] M. Borghini, Spin-temperature model of nuclear dynamic polarization using free radicals, Phys. Rev. Lett. 20 (1968) 419.

[31] J. Heckmann, W. Meyer, E. Radtke, G. Reicherz, S. Goertz, Electron spin resonance and its implication on the maximum nuclear polarization of deuterated solid target materials, Phys. Rev. B 74 (2006), https://doi.org/10.1103/physrevb.74.134418.

[32] S.T. Goertz, The dynamic nuclear polarization process, Nucl. Instrum. Methods Phys. Res. Sect. A 526 (2004) 28–42.

[33] W. Wenckebach, The solid effect, Appl. Magn. Reson. 34 (2008) 227.

[34] Y. Hovav, A. Feintuch, S. Vega, Theoretical aspects of dynamic nuclear polarization in the solid state - the cross effect, J. Magn. Reson. 214 (2012) 29–41.

[35] J. Wolber, F. Ellner, B. Fridlund, A. Gram, H. Johannesson, G. Hansson, L.H. Hansson, M.H. Lerche, S. Mansson, R. Servin, M. Thaning, K. Golman, J.H. Ardenkjaer-Larsen, Generating highly polarized nuclear spins in solution using dynamic nuclear polarization Nucl Instrum Methods, Phys Res. Sect A. 526 (2004) 173–181.

[36] F. Kurdzesau, B. van den Brandt, A. Comment, P. Hautle, S. Jannin, J.J. van der Klink, J.A. Konter, Dynamic nuclear polarization of small labelled molecules in frozen water—alcohol solutions, J. Phys. D Appl. Phys. 41 (2008) 155506.

[37] L. Lumata, M.E. Merritt, Z. Kovacs, Influence of deuteration in the glassing matrix on 13C dynamic nuclear polarization, Phys. Chem. Chem. Phys. 15 (2013) 7032.

[38] C. Parish, P. Niedbalski, Q. Wang, F. Khashami, Z. Hayati, M. Liu, et al., Effects of glassing matrix deuteration on the relaxation properties of hyperpolarized 13C spins and free radical electrons at cryogenic temperatures, J. Chem. Phys. 150 (2019) 234307.

[39] A. Kiswandhi, B. Lama, P. Niedbalski, M. Goderya, J. Long, L. Lumata, The effect of glassing solvent deuteration and Gd3+ Doping on 13C DNP at 5 T, RSC Adv. 6 (2016) 38855–38860.

[40] L. Lumata, Z. Kovacs, C. Malloy, A.D. Sherry, M. Merritt, The effect of 13C enrichment in the glassing matrix on dynamic nuclear Polarization of [1-13C]pyruvate, Phys. Med. Biol. 56 (2011) N85.

[41] N. Bloembergen, On the interaction of nuclear spins in a crystalline lattice, Physica 15 (1949) 410.

[42] P. Niedbalski, C. Parish, A. Kiswandhi, Z. Kovacs, L. Lumata, Influence of 13C isotopic labeling location on dynamic nuclear polarization of acetate, J. Phys. Chem. A 121 (2017) 3227–3233.

[43] C. Parish, P. Niedbalski, A. Kiswandhi, L. Lumata, Dynamic nuclear polarization of carbonyl and methyl 13C spins of acetate using 4-oxo-tempo free radical, J. Chem. Phys. 149 (2018) 054302.

[44] J. Gordon, S. Fain, I. Rowland, Effect of lanthanide ions on dynamic nuclear polarization enhancement and liquid-state T1 relaxation, Magn. Reson. Med. 68 (2012) 1949–1954.

[45] J. Wahsner, E.M. Gale, A. Rodríguez-Rodríguez, P. Caravan, Chemistry of MRI contrast agents: current challenges and new frontiers, Chem. Rev. 119 (2018) 957–1057.

[46] L. Friesen-Waldner, A. Chen, W. Mander, T.J. Scholl, C.A. Mckenzie, Optimisation of dynamic nuclear polarisation of [1-13C] pyruvate by addition of gadolinium-based contrast agents, J. Magn. Reson. 223 (2012) 85–89.

[47] P. Niedbalski, C. Parish, Q. Wang, A. Kiswandhi, Z. Hayati, L. Song, et al., 13C dynamic nuclear polarization using a trimeric Gd3+ complex as an additive, J. Phys. Chem. A 121 (2017) 5127–5135.

[48] A. Kiswandhi, P. Niedbalski, C. Parish, P. Kaur, A. Martins, L. Fidelino, et al., Impact of Ho3+-doping on 13C dynamic nuclear polarization using trityl OX063 free radical, Phys. Chem. Chem. Phys. 18 (2016) 21351–21359.

[49] P. Niedbalski, C. Parish, A. Kiswandhi, L. Fidelino, C. Khemtong, Z. Hayati, et al., Influence of Dy3+ and Tb3+ doping on 13C dynamic nuclear polarization, J. Chem. Phys. 146 (2017) 014303.

[50] A. Capozzi, S. Patel, W.T. Wenckebach, M. Karlsson, M.H. Lerche, J.H. Ardenkjær-Larsen, Gadolinium effect at high-magnetic-field DNP: 70% 13C polarization of [U-13C] glucose using trityl, J. Phys. Chem. Lett. 10 (2019) 3420–3425.

[51] L. Lumata, M.E. Merritt, C.R. Malloy, A.D. Sherry, Z. Kovacs, Impact of Gd3+ on DNP of [1-13C]pyruvate doped with trityl OX063, BDPA, or 4-oxo-TEMPO, J. Phys. Chem. A 116 (2012) 5129–5138.

[52] R.B. Khattri, A.A. Sirusi, E.H. Suh, Z. Kovacs, M.E. Merritt, The influence of Ho3+ doping on 13 C DNP in the presence of BDPA, Phys. Chem. Chem. Phys. 21 (2019) 18629–18635.

[53] A.A. Sirusi, E.H. Suh, Z. Kovacs, M.E. Merritt, The effect of Ho3+ doping on 13C dynamic nuclear polarization at 5 T, Phys. Chem. Chem. Phys. 20 (2018) 728–731.

[54] P. Niedbalski, C. Parish, Q. Wang, Z. Hayati, L. Song, A.F. Martins, et al., Transition metal doping reveals link between electron t1 reduction and 13C dynamic nuclear polarization efficiency, J. Phys. Chem. A 121 (2017) 9221–9228.

[55] A. Cho, R. Eskandari, V.Z. Miloushev, K.R. Keshari, A non-synthetic approach to extending the lifetime of hyperpolarized molecules using D2O solvation, J. Magn. Reson. 295 (2018) 57–62.

[56] P. Miéville, et al., Scavenging free radicals to preserve enhancement and extend relaxation times in NMR using dynamic nuclear polarization, Angew. Chem., Int. Ed. 49 (2010) 6182–6185.

[57] H. Johannesson, S. Macholl, J.H. Ardenkjaer-Larsen, Dynamic nuclear polarization of [1-13C]pyruvic acid at 4.6 tesla, J. Magn. Reson. 197 (2009) 167–175.

[58] S. Jannin, A. Comment, F. Kurdzesau, J.A. Konter, P. Hautle, B. van der Brandt, J.J. van der Klink, A 140 GHz prepolarizer for dissolution dynamic nuclear polarization, J. Chem. Phys. 128 (2008) 241102.

[59] A. Kiswandhi, P. Niedbalski, C. Parish, S. Ferguson, D. Taylor, G. McDonald, et al., Construction and 13C hyperpolarization efficiency of a 180 GHz dissolution dynamic nuclear polarization system, Magn. Reson. Chem. 55 (2017) 828–836.

[60] A. Kiswandhi, P. Niedbalski, C. Parish, Q. Wang, L. Lumata, Assembly and performance of a 6.4 T cryogen-free dynamic nuclear polarization system, Magn. Reson. Chem. 55 (2017) 846–852.

[61] I.H.A. Yoshihara, E. Can, M.H. Karlsson, M. Lerche, J. Schwitter, A. Comment, High-field dissolution dynamic nuclear polarization of [1-13C] pyruvic acid, Phys. Chem. Chem. Phys. 18 (2016) 12409–12413.

[62] J.H. Ardenkjær-Larsen, et al., Cryogen-free dissolution dynamic nuclear polarization polarizer operating at 3.35 T, 6.70 T, and 10.1 T, Magn. Reson. Med. 81 (2019) 2184–2194.

[63] A. Ferrari, J. Peters, M. Anikeeva, et al., Performance and reproducibility of 13C and 15N hyperpolarization using a cryogen-free DNP polarizer, Sci. Rep. 12 (2022) 11694.

[64] M. Baudin, B. Vuichoud, A. Bornet, G. Bodenhausen, S. Jannin, A cryogen-consumption-free system for dynamic nuclear polarization at 9.4 T, J. Magn. Reson. 294 (2018) 115–121.

[65] S. Jannin, A. Bornet, R. Melzi, G. Bodenhausen, High field dynamic nuclear polarization at 6.7 T: carbon-13 polarization above 70 % within 20 min, Chem. Phys. Lett. 549 (2012) 99–102.

[66] A. Bornet, R. Melzi, A.J.P. Linde, P. Hautle, B. van der Brandt, S. Jannin, G. Bodenhausen, Boosting dissolution dynamic nuclear polarization by cross polarization, J. Phys. Chem. Lett. 4 (2013) 111–114.

[67] B. Vuichoud, J. Milani, A. Bornet, R. Melzi, S. Jannin, G. Bodenhausen, Hyperpolarization of deuterated metabolites via remote cross-polarization and dissolution dynamic nuclear polarization, J. Phys. Chem. B 118 (5) (2014) 1411–1415.

[68] A. Bornet, J. Milani, B. Vuichoud, J. Angel, L. Perez, G. Bodenhausen, S. Jannin, Microwave frequency modulation to enhance dissolution dynamic nuclear polarization, Chem. Phys. Lett. 602 (2014) 63–67.

[69] J. Milani, B. Vuichoud, A. Bornet, P. Miéville, R. Mottier, S. Jannin, G. Bodenhausen, A magnetic tunnel to shelter hyperpolarized fluids, Rev. Sci. Instrum. 86 (2015) 024101.

[70] M. Batel, M. Krajewski, K. Weiss, O. With, A. Dapp, A. Hunkeler, M. Gimersky, K.P. Pruessmann, P. Boesiger, B.H. Meier, S. Kozerke, M. Ernst, A multisample 94 GHz dissolution dynamic-nuclear-polarization system, J. Magn. Reson. 214 (2012) 166–174.

[71] Y. Crémillieux, F. Goutailler, B. Montcel, D. Grand, G. Vermeulen, P.E. Wolf, A super-wide bore DNP system for multiple sample preparation: cryogenic performance and polarization at low temperature, Appl. Magn. Reson. 43 (2012) 167–180, 2012.
[72] T. Cheng, A.P. Gaunt, I. Marco-Rius, M. Gehrung, A.P. Chen, J.J. Klink, Arnaud Comment, A multisample 7 T dynamic nuclear polarization polarizer for preclinical hyperpolarized MR, NMR Biomed. 33 (2020) 5, https://doi.org/10.1002/nbm.4264.
[73] S. Bowen, C. Hilty, Rapid sample injection for hyperpolarized NMR spectroscopy, Phys. Chem. Chem. Phys. 12 (2010) 5766–5770.
[74] C. Morgan, C. Olivier, T. El Daraï, F. Samuel, Q.S. Cousin, S. Guibert, S.J. Elliott, A. Bornet, B. Vuichoud, J. Milani, C. Pages, D. Eshchenko, J.G. Kempf, C. Jose, S.A. Lambert, S. Jannin, An automated system for fast transfer and injection of hyperpolarized solutions, Journal of Magnetic Resonance Open 8–9 (2021) 100017.
[75] T. Cheng, M. Mishkovsky, J.A.M. Bastiaansen, O. Ouari, P. Hautle, P. Tordo, B. van der Brandt, A. Comment, Automated transfer and injection of hyperpolarized molecules with polarization measurement prior to in vivo NMR, NMR Biomed. 26 (2013) 1582–1588.

Chapter 3

The chemistry of parahydrogen induced polarization

Francesca Reineri* and Erika Cerutti
Molecular Biotechnology and Health Sciences, University of Turin, Turin, Italy
**Corresponding-author: E-mail: francesca.reineri@unito.it*

The existence of two different kinds of hydrogen (ortho- and parahydrogen, henceforth o-H_2 and p-H_2) was postulated first in 1927 by Heisenberg and Hund [1], and their properties were observed in 1929 by Harteck and Bonhoffer [2,3]. P$-H_2$ is a nuclear spin isomer of the hydrogen molecule, and enrichment in this isomer can be obtained at low temperatures and maintained at higher temperatures in a nonequilibrium condition. In terms of the population of the nuclear spin states, this can be considered a high-spin-order state. In order to obtain hyperpolarized MR signals, the symmetry of hydrogen must be broken by a chemical reaction; therefore, it can be stated that parahydrogen-induced polarization (PHIP) is a chemistry-based hyperpolarization method.

The possibility of transforming the spin order of parahydrogen into nuclear-spin magnetization was theoretically predicted by C.R. Bowers and D. Weitekamp in 1986 [4]. The first experimental observation of this effect was unwittingly obtained in the same year [5] in the hydrogenation of phenylacetylene catalyzed by a Rh complex, but the enhanced antiphase signals observed in the ^{1}H-NMR spectra of vinyl-benzene were interpreted as due to the occurrence of chemically induced dynamic nuclear polarization. A few months later [6], the same authors showed that these results arose from p-H_2, in agreement with the Weitekamp proposal, who, in his turn, observed PHIP experimentally, almost at the same time, in the hydrogenation of acrylonitrile CH_2CHCN to propionitrile CH_3CH_2CN, catalyzed by the Wilkinson catalyst [7].

Because of a major signal enhancement, PHIP provides a highly sensitive NMR tool for the study of reaction mechanisms that involve catalytic activation of H_2, and, at least in the first decade, p-H_2 has been applied mainly to the investigation of hydrogenation mechanisms.

In 2001, the first in-vivo study using a hyperpolarized probe, obtained by means of PHIP, was reported by K. Golman and coworkers [8] and suddenly brought new attention to the field of hyperpolarization.

The application of hyperpolarized probes for biomedical imaging has developed quite quickly in the last 2 decades, making dissolution-dynamic nuclear polarization the gold-standard method for these studies. The versatility of this method, which can, in principle, be applied to any substrate, allows it to hyperpolarize a few molecules that are central to metabolic processes [9–11]. Imaging of metabolic processes in vivo, in real time, is undoubtedly an innovative and powerful tool in the field of diagnostics [12–15].

Among the members of the HP techniques, PHIP is less instrumentally sophisticated and demanding than DNP, which makes it much more affordable in terms of equipment and operational costs. Nevertheless, it is less versatile, and the application of PHIP to the biomedical field has been hindered, for the first 2 decades, by the limited range of suitable substrates. The conventional way to obtain hyperpolarized products relies on the chemical addition of parahydrogen to an unsaturated molecule. It can be referred to as hydrogenative PHIP (h-PHIP), and it is mostly applied to the hyperpolarization of organic substrates that can incorporate two protons, such as alkynes and alkenes. In order to circumvent this issue and to hyperpolarize substrates that cannot incorporate a hydrogen molecule, the PHIP and side arm hydrogenation (PHIP-SAH) method has been introduced [16], and pyruvate, thus hyperpolarized, has been used for metabolic investigations in cells and in vivo [17,18].

The Chemistry of Hyperpolarized Magnetic Resonance Probes. https://doi.org/10.1016/B978-0-323-91842-8.00005-7

The scope of hyperpolarization by means of parahydrogen was significantly increased in 2009 by the PHIP-based technique named signal amplification by reversible exchange (SABRE), which allowed the hyperpolarization of substrates without changing their chemical identity [19,20]. Therefore, this can be considered a nonh-PHIP technique.

It appears clearly that much attention must be paid to the chemistry involved in PHIP-based methods in order to enter this hyperpolarization field.

This chapter focuses on the substrates and catalysts that are used in both h-PHIP and nonh- PHIP. At the beginning, a quick look at the physical properties of parahydrogen, how it can be obtained and detected, and how hyperpolarization derives from parahydrogen will be also given.

3.1 Parahydrogen enrichment and quantification

The existence of the two modifications of hydrogen, o-H_2 and p-H_2, is due to the fact that they obey Fermi-Dirac statistics, and the total wave function describing the hydrogen molecule must be antisymmetric with respect to the exchange of the two atoms. The total symmetry of this wave function depends only on the rotational and nuclear spin contributions.

The latter is given by the four combinations of the $|\alpha\rangle$ and $|\beta\rangle$ states of the two protons and are the following:

$$|T_+\rangle = |\alpha\alpha\rangle, \tag{3.1a}$$

$$|T_0\rangle = \frac{1}{\sqrt{2}}|\alpha\beta + \beta\alpha\rangle, \tag{3.1b}$$

$$|T_-\rangle = |\beta\beta\rangle \tag{3.1c}$$

$$|S_0\rangle = \frac{1}{\sqrt{2}}|\alpha\beta - \beta\alpha\rangle. \tag{3.1d}$$

The first three states are symmetric and are associated with an asymmetric rotational function (odd rotation quantum number $J = 1,3, \ldots$). In the absence of a magnetic field, these states are degenerate (have the same energy) and constitute a triple state (orthohydrogen). The fourth state ($\frac{1}{\sqrt{2}}|\alpha\beta - \beta\alpha\rangle$) is antisymmetric, is associated with symmetric rotational functions (even rotational quantum number $J = 0, 2, \ldots$), and is the p-H_2 state.

The thermodynamical equilibrium between p-H_2 and o-H_2 is governed by Boltzmann's distribution law. At room temperature, all the nuclear spin states are equally populated, and 75% of hydrogen is in the ortho-state, while 25% of hydrogen is p-H_2. We can refer to this mixture as normal-H_2. Being associated with the lowest rotational function ($J = 0$), the singlet state is the most energetically stable. Therefore, the ortho/para ratio changes with temperature, and close to 20K (liquid hydrogen), the para state is the only populated one (Fig. 3.1).

The two modifications of hydrogen, ortho and para, can be considered molecules of two gases differing from each other in certain optical and thermal properties. Spontaneous conversion of these two sets of molecules practically does not take place, and ortho-para transition needs a catalyst, which can be a paramagnetic or diamagnetic solid [22–28], like activated charcoal, iron oxide, or alumina, or paramagnetic gases and ions in solution. They allow them to reach the Boltzmann equilibrium at a particular temperature.

The ortho/para mixture obtained at a low temperature can be maintained at a higher temperature if the conversion catalyst is removed.

To generate hydrogen enriched in the para form, the simplest system has been proposed by Bonhoeffer and Harteck [2,29], who developed an approach to produce a continuous p-H_2 stream. A U-shaped metal tube filled with the conversion catalyst (i.e., activated charcoal) is cooled in a liquid nitrogen bath, and a mixture of 48% ortho and 52% para is obtained (Fig. 3.1). Modern equipment [30,31] allows for higher parahydrogen enrichment by working at a lower temperature, i.e., 36K in the commercial system Bruker parahydrogen generator or even lower (22K) using helium compressors [32].

It must be kept in mind that the same phenomena that allow enrichment in the para-isomer (i.e., adsorption on magnetic surfaces, paramagnetic impurities such as oxygen [32]) bring to the para-ortho equilibration, therefore parahydrogen depletion, at high temperature (RT). Therefore, all these impurities must be carefully removed for the storage of parahydrogen.

3.1.1 Quantification of p-H_2 enrichment

The first setup to measure the p-H_2 enrichment leveraged the different heat capacities of the two isomers, and it was a thermal conductivity cell [33–35]. The thermal conductivity differs by ~20% in para versus ortho hydrogen and reaches

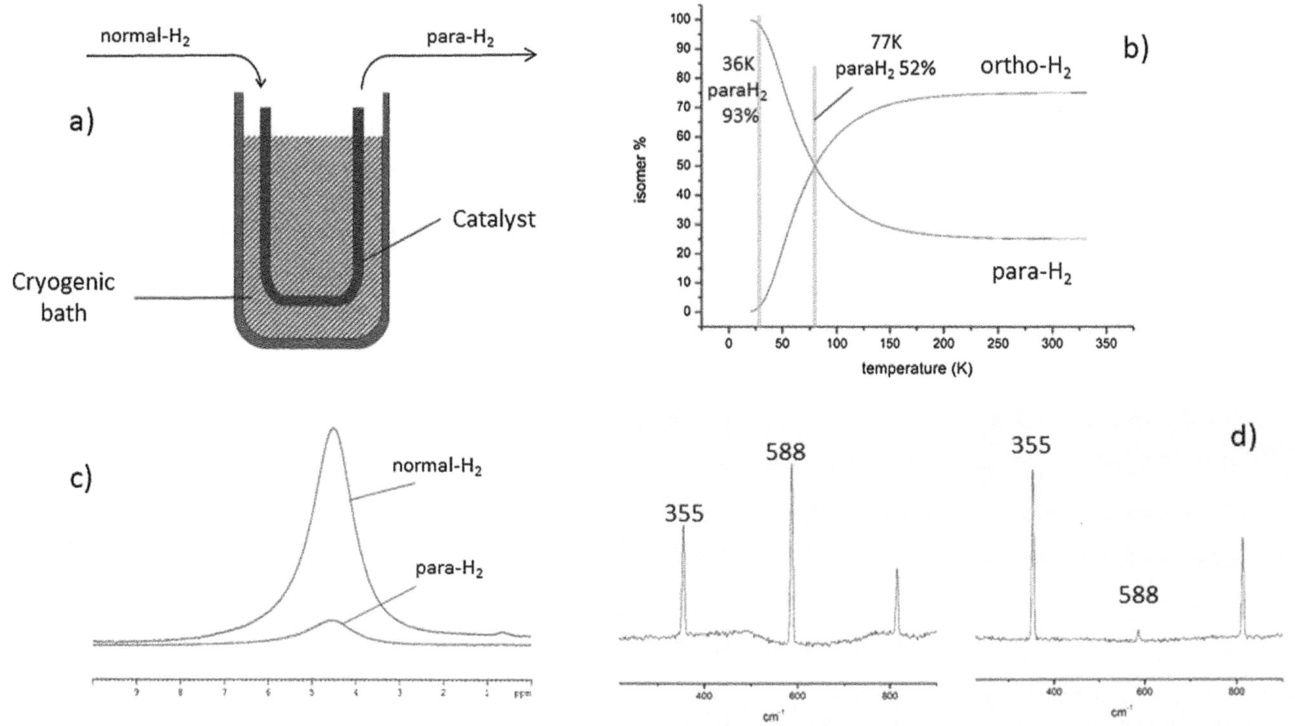

FIGURE 3.1 (A) Schematic drawing of a setup for p-H_2 production: normal-H2 is passed through a U-shaped steel tube containing a catalyst (e.g., charcoal, iron oxide) capable of enabling the o-H_2-p-H_2 conversion. The tube is plunged in a cryogenic bath (e.g., a liquid nitrogen bath). (B) Percentage of o-H_2 and p-H_2, at thermal equilibrium are reported as a function of the temperature. (C) ^{1}H-NMR spectra (128 scans, 50 ppm, repetition time 2s) of p-H_2 and n-H_2 (2 bar) in a gas-tight NMR tube. (D) Raman spectra of hydrogen gas: the left sample contains n-H_2, i.e., 75% ortho (transition $J = 1 \rightarrow J = 3$, 588 cm^{-1}) and 25% para (transition $J = 0 \rightarrow J = 2$, 355 cm^{-1}). The right sample contains 21% ortho and 79% para (equilibrium mixture at T = 50K). Hydrogen was contained in an NMR tube; a Renishaw Raman 200 system was used (laser wavelength 514 nm). The experimental method has been described in Ref. [21].

its maximum at 150 K. The apparatus is made of two cells connected through an electrical circuit that ends with a platinum filament in each cell. Both cells are placed in a liquid nitrogen bath; one cell is filled with normal hydrogen and the other with parahydrogen. The current unbalance between the two filaments is a measure of the para enrichment, and hence it is used to measure the ortho/para ratio.

Being p-H_2 nonmagnetic, it cannot be directly detected by NMR; anyhow, an indirect measurement of the para-enrichment can be obtained by exploiting the kinetics of para-ortho conversion catalyzed by the surface of borosilicate NMR tubes [30]. In practice, the ^{1}H-NMR spectrum of parahydrogen placed in a gas-tight NMR tube is acquired using a sufficiently large spectral window (~50 ppm). The width of the spectral window is due to the linewidth of the ortho-hydrogen signal that, in the gas phase, can be in the order of 1 kHz due to the short T_2 value (less than 0.5 ms) [36]. The ortho-para equilibration is reached in a few hours, thanks to the catalysis of the surface of the NMR tube, and the proton signal of orthohydrogen is acquired again. The difference between the two signals is due to the latent parahydrogen (Fig. 3.1).

A direct measurement of the para and ortho signals can be obtained by means of Raman spectroscopy. The first two Raman transitions of hydrogen, between the rotation states $0 \rightarrow 2$ and $1 \rightarrow 3$, correspond to the first para-ortho transitions, and their intensity, at room temperature, is directly proportional to the percentage of each isomer (Fig. 3.1) [21].

3.2 Hyperpolarization by means of parahydrogen

The spin order of parahydrogen can be exploited to obtain nuclear spin magnetization [2] by breaking the symmetry of the two protons. In h-PHIP, this occurs when the two protons are added to a substrate (organic or inorganic). If the substrate is symmetric and the two protons are chemically equivalent, the symmetry can be broken by asymmetric coupling with a third spin. An additional requirement is that the spin correlation between the two protons is maintained, i.e., they are added in a pairwise fashion.

In nonh-PHIP, termed SABRE, the symmetry of hydrogen is broken by adding the two parahydrogen protons to a metal complex (usually an iridium, Crabtree-like complex [37]) and released to a substrate that is concurrently bound to the metal. The substrate becomes hyperpolarized thanks to the spin-order transfer that occurs through the scalar coupling network between the parahydrogen protons and the spins of the substrate on the metal scaffold. Since at least four spins are involved in SABRE, the description of this way of hyperpolarization is more theoretically challenging than the simplest h-PHIP experiment. A few publications treat the different aspects involved in SABRE hyperpolarization in detail [38–40].

The easiest way to get an understanding of how parahydrogen, which is nonmagnetic, can lead to MR hyperpolarized signals is to consider an h-PHIP experiment in which the two protons are added to a substrate in chemically different positions and are isolated from other magnetically active nuclei.

Using the population model of NMR, we need to find the spin states of the two protons on the product molecule. These are the eigenstates of the Hamiltonian.

$$\mathscr{H} = \frac{1}{2}\omega_\Sigma\left(I_z^{H_1} + I_z^{H_2}\right) + \frac{1}{2}\omega_\Lambda\left(I_z^{H_1} - I_z^{H_2}\right) + \omega_J \boldsymbol{I}^{H_1}\cdot\boldsymbol{I}^{H_2} \tag{3.2}$$

where

$$\omega_\Sigma = \omega_{H_1} + \omega_{H_2},$$

$$\omega_\Lambda = \omega_{H_1} - \omega_{H_2}$$

being ω_{H_1} and ω_{H_2} the Larmor frequencies of the two protons on the product molecule, and

$$\omega_J = 2\pi J_{H_1H_2}$$

where $J_{H_1H_2}$ is their mutual scalar coupling.

Using the Zeeman states ($|\alpha\alpha\rangle$, $|\alpha\beta\rangle$, $|\beta\alpha\rangle$ and $|\beta\beta\rangle$) as a basis, the following eigenstates of the Hamiltonian are obtained:

$$\psi_1 = |\alpha\alpha\rangle \tag{3.3a}$$

$$\psi_2 = \cos\left(\frac{\theta}{2}\right)|\alpha\beta\rangle + sin\left(\frac{\theta}{2}\right)|\beta\alpha\rangle \tag{3.3b}$$

$$\psi_3 = \sin\left(\frac{\theta}{2}\right)|\alpha\beta\rangle - cos\left(\frac{\theta}{2}\right)|\beta\alpha\rangle \tag{3.3c}$$

$$\psi_4 = |\beta\beta\rangle. \tag{3.3d}$$

Where $\tan(\theta) = \frac{\omega_J}{\omega_\Lambda}$.

When the hydrogenation reaction occurs at a low magnetic field, the frequency difference between the two protons is neglectable with respect to their mutual coupling ($\omega_J >> \omega_\Lambda$, strong coupling conditions) (ALTADENA experiment [41]), and the eigenstate ψ_3 is (almost) a singlet state [42].

We can find the population of the spin states (ψ_1, ψ_2, ψ_3 and ψ_4) by calculating the transition probability between the initial state (the parahydrogen state S_0) and each of the product states $|\langle\psi_{1-4}|S_0\rangle|^2$.

In the ALTADENA experiment, the population of parahydrogen is completely transferred to the state ψ_3 (see Fig. 3.2, level population for the ALTADENA experiment) [42]. After the hydrogenation, the sample is transported into the magnet adiabatically, i.e., without any alteration of the populations of the spin states. Therefore, of the states $|\alpha\alpha\rangle$, $|\alpha\beta\rangle$, $|\alpha\beta\rangle$ and $|\beta\beta\rangle$ that are reached at high field, only one remains populated (the state $|\alpha\beta\rangle$ in Fig. 3.2A), and the transitions $|\alpha\beta\rangle \rightarrow |\alpha\alpha\rangle$ and $|\alpha\beta\rangle \leftrightarrow |\beta\beta\rangle$ are hyperpolarized (see Fig. 3.2). In this kind of experiment, one adsorption and one emission signal are observed, i.e., hyperpolarized net magnetization is obtained.

When the hydrogenation reaction occurs in the high magnetic field of the NMR spectrometer (PASADENA experiment [7]), we are in a limit situation in which the frequency difference between the two protons is much bigger than the J coupling (weak coupling $\omega_\Lambda >> \omega_J$) and the nuclear spin states are

$$\psi_1 = |\alpha\alpha\rangle \tag{3.4a}$$

$$\psi_2 = |\alpha\beta\rangle \tag{3.4b}$$

$$\psi_3 = |\beta\alpha\rangle \tag{3.4c}$$

$$\psi_4 = |\beta\beta\rangle. \tag{3.4d}$$

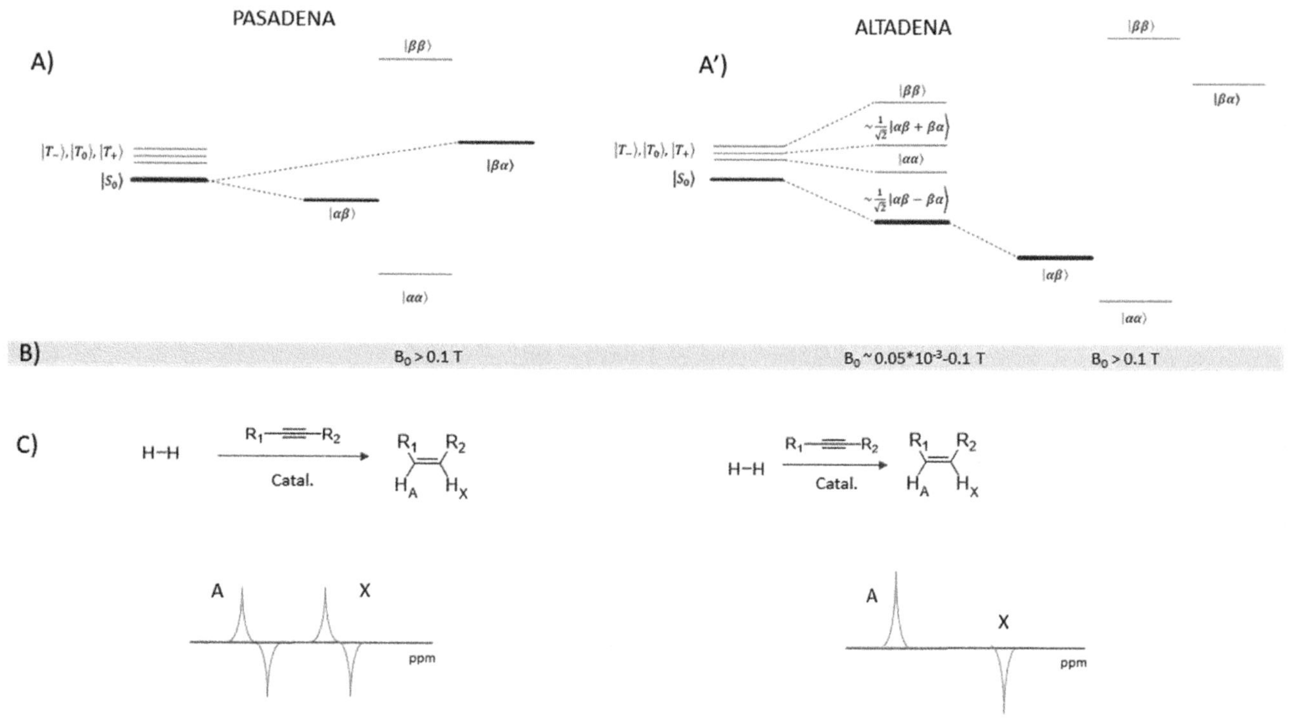

FIGURE 3.2 Representation of the population of the nuclear spin states in a PASADENA and an ALTADENA experiment. In the first kind of experiment, the hydrogenation reaction occurs at a high magnetic field (see B), and the states αβ and βα are suddenly formed, following the hydrogenation. They are populated equally, and the two protons lead to two antiphase signals (bottom part of the figure). In the ALTADENA experiment, the reaction occurs at a low magnetic field (see B), and only one nuclear spin state of the product molecule is populated, which overlaps to the single state. After the adiabatic transport of the sample into the spectrometer, the population of the states is maintained, and both protons give in-phase signals ($+I_z^A$ and $-I_z^X$), as schematized in the bottom part of the figure.

In this case, the two states $|\alpha\beta\rangle$ and $|\beta\alpha\rangle$ are equally populated (see Fig. 3.2A), and four transitions are enhanced:

$$|\alpha\beta\rangle \rightarrow |\alpha\alpha\rangle \text{ and } |\beta\alpha\rangle \rightarrow |\beta\beta\rangle \text{ for the spin X}$$

$$|\alpha\beta\rangle \rightarrow |\beta\beta\rangle \text{ and } |\beta\alpha\rangle \rightarrow |\alpha\alpha\rangle \text{ for the spin A.}$$

In the PASADENA experiments, the selective population of the spin states leads to two antiphase doublets (Fig. 3.2).

3.2.1 Hyperpolarization of heteroatoms

Hyperpolarization is spontaneously transferred (i.e., without the need for the application of RF pulses) from the parahydrogen protons to other spins in the product (protons or heteroatoms). Spin order transfer usually occurs through scalar couplings; the first case has been reported by Bargon and coworkers [43].

Relying on the selective population of the nuclear spin states, we can explain the polarization transfer from the parahydrogen protons to a heteronucleus (X) considering the simplest case of a three-spin system, in which the two protons are isochronous, i.e., have the same Larmor frequency. We have to find the eigenstates of the following Hamiltonian (AA′X system).

$$\mathscr{H} = \omega_H\left(I_z^H + I_z^{H'}\right) + \omega_X I_z^X + 2\pi J_{HH'} \boldsymbol{I}_H \cdot \boldsymbol{I}_{H'} + 2\pi J_{HX} I_z^H I_z^X + 2\pi J_{H'X} I_z^{H'} I_z^X. \tag{3.5}$$

A convenient basis is given by the so-called singlet-triplet-Zeeman basis [44], i.e., the tensor product of the singlet-triplet states ($|T_+\rangle$, $|T_0\rangle$, $|T_-\rangle$, $|S_0\rangle$) for the two protons and the $|\alpha\rangle$ and $|\beta\rangle$ states for the heteroatom [43]. The following eight states are obtained, four of which derive from the mixing of the states $|S_0\alpha\rangle$ and $|T_0\alpha\rangle$ (total spin -½), $|S_0\beta\rangle$ and $|T_0\beta\rangle$ (total spin ½).

$$\psi_1 = |\alpha\alpha\alpha\rangle, \tag{3.6a}$$

$$\psi_2 = |\alpha\alpha\beta\rangle \tag{3.6b}$$

$$\psi_3 = |S_0'\alpha\rangle = \cos\left(\frac{\theta_G}{2}\right)|\beta\alpha\alpha\rangle - sin\left(\frac{\theta_G}{2}\right)|\alpha\beta\alpha\rangle, \tag{3.6c}$$

$$\psi_4 = |T_0'\alpha\rangle = \cos\left(\frac{\theta_G}{2}\right)|\alpha\beta\alpha\rangle + sin\left(\frac{\theta_G}{2}\right)|\beta\alpha\alpha\rangle, \tag{3.6d}$$

$$\psi_5 = |S_0'\beta\rangle = \cos\left(\frac{\theta_G}{2}\right)|\alpha\beta\beta\rangle - sin\left(\frac{\theta_G}{2}\right)|\beta\alpha\beta\rangle, \tag{3.6e}$$

$$\psi_6 = |T_0'\beta\rangle = \cos\left(\frac{\theta_G}{2}\right)|\beta\alpha\beta\rangle + sin\left(\frac{\theta_G}{2}\right)|\alpha\beta\beta\rangle, \tag{3.6f}$$

$$\psi_7 = |\beta\beta\alpha\rangle \tag{3.6g}$$

$$\psi_8 = |\beta\beta\beta\rangle. \tag{3.6h}$$

Where

$$\theta_G = arctan\frac{2\omega_{HH'}}{\omega^{\Delta}_{HH'X}} \text{ with } \omega_{HH'} = 2\pi J_{HH'} \text{ and } \omega^{\Delta}_{HH'X} = 2\pi(J_{HX} - J_{H'X}).$$

The percentage of singlets of the states ψ_3, ψ_4, ψ_5 and ψ_6 is a function of the J-couplings involved in the three-spin system. When the two H-X couplings are equal, the system is symmetric, and the states correspond to pure singlet-Zeeman and triplet-Zeeman states. The heteronuclear transitions $\psi_3 \rightarrow \psi_6$ and $\psi_4 \rightarrow \psi_5$ have nonnull probability and are hyperpolarized because the population differences are out of thermal equilibrium.

Heteronuclear hyperpolarization is of great interest for the application to biomedical investigations and in vivo imaging [45,46]; therefore, much attention has been paid to the development of different methods to obtain net magnetization and to improve the spin order transfer [38,47,48].

3.3 Hydrogenative PHIP (h-PHIP)

Parahydrogen can enhance the intensity of NMR signals of chemical species on which it is incorporated, and, usually, organic molecules that contain double or triple C—C bonds are used as substrates.

In order to be suitable for parahydrogen hyperpolarization, i.e., to allow high polarization on the products and fast hydrogenation, the unsaturation should be well accessible to the catalyst (alkynes are hydrogenated faster than alkenes), and small molecules are usually preferred since the relaxation rate (T_1) becomes shorter on heavier molecules and hyperpolarization decay is faster.

The final aim for the preparation of hyperpolarized molecules is often given by their application as imaging agents, not only for biomedical purposes [17,18,49] but also for imaging of void spaces [50,51]. Quite a few of these molecules can be found in Table 3.1, together with the polarization level that has been reported, the catalyst used, and the solvent in which they have been obtained. This latter feature is particularly relevant when the hyperpolarized product aims at biological applications, since, in that case, aqueous solutions are necessary and the presence of toxic impurities must be avoided. It can also be considered that some of these molecules have been hyperpolarized using different PHIP-based methods.

3.3.1 Substrates for h-PHIP

Among the h-PHIP polarized molecules suitable for biomedical application, the following metabolites must be mentioned: succinate [52], fumarate [53,54], phospholactate [55—57], AA derivatives (e.g., N-Acetyl-phenyl alanine and their esters [58,59]), and N-propargyl-choline [60]. Two nonmetabolic compounds must also be mentioned, in this context, i.e., hydroxyethyl propionate (HEP [61]) and tetrafluoropropyl propionate (TFPP) [62]. HEP has been used in the first in-vivo studies that have been carried out using a hyperpolarized substance [61,63]. The ^{13}C images thus obtained clearly showed the distribution of the agent into the vasculature system of an animal model. This compound can also be considered a kind of reference for testing the PHIP setup [64]. Conversely, TFPP has been used in vivo for imaging atherosclerotic plaques, thanks to its lipophilicity.

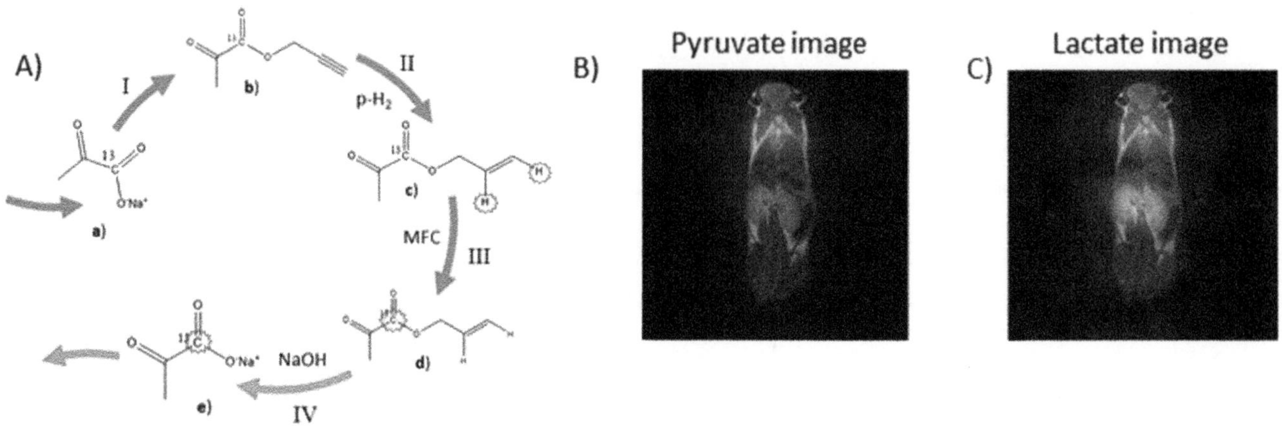

FIGURE 3.3 Left: scheme of the PHIP-SAH method for the hyperpolarization of $[1-^{13}C]$pyruvate. The metabolic transformation of pyruvate into lactate has been observed following the injection of the aqueous solution of the hyperpolarized metabolite in in vivo, in a mouse. The ^{13}C-csi image (single shot) is overlapped with an anatomical image (T_2 weighted fast spin echo) for spatial localization. *Reproduced from Ref. [18].*

Using the so-called PHIP-SAH [16,65,66] technique, a few other metabolites (pyruvate, acetate, and lactate) have been hyperpolarized. In this case, hydrogen is not incorporated into the final hyperpolarized product but is added to a cleavable, unsaturated molecular tag (the so-called sidearm) that is bound to a carboxylate group of the target molecule. The PHIP-SAH method relies on the following steps (Fig. 3.3): (1) Esterification of the carboxylate group with an unsaturated alcohol (propargyl- or vinyl-alcohol) [67]; (2) parahydrogenation of the unsaturated moiety in a hydrophobic solvent; (3) transfer of the spin order from the parahydrogen protons to the ^{13}C carboxylate spin; (4) hydrolysis of the ester using an aqueous base (sodium hydroxide); and (5) extraction of the carboxylate salt into the water phase. The last two steps occur simultaneously, and the phase extraction of the hyperpolarized product in the aqueous solution allows to remove, almost completely, the hydrogenation catalyst, which is retained in the organic solvent [68].

Since the main drawback of hyperpolarization is its limited lifetime, the hyperpolarization of heteroatoms is also an important topic. ^{13}C and ^{15}N have been reported more often, but other nuclei (^{29}Si [69], ^{19}F [70,71]) have been hyperpolarized on h-PHIP products.

Hyperpolarized gases (propane and propane-d_6), obtained from heterogeneous PHIP, have been used for high-resolution MR imaging of void spaces and in microreactors [50,72,73]. The use of MRI and the highly polarized product for the characterization of microreactors is advantageous because the technique is noninvasive and can probe optically opaque media.

3.3.2 Catalysts for h-PHIP

The efficiency of the catalyst, i.e., the amount of substrate that can be hydrogenated in a unit time, as well as the hyperpolarization level of the parahydrogenation product, are fundamental parameters for the production of hyperpolarized molecules that have to be applied for MR imaging.

Although several different metal complexes can be used for h-PHIP, only a few of them are widely applied thanks to their hydrogenation pathway, which makes them particularly efficient for hyperpolarization. Here, these hydrogenation catalysts are grouped by metal.

3.3.2.1 Rhodium

Rhodium complexes are the most widely used for h-PHIP, and the Wilkinson catalyst ($RhCl(PPh_3)_3$) was applied in the first PHIP experiment [7]. In that work, the hydrogenation of acrylonitrile (CH_2CHCN) led to hyperpolarized signals in the ^1H-NMR spectrum of propionitrile (CH_3CH_2CN), and the formation of a dihydride complex $[RhCl(H)_2(Ph_3)_3]$, with significantly enhanced ^1H-NMR signals in the hydride region, was also observed. The reactivity of the Wilkinson catalyst was further investigated, and a number of reaction intermediates were detected thanks to PHIP [103] and D-NMR spectra.

In that hydrogenation pathway, hydrogen exchange occurs on the metal center (Fig. 3.4), thus leading to para-ortho equilibration rather quickly; therefore, the observed hyperpolarization level is relatively weak and disappears in a few minutes. PHIP on the product molecules can be observed until the ortho-para thermal equilibrium has been reestablished [104].

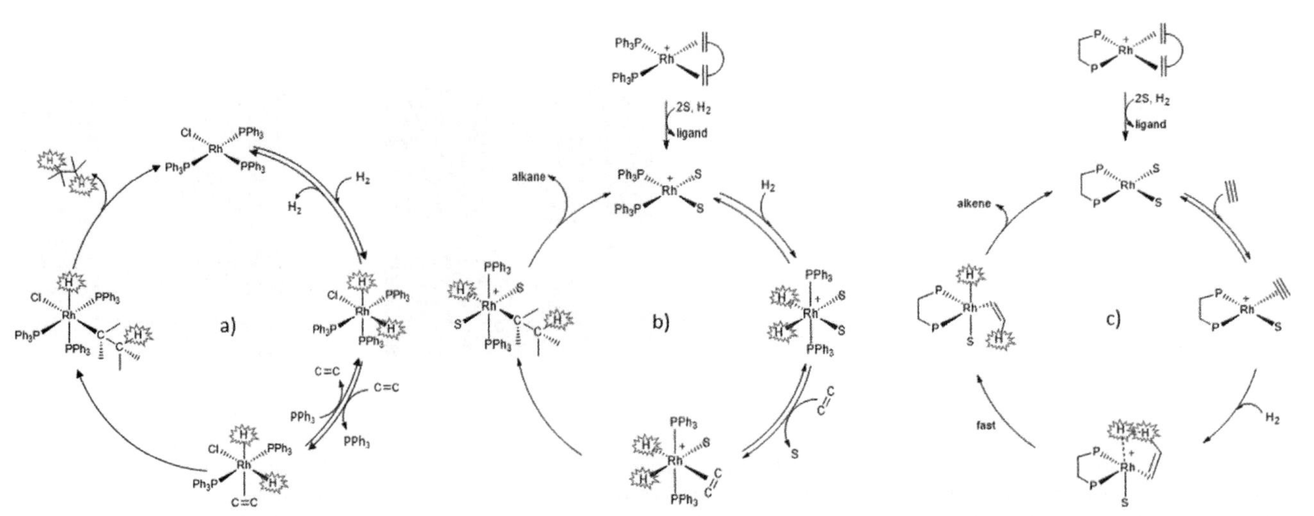

FIGURE 3.4 Scheme of the hydrogenation pathways of three types of rhodium complexes: (A) The Wilkinson complex; (B) Shrock-Osborne Rh(I) complexes; these have mono-dentated phosphines and the hydrogenation pathway is called the "saturated route"; (C) Rh(I) complexes containing a chelating phosphine; in this case, the hydrogenation pathway is called the "unsaturated route" (the unsaturated substrate is coordinated to the metal center before the hydrogen molecule) S = solvent.

Cationic Rh complexes are extensively applied for h-PHIP. These complexes can be distinguished in those containing mono-dentated phosphines, known as Shrock-Osborn catalysts $[Rh(di\text{-}olefine)L_2]^+$ (where L = phosphine ligand) [105—107] and those containing a chelating phosphine [107,108].

When the first group is used, the reaction proceeds through the addition of the hydrogen molecule to the metal center and the formation of a cis dihydride, which is followed by the coordination of the unsaturated substrate according to the known "hydride route" (Fig. 3.4). It has been demonstrated that the ortho-para conversion rate is fast when the amount of substrate is small or the reaction has come to an end [109]. This occurs because rapid equilibration of ortho and para-hydrogen occurs by reversible formation of $[RhH_2(sol)_2L_2]^+$ (sol = solvent). Since the solvent enters the catalytic cycle, the reaction efficiency is related to it [110], and coordinating solvents (methanol, acetone) lead to higher polarization than noncoordinating ones (e.g., chloroform) [110].

The ortho-para equilibration that occurs due to the reversible exchange of hydrogen on the catalyst is not observed when rhodium complexes containing chelating phosphines are used [109]. These complexes, having the general form $[(diphos)Rh(diene)]^+$ (diphos = chelating phosphine, diene = cyclooctadiene or norbornadiene), are also used for selective hydrogenation of alkynes to alkenes. The reaction proceeds through the widely known unsaturated route (Fig. 3.4) [104,107,108]. It starts with: (1) Coordination of the unsaturated substrate to the rhodium center; (2) oxidative addition of hydrogen (rate-determining step); (3) migration of one hydride to the coordinated alkyne, which results in alkenyl asymmetrically coordinated to the Rh(III); and (4) migration of the second hydride that leads to a symmetrically coordinated alkene, as confirmed by using d-DNP-polarized substrates [111].

The rhodium complexes containing a chelating phosphine are particularly efficient in terms of hyperpolarization on products since the lifetime of reaction intermediates is extremely short [112]. Consequently, the singlet state is transferred to the substrate without being mixed with the orthostates. The effect of singlet-triplet mixing, which occurs on reaction intermediates, on the hyperpolarization of product molecules has been described theoretically [113,114].

Modifications of the phosphine ligand are an important means to change the properties of the catalyst, to improve specificity and efficiency [115], to make it soluble in aqueous medium [64,116,117], or to anchor the complex to a solid [50,118,119].

In order to obtain a water-soluble rhodium complex, the chelating phosphine can be varied, introducing hydrophilic substituents (like OCH_3, SO_3H, and THP) [116]. Tetrasulfonated bis(diphenylphosphino)butane (DPPBTS) [116] and 1,4-bis-[(phenyl-3-propanesulfonate)phosphine] butane disodium salt [64] have been used to obtain an aqueous solution of PHIP-polarized products suitable for in vivo application [120].

As far as heterogeneous hydrogenation is concerned, these reactions usually proceed through the chemisorption and dissociation of the hydrogen molecule over the surface of metal clusters. In this case, the spin correlation between protons of the hydrogen molecule is lost, and PHIP could not be observed, at least in principle. Nevertheless, the use of heterogeneous catalysts would allow the easy removal of the catalyst from the products, which is necessary for the biological

application of PHIP-polarized substrates; therefore, the use of heterogeneous catalysts is quite appealing for the production of biocompatible contrast agents for MRI [121].

The use of hydrogenation complexes (such as the Wilkinson catalysts) immobilized on solid surfaces through modified ligand [118] ($RhCl(PPh_3)_2PPh_2(CH_2)_2{-}SiO_2$) has been proposed to circumvent the issue of heterogeneous hydrogenation [50,118,119].

Nevertheless, and quite unexpectedly, PHIP has also been observed on authentic heterogeneous catalysts comprising Rh particles supported on TiO_2 (Rh/TiOs) or alumina (Rh/AlO(OH)) [51,122] (other heterogeneous catalysts based on Pt and Pd will be described in the following section). These have been used for parahydrogenation tests and led to the observation of PHIP-polarized products such as styrene [123], acetate [121], and furane derivatives in the liquid phase or propene gas [51].

The observation of PHIP using a heterogeneous catalyst has been explained considering the isolation or localization of catalytic sites. It seems reasonable that partitioning into smaller surfaces, like those of the smallest metal particles, by various adsorbates should lead to higher yields of the pairwise addition of hydrogen to substrates, which is not necessarily the main reaction mechanism. Localization of catalytic sites on metal surfaces has been tested, and surface diffusion has been limited by ligands that bind to the metal [87].

Anyhow, signal enhancements obtained by means of het-PHIP have been disappointing due to the low pairwise selectivity for the supported metals.

3.3.2.2 Ruthenium

The catalytic semihydrogenation of alkynes to alkenes usually leads to the formation of Z-alkenes because the most widely applied transition metal catalysts deliver the two hydrogen atoms suprafacially to the π-system of the substrate. The isomeric products (E-alkenes) are usually thought to derive from secondary processes.

In a pioneering study using PHIP [124,125], Bargon and coworkers demonstrated that the cationic sorbic acid complex Cp_Ru(η^4-$CH_3CH{=}CHCHCHCO_2H$)][CF_3SO_3] catalyzes the transreduction of internal alkynes such as 2-pentyne or 1-phenyl-1-propyne, while terminal acetylenes were unreactive. The use of PHIP allowed us to demonstrate that the E-isomer is formed directly and not from the isomerization of the Z isomer. At that time, a mechanism proceeding via a binuclear Ru complex was proposed to explain the formation of the (E)-alkenes.

In later studies, the commercial precatalyst [CpRuCl(COD)] [126] led to the formation of the E configuration of the alkene. In that case, more than one Ru-hydride species was identified, although a detailed picture of the operative mechanism could not be established. The active catalyst was generated in situ from [Cp*RuCl(cod)] and AgOTf under hydrogen pressure.

The use of PHIP (together with computational studies of the reaction intermediates) led to more insights into the mechanism of the trans-hydrogenation of internal alkynes by [CpRu] complexes [127]. The cationic complex [Cp*Ru(-MeCN)$_3$]PF_6 or the neutral precatalysts, such as [Cp*RuCL(cod)] or [[Cp*$RuCl_4$], lead to the trans-reduction of internal alkynes, which competes with a pathway in which both H atoms of the H_2 molecule are delivered to the same C atom of the alkyne, i.e., the gem-hydrogenation product. Gem-hydrogenation has already been observed using Rh(I) and Pd(0) catalysts as a result of a pairwise proton exchange mechanism [128].

The unorthodox stereochemical course of hydrogenation catalyzed by [Cp*Ru] complexes has been exploited to obtain hyperpolarized fumarate from the parahydrogenation of acetylendicarboxylic acid. The complex Cp*Ru(MeCN)$_3$]PF_6 (6 mM) [53] in D_2O leads to a trans-cis ratio more than 500/1 when the additive sodium disulfite is added. Although the role of the salt is not clear, this reaction allowed it to reach more than 25% polarization on the ^{13}C carboxylate spin following the application of the magnetic field cycle [54,129], and this hyperpolarized product is a good candidate for biological applications and in vivo investigations. When sodium disulfite is omitted, the trans-cis ratio goes to 3/1 and another product is formed, on which parahydrogen is added into geminal positions to form a CH_2 group [130] (Fig. 3.5). The hyperpolarized singlet order is accumulated in this product, and, being the two protons inequivalent, it can be detected by means of a suitable pulse sequence [130,131].

3.3.2.3 Cobalt

The cobalt complex (^{Mes}CCC)Co(N2)(PPh3) (MesCCC is the monoionic pincer ligand: bis(mesityl-benzimidazol-2-ylidene)phenyl) brought to PHIP polarized signals on the products of catalytic hydrogenation of alkynes to alkenes, together with the cis-trans isomerization [132,133]. The observation of hyperpolarized signals on the product confirms that both atoms of the hydrogen molecule are added in a concerted fashion and shows that the trans-isomer is not directly obtained from parahydrogen addition to the alkyne [134].

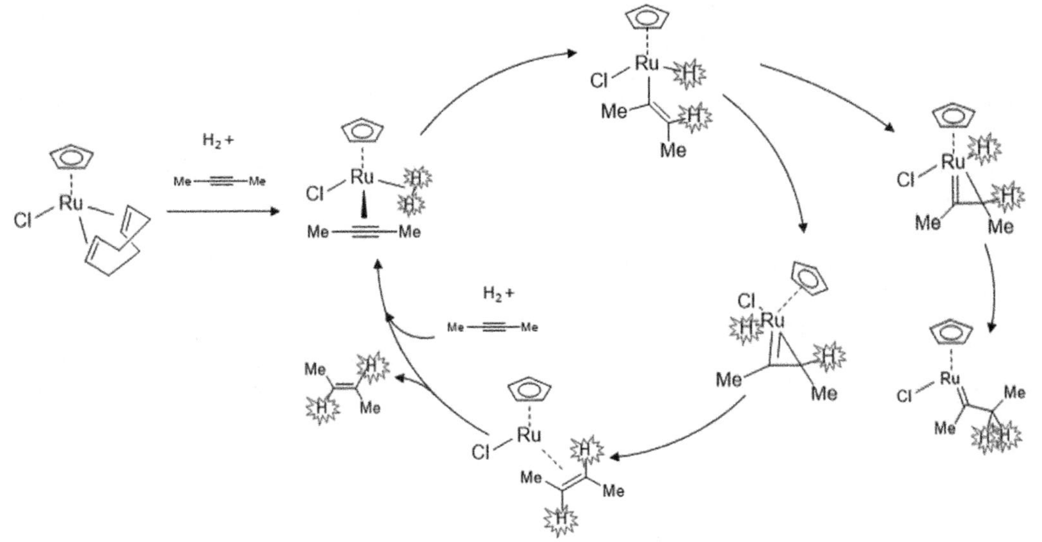

FIGURE 3.5 Scheme of the hydrogenation pathway of the [Ru(Cp)]+ complex that brings to the transhydrogenation of an alkyne to E-alkene. The gem-addition of hydrogen to the same carbon atom is also obtained as a subproduct [128].

Strong ^{13}C hyperpolarization has also been observed following the parahydrogenation of the test substrate, ethyl-propionate, using the cobalt complex (^{Mes}CCC)Co-py [135]. This has been chosen instead of [(Mes-CCC)Co(N_2)PPh_3] because the phosphine inhibits catalysis due to the ligand-substrate exchange (Fig. 3.6).

Hydrogen transfer to the substate has been elucidated thanks to the detection of catalytically relevant intermediates [132]. The semihydrogenation of internal alkynes proceeds with good yields, while the hydrogenation of terminal alkynes (such as phenylacetylene) leads to catalyst decomposition and no hydrogenation of the product.

PHIP on hydroformylated alkenes has also been observed using the cobalt-allyl complex Co(η3-C_3H_5) $(CO)_2$-(PCy_3), deriving from the $HCo(CO)_4$-based complex [136], and reaction intermediates have also been detected thanks to PHIP.

3.3.2.4 Platinum and palladium

These metals have been more widely applied as heterogeneous hydrogenation catalysts. Substantial NMR signal enhancement provided by PHIP depends on the ability of the catalyst to incorporate both hydrogen atoms into the same product molecule, which can be obtained thanks to the isolation or localization of catalytic sites.

The use of PHIP allowed it to be shown that the degree of pairwise H_2 addition in the propene hydrogenation catalyzed by metal Pt and Pd in the gas phase [137] depends on the support of the metal catalyst (Al_2O_3, SiO_2, ZrO_2, and TiO_2 have been used) and on the size of the metal particles. It has also been found that there is an increase toward the pairwise addition for Pt/TiO_2 compared to the other supports, which might be due to the strong metal-support interaction. Anyhow, the pairwise addition of a supported Pt catalyst remains a minor process compared to the addition of random H atoms to the double bond.

FIGURE 3.6 Cocomplexes used for hyperpolarization of substrates by means of PHIP.

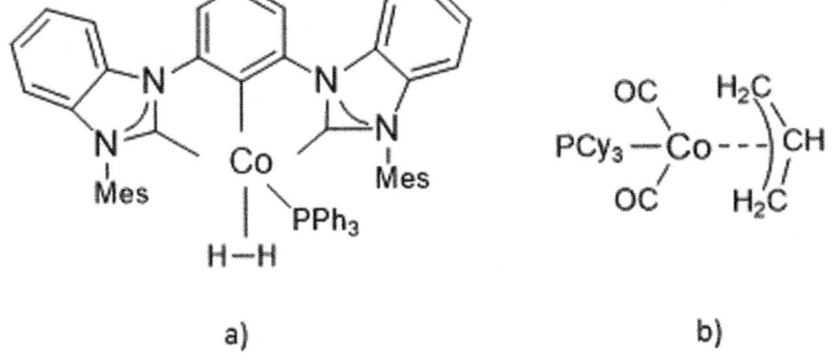

Pt and Pd nanoparticles supported on γ-Al_2O_3 [138] and intermetallic nanoparticles (Pt_3Sn and PtSn) encapsulated in a thermally robust mesoporous silica ($mSiO_2$) have also been tested for HetPHIP [139]. The pairwise selectivity was significantly increased using the intermetallic nanoparticles PtSn@$mSiO_2$, with respect to that observed using a monometallic catalyst. This result was explained in terms of the elimination of the three-fold sites on the Pt(111) surface, which are the sites for dissociative H_2 chemisorption.

Since the efficiency of het-PHIP is related to the isolation of the catalytic sites, the vast metal surface can be partitioned into isolated islands of bare metal by means of other species, and various adsorbates should lead to higher yields of the pairwise addition of hydrogen to substrates. Water-dispersible Pd nanoparticles have been stabilized by means of N-acetylcysteine and L-cysteine (NAC@Pd and Cys1@Pd) [140,141], and this has brought significantly higher polarization on ^{13}C. Platinum nanoparticles capped with amino-acid ligands (glutathione or cysteine) have been applied for the parahydrogenation of different substrates (NAC@Pt and Cys1@Pt) [141,142] (see Table 3.1).

The main advantage of using heterogeneous catalysts is the possibility of filtering the solutions in order to remove the metal, which is necessary for the exploitation of the PHIP-polarized products in in vivo studies. Nevertheless, when nanoparticles are employed, their size (2–2.5 nm) makes the separation of the catalyst quite challenging.

A few homogeneous Pt and Pd complexes have also been explored in hydrogenation reactions using PHIP, among them the complex trans-PtCl(COEt) $(PPh_3)_2$ [143] that led to one-H PHIP of the aldehyde proton of propanal. This is the first case in which PHIP has been reported following the addition of only one proton from parahydrogen.

3.4 Nonhydrogenative PHIP (SABRE)

To achieve hyperpolarization by means of SABRE, a metal center must act as a scaffold to temporarily bring together the substrate and the parahydrogen molecule in the form of hydride ligands. The reversible binding of the substrate and parahydrogen to the metal complex is crucial for the efficiency of SABRE. If the binding interaction is too strong, it can become irreversible and parahydrogen exchange is stopped; vice versa, if the interaction is too weak, the SABRE catalyst can become unstable and parahydrogen is rapidly depleted.

Therefore, the binding affinity of the substrates for the metal center as well as the stability of the complex must be accurately chosen and tuned in order to be able to mediate the exchange on an appropriate timescale.

3.4.1 Catalysts for SABRE

The initial SABRE experiments made use of Crabtree's complex $[Ir(COD)(PCy_3)(py)]BF_4$ (COD = cis,cis-1,5-cyclooctadiene, Cy = cyclohexyl, and py = pyridine) to mediate polarization transfer. In that case, the active complex was $[Ir(H)_2(PCy_3)(py)_3]BF_4$, formed by the oxidative addition of hydrogen in the presence of excess pyridine. The small scalar couplings between the hydride ligands and the orthopyridine protons mediate the spontaneous polarization transfer from parahydrogen protons to the ligand (Fig. 3.7) in a low magnetic field.

The identity of the phosphine on the Crabtree's complex plays an important role in the process due to the interplay between its steric bulk, which promotes substrate loss, and the electron donation that is necessary to promote H_2 exchange. Due to this, the PCy_2Ph catalyst was shown to be superior to those containing PPh_3 or PCy_3 [144].

In the second generation of SABRE catalysts, the phosphine ligand was replaced with N-heterocyclic carbenes, stronger electron-donating ligands. The air-stable complex [IrCl(COD)(IMes)] (IMes = 1,3-bis(2,4,6-trimethylphenyl) imidazole-2-ylidene) yielded much higher polarization; in particular, the polarization level on pyridine protons was about four times that obtained using the phosphine complex [145]. This increase was associated with the higher exchange rate of pyridine on the active catalyst $[Ir(H)_2(IMes)(py)_3]Cl$, 23 s^{-1}, instead of 0.45 s^{-1} when using the phosphine-derived catalyst, as measured using EXSY.

The effect of various substitutions on the carbene ligand has been explored [146], and it has been shown that the polarization on substrates is related to the steric bulk and the electron-donating power of the ligand [146], although it has not been possible to quantify the correlation between these parameters due to the complexity of the SABRE polarization transfer.

As the catalyst ($[Ir(H)_2(IMes)(analyte)_3]Cl$) in a charged system, polar solvents such as D_4-MeOH or D_6-EtOH have been more commonly used. The use of a neutral iridium catalyst containing bidentated NHC [147] ligand allowed the use of nonpolar solvents such as benzene-d_6, CD_2Cl_2, and THF-d_8.

Other SABRE catalysts, such as the bidentated NHC-phenolate [147], an iridium PNP-pincer [148], and an iridium cyclooctadiene complex [149], can undergo the nonhydrogenative transfer of polarization, although the signal gain was lower compared to [IrCl(COD)(IMes)].

TABLE 3.1 Molecules that have been hyperpolarized by means of parahydrogen (hydrogenative and nonhydrogenative).

	Molecule	pH_2	Solvent	[HPmolecule]	Catalyst and PHIP method	^{13}C HP	Ref.
Metabolites	Acetate	92%	D_2O	75 mM	Rh PHIP-SAH	^{13}C Yes, n.d	[16]
	1-^{13}C-acetate	–	CD_2Cl_2	50 mM	Ir SABRE-RELAY	^{13}C enhancement 48 fold@ 9.4 T	[74]
		80%	Methanol-d_4/ethanol-d_6	30–40 mM	Ir SABRE	^{13}C enhancement 108 fold@ 9.4 T	[75]
	Acetate-d_3	92%	Methanol-d_4/D2O	0.9 mM	Rh	^{13}C 19.3%	[76]
	1-^{13}C-acetate-d_3	92%	Acetone-d_6/D_2O or methanol-d_4/D_2O	0.5 mM	Rh PHIP-SAH	^{13}C 3.6%	[77]
	1-^{13}C-fumarate	86%	D_2O	45 mM	Ru (water soluble) h-PHIP	^{13}C 24% @ 14.1 T	[54]
	1-^{13}C-lactate	92%	H_2O	40–60 mM	Rh PHIP-SAH	^{13}C 2.9% @ 14.1 T	[65]
	1-^{13}C-phospholactate-d_2	90%	D_2O	25 mM	Rh (water soluble) h-PHIP	^{13}C 15% @ 47.5 mT	[78] [56]
	1-^{13}C-pyruvate	92%	H_2O	30–40 mM	Rh PHIP-SAH	^{13}C 3.5% @ 14.1 T	[79] [80]
	1-^{13}C-pyruvate	92%	Acetone-d_6/Methanol-d_4/ D_2O	0.75 mM	Rh PHIP-SAH	^{13}C 3.4%	[77]
	1-^{13}C- pyruvate 2-^{13}C-pyruvate	93%	Methanol-d_4	25–30 mM	Ir SABRE	^{13}C 1% @ 9.4T ^{13}C 0.6% @ 9.4T	[81]
	1-^{13}C- pyruvate	–	CD_2Cl_2	50 mM	Ir SABRE-RELAY	^{13}C 0.039% @ 9.4 T (48 fold)	[74]
	1,2-$^{13}C_2$- pyruvate	99%	Methanol-d_4	30 mM	Ir SABRE	^{13}C 1.7% @ 9.4 T (C1 and C2 average)	[82]
	1,2-$^{13}C_2$-pyruvate	93%	Methanol-d_4	25–30 mM	Ir SABRE	^{13}C 1.8% (C1) ^{13}C 1.6% (C2)@ 9.4T	[81]
	1,2-$^{13}C_2$-pyruvate	93%	D_2O/Ethanol-d_6 (70:30)	0.4 mM	Ir SABRE	n.d.	[81]
	Pyruvic acid	50%	Acetone-d_6	100 mM	Rh PHIP-X	<0.1	[83]
	1-^{13}C-succinate-d_2	90%	D_2O	30 mM	Rh (water soluble) h-PHIP	^{13}C 28% @ 47.5 mT	[78]
	1-^{13}C-succinate-d_2	90%	H_2O	5 mM	Rh (water soluble) h-PHIP	^{13}C 11%	[84]
	Urea^{-13}C	–	CD_2Cl_2	50 mM	Ir SABRE-RELAY	^{13}C 0.15% @ 9.4 T (182 fold)	[74]

	Urea–^{13}C–$^{15}N_2$	–	CD_2Cl_2	25 mM	Ir SABRE-RELAY	^{13}C 0.33% on ^{13}C @ 9.4 T (408 fold) ^{15}N HP yes, n.d.	[74]
	Ethyl [1-^{13}C]pyruvate	86%	H_2O/D_2O	80 mM	Rh water soluble PHIP-SAH	^{13}C 4.8% ^{13}C	[66]
Metabolites derivatives	Allyl-glycine	50%	D_2O	40 mM	Rh h-PHIP	^{13}C (n.d.)	[85]
	Propargyl [15N]Choline	50%	Acetone-d_6/MeOD	15 mM	Rh h-PHIP	^{15}N (0.4%)	[60]
	Diethyl-succinate	n.a.	H_2O/D_2O	20 mM	Rh h-PHIP	^{13}C (2.1%)	[86]
Amino acids and their derivatives	1-^{13}C-Glycine	80%	D_2O	0.6 mM	NAC@Rh h-PHIP	^{1}H 1.14% ^{13}C 0.29% @ 7 T, 353K	[87]
	1-^{13}C-alanine	80%	D_2O	0.6 mM	NAC@Rh h-PHIP	^{13}C 0.25% @ 7 T, 353K	[87]
	N-acetyl-AAs (phenylalanine)	50%	Methanol-d_4	40 mM	Rh h-PHIP	^{13}C 1.3% @ 11.7 T, 298K of fully deuterated *N*-acetyl-1-^{13}C-phenylalanine	[59]
	N-unprotected AA derivatives (esters)	95%	D_2O	2 mM approx. (5%–10% of 25 mM)	Rh (water soluble) h-PHIP	^{1}H 0.7% Gly and Ala deriv. ^{1}H 0.5% Gln @ 14.1 T, 353K	[88]
	Glycine-Histidine-Phenylalanine	50%	Methanol-d_4	~2 mM	Ir SABRE	–	[89]
	Alanine, glutamine esters	95%	D_2O	25 mM	h-PHIP	^{1}H 0.5%–1%; ^{13}C 4.4% (deuterated derivative)	[88]
Nitrogenous bases	Adenine Adenosine	95%	Methanol-d_4	10 mM	Ir SABRE	–	[90]
	Adenine	80% –90%	Methanol-d_4	n.d.	Ir SABRE-SHEAT	^{13}C 0.06% ^{15}N @ 8.5 T	[91]
	Adenosine 5′-triphosphate (ATP)	–	CD_2Cl_2	50 mM	Ir SABRE-RELAY	^{13}C 0.063%, 0.042%, 0.026% on ^{31}P @ 9.4 T (48,32,20 fold)	[74]
	Purine		Methanol-d_4	100 mM	Ir SABRE		[19]

Continued

TABLE 3.1 Molecules that have been hyperpolarized by means of parahydrogen (hydrogenative and nonhydrogenative).—cont'd

	Molecule	pH_2	Solvent	[HPmolecule]	Catalyst and PHIP method	^{13}C HP	Ref.
Nutrients	D-glucose	–	Acetone/DMSO	50 mM	Ir SABRE-RELAY	^{1}H 0.13% @ 9.4 T (40 fold)	[74]
	D-glucose-$^{13}C_6$	–	CD_2Cl_2	50 mM	Ir SABRE-RELAY	n.d.	[74]
	D-glucose-$^{13}C_6$ Fructose$^{-13}C_6$-d_7	99%	CD_2Cl_2/DMF	20 mM	Ir SABRE-RELAY	1.1% ^{13}C @ 1 T 0.2% ^{13}C @ 9.4 T	[20]
	Glucose	50%	Acetone-d_6	40 mM	Rh PHIP-X	$^{13}C < 0.1$	[83]
	Nicotinic acid (vitamin B_3)	–	Methanol-d_4	40 mM	Ir SABRE	2.6%/total ^{1}H signal @9.4 T (800 fold)	[92]
	Nicotinamide	50%	Methanol-d_4	n.d	Ir SABRE-SHEAT	3.4% ^{15}N @ 8.5 T	[91]
	Nicotinamide-d_2	–	Methanol-d_4 Ethanol-d_6	20 mM	Ir SABRE	^{1}H 4.1% ^{1}H 2.1%	[93]
	15N-nicotinamide		H_2O	100 mM	**SABRE-SHEATH**	(^{15}N) 0.37%	[94]
Drugs	Alectinib (potential chemotherapeutic)	80% –90%	Methanol-d_4	n.d.	Ir SABRE-SHEAT	^{13}C 1.5% on ^{15}N @ 8.5 T	[91]
	CHCA (chemotherapeutic)	80% –90%	Methanol-d_4	n.d.	Ir SABRE-SHEAT	1.1% on ^{15}N @ 8.5 T	[91]
	Isoniazid (antitubercolosis)	92.5%	Methanol-d_4	4 mM	Ir SABRE	^{13}C 8% @ 16.4T, 319K	[95]
	Metronidazole (antibiotic)	75% –85%	Methanol-d_4	20 mM	Ir SABRE-SHEAT	^{13}C 34% on ^{15}N @ 9.4 T, 300K	[91,96]
	PET tracers O-(2-[^{18}F]fluoroethyl)-L-tyrosine ([^{18}F]FET) [^{18}F]fallypride ([^{18}F]FP)	92%	Acetone-d_6 /D_2O And ethanol-d_6/D_2O MeOD-d_4	12 mM (12% of 100 mM) 8 mM	Rh h-PHIP	1H 4.5%@ 7T ^{13}C 4%–7.4% @ 7 T, 333K 1H 3.5% @ 7 T, 333K ^{13}C 3.1 % @ 7 T, 333 K	[97]
	Pyrazinamide (antitubercolosis)	92.5%	Methanol-d_4	4 mM	Ir SABRE	–	[95]

Other molecules (h-PHIP)	**2-Hydroxyethyl propionate (HEP)** (In-vivo angiography)	95%	H_2O/D_2O	500 mM	h-PHIP	^{13}C 30%	[61]
	Tetrafluoropropyl propionate (TFPP) **Plaque imaging** in vivo	90%	H_2O/D_2O	20 mM	h-PHIP	^{13}C, 17%	[62]
	Propane (HP-gas)	90%	No solvent		HET-PHIP	^{1}H, 1%	[73]
	Trimethylvinyl silane	50	Acetone-d_6	300 mM	h-PHIP	^{29}Si, 0.2%	[98]
	bis(trimethylsilyl) ethene	50%	Acetone-d_6	20 mM	h-PHIP	^{29}Si 8.3%	[69]
Other molecules (SABRE)	Nicotinamide	90%	MeOD or EtOH	20 mM	SABRE	^{1}H: up to 3.4% on deuterated analogs	[93]
	Methy-nicotinate	90%	MeOD or EtOH	20 mM	SABRE	^{1}H: up to 11% on deuterated analogs	
	Propanol	99%	CD_2Cl_2	20 mM	Ir SABRE-RELAY	^{1}H pol 1.8%, ^{13}C 0.2%	[74]
	Other alcohols: n-butanol, n-pentanol, n-hexanol, n-heptanol, n-octanol, isopropanol, tert-butanol)	99%	CD_2Cl_2	10–20 mM	Ir SABRE-RELAY	^{1}H pol 0.6%–1.4%, ^{13}C 0.2%–0.4%	[74]
	Acetonitrile	90%	MeOD	100 mM	SABRE	^{1}H <0.1%	[99]
	Methanol	99%	CD_2Cl_2	40 mM	Ir SABRE-RELAY	^{1}H 1.6%, ^{13}C 0.6%	[74]
	^{15}N-pyridine	90%	MeOD	4 mM	SABRE-SHEAT	^{15}N 9.7%	[100]
	^{15}N-pyridine	90%	D_2O	20 mM	SABRE-SHEAT	^{15}N, 0.3%	[94]
	^{15}N-benzonitrile	90%	D_2O	50 mM	SABRE-SHEAT	^{15}N 0.58%	[94]
	α-cyano-4-hydroxycinnamic acid (CHCA) Schemoterapic	90%	D_2O	50 mM	SABRE-SHEAT	^{15}N <0.1%	[94]
	15N-diaziridine derivatives	90%	MeOD	10 mM	SABRE-SHEAT	^{15}N, 5%	[101]
	Silanols and silanes	90%	CD_2Cl_2	30 mM	SABRE	^{29}Si (0.2–2,3%)	[102]

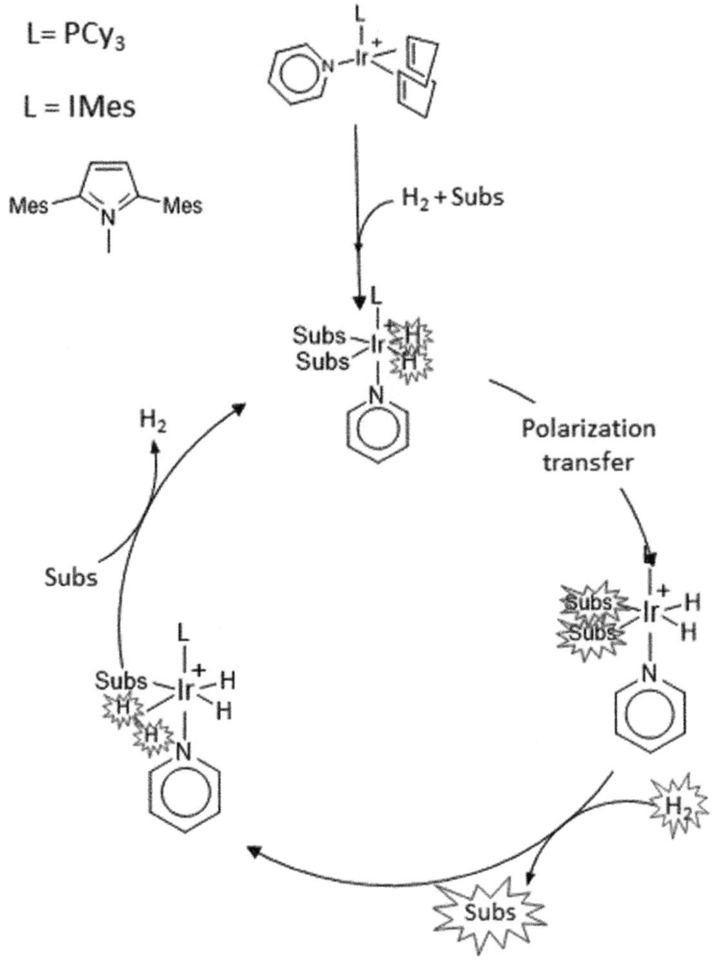

FIGURE 3.7 Scheme of SABRE polarization of a generic substrate (=Subs), pyridine can be a substrate as well. In the initially reported SABRE complex, the ligand L = PCy_3 has been replaced by IMes, bringing to a significantly more efficient complex.

The use of water or ethanol-d_6/water mixtures is desirable for in-vivo applications; therefore, different strategies have been investigated for this purpose.

The solubility of the IMes-Ir complexes in water has increased [94,150,151] through their activation in organic solvents and subsequent addition of water [150]. Water-soluble phosphine ligands [152] or modifications of the IMes [94] have also been tested.

Heterogeneous SABRE catalysts have been obtained through the immobilization of iridium-NHC catalysts on polymer microbeads [153] or silica [154]. Currently, neither heterogeneous nor aqueous SABRE delivers polarization levels comparable to those observed using the most widely applied SABRE complex.

3.4.2 Substrates for SABRE

As can be seen in Table 3.1, nitrogen-containing heterocycles are usually the best candidates for ^{1}H-SABRE polarization. Pyridine and substituted pyridyl [93], nicotinamide, nicotinate, and nicotinic acid (Vit-B3) [92] are often used. Molecules containing multiple ligation sites such as pyrimidine, pyrazine, pyridazine, and phtalazine [155], biologically relevant adenine and adenosine [90], indazole, and imidazole [156] have also been hyperpolarized.

Binding to the SABRE complexes of heteroaromatic compounds, such as nucleobases containing the pyridine, purine, or adenine group [157], nitriles, and amines also found application in the fields of chemosensing and mixture analysis and allowed the detection of picomole amounts of biological substrates that can bind to Ir complexes [157,158]. This is obtained thanks to the formation of hydride signals in a normally unoccupied region of the NMR spectrum. These experiments are usually carried out in methanol or methanol/water mixtures, but urine diluted in MeOH (80/20 vol% methanol/water mixture) [159] has also been used, and α-amino acids have been detected in this complex mixture. The use

of PHIP as a chemosensor for the detection of analytes provides a sensitivity gain that exceeds that of an NMR cryo-probe by at least two orders of magnitude and compares with HPLC-MS.

The affinity of the SABRE complex to substrates can be modified thanks to the effects of coligands. The presence of DMSO, or other analogs, allowed for the assembly of the highly reactive polarization transfer complex ([$Ir(H)_2(\eta^2$-pyruvate)(DMSO)(IMes)] that is able to bind pyruvate [81,82,160], a substrate characterized by poor ligation with the SABRE catalyst.

Other coligands, such as acetonitrile, PPh_3, or PCy_3 [99,156,161], have also been used for SABRE because their ligation to the metal leads to symmetry breaking in complexes of the type [$Ir(H)_2$IMes) (substrate)$^{3+}$. In these cases, hyperpolarization is transferred to the ligands too, which occupies an equatorial position.

The breadth of SABRE has been widely extended by SABRE-RELAY [74], for which the requirement for binding the target molecule to the metal has been overcome. In this case, a carrier containing exchangeable protons (usually ammonia or amines) is hyperpolarized by SABRE [20]. After this first step, proton exchanges allow the relay of polarization into the target substrate. Thanks to this method, several alcohols, spanning from methanol to pentanol to glucose [20], sodium salts (acetate and pyruvate), sodium phosphate, and sodium hydrogen carbonate, have been polarized.

It has already been mentioned that the hyperpolarization of ^{15}N-labeled substrates can be particularly interesting due to the slow relaxation of these signals and the subsequent long lifetime of the hyperpolarized signals. ^{15}N hyperpolarization has been obtained by means of the SABRE-SHEAT experiments [100,162]. In these experiments, SABRE occurs in a magnetic shield (a mu-metal shield) that allows it to lower the geomagnetic filed and reach a few mG. At this field, strong coupling conditions between protons and heteroatoms take place, and spin order has been transferred to ^{15}N on several N-heterocycles [91], Shiff bases [163], and diazirine tags [101,164].

3.5 Summary and outlook

Since its discovery in the early 1980s, PHIP has evolved from a method used mainly for the investigation of hydrogenation pathways to a more flexible tool for the hyperpolarization of substrates for various applications. Among the purposes of PHIP-polarized molecules, their application to biomedical imaging is probably the most relevant one at the present time.

The recent developments in the chemistry of PHIP have significantly expanded the scope of this method. The main limitation of h-PHIP, i.e., the chemical modification of the substrate that incorporates a hydrogen molecule, has been overcome by the introduction of SABRE.

Hyperpolarization of biologically relevant substrates and metabolites has been obtained using different PHIP-based methods. As far as this application is concerned, it must be kept in mind that an aqueous solution of the hyperpolarized probe is necessary and toxic compounds (hydrogenation catalysts, solvents) have to be removed. Several methods have been proposed, from the application of heterogeneous catalysts to the development of fast purification strategies. Among these, liquid-liquid extraction [68,165], precipitation of the hyperpolarized product [166], and catalyst scavenging must be mentioned. The first one is based on the hyperpolarization of the target molecule in an organic hydrophobic solvent, followed by the extraction of the hyperpolarized product in the water phase. Precipitation of the hyperpolarized product, followed by its redissolution in an aqueous phase, has been obtained on fumarate, thanks to the low solubility of its acidic form. Scavenging of the metal catalyst by means of supported ligands appears to be a quite efficient method for the removal of the catalyst in the aqueous phase.

The efficiency of PHIP-based methods in the production of HP products relies not only on the chemistry of the process (i.e., of the hydrogenation reaction) but also on the efficiency of spin order transfer [44,129,167–171] and on the automatization of the entire procedure. For this purpose, different methods have been reported [172] that are based on different ways to perform the hydrogenation reaction and the spin order transfer.

In conclusion, PHIP is a chemistry-based hyperpolarization method that needs the application of different competences, from spin physics to engineering of the entire procedure; nevertheless, the development of new applications and substrates has to take its basis in the chemistry of the hyperpolarization process.

References

[1] F. Hund, Zur Deutung der Molekelspektren. II, Z. Phys. 42 (1927) 93–120, https://doi.org/10.1007/BF01397124.

[2] H.K.F. Bonhoeffer, P. Harteck, Die eigenschaften des parawasserstoffes, Zeitschrift Für Elektrochemie Und Angew. Phys. Chemie. 35 (1929) 621–623, https://doi.org/10.1002/bbpc.192900041.

[3] A. Farkas, Orthohydrogen, Parahydrogen and Heavy Hydrogen, Cambridge University Press, London, 1935.

[4] C. Bowers, Transformation of symmetrization order to nuclear-spin magnetization by chemical reaction and nuclear magnetic resonance, Phys. Rev. Lett. Phys. Rev. Lett. 57 (1986) 2645–2648.

[5] S.I. Hommeltoft, D.H. Berry, R. Eisenberg, A metal-centered radical-pair mechanism for alkyne hydrogenation with a binuclear rhodium hydride complex. CIDNP without organic radicals, J. Am. Chem. Soc. 108 (1986) 5345–5347, https://doi.org/10.1021/ja00277a050.

[6] T.C. Eisenschmid, R.U. Kirss, P.P. Deutsch, S.I. Hommeltoft, R. Eisenberg, J. Bargon, R.G. Lawler, A.L. Balch, Para hydrogen induced polarization in hydrogenation reactions, J. Am. Chem. Soc. 109 (1987) 8089–8091, https://doi.org/10.1021/ja00260a026.

[7] C.R. Bowers, D.P. Weitekamp, Parahydrogen and synthesis allow dramatically enhanced nuclear alignment, J. Am. Chem. Soc. 109 (1987) 5541–5542, https://doi.org/10.1021/ja00252a049.

[8] K. Golman, O. Axelsson, H. Johannesson, S. Månsson, C. Olofsson, J.S. Petersson, Parahydrogen-induced polarization in imaging: subsecond C-13 angiography, Magn. Reson. Med. 46 (2001) 1–5.

[9] J.H. Ardenkjaer-Larsen, B. Fridlund, A. Gram, G. Hansson, L. Hansson, M.H. Lerche, R. Servin, M. Thaning, K. Golman, Increase in signal-to-noise ratio of > 10,000 times in liquid-state NMR, Proc. Natl. Acad. Sci. U. S. A. 100 (2003) 10158–10163, https://doi.org/10.1073/pnas.1733835100.

[10] J.H. Ardenkjaer-Larsen, A.M. Leach, N. Clarke, J. Urbahn, D. Anderson, T.W. Skloss, Dynamic nuclear polarization polarizer for sterile use intent, NMR Biomed. 24 (2011) 927–932, https://doi.org/10.1002/nbm.1682.

[11] J.H. Ardenkjaer-Larsen, On the present and future of dissolution-DNP, J. Magn. Reson. 264 (2016) 3–12, https://doi.org/10.1016/j.jmr.2016.01.015.

[12] K. Golman, R.I.T. Zandt, M.H. Lerche, R. Pehrson, J.H. Ardenkjaer-Larsen, Metabolic imaging by hyperpolarized 13C magnetic resonance imaging for in vivo tumor diagnosis, Cancer Res. 66 (2006) 10855–10860, https://doi.org/10.1158/0008-5472.CAN-06-2564.

[13] F.A. Gallagher, M.I. Kettunen, K.M. Brindle, Biomedical applications of hyperpolarized 13C magnetic resonance imaging, Prog. Nucl. Magn. Reson. Spectrosc. 55 (2009) 285–295, https://doi.org/10.1016/j.pnmrs.2009.06.001.

[14] J. Kurhanewicz, D.B. Vigneron, K. Brindle, E.Y. Chekmenev, R.J. Deberardinis, G.G. Green, R.R. Rizi, B.D. Ross, W.S. Warren, Analysis of cancer metabolism by imaging hyperpolarized nuclei: prospects for translation to clinical research, Neoplasia 13 (2011) 81–97, https://doi.org/10.1593/neo.101102.

[15] K.M. Brindle, Imaging metabolism with hyperpolarized ^{13}C-labeled cell substrates, J. Am. Chem. Soc. 137 (2015) 6418–6427, https://doi.org/10.1021/jacs.5b03300.

[16] F. Reineri, T. Boi, S. Aime, ParaHydrogen Induced Polarization of 13C carboxylate resonance in acetate and pyruvate, Nat. Commun. 6 (2015) 5858, https://doi.org/10.1038/ncomms6858.

[17] E. Cavallari, C. Carrera, S. Aime, F. Reineri, Metabolic studies of tumor cells using [1- 13 C] pyruvate hyperpolarized by means of PHIP-side arm hydrogenation, ChemPhysChem 20 (2019) 318–325, https://doi.org/10.1002/cphc.201800652.

[18] E. Cavallari, C. Carrera, M. Sorge, G. Bonne, A. Muchir, S. Aime, F. Reineri, The 13 C hyperpolarized pyruvate generated by ParaHydrogen detects the response of the heart to altered metabolism in real time, Sci. Rep. 8 (2018) 8366, https://doi.org/10.1038/s41598-018-26583-2.

[19] R.W. Adams, J.A. Aguilar, K.D. Atkinson, M.J. Cowley, P.I.P. Elliott, S.B. Duckett, G.G.R. Green, I.G. Khazal, J. Lopez-Serrano, D.C. Williamson, Reversible interactions with para-hydrogen enhance NMR sensitivity by polarization transfer, Science 323 (2009) 1708–1711, https://doi.org/10.1126/science.1168877.

[20] P.M. Richardson, W. Iali, S.S. Roy, P.J. Rayner, M.E. Halse, S.B. Duckett, Rapid 13C NMR hyperpolarization delivered from: para -hydrogen enables the low concentration detection and quantification of sugars, Chem. Sci. 10 (2019) 10607–10619, https://doi.org/10.1039/c9sc03450a.

[21] P. Kaur, Preparation and Raman Spectroscopy of Parahydrogen, Southampton University, Southampton, UK, 2005.

[22] K.F. Bonhoeffer, A. Farkas, K.W. Rummel, Über die heterogene Katalyse der Parawasserstoffumwandlung, Z. Phys. Chem. 21B (1933) 225–234, https://doi.org/10.1515/zpch-1933-2120.

[23] H.S. Taylor, H. Diamond, The spin-isomerization of hydrogen at paramagnetic and diamagnetic surfaces, J. Am. Chem. Soc. (1935) 1251–1256.

[24] Y.L. Sandler, The adsorption and the magnetic ortho-para conversion of hydrogen on diamagnetic solids. I. Some experiments in surface para-magnetism, J. Phys. Chem. 58 (1954) 54–57, https://doi.org/10.1021/j150511a013.

[25] Y. Sandler, The adsorption and ortho-para conversion of hydrogen on diamagnetic solids. II. The relative adsorbabilities of orthohydrogen and parahydrogen, J. Phys. Chem. (1954) 58–61.

[26] L. Farkas, L. Sandler, On the heterogeneous ortho- paraconversion on paramagnetic crystals, J. Chem. Phys. 8 (1940) 248, https://doi.org/10.1063/1.1750638.

[27] Y. Sandler, The kinetics of the heterogeneous parahydrogen conversion, J. Chem. Phys. 21 (1953) 2243.

[28] D.D. Eley, The conversion of parahydrogen by porphyrin compounds, including hemoglobin, Trans. Farady Soc. 35 (1940) 500–505.

[29] K.F. Bonhoeffer, P. Harteck, Experimente über Para- und Orthowasserstoff, Naturwissenschaften (1929) 182, https://doi.org/10.1007/BF01506559.

[30] B. Feng, A.M. Coffey, R.D. Colon, E.Y. Chekmenev, K.W. Waddell, A pulsed injection parahydrogen generator and techniques for quantifying enrichment, J. Magn. Reson. 214 (2012) 258–262, https://doi.org/10.1016/j.jmr.2011.11.015.

[31] J.B. Hövener, S. Bär, J. Leupold, K. Jenne, D. Leibfritz, J. Hennig, S.B. Duckett, D. Von Elverfeldt, A continuous-flow, high-throughput, high-pressure parahydrogen converter for hyperpolarization in a clinical setting, NMR Biomed. 26 (2013) 124–131, https://doi.org/10.1002/nbm.2827.

[32] S. Wagner, Conversion rate of para-hydrogen to ortho-hydrogen by oxygen: implications for PHIP gas storage and utilization, Magma 27 (2014) 195–199, https://doi.org/10.1007/s10334-013-0399-y.

[33] E.R. Grilly, A simple thermal conductivity analyzer for ortho-para hydrogen, Rev. Sci. Instrum. 24 (1953) 72, https://doi.org/10.1063/1.1770522.

[34] A.T. Stewart, G.L. Squires, Analysis of ortho- and para-hydrogen mixtures by the thermal conductivity method, J. Sci. Instrum. 32 (1955) 26–29, https://doi.org/10.1088/0950-7671/32/1/313.

[35] T.W. Bradshaw, Observations on the use of a thermal conductivity cell to measure the para hydrogen concentration in a mixture of para and ortho hydrogen gas, Rev. Sci. Instrum. (1986) 26, https://doi.org/10.1063/1.1139519.

[36] A. Gamliel, H. Allouche-Arnon, R. Nalbandian, C.M. Barzilay, J.M. Gomori, R. Katz-Brull, An apparatus for production of isotopically and spin-enriched hydrogen for induced polarization studies, Appl. Magn. Reson. 39 (2010) 329–345, https://doi.org/10.1007/s00723-010-0161-9.

[37] K.D. Atkinson, M.J. Cowley, S.B. Duckett, P.I.P. Elliott, G.G.R. Green, J. López-Serrano, I.G. Khazal, A.C. Whitwood, Para-hydrogen induced polarization without incorporation of para-hydrogen into the analyte, Inorg. Chem. 48 (2009) 663–670, https://doi.org/10.1021/ic8020029.

[38] R.A. Green, R.W. Adams, S.B. Duckett, R.E. Mewis, D.C. Williamson, G.G.R. Green, The theory and practice of hyperpolarization in magnetic resonance using parahydrogen, Prog. Nucl. Magn. Reson. Spectrosc. 67 (2012) 1–48, https://doi.org/10.1016/j.pnmrs.2012.03.001.

[39] K.L. Ivanov, A.N. Pravdivtsev, A.V. Yurkovskaya, H.M. Vieth, R. Kaptein, The role of level anti-crossings in nuclear spin hyperpolarization, Prog. Nucl. Magn. Reson. Spectrosc. 81 (2014) 1–36, https://doi.org/10.1016/j.pnmrs.2014.06.001.

[40] D.A. Barskiy, S. Knecht, A.V. Yurkovskaya, K.L. Ivanov, SABRE: chemical kinetics and spin dynamics of the formation of hyperpolarization, Prog. Nucl. Magn. Reson. Spectrosc. 114–115 (2019) 33–70, https://doi.org/10.1016/j.pnmrs.2019.05.005.

[41] M.G. Pravica, D.P. Weitekamp, Net nmr alignment by adiabatic transport of para-hydrogen addition-products to high magnetic-field, Chem. Phys. Lett. 145 (1988) 255–258.

[42] G. Pileio, Singlet NMR methodology in two-spin-1/2 systems, Prog. Nucl. Magn. Reson. Spectrosc. 98–99 (2017) 1–19, https://doi.org/10.1016/j.pnmrs.2016.11.002.

[43] J. Barkemeyer, M. Haake, J. Bargon, Hetero-NMR enhancement via parahydrogen labeling, J. Am. Chem. Soc. 117 (1995) 2927–2928, https://doi.org/10.1021/ja00115a028.

[44] J. Eills, J.W. Blanchard, T. Wu, C. Bengs, J. Hollenbach, D. Budker, M.H. Levitt, Polarization transfer via field sweeping in parahydrogen-enhanced nuclear magnetic resonance, J. Chem. Phys. 150 (2019) 174202, https://doi.org/10.1063/1.5089486.

[45] A.J. Hovener, A.N. Pravdivtsev, B. Kidd, R. Bowers, S. Glöggler, K. V Kovtunov, R. Katz-brull, K. Buckenmaier, F. Reineri, T. Theis, V. Roman, S. Wagner, N.M.M. Zacharias, E.Y. Chekmenev, Parahydrogen-based hyperpolarization for biomedicine, Angew. Chem. Int. Ed. 57 (2018) 11140–11162, https://doi.org/10.1002/anie.201711842.

[46] F. Reineri, E. Cavallari, C. Carrera, S. Aime, Hydrogenative-PHIP polarized metabolites for biological studies, Magn. Reson. Mater. Phys. Biol. Med. 34 (2021) 25–47, https://doi.org/10.1007/s10334-020-00904-x.

[47] J. Natterer, J. Bargon, Parahydrogen induced polarization, Prog. Nucl. Magn. Reson. Spectrosc. 31 (1997) 293–315.

[48] S.B. Duckett, C.J. Sleigh, Applications of the parahydrogen phenomenon: a chemical perspective, Prog. Nucl. Magn. Reson. Spectrosc. 34 (1999) 71–92, https://doi.org/10.1016/S0079-6565(98)00027-2.

[49] E. Cavallari, C. Carrera, G. Di Matteo, O. Bondar, S. Aime, F. Reineri, In-vitro NMR studies of prostate tumor cell metabolism by means of hyperpolarized [1-13C]pyruvate obtained using the PHIP-sah method, Front. Oncol. 10 (2020) 1–9, https://doi.org/10.3389/fonc.2020.00497.

[50] L.-S. Bouchard, S.R. Burt, M.S. Anwar, K. V Kovtunov, I. V Koptyug, A. Pines, NMR imaging of catalytic hydrogenation in microreactors with the use of para-hydrogen, Science 319 (2008) 442–445, https://doi.org/10.1126/science.1151787.

[51] K.V. Kovtunov, D.A. Barskiy, A.M. Coffey, M.L. Truong, O.G. Salnikov, A.K. Khudorozhkov, E.A. Inozemtseva, I.P. Prosvirin, V.I. Bukhtiyarov, K.W. Waddell, E.Y. Chekmenev, I.V. Koptyug, High-resolution 3D proton MRI of hyperpolarized gas enabled by parahydrogen and Rh/TiO2 heterogeneous catalyst, Chem. Eur J. 20 (2014) 11636–11639, https://doi.org/10.1002/chem.201403604.

[52] E.Y. Chekmenev, J. Hövener, V.A. Norton, K. Harris, L.S. Batchelder, P. Bhattacharya, B.D. Ross, D.P. Weitekamp, PASADENA hyperpolarization of succinic acid for MRI and NMR spectroscopy, J. Am. Chem. Soc. 130 (2008) 4212–4213, https://doi.org/10.1021/ja7101218.

[53] B. Ripka, J. Eills, H. Kour, M. Leutzsch, M.H. Levitt, K. Munnemann, Hyperpolarized fumarate via parahydrogen, Chem. Commun. (2018) 12246–12249, https://doi.org/10.1039/C8CC06636A.

[54] J. Eills, E. Cavallari, C. Carrera, D. Budker, S. Aime, F. Reineri, Real-time nuclear magnetic resonance detection of Fumarase Activity using parahydrogen-hyperpolarized [1-13C]fumarate, J. Am. Chem. Soc. 141 (2019) 20209–20214, https://doi.org/10.1021/jacs.9b10094.

[55] R.V. Shchepin, A.M. Coffey, K.W. Waddell, E.Y. Chekmenev, PASADENA hyperpolarized 13C phospholactate, J. Am. Chem. Soc. 134 (2012) 3957–3960, https://doi.org/10.1021/ja210639c.

[56] R. V Shchepin, A.M. Coffey, K.W. Waddell, E.Y. Chekmenev, Parahydrogen induced polarization of 1- 13 C-phospholactate- d 2 for biomedical imaging with >30,000,000-fold NMR signal enhancement in water, Anal. Chem. 86 (2014) 5601–5605, https://doi.org/10.1021/ac500952z.

[57] O.G. Salnikov, R.V. Shchepin, N.V. Chukanov, L. Jaigirdar, W. Pham, K.V. Kovtunov, I.V. Koptyug, E.Y. Chekmenev, Effects of deuteration of 13C-enriched phospholactate on efficiency of parahydrogen-induced polarization by magnetic field cycling, J. Phys. Chem. C 122 (2018) 24740–24749, https://doi.org/10.1021/acs.jpcc.8b07365.

[58] E. Cerutti, A. Viale, C. Nervi, R. Gobetto, S. Aime, The role of the amino protecting group during parahydrogenation of protected dehydroamino acids, J. Phys. Chem. A 119 (2015) 11271–11279, https://doi.org/10.1021/acs.jpca.5b06802.

[59] P.C. Soon, X. Xu, B. Zhang, F. Gruppi, J.W. Canary, A. Jerschow, Hyperpolarization of amino acid precursors to neurotransmitters with parahydrogen induced polarization, Chem. Commun. 49 (2013) 5304–5306, https://doi.org/10.1039/c3cc40426a.

[60] F. Reineri, A. Viale, S. Ellena, D. Alberti, T. Boi, G.B. Giovenzana, R. Gobetto, S.S.D. Premkumar, S. Aime, 15N magnetic resonance hyperpolarization via the reaction of parahydrogen with 15N-propargylcholine, J. Am. Chem. Soc. 134 (2012) 11146–11152, https://doi.org/10.1021/ja209884h.

[61] L.E. Olsson, C.M. Chai, O. Axelsson, M. Karlsson, K. Golman, J.S. Petersson, MR coronary angiography in pigs with intraarterial injections of a hyperpolarized 13C substance, Magn. Reson. Med. 55 (2006) 731–737, https://doi.org/10.1002/mrm.20847.
[62] P. Bhattacharya, E.Y. Chekmenev, W.F. Reynolds, S. Wagner, N. Zacharias, H.R. Chan, R. Bünger, B.D. Ross, Parahydrogen-induced polarization (PHIP) hyperpolarized MR receptor imaging in vivo: a pilot study of 13C imaging of atheroma in mice, NMR Biomed. 24 (2011) 1023–1028, https://doi.org/10.1002/nbm.1717.
[63] M. Goldman, H. Jóhannesson, O. Axelsson, M. Karlsson, Hyperpolarization of 13C through order transfer from parahydrogen: a new contrast agent for MRI, Magn. Reson. Imaging (2005) 153–157, https://doi.org/10.1016/j.mri.2004.11.031.
[64] J.-B. Hövener, E.Y. Chekmenev, K.C. Harris, W.H. Perman, T.T. Tran, B.D. Ross, P. Bhattacharya, Quality assurance of PASADENA hyperpolarization for carbon-13 biomolecules, Magn. Reson. Mater. Phys. 22 (2009) 123–134, https://doi.org/10.1007/s10334-008-0154-y.
[65] E. Çavallari, C. Carrera, S. Aime, F. Reineri, ^{13}C-MR Hyperpolarization of Lactate using ParaHydrogen and metabolic transformation in vitro, Chem. Eur J. 23 (2017) 1200–1204, https://doi.org/10.1002/chem.201605329.
[66] C. Carrera, E. Cavallari, G. Digilio, O. Bondar, S. Aime, F. Reineri, ParaHydrogen polarized ethyl-[1-13C]pyruvate in water, a key substrate for fostering the PHIP-SAH approach to metabolic imaging, ChemPhysChem (2021), https://doi.org/10.1002/cphc.202100062.
[67] R.V. Shchepin, D.A. Barskiy, A.M. Coffey, I.V. Manzanera Esteve, E.Y. Chekmenev, Efficient synthesis of molecular precursors for parahydrogen-induced polarization of ethyl acetate-1-13C and beyond, Angew. Chem. Int. Ed. 55 (2016) 6071–6074, https://doi.org/10.1002/anie.201600521.
[68] F. Reineri, A. Viale, S. Ellena, T. Boi, V. Daniele, R. Gobetto, S. Aime, Use of labile precursors for the generation of hyperpolarized molecules from hydrogenation with parahydrogen and aqueous-phase extraction, Angew. Chem. Int. Ed. 50 (2011) 7350–7353, https://doi.org/10.1002/anie.201101359.
[69] S. Ellena, A. Viale, R. Gobetto, S. Aime, Para-hydrogen induced polarization of Si-29 NMR resonances as a potentially useful tool for analytical applications, Magn. Reson. Chem. 50 (2012) 529–533, https://doi.org/10.1002/mrc.3832.
[70] M. Plaumann, U. Bommerich, T. Trantzschel, D. Lego, S. Dillenberger, G. Sauer, J. Bargon, G. Buntkowsky, J. Bernarding, Parahydrogen-induced polarization transfer to 19 F in perfluorocarbons for 19 F NMR spectroscopy and MRI, Chem. Eur J. 19 (2013) 6334–6339, https://doi.org/10.1002/chem.201203455.
[71] U. Bommerich, T. Trantzschel, S. Mulla-osman, G. Buntkowsky, J. Bargon, J. Bernarding, Variation Enable 19 F-MRI at Low Spin Density, 2010, https://doi.org/10.1039/c001265c.
[72] D.A. Barskiy, A.M. Coffey, P. Nikolaou, D.M. Mikhaylov, B.M. Goodson, R.T. Branca, G.J. Lu, M.G. Shapiro, V.V. Telkki, V.V. Zhivonitko, I.V. Koptyug, O.G. Salnikov, K.V. Kovtunov, V.I. Bukhtiyarov, M.S. Rosen, M.J. Barlow, S. Safavi, I.P. Hall, L. Schröder, E.Y. Chekmenev, NMR hyperpolarization techniques of gases, Chem. Eur J. 23 (2017) 725–751, https://doi.org/10.1002/chem.201603884.
[73] K. V Kovtunov, M.L. Truong, D.A. Barskiy, O.G. Salnikov, V.I. Bukhtiyarov, A.M. Coffey, K.W. Waddell, I. V Koptyug, E.Y. Chekmenev, Propane-d6 heterogeneously hyperpolarized by parahydrogen, J. Phys. Chem. C. Nanomater. Interfaces. 118 (2014) 28234–28243, https://doi.org/10.1021/jp508719n.
[74] W. Iali, P.J. Rayner, S.B. Duckett, Using parahydrogen to hyperpolarize amines, amides, carboxylic acids, alcohols, phosphates, and carbonates, Sci. Adv. 4 (2018) 1–7, https://doi.org/10.1126/sciadv.aao6250.
[75] M.E. Gemeinhardt, M.N. Limbach, T.R. Gebhardt, C.W. Eriksson, S.L. Eriksson, J.R. Lindale, E.A. Goodson, W.S. Warren, E.Y. Chekmenev, B.M. Goodson, "Direct" 13C hyperpolarization of 13C-acetate by MicroTesla NMR signal amplification by reversible exchange (SABRE), Angew. Chem. Int. Ed. 59 (2020) 418–423, https://doi.org/10.1002/anie.201910506.
[76] S. Korchak, S. Mamone, S. Glçggler, Over 50 % 1 H and 13 C polarization for generating hyperpolarized metabolites — a para -hydrogen approach, Chem. Open (2018) 672–676, https://doi.org/10.1002/open.201800086.
[77] S. Korchak, S. Yang, S. Mamone, S. Glçggler, Pulsed magnetic resonance to signal-enhance metabolites within seconds by utilizing para -hydrogen, Chem. Open. (2018) 344–348, https://doi.org/10.1002/open.201800024.
[78] A.M. Coffey, R.V. Shchepin, M.L. Truong, K. Wilkens, W. Pham, E.Y. Chekmenev, Open-source automated parahydrogen hyperpolarizer for molecular imaging using 13C metabolic contrast agents, Anal. Chem. 88 (2016) 8279–8288, https://doi.org/10.1021/acs.analchem.6b02130.
[79] R.F. Cavallari Eleonora, C. Carrera, S. Matteo, B. Gisèle, M. Antoine, S. Aime, The 13C hyperpolarized pyruvate generated by ParaHydrogen detects the response of the heart to altered metabolism in real time, Sci. Rep. 8 (2018) 8366.
[80] E. Cavallari, C. Carrera, S. Aime, F. Reineri, Studies to enhance the hyperpolarization level in PHIP-SAH-produced C13-pyruvate, J. Magn. Reson. 289 (2018) 12–17, https://doi.org/10.1016/j.jmr.2018.01.019.
[81] W. Iali, S.S. Roy, B.J. Tickner, F. Ahwal, A.J. Kennerley, S.B. Duckett, Hyperpolarising pyruvate through signal amplification by reversible exchange (SABRE), Angew. Chem. Int. Ed. 58 (2019) 10271–10275, https://doi.org/10.1002/anie.201905483.
[82] B.J. Tickner, O. Semenova, W. Iali, P.J. Rayner, A.C. Whitwood, S.B. Duckett, Optimisation of pyruvate hyperpolarisation using SABRE by tuning the active magnetisation transfer catalyst, Catal. Sci. Technol. 10 (2020) 1343–1355, https://doi.org/10.1039/c9cy02498k.
[83] K. Them, F. Ellermann, A.N. Pravdivtsev, O.G. Salnikov, I.V. Skovpin, I.V. Koptyug, R. Herges, J.B. Hövener, Parahydrogen-induced polarization relayed via proton exchange, J. Am. Chem. Soc. 143 (2021) 13694–13700, https://doi.org/10.1021/jacs.1c05254.
[84] S. Berner, A.B. Schmidt, M. Zimmermann, A.N. Pravdivtsev, S. Glöggler, J. Hennig, D. von Elverfeldt, J.B. Hövener, SAMBADENA hyperpolarization of 13C-succinate in an MRI: singlet-triplet mixing causes polarization loss, ChemistryOpen 8 (2019) 728–736, https://doi.org/10.1002/open.201900139.

[85] T. Trantzschel, M. Plaumann, J. Bernarding, D. Lego, T. Ratajczyk, S. Dillenberger, G. Buntkowsky, J. Bargon, U. Bommerich, Application of parahydrogen-induced polarization to unprotected dehydroamino carboxylic acids, Appl. Magn. Reson. 44 (2013) 267–278, https://doi.org/10.1007/s00723-012-0391-0.

[86] N.M. Zacharias, H.R. Chan, N. Sailasuta, B.D. Ross, P. Bhattacharya, Real-time molecular imaging of tricarboxylic acid cycle metabolism in vivo by hyperpolarized 1- 13C diethyl succinate, J. Am. Chem. Soc. 134 (2012) 934–943, https://doi.org/10.1021/ja2040865.

[87] L. Kaltschnee, A.P. Jagtap, J. McCormick, S. Wagner, L.S. Bouchard, M. Utz, C. Griesinger, S. Glöggler, Hyperpolarization of amino acids in water utilizing parahydrogen on a rhodium nanocatalyst, Chem. Eur J. 25 (2019) 11031–11035, https://doi.org/10.1002/chem.201902878.

[88] S. Glöggler, S. Wagner, L.-S. Bouchard, Hyperpolarization of amino acid derivatives in water for biological applications, Chem. Sci. 6 (2015) 4261–4266, https://doi.org/10.1039/C5SC00503E.

[89] S. Glöggler, R. Müller, J. Colell, M. Emondts, M. Dabrowski, B. Blümich, S. Appelt, Para-hydrogen induced polarization of amino acids, peptides and deuterium-hydrogen gas, Phys. Chem. Chem. Phys. 13 (2011) 13759–13764, https://doi.org/10.1039/c1cp20992b.

[90] J.-B. Hövener, N. Schwaderlapp, T. Lickert, S.B. Duckett, R.E. Mewis, L. a R. Highton, S.M. Kenny, G.G.R. Green, D. Leibfritz, J.G. Korvink, J. Hennig, D. von Elverfeldt, A hyperpolarized equilibrium for magnetic resonance, Nat. Commun. 4 (2013) 2946, https://doi.org/10.1038/ncomms3946.

[91] J.F.P. Colell, A.W.J. Logan, Z. Zhou, R.V. Shchepin, D.A. Barskiy, G.X. Ortiz, Q. Wang, S.J. Malcolmson, E.Y. Chekmenev, W.S. Warren, T. Theis, Generalizing, extending, and maximizing nitrogen-15 hyperpolarization induced by parahydrogen in reversible exchange, J. Phys. Chem. C 121 (2017) 6626–6634, https://doi.org/10.1021/acs.jpcc.6b12097.

[92] A.M. Olaru, M.J. Burns, G.G.R. Green, S.B. Duckett, SABRE hyperpolarisation of vitamin B3 as a function of pH, Chem. Sci. 8 (2017) 2257–2266, https://doi.org/10.1039/c6sc04043h.

[93] P.J. Rayner, M.J. Burns, A.M. Olaru, P. Norcott, M. Fekete, G.G.R. Green, L.A.R. Highton, R.E. Mewis, S.B. Duckett, Delivering strong 1H nuclear hyperpolarization levels and long magnetic lifetimes through signal amplification by reversible exchange, Proc. Natl. Acad. Sci. U.S.A. 114 (2017) E3188–E3194, https://doi.org/10.1073/pnas.1620457114.

[94] J.F.P. Colell, M. Emondts, A.W.J. Logan, K. Shen, J. Bae, R.V. Shchepin, G.X. Ortiz, P. Spannring, Q. Wang, S.J. Malcolmson, E.Y. Chekmenev, M.C. Feiters, F.P.J.T. Rutjes, B. Blümich, T. Theis, W.S. Warren, Direct hyperpolarization of nitrogen-15 in aqueous media with parahydrogen in reversible exchange, J. Am. Chem. Soc. 139 (2017) 7761–7767, https://doi.org/10.1021/jacs.7b00569.

[95] H. Zeng, J. Xu, J. Gillen, M.T. McMahon, D. Artemov, J.M. Tyburn, J.A.B. Lohman, R.E. Mewis, K.D. Atkinson, G.G.R. Green, S.B. Duckett, P.C.M. Van Zijl, Optimization of SABRE for polarization of the tuberculosis drugs pyrazinamide and isoniazid, J. Magn. Reson. 237 (2013) 73–78, https://doi.org/10.1016/j.jmr.2013.09.012.

[96] B.E. Kidd, J.L. Gesiorski, M.E. Gemeinhardt, R.V. Shchepin, K.V. Kovtunov, I.V. Koptyug, E.Y. Chekmenev, B.M. Goodson, Facile removal of homogeneous SABRE catalysts for purifying hyperpolarized metronidazole, a potential hypoxia sensor, J. Phys. Chem. C 122 (2018) 16848–16852, https://doi.org/10.1021/acs.jpcc.8b05758.

[97] M. Braun, S. Häseli, F. Rösch, M. Piel, K. Münnemann, NMR hyperpolarization of established PET tracers, ChemistrySelect 3 (2018) 5176–5184, https://doi.org/10.1002/slct.201800364.

[98] M. Haake, J. Natterer, J. Bargon, Efficient NMR pulse sequences to transfer the parahydrogen-induced polarization to hetero nuclei, J. Am. Chem. Soc. 118 (1996) 8688–8691, https://doi.org/10.1021/ja960067f.

[99] R.E. Mewis, R.A. Green, M.C.R. Cockett, M.J. Cowley, S.B. Duckett, G.G.R. Green, R.O. John, P.J. Rayner, D.C. Williamson, Strategies for the hyperpolarization of acetonitrile and related Ligands by SABRE, J. Phys. Chem. B 119 (2015) 1416–1424, https://doi.org/10.1021/jp511492q.

[100] M.L. Truong, T. Theis, A.M. Coffey, R.V. Shchepin, K.W. Waddell, F. Shi, B.M. Goodson, W.S. Warren, E.Y. Chekmenev, ^{15}N hyperpolarization by reversible exchange using SABRE-SHEATH, J. Phys. Chem. C 119 (2015) 8786–8797, https://doi.org/10.1021/acs.jpcc.5b01799.

[101] T. Theis, G.X. Ortiz, A.W.J. Logan, K.E. Claytor, Y. Feng, W.P. Huhn, V. Blum, S.J. Malcolmson, E.Y. Chekmenev, Q. Wang, W.S. Warren, Direct and cost-efficient hyperpolarization of long-lived nuclear spin states on universal 15N2-diazirine molecular tags, Sci. Adv. 2 (2016), https://doi.org/10.1126/sciadv.1501438.

[102] P.J. Rayner, P.M. Richardson, S.B. Duckett, The detection and reactivity of silanols and silanes using hyperpolarized 29Si nuclear magnetic resonance, Angew. Chem. Int. Ed. 59 (2020) 2710–2714, https://doi.org/10.1002/anie.201915098.

[103] S.B. Duckett, C.L. Newell, R. Eisenberg, Observation of new intermediates in hydrogenation catalyzed by Wilkinson catalyst, RhCl(PPh3)3, using parahydrogen-induced polarization, J. Am. Chem. Soc. (1994) 10548–10556.

[104] R.U. Kirss, R. Eisenberg, Para-hydrogen-induced polarization in rhodium complex-catalyzed hydrogenation reactions, J. Organomet. Chem. 359 (1989) C22–C26.

[105] R.R. Schrock, J.A. Osborn, Catalytic hydrogenation using cationic rhodium complexes. I. Evolution of the catalytic system and the hydrogenation of olefins, J. Am. Chem. Soc. (1976) 2134–2143.

[106] R.R. Schrock, J.A. Osborn, Catalytic hydrogenation using cationic rhodium complexes. II. The selective hydrogenation of alkynes to cis olefins, J. Am. Chem. Soc. (1976) 2143–2147.

[107] R.A. Sánchez-Delgado, M. Rosales, Kinetic studies as a tool for the elucidation of the mechanisms of metal complex-catalyzed homogeneous hydrogenation reactions, Coord. Chem. Rev. 196 (2000) 249–280, https://doi.org/10.1016/S0010-8545(99)00168-X.

[108] J. Halpern, Mechanism and stereoselectivity of asymmetric hydrogenation, Science 217 (1982) 401–407.

[109] J.M. Brown, L.R. Canning, A.J. Downs, Equilibration of ortho- and para-hydrogen by homogeneous hydrogenation catalysts in solution; A test for the reversibility of hydrogen addition using Raman spectroscopy, J. Organomet. Chem. (1983) 103–111.

[110] O. Bondar, E. Cavallari, C. Carrera, S. Aime, F. Reineri, Effect of the hydrogenation solvent in the PHIP-SAH hyperpolarization of [1-13C] pyruvate, Catal. Today 397–399 (2021) 94–102, https://doi.org/10.1016/j.cattod.2021.11.030.

[111] P.A. Boeg, J. Duus, J.H. Ardenkjær-larsen, M. Karlsson, S. Mossin, Real-time detection of intermediates in rhodium-catalyzed hydrogenation of alkynes and alkenes by dissolution DNP, J. Phys. Chem. C 123 (2019) 9949–9956, https://doi.org/10.1021/acs.jpcc.9b01376.

[112] R. Giernoth, H. Heinrich, N.J. Adams, R.J. Deeth, J. Bargon, J.M. Brown, PHIP detection of a transient rhodium dihydride intermediate in the homogeneous hydrogenation of dehydroamino acids, J. Am. Chem. Soc. 122 (2000) 12381–12382, https://doi.org/10.1021/ja002516o.

[113] J. Natterer, O. Schedletzky, J. Barkemeyer, J. Bargon, S. Glaser, Investigating catalytic processes with parahydrogen: evolution of zero-quantum coherence in AA'X spin systems, J. Magn. Reson. 133 (1998) 92–97, https://doi.org/10.1006/jmre.1998.1421.

[114] M. Haake, J. Barkemeyer, J. Bargon, Symmetry breakdown during parahydrogen-labeling of symmetric substrates : proton spin polarization and singlet/triplet mixing due to 13C-isotopes in natural abundance, J. Phys. Chem. 99 (1995) 17539–17543, https://doi.org/10.1021/j100049a012.

[115] M. Itoda, Y. Naganawa, Structural exploration of rhodium catalysts and their kinetic studies for e ffi cient parahydrogen- induced polarization by side arm hydrogenation, RSC Adv. 9 (2019) 18183–18190, https://doi.org/10.1039/C9RA02580D.

[116] M. Ahlquist, M. Gustafsson, M. Karlsson, M. Thaning, O. Axelsson, O.F. Wendt, Rhodium(I) hydrogenation in water: kinetic studies and the detection of an intermediate using 13C{1H} PHIP NMR spectroscopy, Inorg. Chim. Acta. 360 (2007) 1621–1627, https://doi.org/10.1016/j.ica.2006.08.042.

[117] R.V. Shchepin, A.M. Coffey, K.W. Waddell, E.Y. Chekmenev, Parahydrogen-induced polarization with a Rh-based monodentate ligand in water, J. Phys. Chem. Lett. 3 (2012) 3281–3285, https://doi.org/10.1021/jz301389r.

[118] I.V. Koptyug, K.V. Kovtunov, S.R. Burt, M.S. Anwar, C. Hilty, S.-I. Han, A. Pines, R.Z. Sagdeev, Para-hydrogen-induced polarization in heterogeneous hydrogenation reactions, J. Am. Chem. Soc. 129 (2007) 5580–5586, https://doi.org/10.1021/ja068653o.

[119] L.S. Bouchard, K.V. Kovtunov, S.R. Burt, M.S. Anwar, I.V. Koptyug, R.Z. Sagdeev, A. Pines, Para-hydrogen-enhanced hyperpolarized gas-phase magnetic resonance imaging, Angew. Chem. Int. Ed. 46 (2007) 4064–4068, https://doi.org/10.1002/anie.200700830.

[120] A.B. Schmidt, S. Berner, M. Braig, M. Zimmermann, J. Hennig, D. Von Elverfeldt, J.B. Hövener, In vivo 13C-MRI using SAMBADENA, PLoS One 13 (2018) 1–15, https://doi.org/10.1371/journal.pone.0200141.

[121] K. V Kovtunov, D.A. Barskiy, R. V Shchepin, O.G. Salnikov, I.P. Prosvirin, A. V Bukhtiyarov, L.M. Kovtunova, V.I. Bukhtiyarov, I. V Koptyug, E.Y. Chekmenev, Production of pure aqueous 13 C-hyperpolarized acetate by heterogeneous parahydrogen-induced polarization, Chem. Eur J. 22 (2016) 16446–16449, https://doi.org/10.1002/chem.201603974.

[122] O.G. Salnikov, K.V. Kovtunov, D.A. Barskiy, A.K. Khudorozhkov, E.A. Inozemtseva, I.P. Prosvirin, V.I. Bukhtiyarov, I.V. Koptyug, Evaluation of the mechanism of heterogeneous hydrogenation of α,β-unsaturated carbonyl compounds via pairwise hydrogen addition, ACS Catal. 4 (2014) 2022–2028, https://doi.org/10.1021/cs500426a.

[123] I.V. Koptyug, V.V. Zhivonitko, K.V. Kovtunov, New perspectives for parahydrogen-induced polarization in liquid phase heterogeneous hydrogenation: an aqueous phase and ALTADENA study, ChemPhysChem 11 (2010) 3086–3088, https://doi.org/10.1002/cphc.201000407.

[124] D. Schleyer, H.G. Niessen, J. Bargon, In situ 1H-PHIP-NMR studies of the stereoselective hydrogenation of alkynes to (E)-alkenes catalyzed by a homogeneous [Cp*Ru]+ catalyst, New J. Chem. 25 (2001) 423–426, https://doi.org/10.1039/b007201j.

[125] H.G. Niessen, D. Schleyer, S. Wiemann, J. Bargon, S. Steines, B. Driessen-Hoelscher, In situ PHIP-NMR studies during the stereoselective hydrogenation of sorbic acid with a [Cp*Ru]+ catalyst, Magn. Reson. Chem. 38 (2000) 747–750, https://doi.org/10.1002/1097-458X(200009)38:9<747::AID-MRC715>3.0.CO;2-2.

[126] K. Radkowski, B. Sundararaju, A. Fürstner, A functional-group-tolerant catalytic trans hydrogenation of alkynes, Angew. Chem. Int. Ed. 52 (2013) 355–360, https://doi.org/10.1002/anie.201205946.

[127] M. Leutzsch, L.M. Wolf, P. Gupta, M. Fuchs, W. Thiel, C. Farès, A. Fürstner, Formation of ruthenium carbenes by gem -hydrogen transfer to internal alkynes: implications for alkyne trans -hydrogenation, Angew. Chem. 127 (2015) 12608–12613, https://doi.org/10.1002/ange.201506075.

[128] A. Harthun, R. Giernoth, C.J. Elsevier, J. Bargon, Rhodium- and palladium-catalysed proton exchange in styrene detected in situ by para-hydrogen induced polarization, Chem. Commun. (1996) 2483–2484, https://doi.org/10.1039/cc9960002483.

[129] H. Jóhannesson, O. Axelsson, M. Karlsson, Transfer of para-hydrogen spin order into polarization by diabatic field cycling, Compt. Rendus Phys. 5 (2004) 315–324, https://doi.org/10.1016/j.crhy.2004.02.001.

[130] L. Dagys, B. Ripka, M. Leutzsch, G.A.I. Moustafa, J. Eills, J.F.P. Colell, M.H. Levitt, Geminal parahydrogen-induced polarization: accumulating long-lived singlet order on methylene proton pairs, Magnes. Res. 1 (2020) 175–186, https://doi.org/10.5194/mr-1-175-2020.

[131] G. Pileio, M.H. Levitt, Theory of long-lived nuclear spin states in solution nuclear magnetic resonance. II. Singlet spin locking, J. Chem. Phys. 130 (2009), https://doi.org/10.1063/1.3139064.

[132] K. Tokmic, A.R. Fout, Alkyne semihydrogenation with a well-defined nonclassical Co-H2 catalyst: a H2 spin on isomerization and E-selectivity, J. Am. Chem. Soc. 138 (2016) 13700–13705, https://doi.org/10.1021/jacs.6b08128.

[133] A. Guthertz, M. Leutzsch, L.M. Wolf, P. Gupta, S.M. Rummelt, R. Goddard, C. Farès, W. Thiel, A. Fürstner, Half-sandwich ruthenium carbene complexes link trans-hydrogenation and gem-hydrogenation of internal alkynes, J. Am. Chem. Soc. 140 (2018) 3156–3169, https://doi.org/10.1021/jacs.8b00665.

[134] K. Tokmic, B.J. Jackson, A. Salazar, T.J. Woods, A.R. Fout, Cobalt-catalyzed and lewis acid-assisted nitrile hydrogenation to primary amines: a combined effort, J. Am. Chem. Soc. 138 (2017) 13554–13561, https://doi.org/10.1021/jacs.7b07368.

[135] K. Tokmic, R.B. Greer, L. Zhu, A.R. Fout, C NMR signal enhancement using parahydrogen-induced polarization mediated by a cobalt hydrogenation catalyst, J. Am. Chem. Soc. 140 (2018) 14844–14850, https://doi.org/10.1021/jacs.8b08614.

[136] C. Godard, S.B. Duckett, S. Polas, R. Tooze, A.C. Whitwood, Detection of intermediates in cobalt-catalyzed hydroformylation using para-hydrogen-induced polarization, J. Am. Chem. Soc. 127 (2005) 4994–4995, https://doi.org/10.1021/ja0434533.

[137] V.V. Zhivonitko, K.V. Kovtunov, I.E. Beck, A.B. Ayupov, V.I. Bukhtiyarov, I.V. Koptyug, Role of different active sites in heterogeneous alkene hydrogenation on platinum catalysts revealed by means of parahydrogen-induced polarization, J. Phys. Chem. C 115 (2011) 13386–13391, https://doi.org/10.1021/jp203398j.

[138] K.V. Kovtunov, I.E. Beck, V.I. Bukhtiyarov, I.V. Koptyug, Observation of parahydrogen-induced polarization in heterogeneous hydrogenation on supported metal catalysts, Angew. Chem. Int. Ed. 47 (2008) 1492–1495, https://doi.org/10.1002/anie.200704881.

[139] E.W. Zhao, R. Maligal-Ganesh, C. Xiao, T.W. Goh, Z. Qi, Y. Pei, H.E. Hagelin-Weaver, W. Huang, C.R. Bowers, Silica-encapsulated Pt-Sn intermetallic nanoparticles: a robust catalytic platform for parahydrogen-induced polarization of gases and liquids, Angew. Chem. Int. Ed. 56 (2017) 3925–3929, https://doi.org/10.1002/anie.201701314.

[140] J. McCormick, A.M. Grunfeld, Y.N. Ertas, A.N. Biswas, K.L. Marsh, S. Wagner, S. Glöggler, L.S. Bouchard, Aqueous ligand-stabilized palladium nanoparticle catalysts for parahydrogen-induced 13C hyperpolarization, Anal. Chem. 89 (2017) 7190–7194, https://doi.org/10.1021/acs.analchem.7b01363.

[141] S. Glöggler, A.M. Grunfeld, Y.N. Ertas, J. McCormick, S. Wagner, L.S. Bouchard, Surface ligand-directed pair-wise hydrogenation for heterogeneous phase hyperpolarization, Chem. Commun. 52 (2016) 605–608, https://doi.org/10.1039/c5cc08648e.

[142] S. Gloggler, A.M. Grunfeld, Y.N. Ertas, J. McCormick, S. Wagner, P.P.M. Schleker, L.S. Bouchard, A nanoparticle catalyst for heterogeneous phase para-hydrogen-induced polarization in water, Angew. Chem. Int. Ed. 54 (2015) 2452–2456, https://doi.org/10.1002/anie.201409027.

[143] A.B. Permin, R. Eisenberg, One-hydrogen polarization in hydroformylation promoted by platinum-tin and iridium carbonyl complexes: a new type of parahydrogen-induced effect, J. Am. Chem. Soc. 124 (2002) 12406–12407, https://doi.org/10.1021/ja026698t.

[144] K.D. Atkinson, M.J. Cowley, P.I.P. Elliott, S.B. Duckett, G.G.R. Green, J. López-Serrano, A.C. Whitwood, Spontaneous transfer of parahydrogen derived spin order to pyridine at low magnetic field, J. Am. Chem. Soc. 131 (2009) 13362–13368, https://doi.org/10.1021/ja903601p.

[145] J. Cowley, W. Ralph, Iridium N-heterocyclic carbene complexes as efficient catalysts for magnetization transfer from para-hydrogen, J. Am. Chem. Soc. 133 (2011) 6134–6137.

[146] L.S. Lloyd, A. Asghar, M.J. Burns, A. Charlton, S. Coombes, M.J. Cowley, G.J. Dear, S.B. Duckett, G.R. Genov, G.G.R. Green, L.A.R. Highton, A.J.J. Hooper, M. Khan, I.G. Khazal, R.J. Lewis, R.E. Mewis, A.D. Roberts, A.J. Ruddlesden, Hyperpolarisation through reversible interactions with parahydrogen, Catal. Sci. Technol. 4 (2014) 3544–3554, https://doi.org/10.1039/c4cy00464g.

[147] A.J. Ruddlesden, R.E. Mewis, G.G.R. Green, A.C. Whitwood, S.B. Duckett, Catalytic transfer of magnetism using a neutral iridium phenoxide complex, Organometallics 34 (2015) 2997–3006, https://doi.org/10.1021/acs.organomet.5b00311.

[148] A.J. Holmes, P.J. Rayner, M.J. Cowley, G.G.R. Green, A.C. Whitwood, S.B. Duckett, The reaction of an iridium PNP complex with parahydrogen facilitates polarisation transfer without chemical change, Dalton Trans. 44 (2014) 1077–1083, https://doi.org/10.1039/c4dt03088e.

[149] W. Iali, G.G.R. Green, S.J. Hart, A.C. Whitwood, S.B. Duckett, Iridium cyclooctene complex that forms a hyperpolarization transfer catalyst before converting to a binuclear C-H bond activation product responsible for hydrogen isotope exchange, Inorg. Chem. 55 (2016) 11639–11643, https://doi.org/10.1021/acs.inorgchem.6b02560.

[150] M.L. Truong, F. Shi, P. He, B. Yuan, K.N. Plunkett, A.M. Coffey, R.V. Shchepin, D.A. Barskiy, K.V. Kovtunov, I.V. Koptyug, K.W. Waddell, B.M. Goodson, E.Y. Chekmenev, Irreversible catalyst activation enables hyperpolarization and water solubility for NMR signal amplification by reversible exchange, J. Phys. Chem. B 118 (2014) 13882–13889, https://doi.org/10.1021/jp510825b.

[151] S.K. Hasnip, S.B. Duckett, D.R. Taylor, K. Barlow, M.J. Taylor, Kinetic and mechanistic examination of NBu 4 [IrH 2 (CO) 2 I 2] and NBu 4 [RhH 2 (CO) 2 I 2] via para -hydrogen enhanced NMR spectroscopy, Chem. Commun. 2 (1999) 889–890.

[152] M. Fekete, C. Gibard, G.J. Dear, G.G.R. Green, A.J.J. Hooper, A.D. Roberts, F. Cisnetti, S.B. Duckett, Utilisation of water soluble iridium catalysts for signal amplification by reversible exchange, Dalton Trans. 44 (2015) 7870–7880, https://doi.org/10.1039/c5dt00311c.

[153] F. Shi, A.M. Coffey, K.W. Waddell, E.Y. Chekmenev, B.M. Goodson, Heterogeneous solution NMR signal amplification by reversible exchange, Angew. Chem. Int. Ed. 53 (2014) 7495–7498, https://doi.org/10.1002/anie.201403135.

[154] K.V. Kovtunov, L.M. Kovtunova, M.E. Gemeinhardt, A.V. Bukhtiyarov, J. Gesiorski, V.I. Bukhtiyarov, E.Y. Chekmenev, I.V. Koptyug, B.M. Goodson, Heterogeneous microtesla SABRE enhancement of 15N NMR signals, Angew. Chem. Int. Ed. 56 (2017) 10433–10437, https://doi.org/10.1002/anie.201705014.

[155] K.M. Appleby, R.E. Mewis, A.M. Olaru, G.G.R. Green, I.J.S. Fairlamb, S.B. Duckett, Investigating pyridazine and phthalazine exchange in a series of iridium complexes in order to define their role in the catalytic transfer of magnetisation from para-hydrogen, Chem. Sci. 6 (2015) 3981–3993, https://doi.org/10.1039/c5sc00756a.

[156] M. Fekete, P.J. Rayner, G.G.R. Green, S.B. Duckett, Harnessing polarisation transfer to indazole and imidazole through signal amplification by reversible exchange to improve their NMR detectability, Magn. Reson. Chem. 55 (2017) 944–957, https://doi.org/10.1002/mrc.4607.

[157] N.J. Wood, J.A. Brannigan, S.B. Duckett, S.L. Heath, J. Wagstaff, Detection of picomole amounts of biological substrates by para-hydrogen-enhanced NMR methods in conjunction with a suitable receptor complex, J. Am. Chem. Soc. 129 (2007) 11012–11013, https://doi.org/10.1021/ja074286k.

[158] N.K.J. Hermkens, R.L.E.G. Aspers, M.C. Feiters, F.P.J.T. Rutjes, M. Tessari, Trace analysis in water-alcohol mixtures by continuous p-H2 hyperpolarization at high magnetic field, Magn. Reson. Chem. 56 (2018) 633–640, https://doi.org/10.1002/mrc.4692.

[159] L. Sellies, R.L.E.G. Aspers, M.C. Feiters, F.P.J.T. Rutjes, M. Tessari, Parahydrogen hyperpolarization allows direct NMR detection of α-amino acids in complex (Bio)mixtures, Angew. Chem. Int. Ed. 60 (2021) 26954–26959, https://doi.org/10.1002/anie.202109588.

[160] B.J. Tickner, P.J. Rayner, S.B. Duckett, Using SABRE hyperpolarized 13C NMR spectroscopy to interrogate organic transformations of pyruvate, Anal. Chem. 92 (2020) 9095–9103, https://doi.org/10.1021/acs.analchem.0c01334.

[161] M. Fekete, O. Bayfield, S.B. Duckett, S. Hart, R.E. Mewis, N. Pridmore, P.J. Rayner, A. Whitwood, Iridium(III) hydrido N-heterocyclic carbene-phosphine complexes as catalysts in magnetization transfer reactions, Inorg. Chem. 52 (2013) 13453–13461, https://doi.org/10.1021/ic401783c.

[162] T. Theis, M.L. Truong, A.M. Coffey, R.V. Shchepin, K.W. Waddell, F. Shi, B.M. Goodson, W.S. Warren, E.Y. Chekmenev, Microtesla SABRE enables 10% nitrogen-15 nuclear Spin polarization, J. Am. Chem. Soc. 137 (2015) 1404–1407, https://doi.org/10.1021/ja512242d.

[163] A.W.J. Logan, T. Theis, J.F.P. Colell, W.S. Warren, S.J. Malcolmson, Hyperpolarization of nitrogen-15 schiff bases by reversible exchange catalysis with para-hydrogen, Chem. Eur J. 22 (2016) 10777–10781, https://doi.org/10.1002/chem.201602393.

[164] K. Shen, A.W.J. Logan, J.F.P. Colell, J. Bae, G.X. Ortiz, T. Theis, W.S. Warren, S.J. Malcolmson, Q. Wang, Diazirines as potential molecular imaging tags: probing the requirements for efficient and long-lived SABRE-induced hyperpolarization, Angew. Chem. Int. Ed. 56 (2017) 12112–12116, https://doi.org/10.1002/anie.201704970.

[165] W. Iali, A.M. Olaru, G.G.R. Green, S.B. Duckett, Achieving high levels of NMR-hyperpolarization in aqueous media with minimal catalyst contamination using SABRE, Chem. Eur J. 23 (2017) 10491–10495, https://doi.org/10.1002/chem.201702716.

[166] S. Knecht, J.W. Blanchard, D. Barskiy, E. Cavallari, L. Dagys, E. van Dyke, M. Tsukanov, B. Bliemel, K. Münnemann, S. Aime, F. Reineri, M.H. Levitt, G. Buntkowsky, A. Pines, P. Blümler, D. Budker, J. Eills, Rapid hyperpolarization and purification of the metabolite fumarate in aqueous solution, Proc. Natl. Acad. Sci. U. S. A. 118 (2021) 1–6, https://doi.org/10.1073/pnas.2025383118.

[167] M. Goldman, H. Jóhannesson, Conversion of a proton pair para order into C-13 polarization by rf irradiation, for use in MRI, Compt. Rendus Phys. 6 (2005) 575–581, https://doi.org/10.1016/j.crhy.2005.03.002.

[168] J. Eills, G. Stevanato, C. Bengs, S. Glöggler, S.J. Elliott, J. Alonso-valdesueiro, G. Pileio, M.H. Levitt, Singlet order conversion and parahydrogen-induced hyperpolarization of 13 C nuclei in near-equivalent spin systems, J. Magn. Reson. 274 (2017) 163–172, https://doi.org/10.1016/j.jmr.2016.11.010.

[169] G. Stevanato, J. Eills, C. Bengs, G. Pileio, A pulse sequence for singlet to heteronuclear magnetization transfer: S-spin I-spins I-spins τ τ τ τ S-spin τ τ, J. Magn. Reson. 277 (2017) 169–178, https://doi.org/10.1016/j.jmr.2017.03.002.

[170] S. Bär, T. Lange, D. Leibfritz, J. Hennig, D. Von Elverfeldt, J.B. Hövener, On the spin order transfer from parahydrogen to another nucleus, J. Magn. Reson. 225 (2012) 25–35, https://doi.org/10.1016/j.jmr.2012.08.016.

[171] C. Cai, A.M. Coffey, R.V. Shchepin, E.Y. Chekmenev, K.W. Waddell, Efficient transformation of parahydrogen spin order into heteronuclear magnetization, J. Phys. Chem. B 117 (2013) 1219–1224, https://doi.org/10.1021/jp3089462.

[172] A.B. Schmidt, C.R. Bowers, K. Buckenmaier, E.Y. Chekmenev, H. de Maissin, J. Eills, F. Ellermann, S. Glöggler, J.W. Gordon, S. Knecht, I.V. Koptyug, J. Kuhn, A.N. Pravdivtsev, F. Reineri, T. Theis, K. Them, J.-B. Hövener, Instrumentation for hydrogenative parahydrogen-based hyperpolarization techniques, Anal. Chem. 94 (2022) 479–502, https://doi.org/10.1021/acs.analchem.1c04863.

Chapter 4

Beyond ^{13}C-pyruvate: Prospects for biomedical applications of alternative hyperpolarized probes and isotopes

Mor Mishkovsky[1,*] and Hikari A.I. Yoshihara[1,a]

[1]*Laboratory of Functional and Metabolic Imaging, Ecole Polytechnique Fédérale de Lausanne (EPFL), Lausanne, Switzerland*

**Corresponding author: E-mail: mor.mishkovsky@epfl.ch, hikari.yoshihara@uzh.ch*

4.1 Introduction

Metabolism is universal to living systems, encompassing all the chemical processes that enable cells to produce the energy required to maintain vital processes and synthesize new organic material. Magnetic resonance (MR) spectroscopic and imaging methods enable one to interrogate metabolic processes in vivo noninvasively without ionizing radiation.

4.1.1 Thermally polarized MR modalities to interrogate metabolism

Proton (^{1}H) MR spectroscopy (MRS), in combination with efficient suppression of the otherwise overwhelmingly large water signal, can detect endogenous metabolites. The specific ^{1}H resonance frequencies associated with each chemical species, combined with spectral fitting software [1,2] enable one to quantitate the various constituents and define the so-called neurochemical profile [3]. However, the limited ^{1}H chemical shift dispersion, together with the relatively broad lines typically recorded in vivo, leads to crowded spectra with multiple overlapping peaks. Only the most abundant metabolites (concentrations typically 1 mM and higher) can be detected, and spatial and temporal resolution is limited. Despite recent developments, the applicability of ^{1}H MRS techniques to clinical diagnostics remains challenging, as a good-quality ^{1}H MRS spectrum requires high magnetic field homogeneity. Moreover, the measurement of steady-state metabolite concentrations cannot provide direct information on metabolic fluxes.

The MRS detection of nonhyperpolarized (i.e., thermally polarized) ^{13}C-labeled molecules (particularly glucose, acetate, and lactate) provides important kinetic information on metabolism. However, its low sensitivity necessitates long acquisition times and long infusion times, which encumber its application to human studies and limit its suitability for clinical applications. Nonetheless, it is an invaluable preclinical research tool, providing extensive biochemical information on metabolism in vivo. The renewed interest in deuterated substrates to follow metabolism in vivo provides an alternative to the traditional ^{13}C methodologies. MR spectroscopic imaging following the administration of nonhyperpolarized [6,6′-$^2H_2$]glucose has been recently demonstrated, and it shows great potential to image glucose metabolism in the human brain [4]. In particular, the shorter longitudinal relaxation times of ^{2}H compared to those of ^{1}H or ^{13}C enable higher sensitivity per unit of scanner time [5]. Nonetheless, in protocols requiring prolonged substrate infusion, ranging from tens of minutes to hours, one should consider the potential systemic effects on the metabolic information being measured [6,7].

---

[a] Current affiliation: Institute for Biomedical Engineering, ETH Zurich and University of Zurich, Zurich, Switzerland.

The Chemistry of Hyperpolarized Magnetic Resonance Probes. https://doi.org/10.1016/B978-0-323-91842-8.00001-X

4.1.2 Hyperpolarized MRS

Hyperpolarization of small molecules temporarily enhances spin polarization up to five orders of magnitude [8]. This increase in sensitivity has opened a new class of MR-based applications to study biochemical reactions as they happen, both in vitro and in vivo settings. Generally, a compound with biomedical relevance is labeled with an NMR active isotope, and its signal is enhanced by one of the hyperpolarization techniques described in the preceding chapters. In most applications, the hyperpolarized sample is dissolved to create a biocompatible solution of the hyperpolarized probe and then infused as a bolus into the living system to be studied. Labeled metabolites are detected by their chemical shift difference from the hyperpolarized precursor, and their identity and rate of appearance report on the metabolic process being interrogated.

4.1.3 Disadvantages of nonrenewable magnetization

The lifetime of the hyperpolarized signal is finite, and regenerating a new hyperpolarized signal usually demands a time-consuming polarization process. Therefore, the hyperpolarized signal is considered nonrenewable. As such, the hyperpolarized nuclear spins need a relatively long longitudinal relaxation time (T_1) for maintaining the large out-of-equilibrium polarization on their journey, starting from the end of the polarization process to its delivery to the target tissue and reporting on the biological process under study. Hence, only metabolic processes with relatively high uptake and fast kinetics on the time scale of the precursor's T_1 can be interrogated. In slower processes, the signal of the hyperpolarized substrate completely decays before uptake and/or any detectable metabolic reaction happen. Note that for some probes, useful information can be obtained even if no metabolic transformation occurs—monitoring perfusion for instance. For metabolized probes, suitable substrates are labeled such that their chemical shift is sufficiently different from their metabolic products. An efficient acquisition scheme is required to make the best use of the nonrenewable hyperpolarized signal. To probe a time-evolving process, only a small portion of the signal is excited at a time (Fig. 4.1). The pulse sequence considerations for efficient detection of hyperpolarized probes are beyond the scope of this chapter but are described elsewhere [9−12].

4.1.4 Advantages of nonrenewable magnetization

While the nonrenewable nature of the hyperpolarized signal brings several constraints for its application, it also presents some advantages. It provides new insight into metabolic processes compared to its thermally polarized counterpart's methods. It is rapid and, thus, has more potential to be translated into clinical diagnostics. It requires much lower doses of substrate than thermally polarized X-nuclei MRS measurements, which typically require prolonged infusions. In metabolic imaging, it enables one to measure information that is closely related to enzyme activity. The limited lifetime of the probe ensures that we do not measure the accumulation of the product and minimizes the influence of systemic effects on the observed metabolic process. The dramatic increase in sensitivity enables the in vivo detection of metabolic intermediates that are inaccessible using thermally polarized techniques [13−15].

While not every metabolite or labeling site is suitable for hyperpolarization, there are many substrates that can be polarized and used to study function and metabolism [16]. In particular, ^{13}C labeling at sites without directly bonded protons, such as carboxylic acids or carbonyls, has relatively long T_1 values ranging from tens of seconds to minutes and is particularly interesting as a potential hyperpolarized contrast agent. This chapter focuses on the potential applications of these probes for monitoring biochemical processes.

4.2 Hyperpolarized [1-^{13}C]Pyruvate

The vast majority of hyperpolarized MR preclinical research has employed [1-^{13}C]pyruvate for real-time molecular imaging, and it is the first hyperpolarized ^{13}C-labeled metabolic contrast agent translated into clinical studies [17]. This is due to the exceptionally excellent properties of [1-^{13}C]pyruvate as a hyperpolarized precursor. It has a long T_1 (about 60 s at 3 T) that enables the maintenance of a large portion of the hyperpolarized signal. Moreover, it is ideally suited for the dissolution dynamic nuclear polarization technique; neat pyruvic acid forms a glassy solid when frozen rapidly, and doped with a sufficient amount of polarizing agent, it can be hyperpolarized to high levels (more than 60%) [18,19]. Moreover, pyruvate is an α-keto acid, and ultraviolet irradiation of frozen pyruvic acid in liquid nitrogen generates nonpersistent radicals that can act as the polarizing agent for the DNP process. These nonpersistent radicals are quenched upon dissolution, yielding a radical-free solution of hyperpolarized pyruvate [20]. The parahydrogen-induced polarization hyperpolarization process with sidearm hydrogenation provides an alternative route to produce hyperpolarized pyruvate [21], and it can yield a [1-^{13}C]pyruvate polarization level sufficient for preclinical imaging within a few minutes [22].

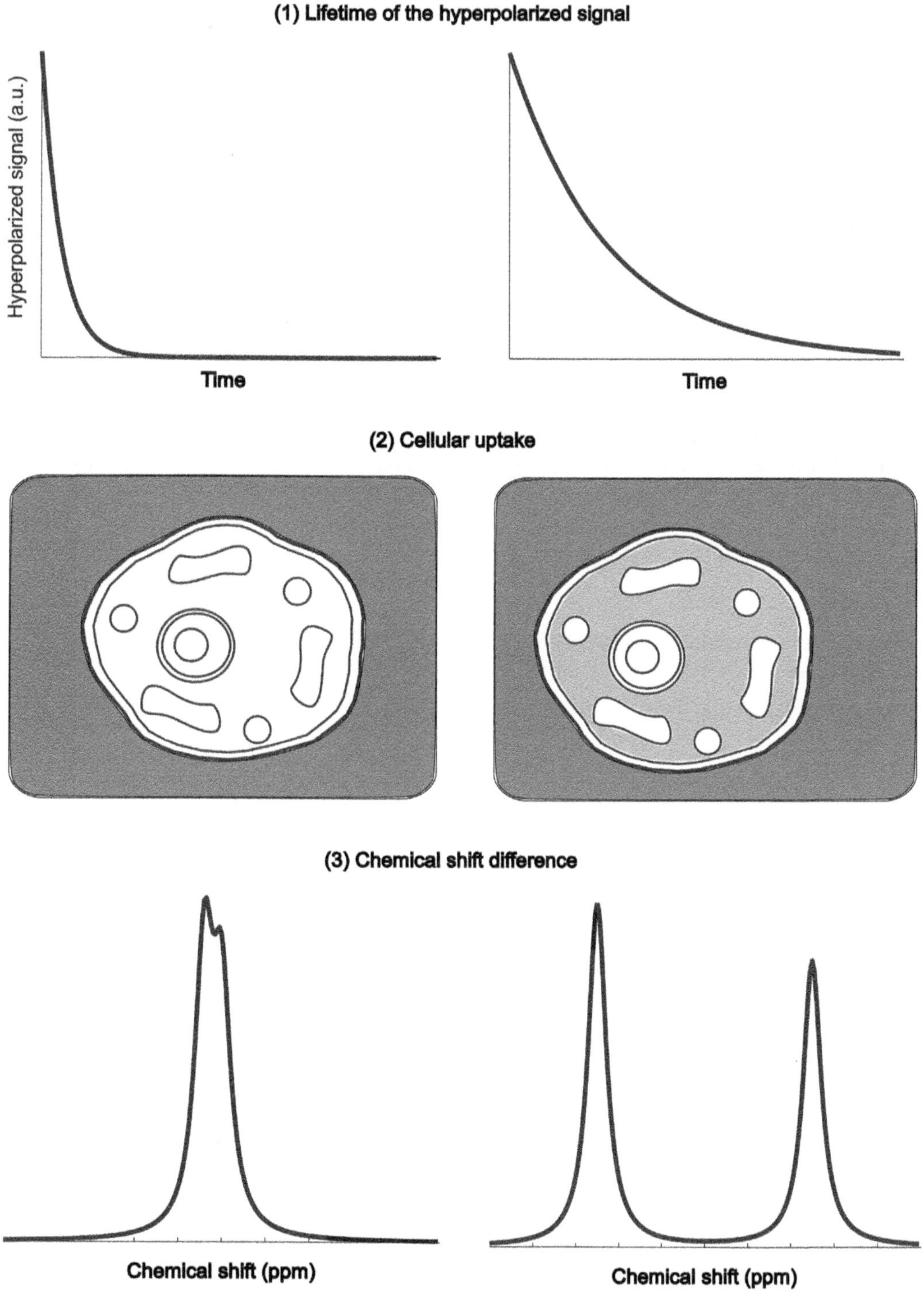

FIGURE 4.1 Main characteristics of good hyperpolarized ^{13}C agents. (1) The ideal agent will relax slowly in vivo, prolonging the lifetime of the hyperpolarized state. (2) Cellular uptake in the target compartment will be rapid and specific, with sufficient enrichment and metabolic conversion to detect products. (3) The chemical shift difference between the probe and its metabolites should be well resolved with the magnetic field homogeneities typically achieved in clinical and preclinical scanners.

4.2.1 Pyruvate metabolism

Pyruvate is the end product of glycolysis, and it is located at an important metabolic crossroads. It has several fates, governed by various cell-intrinsic factors (such as cell type, redox state, ATP content, and the activities of other metabolic

pathways) and extrinsic factors (including oxygen concentration, pH, and nutrient levels). Pyruvate can be reduced to lactate by lactate dehydrogenase (LDH), and it can be transaminated to alanine by alanine aminotransferase (ALT) (Fig. 4.2). Pyruvate can be transported into mitochondria and oxidized to acetyl-CoA via the pyruvate dehydrogenase (PDH) complex, followed by entry into the tricarboxylic acid (TCA) cycle. Alternatively, it can enter the TCA cycle following conversion to oxaloacetate by pyruvate carboxylase (PC).

4.2.2 Lactate dehydrogenase

Lactate is the main hyperpolarized pyruvate metabolite, and its production by LDH has been extensively studied and used for imaging metabolism. A bolus injection of hyperpolarized [1-^{13}C]pyruvate results in the rapid exchange of the ^{13}C label from pyruvate to lactate, with relatively little net conversion of pyruvate to lactate [23]. With sufficient cellular uptake of pyruvate through monocarboxylic acid transporters (MCT1) [24], the amount of lactate that becomes labeled is largely dependent on the endogenous lactate pool size. Endogenous lactate concentrations are typically in the millimolar range and are elevated in many pathologies, whereas the pyruvate pool size is normally submillimolar. Consequently, much of the labeled pyruvate exchanges into the larger lactate pool, and the lactate signals persist with a high signal-to-noise ratio (SNR). Additionally, the 12.5 ppm difference in chemical shift between [1-^{13}C]pyruvate and [1-^{13}C]lactate allows for their reliable detection. All these characteristics provide favorable conditions for imaging the conversion of hyperpolarized [1-^{13}C]pyruvate to [1-^{13}C]lactate.

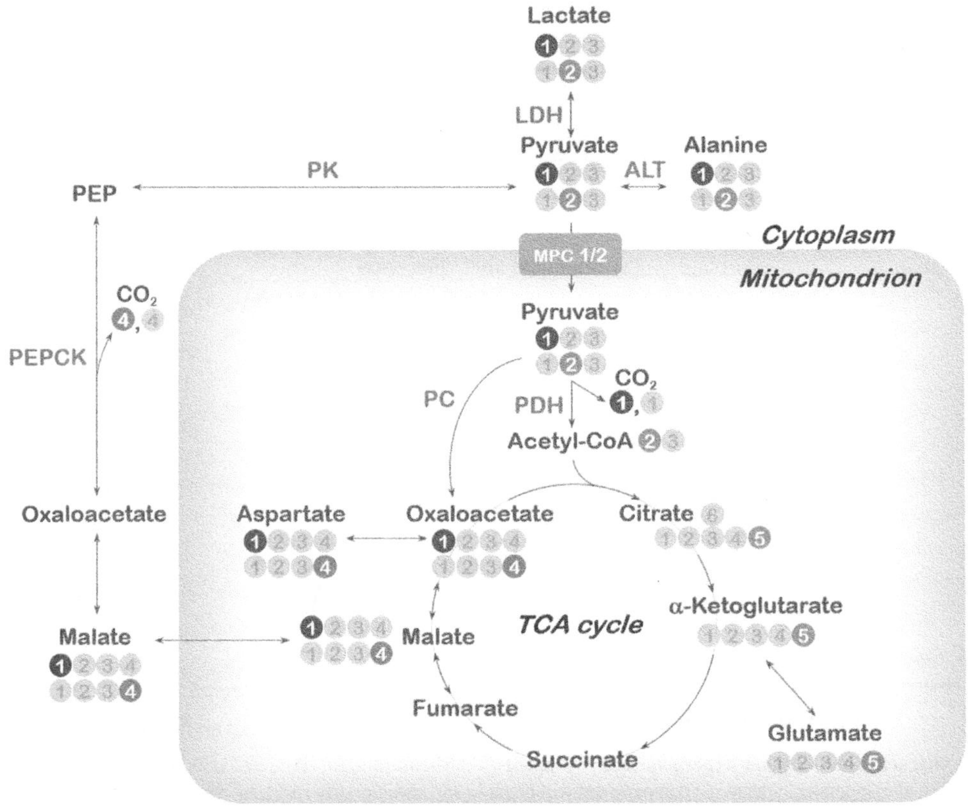

FIGURE 4.2 Metabolic network of hyperpolarized ^{13}C pyruvate. Filled circles represent ^{13}C-enriched carbons, with dark gray circles corresponding to labeling by [1-^{13}C]pyruvate and labeling by [2-^{13}C]pyruvate represented by gray circles. Pyruvate, lactate, and alanine are reversibly interconverted in the cell by lactate dehydrogenase (LDH) and alanine transaminase (ALT). Pyruvate is transported into the mitochondria by the mitochondrial pyruvate carrier (MPC1/2). Oxidation of [1-^{13}C]pyruvate by pyruvate dehydrogenase (PDH) yields labeled carbon dioxide, while PDH oxidation of [2-^{13}C]pyruvate yields [1-^{13}C]acetyl-CoA. Metabolism of [1-^{13}C]acetyl-CoA in the tricarboxylic acid (TCA) cycle yields observable [5-^{13}C]citrate as well as α-keto [5-^{13}C]glutarate, which interconverts with [5-^{13}C]glutamate. The TCA cycle is also fed by pyruvate carboxylase (PC), which converts [1-^{13}C]pyruvate to [1-^{13}C]oxaloacetate. Reversible conversion of [1-^{13}C]oxaloacetate to [1-^{13}C]malate and [1/4-^{13}C]fumarate results in labeling at C4 (light gray), which can lead to labeled carbon dioxide by phosphoenolpyruvate carboxykinase (PEPCK), a key early step of gluconeogenesis.

4.2.3 Metabolic contrast by pyruvate-to-lactate exchange

Endogenous lactate concentration is partially determined by the rate of glycolysis; therefore, the amount of lactate labeling, reflecting the intracellular lactate pool, provides an indirect evaluation of glycolytic activity. Hyperactive glucose metabolism and increased secretion of lactate are considered hallmarks of cancer via the so-called "Warburg Effect" [25]. Golman et al. were the first to demonstrate how hyperpolarized [1-^{13}C]pyruvate to [1-^{13}C]lactate conversion can be used to characterize tumors [26]. Thereafter, many studies using hyperpolarized [1-^{13}C]pyruvate investigated pyruvate-to-lactate exchange as a new metabolic contrast agent for imaging tumor metabolism [27–32]. The first clinical study with hyperpolarized [1-^{13}C]pyruvate underscored the potential diagnostic power of the technique, where pyruvate-to-lactate exchange could detect tumors where other MR modalities failed [17].

Other pathologies are associated with increased activity of LDH-mediated pyruvate exchange. For instance, activated immune cells have increased glucose catabolism [33], and this metabolic shift leads to increased lactate production. Therefore, hyperpolarized [1-^{13}C]pyruvate can potentially enable one to image inflammation via increased [1-^{13}C]lactate labeling [34–36]. PDH activity is inhibited after traumatic brain injury (TBI), with mitochondrial dysfunction starting within 4 h postinjury [37]. Elevated lactate-to-pyruvate ratios were observed in TBI models, and these findings may improve our understanding of the degree of injury sustained [38,39].

One should take caution before directly relating the lactate-to-pyruvate ratio to the glycolytic rate, as other factors may influence this labeling. The rate of [1-^{13}C]lactate formation is dependent on pyruvate delivery and uptake, a function of perfusion and transporter expression, as well as LDH activity, NADH coenzyme concentrations, and pyruvate flux into other competing metabolic pathways, such as transamination, oxidation, and carboxylation.

4.2.4 Transamination

Conversion of pyruvate to alanine results from the transamination reaction between pyruvate and the amino donor glutamate, also yielding α-ketoglutarate. This reaction is catalyzed by ALT/glutamic-pyruvic transaminase and is reliably observed in muscles, kidneys, and most prominently in the liver. A lower normalized liver alanine signal is noted in fasted rats infused with [1-^{13}C]pyruvate [40,41]. This reaction may also be a marker of tissue remodeling in the kidney. Nielsen et al. showed a relationship between elevated hyperpolarized [1-^{13}C]pyruvate-to-[1-^{13}C]alanine conversion and renal fibrosis following ischemia-reperfusion injury in diabetic kidneys [42].

4.2.5 Pyruvate dehydrogenase and carboxylase

Metabolism of [1-^{13}C]pyruvate PDH flux is measured via ^{13}C-labeled CO_2 production and its rapid equilibration with the cellular bicarbonate pool by carbonic anhydrase. Bicarbonate production from hyperpolarized [1-^{13}C]pyruvate by PDH activity was reported in the heart [43,44] and brain [45,46]. Moreover, the equilibrium reaction between $^{13}CO_2$ and ^{13}C-bicarbonate brings the possibility to evaluate intracellular pH by analyzing the ratio between the ^{13}C signals [47,48]. Hyperpolarized aspartate and malate, metabolites downstream of oxaloacetate, the product of PC, can be detected in the liver [49,50], kidney [51], and fasted long-chain acyl-CoA dehydrogenase knockout mouse heart [52]. It was demonstrated that the labeled bicarbonate seen in fasted rat and mouse livers [49,50] is absent after inhibiting or knocking out phosphoenolpyruvate carboxykinase (PEPCK) activity, indicating that it results from gluconeogenic flux through PC and PEPCK rather than PDH. In the fed state, bicarbonate is produced in the liver even after inhibiting PEPCK, consistent with PDH flux [49]. In the kidney, gluconeogenesis takes place in the convoluted proximal tubule, while the distal tubule has predominantly glycolytic metabolism, and flux through both PC and PDH is observed [51]. Unlike the liver, pyruvate-to-bicarbonate conversion in the kidney is mainly due to PDH flux in both fed and fasted states, and the [1-^{13}C]aspartate signal, which increases in the fasted state and is greatly diminished with PEPCK inhibition, is a candidate marker of renal gluconeogenic flux.

4.3 Lactate and alanine as pyruvate precursors

The transformation of hyperpolarized [1-^{13}C]pyruvate to [1-^{13}C]lactate or to [1-^{13}C]alanine enables the measurement of flux or exchange through a single enzymatic step. However, a complementary view of the same metabolic pathways can be attained by starting from hyperpolarized [1-^{13}C]lactate or [1-^{13}C]alanine. The C1 position in lactate and alanine is a carboxyl group and thus has a relatively long T_1 necessary for preserving the hyperpolarized state; nonetheless, polarization levels in the C1 of lactate and alanine are lower than those achieved with pyruvate [53].

4.3.1 Hyperpolarized lactate

Plasma levels of lactate are higher than most other metabolites [54], and even with a concentrated bolus, one can detect hyperpolarized [1-^{13}C]lactate metabolism close to its physiological concentrations. The conversion of hyperpolarized ^{13}C lactate into the unlabeled pyruvate pool provides LDH activity in vivo in different organs [55–57]. Takado et al. demonstrated that the blood-brain barrier (BBB) plays a rate-limiting role in the cerebral metabolism of hyperpolarized [1-^{13}C] lactate [57], with the [1-^{13}C]pyruvate labeling reflecting the amount of lactate that crossed the BBB. It is well established that the lactate pool increases after ischemic stroke, and interestingly, preclinical studies show that providing additional lactate has a neuroprotective effect [58]. Building on that result, the real-time metabolism of hyperpolarized [1-^{13}C]lactate was investigated in a transient focal cerebral ischemia. This study aimed to demonstrate the potential of hyperpolarized lactate as a theranostic contrast agent, providing neuroprotection along with metabolic information [59]. Further metabolism of hyperpolarized lactate to alanine and bicarbonate is also noted in skeletal muscle [55], heart [60], and liver [61], with the expected decrease in PDH flux to bicarbonate under fasting conditions. Additionally, the conversion of lactate to aspartate via PC has been observed in the liver and, as with pyruvate infusion, is increased with fasting.

4.3.2 Hyperpolarized alanine

Hyperpolarized [1-^{13}C]alanine provides another route for measuring pyruvate-to-lactate conversion with the advantage of detecting intracellular [1-^{13}C]pyruvate and [1-^{13}C]lactate ratios that reflect the actual physiological values [62]. Being proportional to free cytosolic [NADH]/[NAD^+], the intracellular lactate-to-pyruvate concentration ratio is a commonly used metric of the redox state. Consequently, the implementation of hyperpolarized [1-^{13}C]alanine for assessing in vivo redox levels was demonstrated in the rat liver [63]. It was recently demonstrated that the differential metabolism of hyperpolarized [1-^{13}C]alanine to either [1-^{13}C]pyruvate or [1-^{13}C]lactate can be employed to distinguish between different telomere-maintenance mechanisms (a universal hallmark of cancer) in both genetically-engineered and patient-derived glioma cell lines and tumors [64]. Although D-amino acids are less abundant than L-amino acids, they do take part in normal mammalian metabolism. Radaelli et al. have recently reported the conversion of hyperpolarized D-[1-^{13}C]alanine to pyruvate, lactate, and bicarbonate in the rat kidney [65], demonstrating the possibility of directly monitoring in vivo D-amino acid oxidase activity.

4.4 Glucose metabolism and related pathways

Because dysregulation of glucose metabolism is a key factor in many disorders, directly monitoring glycolytic flux with hyperpolarized ^{13}C-glucose could be a useful application. However, unlike the carboxyl carbons in pyruvate, lactate, and alanine, the glucose carbons have very short longitudinal relaxation times (T_1), due to the dipole-dipole ^{13}C relaxation by the neighboring protons. Perdeuteration of the molecule can extend the T_1s, allowing the detection of the glycolytic pathway in cells in vitro [66]. This approach has also been employed in vivo, enabling the detection of the pentose phosphate pathway (PPP) intermediate 6PG in a lymphoma mouse model [13] and TERT expression in brain tumors [67]. The detection of the glycolytic intermediate 3PG has become possible in the naïve mouse brain [15] following a bolus of hyperpolarized [$^2H_7$,$^{13}C_6$]glucose. By hyperpolarizing deuterated glucose, specifically ^{13}C labeled at the C3 and C4 positions, the SNR of the experiment increases due to the absence of ^{13}C-^{13}C splitting in the detected metabolic products. The above-mentioned studies [13,15,67] have also demonstrated the possibility of monitoring lactate production shortly after injection of hyperpolarized D-[$^2H_7$,$^{13}C_6$]glucose (Fig. 4.3).

4.4.1 Lactate production from glucose

While lactate formed from hyperpolarized pyruvate relies on cellular uptake via the MCTs and a single enzymatic step (Fig. 4.2), the lactate produced from hyperpolarized ^{13}C-glucose results from 12 biochemical steps, including glucose transport, the 10 enzymatic steps of glycolysis, and LDH (Fig. 4.3). A study performed on mouse models of brain tumors has reported a discrepancy between the pool size of lactate (measured by ^{1}H MRS) and the ^{13}C labeling of lactate from hyperpolarized [$^2H_7$,$^{13}C_6$]glucose [68]. Other work that investigated the effect of different anesthetics on cerebral glucose metabolism demonstrated that, while the conversion of [6,6-$^2H_2$]glucose to [3-^{2}H]lactate measured by thermally-polarized ^{2}H MRS was consistent with the lactate pool size detected by ^{1}H MRS, lactate production from hyperpolarized ^{13}C-glucose is not sensitive to the endogenous pool [6]. Those findings indicate that hyperpolarized glucose metabolism enables monitoring the net ^{13}C lactate production and is related to the cerebral metabolic rate of glycolysis [6,68]. Nonetheless,

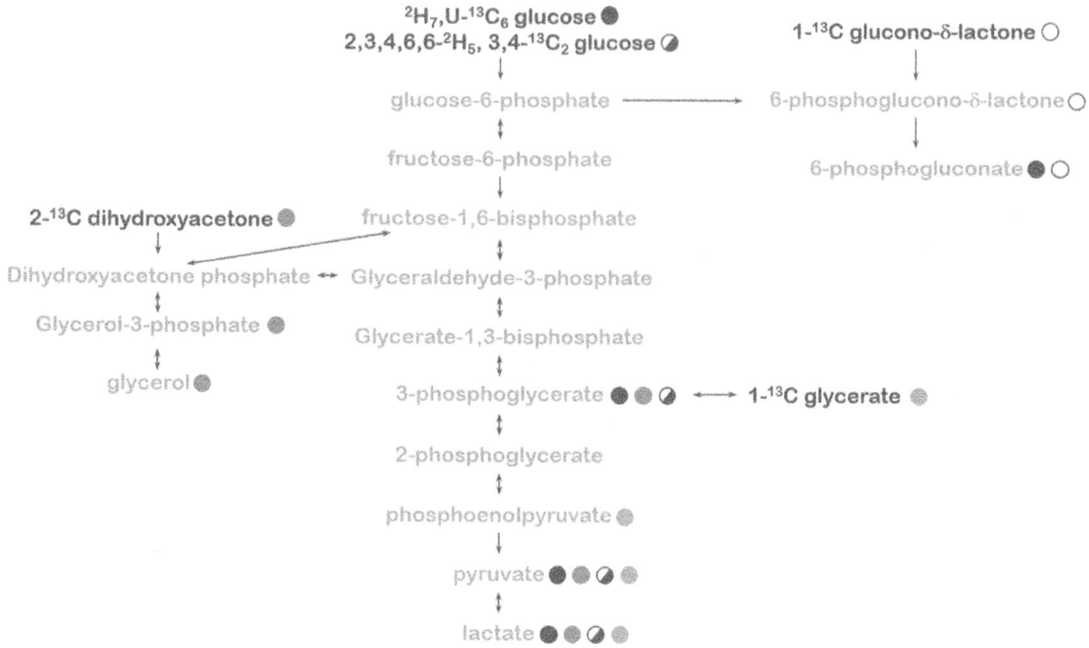

FIGURE 4.3 **Hyperpolarized probes of the glycolytic metabolic network and PPP.** Hyperpolarized agents are in dark gray. In-vivo detectable $^{2}H_{7}$,U-$^{13}C_{6}$ glucose metabolites are noted with dark gray circles; post 2,3,4,6,6-$^{2}H_{5}$, 3,4-$^{13}C_{2}$ glucose with half-filled dark gray; PPP metabolites of 1-^{13}C glucono-δ-lactone with empty circles; 2-^{13}C dihydroxyacetone metabolites with gray circles; and 1-^{13}C glycerate metabolism with light gray circles.

technical challenges need to be overcome before introducing hyperpolarized glucose for clinical diagnostics, in particular the limitations related to the short lifetime of the hyperpolarized state of glucose. Recent developments around photo-induced radicals may open opportunities to reduce the delay between the preparation and injection of hyperpolarized substrates, possibly allowing the translation of hyperpolarized ^{13}C-glucose to clinical applications [69,70].

4.4.2 Other probes to monitor in vivo glycolysis

Glucose metabolism is regulated at multiple points within the glycolytic pathway, and other molecules have been proposed as probes of specific parts of the pathway. Hyperpolarized [1-^{13}C]glycerate was employed to monitor glycolytic activity in the rat liver, where the detected metabolic products, [1-^{13}C]pyruvate and [1-^{13}C]lactate, originated from glycolysis. The levels of production and relative intensities of these metabolites are different between the fasted and fed states, which modulate glycolytic metabolism in the liver [71]. In normal conditions, hydrophilic glycerate should not appreciably cross the BBB, but following BBB opening with focused ultrasound, the metabolism of hyperpolarized [1-^{13}C]glycerate through glycolysis could be detected in the injured area [72]. Another hyperpolarized probe [2-^{13}C]dihydroxyacetone (DHAc) has been shown to report on glucose metabolism in the liver [73–75] and kidney [74]. DHAc is a precursor of dihydroxyacetone phosphate (DHAP), located at the midpoint of glycolysis, and it may probe the delicate balance between glycolysis, gluconeogenesis, and lipid synthesis. The conversion of DHAP to fructose-1,6-bisphosphate and then to glucose could inform on gluconeogenesis. Alternatively, isomerization of DHAP to glyceraldehyde-3-phosphate and conversion to glycerate-1,3-bisphosphate toward pyruvate probe glycolysis, and DHAP is also reduced and dephosphorylated to yield glycerol (Fig. 4.3).

4.4.3 Pentose phosphate pathway

The PPP is required for the synthesis of ribonucleotides and is a major source of nicotinamide adenine dinucleotide phosphate (NADPH). NADPH is necessary for fatty acid synthesis and scavenging reactive oxygen species (ROS), and the PPP is therefore another important metabolic fate of glucose in cancer cells, helping meet anabolic demands and combat oxidative stress. The PPP branches off from glycolysis at glucose-6-phosphate (Fig. 4.3), and hyperpolarized δ-[1-^{13}C]gluconolactone metabolism can be detected in the perfused mouse liver [76]. Its predominant metabolite is 6-phospho-[1-^{13}C]gluconate (6PG), resulting from phosphorylation by gluconokinase and hydrolysis by 6P-gluconolactonase, and the

oxidation of 6PG by 6-phosphogluconate dehydrogenase is detected as [^{13}C]bicarbonate. A twofold higher conversion to bicarbonate is seen following H_2O_2 treatment in glucose- and insulin-perfused livers, indicating a higher PPP flux with oxidative stress.

Hyperpolarized δ-[1-^{13}C]gluconolactone metabolism to 6PG is also detected in cultured glioblastoma cells and is higher in rat orthotopic glioblastoma xenografts compared to tumor-free brain [77]. Additionally, 6PG production is lower in U251 cell tumors following chemotherapeutic treatment with temozolomide, demonstrating the potential of δ-[1-^{13}C] gluconolactone as a probe of tumor metabolism and to monitor treatment response.

4.5 Direct detection of TCA cycle metabolites

The mitochondrial TCA cycle is central to energy homeostasis [78], and it is the final common path for aerobic oxidation of any metabolite that can be transformed into acetyl-CoA. The cycle is also a source of precursors for some amino acids. A difficulty in studying the regulation of TCA-cycle activity in vivo is rooted in the fact that the concentrations of TCA-cycle intermediates are too small to be directly detected using established thermally-polarized methods [79]. Many substrates that feed the TCA cycle can be labeled at sites with a relatively long T_1 and are suitable for hyperpolarization, enabling the directional detection of some TCA cycle intermediates.

4.5.1 Carbohydrate metabolism

As noted earlier, sugars are generally not ideal substrates for hyperpolarization due to their relatively short T_1s; however, pyruvate, as the product of glycolysis, can be used instead to probe the downstream pathways of carbohydrate metabolism. While labeling of ^{13}C-bicarbonate from [1-^{13}C]pyruvate can in many cases be related to PDH flux and TCA-cycle activity, [2-^{13}C]pyruvate is metabolized by PDH to [1-^{13}C]acetyl-CoA and directly enters the TCA cycle or is buffered as [1-^{13}C] acetylcarnitine. The detectable hyperpolarized [2-^{13}C]pyruvate metabolites in the rat heart include the TCA cycle intermediate [5-^{13}C]citrate and [5-^{13}C]glutamate labeled via TCA-cycle activity [80]. The brain metabolism of hyperpolarized [2-^{13}C]pyruvate has been investigated under several different anesthetic conditions, and the formation of the TCA-cycle intermediate α-keto[5-^{13}C]glutarate was observed under a light anesthetic condition [81]. The conversion of hyperpolarized [2-^{13}C]pyruvate to [5-^{13}C]glutamate was also demonstrated in the human brain [82]. The [5-^{13}C]glutamate and α-keto[5-^{13}C]glutarate resonances vary in chemical shift by only 0.02 ppm [83] and cannot be distinguished in vivo on the basis of labeling at this site alone. With glutamate being present in tissue at much higher concentrations, it is generally assumed that [5-^{13}C]glutamate produced by transamination of α-keto[5-^{13}C]glutarate is responsible for much of this signal.

4.5.2 Fatty acid metabolism

Fatty acids are important energy substrates in several major organs and are broken down by β-oxidation to acetyl-CoA units that are further oxidized in the TCA cycle. The carboxyl carbon in various small-to-medium-chain fatty acids can be hyperpolarized to a level sufficient to detect TCA cycle activity in vivo. The heart is normally fueled primarily by fatty acids, and it is the main organ that has been used to study hyperpolarized fatty acid metabolism by TCA-cycle activity. As with [2-^{13}C]pyruvate, the most prominent hyperpolarized fatty acid metabolites downstream of [1-^{13}C]acetyl-CoA are [1-^{13}C]acetylcarnitine, [5-^{13}C]citrate, and [5-^{13}C]glutamate. [1-^{13}C]Acetate, the simplest fatty acid, is readily converted to acetylcarnitine in the liver, skeletal muscle, heart, and kidney [84–86]. The dynamics of acetyl-CoA synthetase activity were quantified in skeletal muscle by the conversion of hyperpolarized [1-^{13}C]acetate to [1-^{13}C]acetylcarnitine [87]. In the heart, the time evolution of [5-^{13}C]citrate production was used to estimate TCA-cycle flux [88]. A more complex view of myocardial metabolism could be probed with the longer-chain fatty acids [1-^{13}C]butyrate [89,90] and [1-^{13}C]octanoate [91], which require β-oxidation to generate [1-^{13}C]acetyl-CoA. The carboxylate carbons of butyrate and octanoate have a higher chemical shift than that of acetate and interfere less with the detection of the adjacent [5-^{13}C]glutamate metabolite peak. However, they have shorter in vivo T_1s than acetate, which limits their sensitivity and suitability for translation into clinical applications.

4.5.3 Acetate in brain

Although fatty acids are not the brain's main fuel, acetate is an interesting substrate for studying cerebral metabolism due to its preferential uptake by glial cells [92,93]. Acetate enters the brain via the MCT transporters, and it has been employed to investigate glial TCA-cycle metabolism using thermally polarized ^{13}C MRS [94,95]. Similarly,

hyperpolarized [1-^{13}C]acetate was used to interrogate the glial TCA cycle and led to the first observation in the brain of a TCA cycle intermediate, α-keto[5-^{13}C]glutarate [14], a metabolite related to both energy metabolism and neuronal activity. Later work has demonstrated that partial inhibition of TCA-cycle activity results in a reduction of the α-ketoglutarate signal [96].

4.5.4 Ketone bodies

Ketone-body catabolism generates acetyl-CoA units that can be oxidized within the TCA cycle, and it is influenced by nutritional and disease states. The ketone bodies acetoacetate and β-hydroxybutyrate have been proposed as hyperpolarized ^{13}C agents. Both C1 and C3 carboxyl carbons of acetoacetate can be hyperpolarized; however, it has low chemical stability and must be prepared on-site prior to use [97–99]. The more stable β-hydroxy[1-^{13}C]butyrate shows lower polarization levels and a lower extent of metabolism in vivo [98]. Myocardial hyperpolarized [1-^{13}C]acetoacetate metabolism is changed between the fed and fasted states, resulting in decreased conversion to [1-^{13}C]acetylcarnitine and increased [5-^{13}C]glutamate with fasting [98]. Hyperpolarized [3-^{13}C]acetoacetate was employed to monitor the effect of a ketogenic diet on cardiac ketone-body metabolism [100]. Acetoacetate and β-hydroxybutyrate are a redox pair, and the formation of β-hydroxybutyrate from acetoacetate depends on the mitochondrial NADH/NAD^{+} ratio. As such, the reduction of hyperpolarized acetoacetate to β-hydroxybutyrate, a redox marker, has also been studied in a diabetic rat model [97], in perfused hearts [101,102], kidneys [99], naïve brains, and U87 tumors [103].

4.6 Nitrogen balance and amino-acid metabolism

Nitrogen is critical for all living organisms, as it is an essential component of proteins and nucleic acids. In mammals, ammonia, produced mainly from amino-acid degradation, is detoxified and excreted as urea. Glutamine degradation releases ammonia, and its synthesis from glutamate conversely consumes ammonia. Studies of cerebral hyperpolarized glutamate metabolism show that hyperpolarized [1-^{13}C]glutamate conversion to [1-^{13}C]glutamine can be detected after temporary BBB disruption in the rat brain, possibly providing an alternative means to interrogate glutamate–glutamine synthesis [104].

4.6.1 Glutamine metabolism in tumors

With the high consumption of glutamine by some tumors, glutaminolysis (degradation of glutamine) is considered a hallmark of tumor cell metabolism [105], and several groups have investigated hyperpolarized glutamine to monitor tumor metabolism [106–109]. Hyperpolarized [5-^{13}C]glutamine conversion to [5-^{13}C]glutamate is seen in rats bearing McA-RH7777 hepatoma cell tumors, but not in control rats [110]. Hyperpolarized [1-^{13}C]glutamine has been used to noninvasively measure the production of glutamine-derived 2-hydroxy[1-^{13}C]glutarate (2-HG) [111], a signature oncometabolite that accumulates as a result of gain-of-function mutations in isocitrate dehydrogenase (IDH) and a decreased activity of branched-chain amino acid transaminase (BCAT). While the preclinical findings with hyperpolarized glutamine are promising, the main obstacles for clinical application are the instability of glutamine in solution, particularly its cyclization to pyroglutamate [110], and the relatively slow cellular uptake of glutamine, even in tumor cells [112]. Additionally, since glutamine is the most abundant amino acid in the bloodstream, the apparent metabolism of labeled glutamine is limited by low fractional enrichment. In glioma tumors, hyperpolarized α-keto[1-^{13}C]glutarate is also able to probe 2-HG production resulting from the IDH1 mutation [113], along with lower conversion to [1-^{13}C]glutamate by transamination [114].

4.6.2 KIC transamination

Like the interconversion of pyruvate and alanine by transamination, α-ketoisocaproic acid (KIC) is transaminated to leucine by BCAT, with the concomitant conversion of glutamine to α-ketoglutarate. In the brain, BCAT has an important role in nitrogen shuttling and glutamate metabolism. Like pyruvic acid, neat KIC readily glasses when frozen rapidly, and it polarizes well without dilution. Butt et al. [115] demonstrated that hyperpolarized [1-^{13}C]KIC rapidly enters the brain and is metabolized to [1-^{13}C]leucine by BCAT. As this reaction also requires glutamate, this probe has the potential for evaluating cerebral BCAT activity in conjunction with neurological disease (Fig. 4.4). Higher levels of isoflurane anesthesia cause vasodilation and increase the brain KIC signal, while the leucine signal remains unchanged, suggesting that cellular uptake is the limiting factor in these conditions [116]. Increased BCAT activity in EL4 lymphoma xenografts is

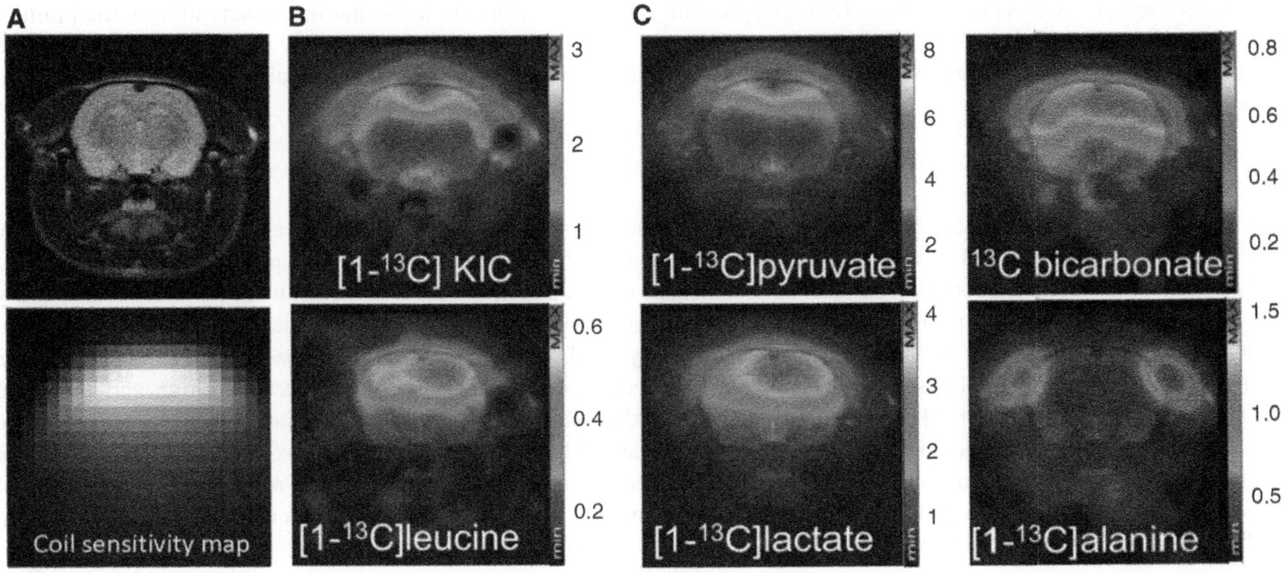

FIGURE 4.4 Imaging of branched chain amino acid transaminase (BCAT) activity in vivo *in rat brain*. (A) Anatomical proton image and the surface coil sensitivity profile are shown. (B) Metabolic maps of [1-^{3}C]KIC and [1-^{13}C]leucine postbolus of hyperpolarized [1-^{13}C]KIC injection. Hyperpolarized [1-^{13}C]leucine is specifically observed inside the brain. (C) Hyperpolarized [1-^{13}C]pyruvate was used as a reference substrate, and its metabolic map is shown together with [1-^{13}C]lactate, [1-^{13}C]bicarbonate, and [1-^{13}C]alanine. *Reproduced with permission from [115].*

detectable by the conversion of hyperpolarized KIC to leucine, which was not apparent in surrounding tissue or in R3230 adenocarcinoma xenografts, which have much lower BCAT activity [117].

4.6.3 KIC oxidation and reduction

Oxidation of hyperpolarized KIC by the branched-chain α-ketoacid dehydrogenase complex (BCKDC) is detectable in a rat glioma model tumor but not in a normal-appearing brain [118]. Distinguishing brain tumors based on the KIC-metabolite bicarbonate/leucine ratio or similar metrics may be a particularly compelling feature for this probe. KIC metabolism by BCAA and BCKDC is also seen in the mouse liver, and mice with carcinogen-induced tumors show decreased flux through BCKDC [119]. KIC is additionally a substrate of LDH-X, a version of the enzyme expressed only in the testes. The limited normal expression of LDH-X makes it suitable for use as a reporter gene, with its conversion of hyperpolarized KIC to α-hydroxyisocaproate as a readout [120].

4.7 Hyperpolarized probes of other enzymes

While the majority of work with hyperpolarized probes has been focused on energy metabolism, a number of substrates of other enzymes have been described, and they may be useful for more specialized applications.

4.7.1 α-KB

α-ketobutyrate (α-KB) is structurally similar to pyruvate, containing a 2-carbon ethyl chain instead of a methyl group. α-KB is also a substrate of LDH, but the isoforms of LDH show different kinetic properties. α-KB is a better substrate of LDH-5 (LDHB homotetramer), the form predominant in the heart, and higher conversion to α-hydroxybutyrate is seen in the rat heart as compared to the liver and kidney [121].

4.7.2 γ-Glutamyl(1-^{13}C)glycine

γ-Glutamyltransferase (GGT) is an enzyme expressed on the cell surface that plays an important role in facilitating intracellular glutathione production. GGT is highly expressed on the luminal surface of renal proximal tubules and is overexpressed in some tumors, where it plays a role in combating oxidative stress. The hyperpolarized substrate γ-glutamyl(1-^{13}C)glycine is rapidly converted by renal GGT to [1-^{13}C]glycine at a high level (>10%) [122], and it can

detect increased brain tumor GGT activity in a glioma model [123]. GGT and related enzymes are attractive targets as infused hyperpolarized substrates are not subject to the potentially rate-limiting effect of cellular uptake.

4.7.3 Substrate esters

With transport into cells being, in many instances, a factor limiting the access of the substrate to the target enzyme, esterification has been shown in a number of instances to improve the uptake and metabolism of hyperpolarized probes. Esterification, a strategy commonly used in prodrugs, increases the lipophilicity and cell permeability of polar compounds bearing organic acids, and ester functionalities are generally labile in cells such that the active form is rapidly generated upon entry. Hyperpolarized ethyl (1-^{13}C)pyruvate was tested as a precursor to pyruvate with better permeability across the BBB. In the brain, it shows greater conversion to lactate compared to unesterified pyruvate [124,125], and it may be useful in cases where the highest enhancement of brain pyruvate metabolite signals is needed. Similarly, hyperpolarized ethyl L-(1-^{13}C)alanine shows higher conversion to lactate in the liver compared to unesterified alanine. Additionally, esterification lowers the pK_a of alanine's amino group to 8, and pH-dependent changes in chemical shifts associated with intracellular and extracellular compartments can be resolved [126]. Esterification of α-keto[1,2-$^{13}C_2$]glutarate also improves its liver uptake, such that its oxidation in vivo by α-ketoglutarate dehydrogenase is detectable [127]. Ethyl [1,3-$^{13}C_2$]acetoacetate is a chemically stable precursor to [1,3-$^{13}C_2$]acetoacetate, and its deesterification reaction may itself provide useful metabolic information. Whereas it is rapidly converted to [1,3-$^{13}C_2$]acetoacetate in the healthy liver, this reaction is significantly slower in the tumors of rats bearing Morris hepatocellular carcinoma [128].

4.8 Assessment of chemotherapy treatment response and detecting necrosis

The ability to monitor tumor cell death following treatment is an essential clinical tool for monitoring disease progression and predicting treatment outcomes. Studies with hyperpolarized [1-^{13}C]pyruvate have demonstrated reduced LDH-flux to lactate in a mouse xenograft EL4 tumor model within 24 h of chemotherapy [28]. Lodi et al. compared the treatment response in MCF-7 breast cancer cells and PC3 prostate cancer cells. Interestingly, while the breast cancer cells showed a decrease in the level of lactate upon treatment, the prostate cancer cells showed an increased level of lactate generation [129]. The authors attributed the decreased lactate production in the MCF-7 cells to the downregulation of MCT1 expression and lower pyruvate uptake. These findings suggest the possible application of decreased hyperpolarized [1-^{13}C] pyruvate to lactate conversion as a marker of tumor treatment response. The increased labeled lactate production in PC3 cells following treatment demonstrates that the response is not the same in every tumor type, and one must bear in mind the multiple factors that can influence hyperpolarized pyruvate metabolism, including uptake, LDH expression, and redox state.

Alternatively, hyperpolarized [1,4-$^{13}C_2$]fumarate has been proposed as a positive hyperpolarized contrast agent of cellular necrosis. The conversion of [1,4-$^{13}C_2$]fumarate to malate by fumarase in the cytosol and mitochondria is greatly enhanced in necrotic tissue. In normal conditions, little or no malate production from hyperpolarized [1,4-$^{13}C_2$]fumarate is detectable in vivo due to the limited transport of fumarate across the cell and mitochondrial membranes relative to the lifetime of the hyperpolarized signal. However, in the case of cellular necrosis, with the loss of plasma- and mitochondrial-membrane integrity, fumarate can more easily enter and access fumarase. The hydration reaction of fumarase greatly favors malate formation, and it does not depend on other metabolic processes. It can therefore proceed in a dead necrotic cell, and the elevated appearance of malate indicates cell death. It was demonstrated as a detector of cell death in tumors following chemotherapy [130,131] (Fig. 4.5), to differentiate between kidney pathologies in animal disease models [132,133], to imaging myocardial necrosis [134] (Fig. 4.6) and to detect necrosis in the liver following treatment with carbon tetrachloride [135]. The medically relevant information provided by hyperpolarized fumarate, combined with the relatively high SNR of the produced malate, makes it a promising probe to be investigated for clinical applications.

4.9 pH sensors

Acidity, measured as pH ($-\log_{10}[H^+]$), has a broad influence on a variety of biochemical processes. pH is a tightly regulated physiological parameter, particularly in the intracellular environment, and it varies between different compartments. The lower pH characteristic of the tumor extracellular environment and the association of lower tumor pH values with poorer prognosis and increased invasiveness [136,137] provide a strong rationale for the development of hyperpolarized tumor pH imaging to complement metabolic imaging with pyruvate or other agents. More generally, minimally-invasive methods to image pH in organs and tissue are also useful in studies of physiology and metabolism.

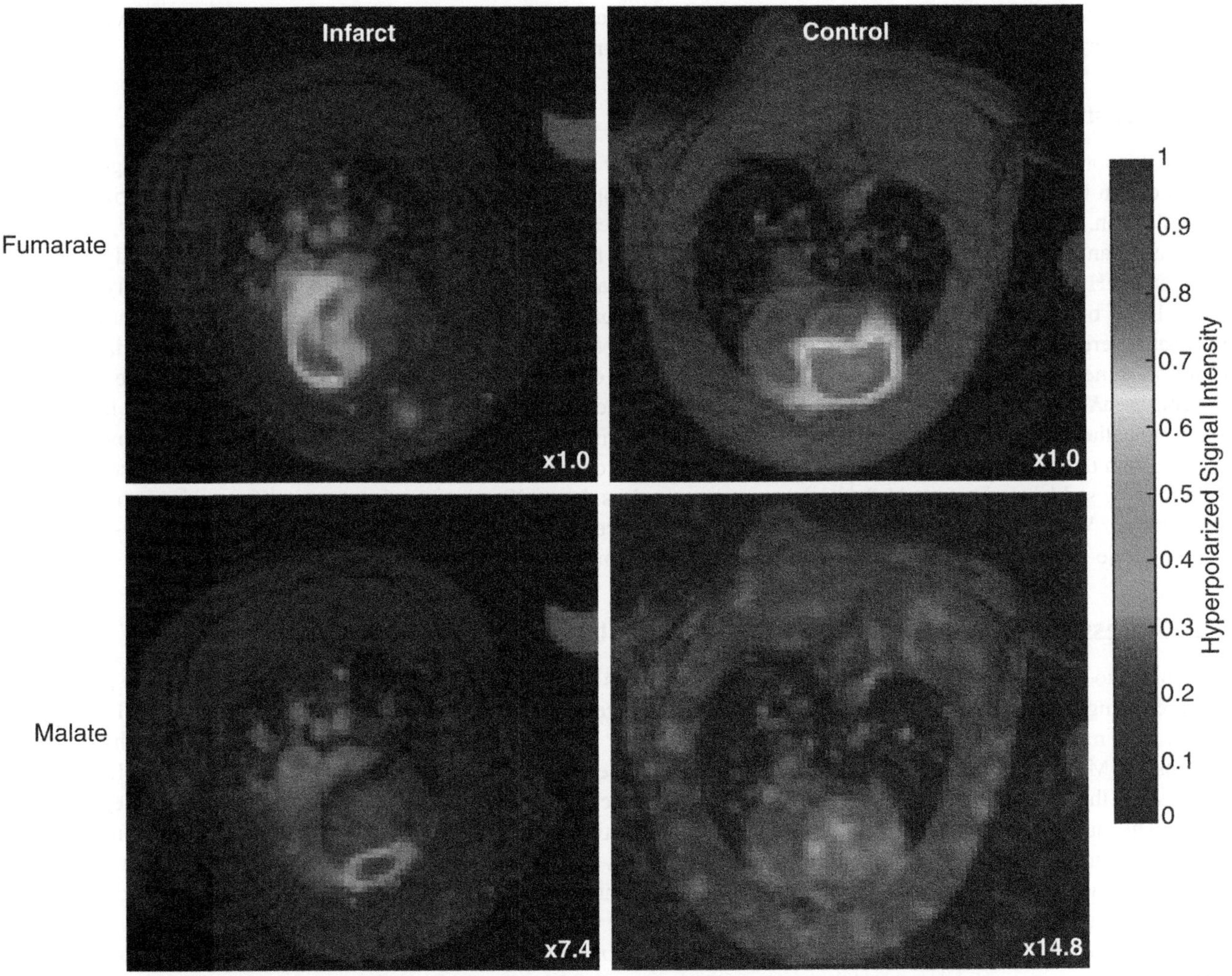

FIGURE 4.5 Monitoring cell death with hyperpolarized fumarate in infarcted heart. Hyperpolarized [1,4-$^{13}C_2$]fumarate metabolism to [1,4-$^{13}C_2$] malate is visible only in myocardial-infarcted animals. *Reproduced with permission from [134].*

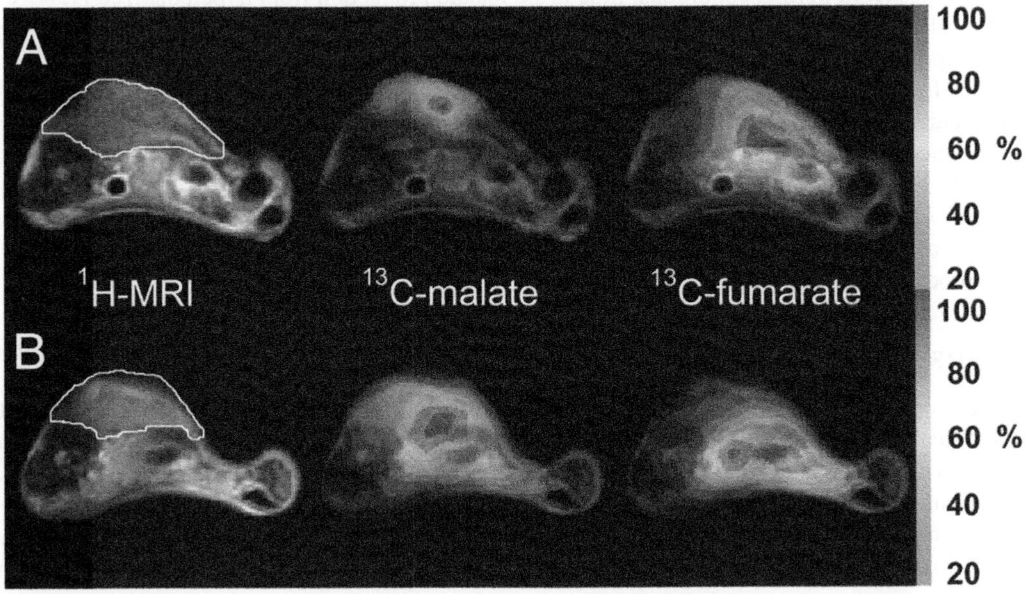

FIGURE 4.6 Monitoring cell necrosis and treatment response with hyperpolarized fumarate in lymphoma tumors. Hyperpolarized [1,4-$^{13}C_2$] fumarate to [1,4-$^{13}C_2$]malate conversion in untreated (A) and etoposide-treated (B) mice with subcutaneously implanted lymphoma tumors. A large increase in the malate signal can be observed after treatment. *Reproduced with permission from [130].*

Several MR methods are used to measure tissue pH, including ones based on ^{31}P MRS, chemical-exchange saturation transfer, and contrast agents with pH-dependent relaxivity effects [138,139]. These methods can complement each other by targeting different compartments. Additionally, a number of hyperpolarized ^{13}C-based pH sensors have been reported. Since the short-lived hyperpolarized signal both enables and requires the rapid acquisition of data, these methods allow for short scan times and the possibility of acquiring dynamic pH maps with good temporal resolution.

4.9.1 Bicarbonate

Most tissue pH measurements using hyperpolarized ^{13}C have thus far used ^{13}C-bicarbonate and its pH-dependent equilibrium with carbonic acid and carbon dioxide. The large 35 ppm chemical shift difference between ^{13}C-bicarbonate and $^{13}CO_2$ allows both species to be detected, and the pH can be calculated using the ratio of their intensities and the Henderson-Hasselbalch equation ($pH = pK_a + \log_{10}(Bic/CO_2)$). The carbonic acid p$K_a$ varies with ionic strength and protein concentration [140], but a value of 6.17 is typically used for in vivo experiments [107,108]. The proof of principle for hyperpolarized ^{13}C-bicarbonate to measure lower extracellular pH in tumors was demonstrated in a mouse EL4 xenograft model [107,108], along with the need for carbonic anhydrase activity to promote the reversible hydration of CO_2 to ensure that the CO_2/bicarbonate ratio reflects their equilibrium values. Bicarbonate has also been used to image tumors in a TRAMP mouse prostate cancer model and show the relationship between lower tumor pH and higher tumor grade [141]. This latter study relied on polarizing a precursor, 1,2-glycerol (^{13}C)carbonate, followed by hydrolysis, in order to generate a more concentrated bicarbonate solution with a good polarization level [142]. Sodium bicarbonate has limited solubility in water and glassing solvents such as glycerol, and a variety of methods have been described to generate polarized sodium ^{13}C-bicarbonate suitable for imaging. These include polarizing the more soluble cesium bicarbonate, followed by ion exchange [107,108], the oxidation of polarized ^{13}C-pyruvate with hydrogen peroxide (H_2O_2) [143], and the enzymatic hydrolysis of ethyl acetyl (^{13}C)carbonate (^{13}C-EAC) [144]. Additionally, in tissues with a high level of pyruvate dehydrogenase (PDH) flux, such as the heart, it is possible to detect both hyperpolarized ^{13}C-bicarbonate and $^{13}CO_2$ resulting from the metabolism of [1-^{13}C]pyruvate [145]. These signals have been used to image intracellular pH in the rat heart [47,48], and show a decrease in tissue pH associated with necrosis following an ischemic injury [146].

4.9.2 Other hyperpolarized MR pH probes

While most preclinical work to image tissue pH using hyperpolarized ^{13}C has used bicarbonate/CO_2, a number of other pH-sensitive methods have been reported, and some of them have been demonstrated to work in vivo. Unlike the ratiometric approach typically used with bicarbonate, all of these probes measure pH by the change in chemical shift. This has some potential advantages in having the pH indicated by a single signal, but it requires an acquisition scheme sensitive to a range of chemical shifts rather than to two specific frequencies.

4.9.2.1 Zymonic acid

(1,5-$^{13}C_2$)Zymonic acid is perhaps the best-studied of these alternative probes. It occurs naturally in yeast from the reaction of two molecules of pyruvate, and the labeled version is prepared from (1-^{13}C)pyruvic acid [147]. It has two ionizable groups with pK_a values of 2.35 and 6.90. Between pH 5 and pH 8, the C1 chemical shift changes by 5.13 ppm, while the C5 chemical shift changes by 2.57 ppm. It has been used to image tumor pH (Fig. 4.7), and it can also detect multiple compartments with different pH values in the kidney. Zymonic acid C1 and C5 have longer T_1s than (^{13}C)bicarbonate, and they are extended 39% by deuteration [148].

4.9.2.2 Amino acids and derivatives as pH sensors

Ethyl esterification of L-(1-^{13}C)alanine to improve its hepatic uptake (also discussed in section 7.3) is noted to lower the α amine pK_a, and the liver signal corresponding to pH 7.4 was assigned to extracellular blood, while a signal at pH 7.0 was assigned to the liver intracellular compartment. Similarly, esterified (1-^{13}C)lysine with a pK_a of 7.18 has been proposed as a pH sensor [149].

Amidation of amino acids has also been described as a strategy to lower α-amine pK_a into the physiologically useful range. Serine amide, with a pK_a of 7.35 (at 25°C), may be a particularly suitable candidate [150]. Amidation is potentially disadvantageous as quadrupolar relaxation of a directly-bonded ^{14}N results in a shortened ^{13}C T_2 [151], which limits spatial

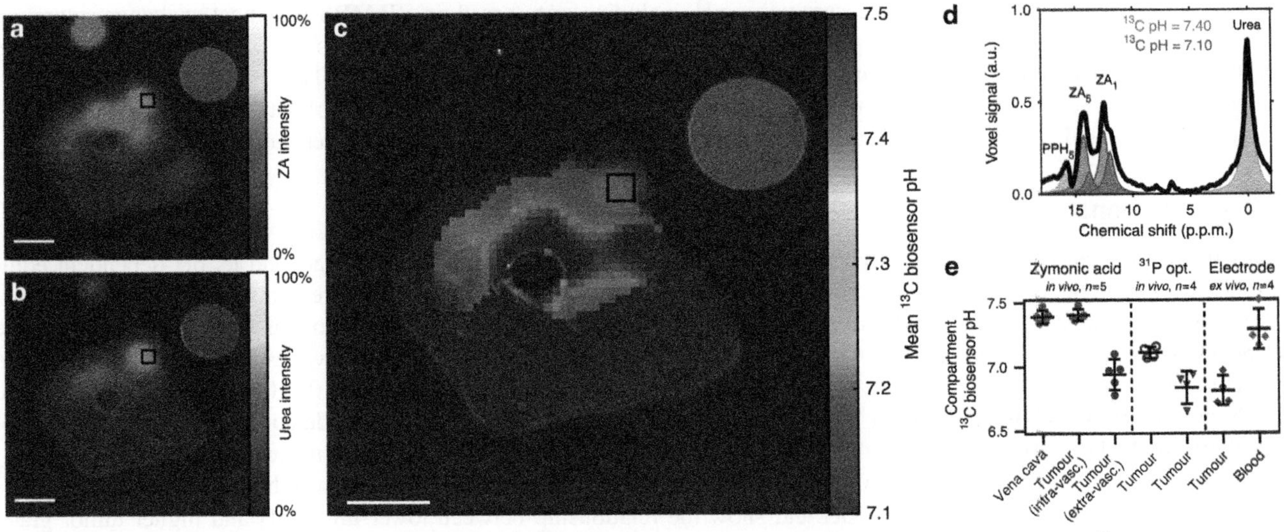

FIGURE 4.7 Imaging tumor pH with hyperpolarized ^{13}C-zymonic acid. Representative tumor data from a hyperpolarized ^{13}C measurement (colored) in an axial slice overlayed on anatomical proton images (greyscale). A calibration phantom containing ^{13}C urea and the catheter used for injection are visible. Both (A) zymonic acid (ZA) and (B) urea accumulate in the MAT B III tumor. (C) The mean pH map shows a lower pH value in the tumor compared to the surrounding tissue. For all ^{13}C images, intensity windows are based on sufficiently high signal levels for either (intensity images) or both (pH images) ZA and urea. Spectra from tumor voxels (D, shown for one representative animal) are best fitted with two pairs of ZA peaks (for ZA_5 and ZA_1) and consistently show different pH values in the tumor in comparison to the vena cava, demonstrated in five animals (E, individual data points and mean ± s.d.) and compared with three other interstitial/extracellular pH measurements in four animals, namely ^{31}P MRS in vivo using 3-APP, a needle-type optical sensor in vivo, and an ex vivo tissue electrode pH measurement. Scale bar, 1 cm. *Reproduced from [147].*

resolution in imaging sequences, and substitution with ^{15}N results in scalar coupling that splits the ^{13}C resonance into a doublet. Some amino acids with α amino pK_as in the physiological range, such as 2,3-diaminopropionic acid, may therefore provide an alternative without the disadvantages resulting from amidation.

More recently, hyperpolarized L-(1-^{13}C)alaninamide was shown to function as an in vivo sensor of pH, aminopeptidase-N activity, and also dissolved CO_2 [152]. With a pK_a of 7.9 at 37°C, it is sensitive to pH values as low as 6, and it readily enters cells to probe pH in both intracellular and extracellular compartments. L-(1-^{13}C)alaninamide also provides sufficient spectral and spatial resolution in the rat kidney to detect at least three compartments with different pH, with an acidic compartment (pH 6.58 ± 0.05) concentrated in the inner medulla that does not correspond to the urine pH. A small but significant increase in the pH of this compartment is seen following treatment with the diuretic acetazolamide, and its increase relative to the other compartments demonstrate the potential for the application of pH-sensitive hyperpolarized probes in studies of renal physiology.

4.9.2.3 Other pH sensors

A hyperpolarized ^{13}C pH sensor based on the Good buffer ACES (N-(2-acetamido)-2-aminoethanesulfonic acid) with a pK_a of 6.58 at 37°C has been described, and its imaging performance has been demonstrated in vitro [153]. Another class of pH-sensitive ^{13}C probes are the malonate derivatives (2-^{13}C,d_{10})diethylmalonic acid and (2-^{13}C,d_4)cyclopropane-1,1-dicarboxylic acid [154]. Between pH 6.5 and 7.4, their C2 chemical shift changes by 1.66 and 2.99 ppm, respectively. Being labeled at quaternary carbons, their ^{13}C T_1s are >70s at 11.7T and appear to be less affected by chemical shift anisotropy, making them particularly well suited to high-field applications.

4.10 Redox

Changes in the balance of reduced and oxidized metabolites in particular organs or compartments are associated with various disease states, and the oxidative stress and upregulation of antioxidant pathways seen in many tumors make noninvasive methods of imaging tissue redox states particularly relevant. Several hyperpolarized probes targeting important redox pathways have been described, and the redox state also influences the metabolism of more common probes such as pyruvate. ROS are important participants in redox reactions, and probes of H_2O_2 are discussed further in section 11.1.

4.10.1 DHA/AA

Ascorbic acid, or vitamin C, is an important physiological antioxidant, and it is regenerated from dehydroascorbate (DHA), its oxidized form, by NADPH- or glutathione-dependent enzymes, including thioredoxin reductase and glutathione dehydrogenase. DHA can enter cells via glucose transporters and is taken up more rapidly compared to vitamin C, and this feature, along with the higher intracellular abundance of vitamin C, results in the substantial conversion of infused hyperpolarized [1-^{13}C]DHA to vitamin C in the mouse brain, liver, and kidney [155]. By contrast, little to no conversion of hyperpolarized [1-^{13}C]vitamin C is seen in vivo [155,156]. Conversion of DHA to vitamin C is seen in xenografts [156], and higher conversion is noted in mouse prostate tumors compared to normal prostate [155]. In the rat brain, the conversion of hyperpolarized DHA to vitamin C is decreased following depletion of GSH by treatment with diethyl maleate, demonstrating its sensitivity to glutathione redox conditions. DHA is also sensitive to oxidative stress in the liver in a rat model of steatohepatitis [157]. Bolus infusions of DHA can trigger strong physiological reactions, including elevated blood pressure and apnea, but pretreatment with DHA moderates these effects.

N-Acetyl(1-^{13}C)cysteine (NAC) is also a probe of the glutathione redox system via the formation of NAC-GSH mixed disulfide [158], presumably through thiol-disulfide exchange with the oxidized form, GSSG. While no NAC-GSH formation was apparent in healthy major organs, most of the infused hyperpolarized NAC was rapidly converted to the mixed disulfide in mouse xenograft tumors, suggesting high GSSG levels and/or oxidizing conditions in the compartments probed.

4.10.2 Oxidoreductases and NADH/NAD$^+$ ratio

Oxidoreductases such as LDH and β-hydroxybutyrate dehydrogenase transfer a hydride equivalent to the nicotinamide coenzyme NAD$^+$, and the equilibrium values of the lactate/pyruvate and β-hydroxybutyrate/acetoacetate ratios depend on the NADH/NAD$^+$ ratio. The conversion of these metabolites is therefore influenced by the NAD$^+$ redox state, and hyperpolarized substrates converted to intracellular pyruvate and lactate provide a measure of the cytoplasmic NADH/NAD$^+$ ratio, while acetoacetate reduction by β-hydroxybutyrate dehydrogenase reflects the redox state in the mitochondrion. As mentioned in section 4.2, the pyruvate/lactate signals derived from hyperpolarized [1-^{13}C]alanine in the liver have been used to detect changes in NADH levels. Hyperpolarized [$^{13}C_6$,$^{2}H_7$]glucose can similarly serve as a pyruvate and lactate precursor, and differences in the pyruvate and lactate signals in PC3 and MCF7 cells can be related to their different NADH/NAD$^+$ ratios and rates of lactate efflux [159]. While using infused hyperpolarized pyruvate does not allow the intracellular pyruvate signal to be measured, its conversion to lactate has been related to perturbations that increase NADH levels. Accordingly, ethanol treatment results in increased pyruvate-to-lactate conversion in the liver [116], and metformin treatment has similar effects in the kidney [160]. Also, as noted in section 5.4, changes in the conversion of hyperpolarized [1,3-$^{13}C_2$]acetoacetate to β-hydroxybutyrate have been related to the NADH/NAD$^+$ redox balance in vivo.

4.11 Other chemical probes

While pH probes are likely the most promising sensors using hyperpolarized ^{13}C, a number of other chemical sensors have been described in recent years.

4.11.1 Peroxide

The medical and biological relevance of ROS as signaling molecules and markers of oxidative stress has motivated the development of hyperpolarized sensors of H_2O_2. α-Ketoacids are oxidatively decarboxylated by peroxides, and ^{13}C-labeled benzoylformic acid (BFA) has been proposed as a sensor of H_2O_2 [161]. BFA is converted to benzoic acid with good specificity by H_2O_2 and is sensitive to peroxide levels as low as 100 μM in vitro with NMR detection, which approaches the highest concentrations that may be found in vivo [162]. An alternative peroxide probe, ^{13}C-thiourea, is selectively converted to thiourea dioxide by peroxide, and oxidation products of hyperpolarized ^{13}C-thiourea can be detected in vivo when coinjected intraperitoneally with H_2O_2 [163]. A trimethylphenylammonium (TMPA)-based hyperpolarized ^{15}N probe bearing an aryl boronic ester also shows modest sensitivity to H_2O_2 [164], but, overall, hyperpolarized probes are not likely to have sufficient sensitivity to detect endogenous H_2O_2 in vivo.

4.11.2 Metal ions

Several hyperpolarized ^{13}C- or ^{15}N-labeled chelating sensors of divalent metal ions have been described. Depending on the dynamics of binding interactions, the read-out of ion concentration is based either on the chemical shift change in the case of rapid complex formation and dissociation or on the relative intensities of the signals of the free and complexed forms of the chelator in the case of slow dissociation. Hyperpolarized ^{15}N sensors based on the chelator *o*-aminophenol-N,N,O-triacetic acid show chemical shift changes associated with Ca^{2+} binding [164,165] and are capable of imaging Ca^{2+} in a blood sample. Hyperpolarized ^{13}C-labeled EDTA and EGTA are sensitive to several different metal ions, with each complex having a distinct chemical shift [166], allowing Ca^{2+}-bound EDTA versus unbound EDTA and Zn^{2+}- versus Mg^{2+}-bound EDTA to be selectively imaged. Perhaps the most developed hyperpolarized agents for metal-ion imaging are for Zn^{2+}, which is normally present at high levels in the healthy prostate gland [167]. ^{13}C-labeled cysteine and imino-diacetic acid can both provide a read-out of Zn^{2+} concentrations and Zn^{2+} levels in tissue samples imaged with hyper-polarized ^{13}C-cysteine, which agrees well with those measured using a conventional fluorometric assay [168]. The tris(2-pyridylmethyl) (^{15}N)amine chelator is another Zn^{2+} sensor, and deuteration of the methylene units adjacent to the labeled nitrogen substantially extends the spin polarization lifetime [169].

4.12 Perfusion

Given the unique properties of nuclear spin hyperpolarization to enable rapid, sensitive detection of chemical changes in intact subjects, most preclinical and clinical applications of hyperpolarized ^{13}C have used metabolized or chemically-transformed probes. Chemically-inert compounds with good polarization properties nonetheless have utility, particularly as markers of perfusion.

4.12.1 (^{13}C)Urea and ($^{13}C,^{15}N_2$)Urea

Waste ammonia is converted to urea in the liver and excreted via the kidneys. Lipid membranes have very low permeability to urea, and urea transporters facilitate its exchange across the cell membrane [170]. The limited urea uptake of most tissues allows infused hyperpolarized [^{13}C]urea to be used as an angiography and perfusion imaging agent, as was demonstrated in the first report of hyperpolarized ^{13}C MRI [171]. Hyperpolarized urea has been used to image perfusion or extravasation in the heart, kidney, and tumors in preclinical models and is, after pyruvate, the second most commonly used in vivo hyperpolarized compound.

In the kidney, the dynamics of hyperpolarized urea uptake and distribution have been described in several different disease models and physiological conditions. The infused urea signal is initially found mainly in the cortical, or outer, parts of the kidney and tends to concentrate more in the medullar, or inner, regions over the first minute, and the degree of the intensity of the medullar signal appears to be related to the normal urine-concentrating function of the kidney. The endogenous urea concentration is higher in the medulla, and a higher medullar signal from hyperpolarized urea develops in rats deprived of food and water [172]. Conversely, the hyperpolarized urea-cortico-medullary gradient is less pronounced during diuresis, where urine is less concentrated. This is seen both in furosemide-treated pigs [173] and diabetic rats [174], and the gradient within the kidney is similarly affected in an ischemia-reperfusion model [175].

Most current applications of hyperpolarized ^{13}C urea make use of the ^{15}N-labeled isotopologue ($^{13}C,^{15}N_2$)urea, which yields images with higher SNR and spatial resolution by avoiding the T_2-shortening quadrupolar relaxation caused by ^{14}N [151]. With a T_2 in plasma increased to ~11 s at 3T, it is possible to measure the spatial dependence of its relaxation and construct ($^{13}C,^{15}N_2$)urea T_2 maps of the kidney showing much slower relaxation in the inner medulla compared to the rest of the kidney [151]; p. 2 (Fig. 4.8). In a rat ischemia-reperfusion model of acute kidney injury, a lower urea ^{13}C T_2 is observed, along with a more uniform anatomical distribution in T_2 relaxation [176], consistent with the impairment of function. Similarly, diabetes induced by streptozotocin in rats results in overall lower urea T_2 values, which was attributed to lower oxygen saturation values resulting from higher metabolic activity, and faster relaxation was observed in the cortex versus the pelvis [177].

Urea has also been used to image cardiac perfusion. In this case, it is necessary to suppress the strong signal from the ventricular blood pool. One approach is to use bipolar gradient pulses to dephase fast-flowing spins while preserving the signal from the slower-moving spins in the myocardium. This method yields images of the urea in the myocardial microvasculature, and it is able to detect increased perfusion following treatment with adenosine [179]. This method has also been extended with the copolarization of (^{13}C)urea and (1-^{13}C)pyruvate to simultaneously image both myocardial

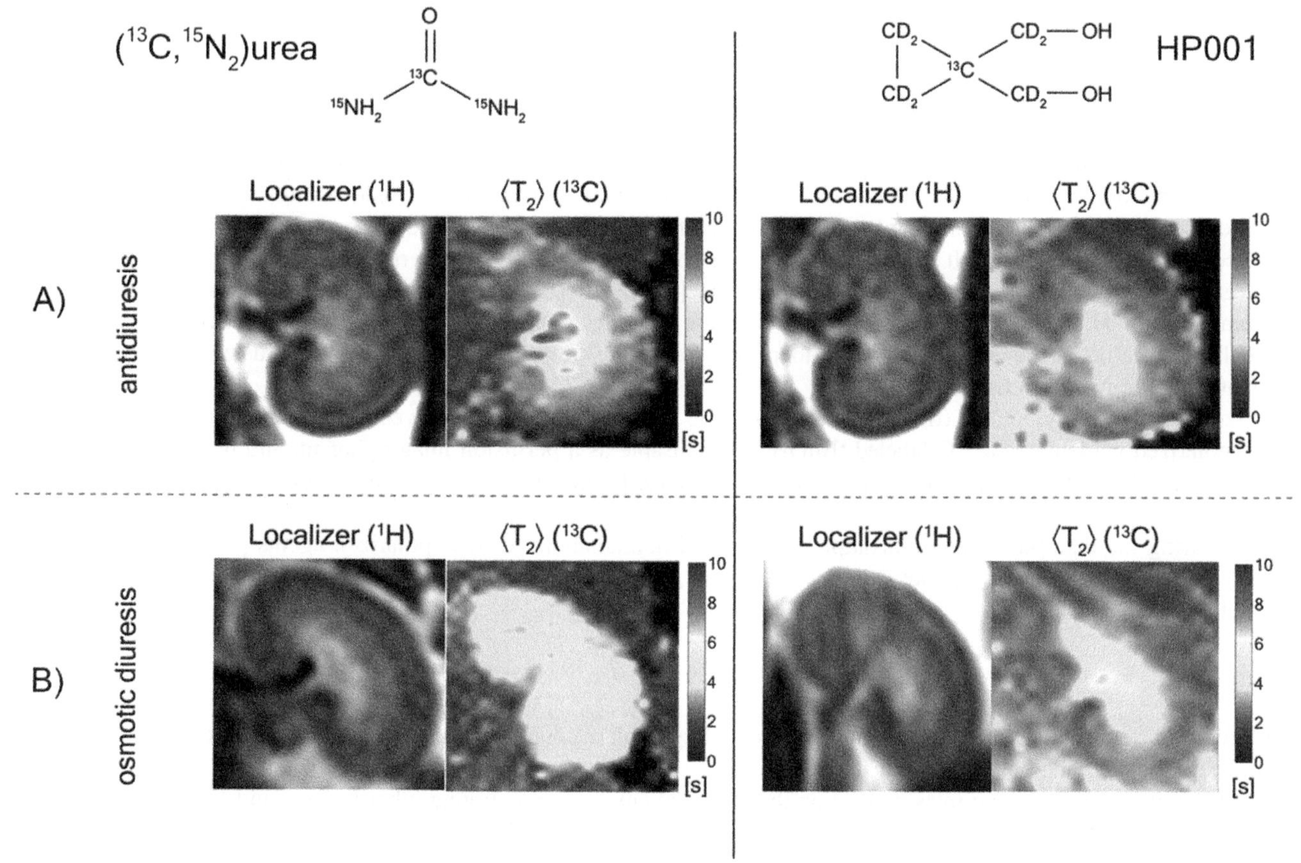

FIGURE 4.8 Hyperpolarized [^{13}C,$^{15}N_2$]urea and HP001 relaxometry during antidiuresis and diuresis. Hyperpolarized [^{13}C,$^{15}N_2$]urea and HP001 images are shown in the left and right columns, respectively. HMCP showed increased medullary T_2, but the effect did not vary between antidiuresis (A) and diuresis (B) states. Decay curves are selected from a single pixel in the inner medulla (IM). The statistically significant change observed was the [^{13}C,$^{15}N_2$] urea T_2 in the IM, consistent with the urea transporter (UT)-A1 and UT-A3 distributions. *Adapted from [178].*

metabolism and perfusion. In this case, the kinetic profiles of the pyruvate and urea signals in the left ventricle and myocardium corresponded closely to each other [47,48]. A similar correspondence between the urea and pyruvate signals is seen in the kidney. An alternative approach uses velocity-sensitive excitation to preferentially excite hyperpolarized urea in the myocardium, and this method is able to detect perfusion defects due to myocardial infarction with performance comparable to conventional Gd^{3+} dynamic contrast enhancement ^1H imaging [180].

In tumors, differences in the delivery and uptake of the hyperpolarized [^{13}C]urea signal have been related to tumor grade and response to treatment. Increased perfusion by pyruvate relative to urea has been noted in mouse liver and prostate tumors compared to healthy liver and kidneys, potentially indicating higher pyruvate uptake in tumor tissue [181]. The utility of hyperpolarized [^{13}C]urea in grading tumors and assessing treatment response has been most extensively studied in the TRAMP transgenic mouse model of prostate cancer. Using a copolarized pyruvate and urea and a dynamic imaging scheme, high-grade tumors have a lower urea signal area under the curve compared to low-grade tumors and a higher k_{trans}, which is consistent with lower perfusion and higher extravasation [182]. Similarly, single-timepoint imaging experiments show a lower urea signal in higher-grade TRAMP tumors, and the lactate/urea ratio metric may also be useful in distinguishing high-grade tumors from low-grade tumors [183]. Radiation treatment [184] and high-intensity focused ultrasound [185] also affect the urea signal in TRAMP tumors. In the case of radiation treatment, the urea signal was clearly increased 1 week posttreatment in responders, whereas it decreased in nonresponders. On the other hand, more modest differences in tumor perfusion compared to healthy tissue were noted using urea in a mouse model of hepatocellular carcinoma following treatment by transcatheter arterial embolization [186]. Formulations of (^{13}C)urea and (1-^{13}C)pyruvic acid suitable for copolarization in clinical studies have been developed [187], and this dual-agent approach is currently being tested in a prostate tumor clinical trial (https://clinicaltrials.gov/ct2/show/NCT02526368).

4.12.2 Alternative ^{13}C perfusion agents: HP001 and tert-(1-^{13}C)butanol

1,1-Bis(hydroxymethyl-(1-^{13}C)cyclopropane-d_8 (HMCP or HP001) (Fig. 4.8), with a quaternary carbon well isolated from nearby protons, is optimized for long-lived spin polarization, and it shows in vivo T_1s in the range of 37−38s at 2.35−3T [188,189]. It has been tested as an angiography [189] and perfusion imaging agent in a number of studies. Overall, the performance of HP001 is similar to that of (^{13}C,$^{15}N_2$)urea, with a time-dependent accumulation of signal in the inner medulla of the kidney [190]. Unlike urea, no change in the T_2 or renal distribution is seen during antidiuresis and diuresis, likely due to its accumulation resulting from general renal function and not specific urine-concentrating mechanisms [178]. In mice with MB-231 xenograft tumors scanned with copolarized HP001 and pyruvate, a stronger correlation is seen between HP001 and lactate signals than with pyruvate [179]. HP001 perfusion in a glioblastoma model shows good agreement with ^1H Gd^{3+} dynamic susceptibility contrast imaging, with both showing increased maximum peak signal, normalized urea-under-the-curve, and delay to peak time in the tumor, reflecting greater perfusion and extravasation compared to normal brain tissue.

Like HP001, tertiary butanol (tBuOH) is water-soluble and contains a quaternary carbon suitable for hyperpolarization. Hyperpolarized perdeuterated ^{13}C-labeled tBuOH is also usable as a perfusion imaging agent, and being more lipophilic than urea or HP001, it is able to readily cross the BBB [191,192]. This property may make it suitable for studies of cerebral blood flow.

Measuring the distribution and relaxation dynamics of nonmetabolized hyperpolarized agents provides physiological information on perfusion and uptake dynamics, and (^{13}C,$^{15}N_2$)urea alone or in combination with metabolized probes is generally suited for clinical and preclinical use.

4.13 ^{15}N

Nitrogen-15 offers an alternative to carbon-13 for some hyperpolarized applications. Nitrogen, being a constituent of many biomolecules, offers the possibility of using endogenous substances as metabolic probes and imaging agents. As for ^{13}C, the ^{15}N longitudinal relaxation rates depend on the chemical environment, among other factors, and ^{15}N T_1 values can be substantially longer than their ^{13}C counterparts in ideal cases, with reported values as high as 1128s [164] in vitro and 160s in vivo [193]. Long-lived hyperpolarized ^{15}N signals are readily achieved in the quaternary amino groups of choline [194] and carnitine [193], as well as nonendogenous compounds such as α-trideuteromethyl(^{15}N) glutamine [195] and TMPA derivatives [164]. Substitution of deuterium for protium on adjacent carbon atoms serves to further isolate the ^{15}N nucleus from dipolar interactions that contribute to relaxation and thereby further enhance hyperpolarization lifetime.

The relative isolation of the quaternary amine nitrogen in choline and carnitine has the disadvantageous property of their metabolites having only negligibly different ^{15}N chemical shifts, making them ill-suited as hyperpolarized metabolic probes. In the case of TMPA compounds, chemical changes associated with covalent reactions or ion-binding interactions of substituents on the aromatic ring are electronically transmitted through the ring, resulting in larger, more readily detected chemical shift changes [164,165].

Among the hyperpolarized ^{15}N compounds imaged in vivo, α-trideuteromethyl(^{15}N)glutamine and L-[^{15}N]carnitine-d_9 both show long-lived polarization and a strong renal signal, which likely reflects their excretion. While the long-lived spin polarization offers the appealing possibility of following an infused hyperpolarized probe over several minutes, as opposed to the <1 min usable lifetimes with ^{13}C, the potential for translation with ^{15}N is not immediately clear. This may, however, change upon the demonstration of clear use cases for ^{15}N perfusion probes and metabolic or physiological sensors.

4.14 Hyperpolarized gas imaging

The noble gas isotopes ^3He and ^{129}Xe have a nuclear spin of 1/2 and are also suitable for hyperpolarized MR imaging and spectroscopy. Both can be hyperpolarized in the gas phase by spin-exchange optical pumping (SEOP) [196] by collisions with rubidium gas vapor bearing an unpaired electron polarized by irradiation with circularly polarized light. Dissolution/sublimation DNP is also able to produce useful amounts of polarized ^{129}Xe [197−199]. By far, the main application of hyperpolarized ^3He and ^{129}Xe has been to image lung function.

4.14.1 ^3He

^3He imaging can reveal ventilation defects in asthma and chronic obstructive pulmonary disease (COPD), and its diffusion properties also reveal differences in the alveolar airway space in these conditions compared to controls [200]. Compared to

^{129}Xe, ^{3}He has the advantages of being easier to polarize to a high level in sufficient quantities by SEOP, and its 2.754-fold higher gyromagnetic ratio provides higher MR sensitivity. The main disadvantage of ^{3}He is its scarcity and high cost. Its limited supply is due to its production by the β-decay of tritium ^{3}H, a byproduct of the nuclear industry. With improvements in polarizer hardware enabling the clinical use of hyperpolarized ^{129}Xe, gas imaging studies are increasingly using this naturally occurring isotope, which doesn't have the same supply constraints.

4.14.2 ^{129}Xe

Like ^{3}He, it is possible with ^{129}Xe to image ventilation defects and increased diffusion times in asthma and COPD [201], though the results are not identical due to the different atomic masses and diffusion properties of the gases. Xe is soluble in blood, and it interacts with biomolecules, resulting in different ^{129}Xe chemical shifts characteristic of their particular environment, with spectral peaks (referenced to the gas-phase ^{129}Xe) at 191, 199, and 213 ppm, respectively, corresponding to ^{129}Xe dissolved in plasma, perialveolar tissue, and red blood cell (RBC) compartments [202].

This sensitivity to the environment enables ^{129}Xe to image several additional aspects of pulmonary function. Dixon-type chemical-shift-sensitive imaging methods allow the simultaneous acquisition of images of the different dissolved Xe phases [201]. Comparing the relative intensities of the signals from the different compartments has revealed features associated with different lung diseases. In idiopathic pulmonary fibrosis (IPF), a higher signal from the "barrier" or tissue compartment is seen compared to controls, and the RBC/barrier ratio distinguishes the IPF group well [203]. Similarly, the RBC/tissue ratio is lower in patients with COPD, and a good correlation is observed between the RBC/gas-phase ratio and the diffusing capacity of the lung for carbon monoxide per unit alveolar volume (DL_{CO}/V_a) [204]. Spectroscopic measurements also detect functional differences in the cariogenic RBC signal oscillations. Distinct patterns of the RBC signal amplitude and resonant-frequency variations associated with each heartbeat are seen for COPD and IPF, as well as left heart failure and pulmonary artery hypertension [205] (Fig. 4.9), underscoring the sensitivity of ^{129}Xe to probe a variety of physiological processes.

In addition to lung imaging, the environmental sensitivity of ^{129}Xe has spawned the development of an array of complementary macro- and supramolecular structures whose presence and chemical state can be probed by virtue of their interactions with Xe [206]. The absorption and transport of inhaled Xe also allow its detection in highly perfused organs, including the brain [207–209] and kidney [210], with the possibility of functional imaging applications.

With the production of hyperpolarized ^{129}Xe for lung imaging nearing regulatory approval and a large number of clinical trials currently underway, it is in an advanced state of development, and a common suite of acquisition protocols has recently been proposed for use in multi-center clinical trials [211].

4.15 Summary and conclusions

Molecular imaging with hyperpolarized substrates complements other imaging modalities by providing temporally- and spatially-resolved metabolic information on the presteady-state conditions immediately following substrate administration. Over the years, many probes have been proposed, and there is now good coverage of the major energy pathways as well as a variety of other processes. The use of multiple hyperpolarized probes to interrogate different aspects of the same biochemical pathway may improve our understanding of both normal function as well as dysregulated metabolism in diseases. In biological and medical applications, the uptake of the probe plays a critical role in determining the extent of the observed metabolism, and this effect must be taken into consideration during data analysis and study design. Uptake is probe- and organ-specific and a particularly important consideration for the brain. Although hyperpolarization improves measurement sensitivity, it does not provide cellular-level resolution. For instance, in organs such as the brain, kidney, and liver, which are composed of multiple different cell types, the same substrate may have different metabolic fates in different parts of the organ, and metabolic compartmentation must also be taken into account.

Having particularly suitable physical and biochemical properties, hyperpolarized [1-^{13}C]pyruvate is the first probe to have entered clinical studies, with promising results for its application in clinical diagnostics. Among the other established hyperpolarized substrates, KIC, fumarate, urea, and α-ketoglutarate have sufficiently high potential to be translated into clinical studies. These substrates meet the criteria of a relatively long lifetime and high polarization level, are safe, and can yield useful medical information. Probes of extracellular pH with suitable performance and safety would also be strong candidates for clinical applications. For gas imaging, ^{129}Xe is becoming increasingly established as the main hyperpolarized agent for functional lung imaging. With numerous clinical trials underway and recent regulatory approval, ^{129}Xe also has the potential for more routine use in clinical applications and physiological studies.

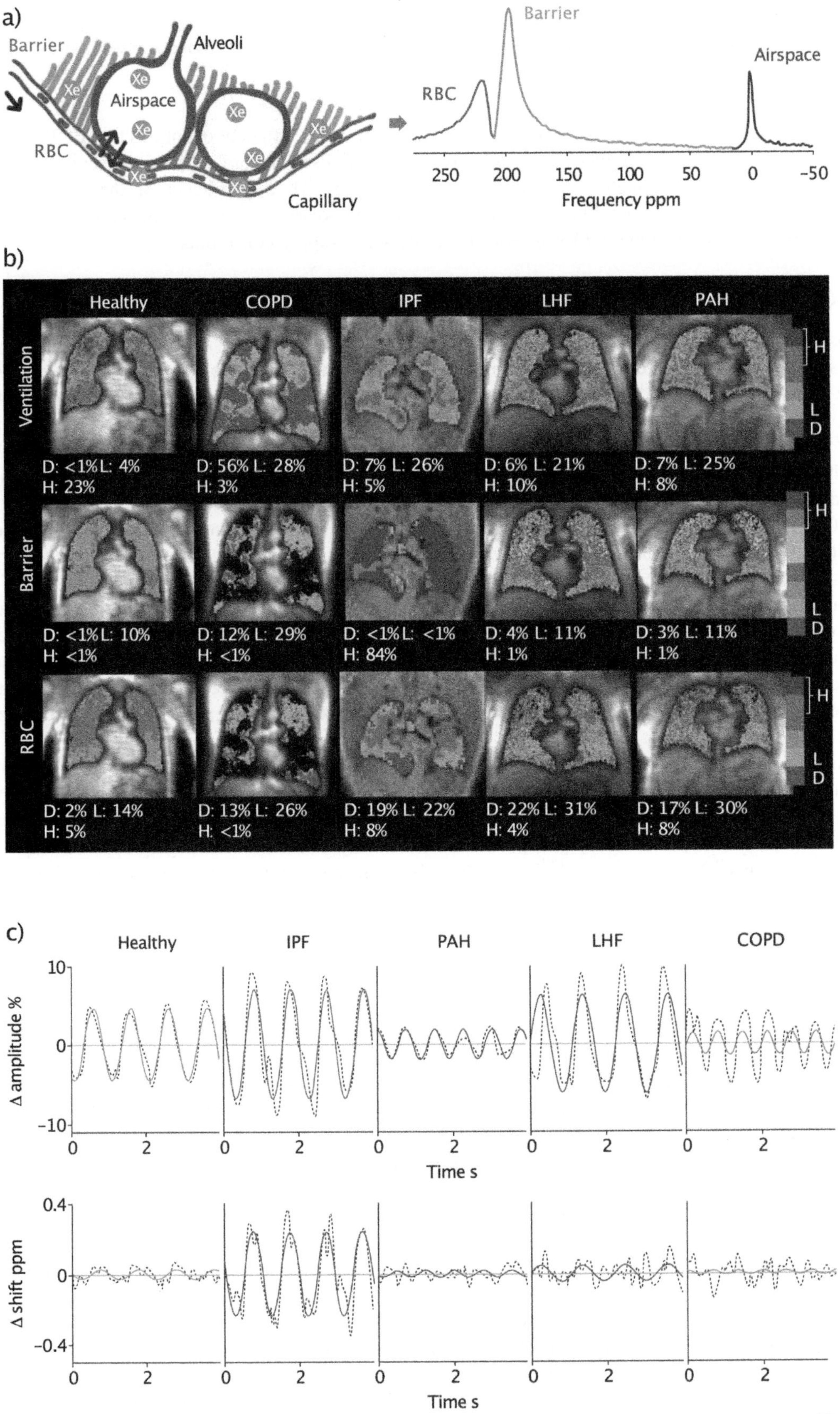

FIGURE 4.9 **Functional imaging of the lung with hyperpolarized ^{129}Xe.** (A) Alveolar–capillary model and ^{129}Xe spectrum. ^{129}Xe freely diffuses from airspace to interstitial barrier tissue to red blood cells (RBCs). In these compartments, the ^{129}Xe atom exhibits distinct frequency shifts of 0, 198 and 217 ppm, respectively. (B) Ventilation, normalized barrier uptake and RBC transfer maps of representative subjects from each cohort. COPD: chronic obstructive pulmonary disease; IPF: idiopathic pulmonary fibrosis; LHF: left heart failure; PAH: pulmonary arterial hypertension. The color bins represent signal intensity, with gray for the lowest, dark gray in ventilation and barrier for the highest and light gray representing voxels in the healthy reference range. Each map is quantified by the percentage of defect (D), low (L) and high (H), calculated as the voxel fraction of the lowest, second lowest and the highest two bins for each map, respectively. The voxels with ventilation defect were excluded from the analysis of barrier uptake and RBC transfer maps. (C) Red blood cell (RBC) signal amplitude and frequency shift demonstrate cardiogenic oscillations in representative patients. The IPF and LHF patients exhibit enhanced RBC amplitude oscillations. By contrast, RBC signal oscillations are diminished in the PAH and COPD patients. Only the IPF patient exhibits oscillations in the RBC shift. *Adapted from [205].*

References

[1] S.W. Provencher, Estimation of metabolite concentrations from localized in vivo proton NMR spectra, Magn. Reson. Med. 30 (1993) 672–679, https://doi.org/10.1002/mrm.1910300604.

[2] L. Vanhamme, A. van den Boogaart, S. Van Huffel, Improved method for accurate and efficient quantification of MRS data with use of prior knowledge, J. Magn. Reson. 129 (1997) 35–43, https://doi.org/10.1006/jmre.1997.1244.

[3] J.M.N. Duarte, H. Lei, V. Mlynárik, R. Gruetter, The neurochemical profile quantified by in vivo 1H NMR spectroscopy, NeuroImage 61 (2012) 342–362, https://doi.org/10.1016/j.neuroimage.2011.12.038. Neuroimaging: Then, Now and the Future.

[4] H.M.D. Feyter, K.L. Behar, Z.A. Corbin, R.K. Fulbright, P.B. Brown, S. McIntyre, T.W. Nixon, D.L. Rothman, R.A. de Graaf, Deuterium metabolic imaging (DMI) for MRI-based 3D mapping of metabolism in vivo, Sci. Adv. 4 (2018) eaat7314, https://doi.org/10.1126/sciadv.aat7314.

[5] M. Lu, X.-H. Zhu, Y. Zhang, G. Mateescu, W. Chen, Quantitative assessment of brain glucose metabolic rates using in vivo deuterium magnetic resonance spectroscopy, J. Cerebr. Blood Flow Metabol. 37 (2017) 3518–3530, https://doi.org/10.1177/0271678X17706444.

[6] E. Flatt, B. Lanz, Y. Pilloud, A. Capozzi, M.H. Lerche, R. Gruetter, M. Mishkovsky, Measuring glycolytic activity with hyperpolarized [2H7, U-13C6] D-glucose in the naive mouse brain under different anesthetic conditions, Metabolites 11 (2021) 413, https://doi.org/10.3390/metabo11070413.

[7] R. Mahar, H. Zeng, A. Giacalone, M. Ragavan, T.H. Mareci, M.E. Merritt, Deuterated water imaging of the rat brain following metabolism of [2H7]glucose, Magn. Reson. Med. 85 (2021) 3049–3059, https://doi.org/10.1002/mrm.28700.

[8] J.H. Ardenkjær-Larsen, B. Fridlund, A. Gram, G. Hansson, L. Hansson, M.H. Lerche, R. Servin, M. Thaning, K. Golman, Increase in signal-to-noise ratio of >10,000 times in liquid-state NMR, Proc. Natl. Acad. Sci. USA 100 (2003) 10158–10163, https://doi.org/10.1073/pnas.1733835100.

[9] J.W. Gordon, J.J. Miller, 3—HP acquisition methods: pulse sequences, reconstruction, and RF coils, in: P.E.Z. Larson (Ed.), Advances in Magnetic Resonance Technology and Applications, Hyperpolarized Carbon-13 Magnetic Resonance Imaging and Spectroscopy, Academic Press, 2021, pp. 49–74, https://doi.org/10.1016/B978-0-12-822269-0.00005-1.

[10] R.E. Hurd, A.P. Chen, Sampling strategies in dynamic hyperpolarized NMR, in: T. Jue, D. Mayer (Eds.), Dynamic Hyperpolarized Nuclear Magnetic Resonance, Handbook of Modern Biophysics, Springer International Publishing, Cham, 2021, pp. 77–102, https://doi.org/10.1007/978-3-030-55043-1_4.

[11] P.E.Z. Larson, J.W. Gordon, Hyperpolarized metabolic MRI—acquisition, reconstruction, and analysis methods, Metabolites 11 (2021) 386, https://doi.org/10.3390/metabo11060386.

[12] G.J. Topping, C. Hundshammer, L. Nagel, M. Grashei, M. Aigner, J.G. Skinner, R.F. Schulte, F. Schilling, Acquisition strategies for spatially resolved magnetic resonance detection of hyperpolarized nuclei, Magn. Reson. Mater. Phys. Biol. Med. 33 (2020) 221–256, https://doi.org/10.1007/s10334-019-00807-6.

[13] T.B. Rodrigues, E.M. Serrao, B.W.C. Kennedy, D.-E. Hu, M.I. Kettunen, K.M. Brindle, Magnetic resonance imaging of tumor glycolysis using hyperpolarized 13C-labeled glucose, Nat. Med. 20 (2014) 93–97, https://doi.org/10.1038/nm.3416.

[14] M. Mishkovsky, A. Comment, R. Gruetter, In vivo detection of brain Krebs cycle intermediate by hyperpolarized magnetic resonance, J. Cerebr. Blood Flow Metabol. 32 (2012) 2108–2113, https://doi.org/10.1038/jcbfm.2012.136.

[15] M. Mishkovsky, B. Anderson, M. Karlsson, M.H. Lerche, A.D. Sherry, R. Gruetter, Z. Kovacs, A. Comment, Measuring glucose cerebral metabolism in the healthy mouse using hyperpolarized 13C magnetic resonance, Sci. Rep. 7 (2017) 11719, https://doi.org/10.1038/s41598-017-12086-z.

[16] K.R. Keshari, D.M. Wilson, Chemistry and biochemistry of 13C hyperpolarized magnetic resonance using dynamic nuclear polarization, Chem. Soc. Rev. 43 (2014) 1627–1659, https://doi.org/10.1039/C3CS60124B.

[17] S.J. Nelson, J. Kurhanewicz, D.B. Vigneron, P.E.Z. Larson, A.L. Harzstark, M. Ferrone, M. van Criekinge, J.W. Chang, R. Bok, I. Park, G. Reed, L. Carvajal, E.J. Small, P. Munster, V.K. Weinberg, J.H. Ardenkjaer-Larsen, A.P. Chen, R.E. Hurd, L.-I. Odegardstuen, F.J. Robb, J. Tropp, J.A. Murray, Metabolic imaging of patients with prostate cancer using hyperpolarized [1-13C]pyruvate, Sci. Transl. Med. 5 (2013) 198ra108, https://doi.org/10.1126/scitranslmed.3006070.

[18] J.H. Ardenkjaer-Larsen, S. Bowen, J.R. Petersen, O. Rybalko, M.S. Vinding, M. Ullisch, N.C. Nielsen, Cryogen-free dissolution dynamic nuclear polarization polarizer operating at 3.35 T, 6.70 T, and 10.1 T, Magn. Reson. Med. 81 (2019) 2184–2194, https://doi.org/10.1002/mrm.27537.

[19] H.A.I. Yoshihara, E. Can, M. Karlsson, M.H. Lerche, J. Schwitter, A. Comment, High-field dissolution dynamic nuclear polarization of [1-13C] pyruvic acid, Phys. Chem. Chem. Phys. 18 (2016) 12409–12413, https://doi.org/10.1039/C6CP00589F.

[20] T.R. Eichhorn, Y. Takado, N. Salameh, A. Capozzi, T. Cheng, J.-N. Hyacinthe, M. Mishkovsky, C. Roussel, A. Comment, Hyperpolarization without persistent radicals for in vivo real-time metabolic imaging, Proc. Natl. Acad. Sci. U. S. A 110 (2013) 18064–18069, https://doi.org/10.1073/pnas.1314928110.

[21] F. Reineri, T. Boi, S. Aime, Parahydrogen induced polarization of 13C carboxylate resonance in acetate and pyruvate, Nat. Commun. 6 (2015) 5858, https://doi.org/10.1038/ncomms6858.

[22] E. Cavallari, C. Carrera, M. Sorge, G. Bonne, A. Muchir, S. Aime, F. Reineri, The 13C hyperpolarized pyruvate generated by ParaHydrogen detects the response of the heart to altered metabolism in real time, Sci. Rep. 8 (2018) 8366, https://doi.org/10.1038/s41598-018-26583-2.

[23] T.H. Witney, M.I. Kettunen, K.M. Brindle, Kinetic modeling of hyperpolarized 13C label exchange between pyruvate and lactate in tumor cells, J. Biol. Chem. 286 (2011) 24572–24580, https://doi.org/10.1074/jbc.M111.237727.

[24] Y. Rao, S. Gammon, N.M. Zacharias, T. Liu, T. Salzillo, Y. Xi, J. Wang, P. Bhattacharya, D. Piwnica-Worms, Hyperpolarized [1-13C]pyruvate-to-[1-13C]lactate conversion is rate-limited by monocarboxylate transporter-1 in the plasma membrane, Proc. Natl. Acad. Sci. USA 117 (2020) 22378–22389, https://doi.org/10.1073/pnas.2003537117.

[25] O. Warburg, On the origin of cancer cells, Science 123 (1956) 309–314, https://doi.org/10.1126/science.123.3191.309.

[26] K. Golman, R. in't Zandt, M. Lerche, R. Pehrson, J.H. Ardenkjaer-Larsen, Metabolic imaging by hyperpolarized 13C magnetic resonance imaging for in vivo tumor diagnosis, Cancer Res. 66 (2006) 10855–10860, https://doi.org/10.1158/0008-5472.CAN-06-2564.

[27] M.J. Albers, R. Bok, A.P. Chen, C.H. Cunningham, M.L. Zierhut, V.Y. Zhang, S.J. Kohler, J. Tropp, R.E. Hurd, Y.-F. Yen, S.J. Nelson, D.B. Vigneron, J. Kurhanewicz, Hyperpolarized 13C lactate, pyruvate, and alanine: noninvasive biomarkers for prostate cancer detection and grading, Cancer Res. 68 (2008) 8607–8615, https://doi.org/10.1158/0008-5472.CAN-08-0749.

[28] S.E. Day, M.I. Kettunen, F.A. Gallagher, D.-E. Hu, M. Lerche, J. Wolber, K. Golman, J.H. Ardenkjaer-Larsen, K.M. Brindle, Detecting tumor response to treatment using hyperpolarized 13C magnetic resonance imaging and spectroscopy, Nat. Med. 13 (2007) 1382–1387, https://doi.org/10.1038/nm1650.

[29] S. Hu, A. Balakrishnan, R.A. Bok, B. Anderton, P.E.Z. Larson, S.J. Nelson, J. Kurhanewicz, D.B. Vigneron, A. Goga, 13C-Pyruvate imaging reveals alterations in glycolysis that precede c-Myc-induced tumor formation and regression, Cell Metabol. 14 (2011a) 131–142, https://doi.org/10.1016/j.cmet.2011.04.012.

[30] K.R. Keshari, R. Sriram, M. Van Criekinge, D.M. Wilson, Z.J. Wang, D.B. Vigneron, D.M. Peehl, J. Kurhanewicz, Metabolic reprogramming and validation of hyperpolarized 13C lactate as a prostate cancer biomarker using a human prostate tissue slice culture bioreactor, Prostate 73 (2013) 1171–1181, https://doi.org/10.1002/pros.22665.

[31] J. Kurhanewicz, D.B. Vigneron, J.H. Ardenkjaer-Larsen, J.A. Bankson, K. Brindle, C.H. Cunningham, F.A. Gallagher, K.R. Keshari, A. Kjaer, C. Laustsen, D.A. Mankoff, M.E. Merritt, S.J. Nelson, J.M. Pauly, P. Lee, S. Ronen, D.J. Tyler, S.S. Rajan, D.M. Spielman, L. Wald, X. Zhang, C.R. Malloy, R. Rizi, Hyperpolarized 13C MRI: path to clinical translation in oncology, Neoplasia 21 (2019) 1–16, https://doi.org/10.1016/j.neo.2018.09.006.

[32] N.V. Rajeshkumar, P. Dutta, S. Yabuuchi, R.F. de Wilde, G.V. Martinez, A. Le, J.J. Kamphorst, J.D. Rabinowitz, S.K. Jain, M. Hidalgo, C.V. Dang, R.J. Gillies, A. Maitra, Therapeutic targeting of the Warburg effect in pancreatic cancer relies on an absence of p53 function, Cancer Res. 75 (2015) 3355–3364, https://doi.org/10.1158/0008-5472.CAN-15-0108.

[33] S. Galván-Peña, L.A.J. O'Neill, Metabolic reprograming in macrophage polarization, Front. Immunol. 5 (2014) 420, https://doi.org/10.3389/fimmu.2014.0042.

[34] E. Can, M. Mishkovsky, H.A.I. Yoshihara, N. Kunz, D.-L. Couturier, U. Petrausch, M.-A. Doucey, A. Comment, Noninvasive rapid detection of metabolic adaptation in activated human T lymphocytes by hyperpolarized 13C magnetic resonance, Sci. Rep. 10 (2020) 200, https://doi.org/10.1038/s41598-019-57026-1.

[35] L.M. Le Page, C. Guglielmetti, C.F. Najac, B. Tiret, M.M. Chaumeil, Hyperpolarized 13C magnetic resonance spectroscopy detects toxin-induced neuroinflammation in mice, NMR Biomed. 32 (2019) e4164, https://doi.org/10.1002/nbm.4164.

[36] J.D. MacKenzie, Y.-F. Yen, D. Mayer, J.S. Tropp, R.E. Hurd, D.M. Spielman, Detection of inflammatory arthritis by using hyperpolarized 13C-pyruvate with MR imaging and spectroscopy, Radiology 259 (2011) 414–420, https://doi.org/10.1148/radiol.10101921.

[37] C.L. Robertson, M. Saraswati, G. Fiskum, Mitochondrial dysfunction early after traumatic brain injury in immature rats, J. Neurochem. 101 (2007) 1248–1257, https://doi.org/10.1111/j.1471-4159.2007.04489.x.

[38] S.J. DeVience, X. Lu, J. Proctor, P. Rangghran, E.R. Melhem, R. Gullapalli, G.M. Fiskum, D. Mayer, Metabolic imaging of energy metabolism in traumatic brain injury using hyperpolarized [1-13C]pyruvate, Sci. Rep. 7 (2017) 1907, https://doi.org/10.1038/s41598-017-01736-x.

[39] C. Guglielmetti, A. Chou, K. Krukowski, C. Najac, X. Feng, L.-K. Riparip, S. Rosi, M.M. Chaumeil, In vivo metabolic imaging of traumatic brain injury, Sci. Rep. 7 (2017) 17525, https://doi.org/10.1038/s41598-017-17758-4.

[40] S. Hu, A.P. Chen, M.L. Zierhut, R. Bok, Y.-F. Yen, M.A. Schroeder, R.E. Hurd, S.J. Nelson, J. Kurhanewicz, D.B. Vigneron, Vivo carbon-13 dynamic MRS and MRSI of normal and fasted rat liver with hyperpolarized ^{13}C-pyruvate, Mol. Imaging Biol. 11 (6) (2009) 399–407, https://doi.org/10.1007/s11307-009-0218-z.

[41] C. von Morze, G.-Y. Chang, P.E.Z. Larson, H. Shang, P.K.R. Allu, R.A. Bok, J.C. Crane, M.P. Olson, C.T. Tan, I. Marco-Rius, S.J. Nelson, J. Kurhanewicz, D. Pearce, D.B. Vigneron, Detection of localized changes in the metabolism of hyperpolarized gluconeogenic precursors ^{13}C-lactate and ^{13}C-pyruvate in kidney and liver, Magn. Reson. Med. 77 (4) (2017) 1429–1437, https://doi.org/10.1002/mrm.26245.

[42] P.M. Nielsen, C.Ø. Mariager, M. Mølmer, N. Sparding, F. Genovese, M.A. Karsdal, R. Nørregaard, L.B. Bertelsen, C. Laustsen, Hyperpolarized [1-13C] alanine production: a novel imaging biomarker of renal fibrosis, Magn. Reson. Med. 84 (2020) 2063–2073, https://doi.org/10.1002/mrm.28326.

[43] O.J. Rider, A. Apps, J.J.J.J. Miller, J.Y.C. Lau, A.J.M. Lewis, M.A. Peterzan, M.S. Dodd, A.Z. Lau, C. Trumper, F.A. Gallagher, J.T. Grist, K.M. Brindle, S. Neubauer, D.J. Tyler, Noninvasive in vivo assessment of cardiac metabolism in the healthy and diabetic human heart using hyperpolarized 13C MRI, Circ. Res. 126 (2020) 725–736, https://doi.org/10.1161/CIRCRESAHA.119.316260.

[44] M.A. Schroeder, L.E. Cochlin, L.C. Heather, K. Clarke, G.K. Radda, D.J. Tyler, In vivo assessment of pyruvate dehydrogenase flux in the heart using hyperpolarized carbon-13 magnetic resonance, Proc. Natl. Acad. Sci. USA 105 (2008) 12051–12056, https://doi.org/10.1073/pnas.0805953105.

[45] J.T. Grist, M.A. McLean, F. Riemer, R.F. Schulte, S.S. Deen, F. Zaccagna, R. Woitek, C.J. Daniels, J.D. Kaggie, T. Matys, I. Patterson, R. Slough, A.B. Gill, A. Chhabra, R. Eichenberger, M.-C. Laurent, A. Comment, J.H. Gillard, A.J. Coles, D.J. Tyler, I. Wilkinson, B. Basu, D.J. Lomas, M.J. Graves, K.M. Brindle, F.A. Gallagher, Quantifying normal human brain metabolism using hyperpolarized [1−13C]pyruvate and magnetic resonance imaging, Neuroimage 189 (2019) 171−179, https://doi.org/10.1016/j.neuroimage.2019.01.027.

[46] M. Marjańska, I. Iltis, A.A. Shestov, D.K. Deelchand, C. Nelson, K. Uğurbil, P.-G. Henry, In vivo 13C spectroscopy in the rat brain using hyperpolarized [1-13C]pyruvate and [2-13C]pyruvate, J. Magn. Reson. 206 (2010) 210−218, https://doi.org/10.1016/j.jmr.2010.07.006.

[47] A.Z. Lau, J.J. Miller, M.D. Robson, D.J. Tyler, Simultaneous assessment of cardiac metabolism and perfusion using copolarized [1-13 C]pyruvate and 13 C-urea, Magn. Reson. Med. 77 (2017) 151−158, https://doi.org/10.1002/mrm.26106.

[48] A.Z. Lau, J.J. Miller, D.J. Tyler, Mapping of intracellular pH in the in vivo rodent heart using hyperpolarized [1-13C]pyruvate, Magn. Reson. Med. 77 (2017) 1810−1817, https://doi.org/10.1002/mrm.26260.

[49] E. Can, J.A.M. Bastiaansen, D.-L. Couturier, R. Gruetter, H.A.I. Yoshihara, A. Comment, [13C]bicarbonate labelled from hyperpolarized [1-13C] pyruvate is an in vivo marker of hepatic gluconeogenesis in fasted state, Commun. Biol. 5 (2022) 1−9, https://doi.org/10.1038/s42003-021-02978-2.

[50] M.E. Merritt, C. Harrison, A.D. Sherry, C.R. Malloy, S.C. Burgess, Flux through hepatic pyruvate carboxylase and phosphoenolpyruvate carboxykinase detected by hyperpolarized 13C magnetic resonance, Proc. Natl. Acad. Sci. USA 108 (2011) 19084−19089, https://doi.org/10.1073/pnas.1111247108.

[51] H.A.I. Yoshihara, A. Comment, J. Schwitter, Assessment of aspartate and bicarbonate produced from hyperpolarized [1-13C]pyruvate as markers of renal gluconeogenesis, Front. Physiol. 12 (2021) 792769, https://doi.org/10.3389/fphys.2021.792769.

[52] A.J. Bakermans, M.S. Dodd, K. Nicolay, J.J. Prompers, D.J. Tyler, S.M. Houten, Myocardial energy shortage and unmet anaplerotic needs in the fasted long-chain acyl-CoA dehydrogenase knockout mouse, Cardiovasc. Res. 100 (3) (2013) 441−449, https://doi.org/10.1093/cvr/cvt212.

[53] M. Karlsson, P.R. Jensen, J.Ø. Duus, S. Meier, M.H. Lerche, Development of dissolution DNP-MR substrates for metabolic research, Appl. Magn. Reson. 43 (2012) 223−236, https://doi.org/10.1007/s00723-012-0336-7.

[54] M.L. Goodwin, J.E. Harris, A. Hernández, L.B. Gladden, Blood lactate measurements and analysis during exercise: a guide for clinicians, J. Diabetes Sci. Technol. 1 (2007) 558−569, https://doi.org/10.1177/193229680700100414.

[55] J.A.M. Bastiaansen, H.A.I. Yoshihara, Y. Takado, R. Gruetter, A. Comment, Hyperpolarized 13C lactate as a substrate for in vivo metabolic studies in skeletal muscle, Metabolomics 10 (2014) 986−994, https://doi.org/10.1007/s11306-014-0630-5.

[56] A.P. Chen, J. Kurhanewicz, R. Bok, D. Xu, D. Joun, V. Zhang, S.J. Nelson, R.E. Hurd, D.B. Vigneron, Feasibility of using hyperpolarized [1-13C] lactate as a substrate for in vivo metabolic 13C MRSI studies, Magn. Reson. Imaging 26 (2008) 721−726, https://doi.org/10.1016/j.mri.2008.01.002.

[57] Y. Takado, T. Cheng, J.A.M. Bastiaansen, H.A.I. Yoshihara, B. Lanz, M. Mishkovsky, S. Lengacher, A. Comment, Hyperpolarized 13C magnetic resonance spectroscopy reveals the rate-limiting role of the blood−brain barrier in the cerebral uptake and metabolism of l-lactate in vivo, ACS Chem. Neurosci. 9 (2018) 2554−2562, https://doi.org/10.1021/acschemneuro.8b00066.

[58] C. Berthet, X. Castillo, P.J. Magistretti, L. Hirt, New evidence of neuroprotection by lactate after transient focal cerebral ischaemia: extended benefit after intracerebroventricular injection and efficacy of intravenous administration, Cerebrovasc. Dis. 34 (2012) 329−335, https://doi.org/10.1159/000343657.

[59] J.-N. Hyacinthe, L. Buscemi, T.P. Lê, M. Lepore, L. Hirt, M. Mishkovsky, Evaluating the potential of hyperpolarised [1-13C] L-lactate as a neuroprotectant metabolic biosensor for stroke, Sci. Rep. 10 (2020) 5507, https://doi.org/10.1038/s41598-020-62319-x.

[60] A.P. Chen, J.Y.C. Lau, R.D.A. Alvares, C.H. Cunningham, Using [1-(13) C]lactic acid for hyperpolarized (13) C MR cardiac studies, Magn. Reson. Med. 73 (2015) 2087−2093, https://doi.org/10.1002/mrm.25354.

[61] J. Chen, E.P. Hackett, Z. Kovacs, C.R. Malloy, J.M. Park, Assessment of hepatic pyruvate carboxylase activity using hyperpolarized [1-^{13}C]-l-lactate, Magn. Reson. Med. 85 (2021) 1175−1182, https://doi.org/10.1002/mrm.28489.

[62] S. Hu, M. Zhu, H.A.I. Yoshihara, D.M. Wilson, K.R. Keshari, P. Shin, G. Reed, C. von Morze, R. Bok, P.E.Z. Larson, J. Kurhanewicz, D.B. Vigneron, In vivo measurement of normal rat intracellular pyruvate and lactate levels after injection of hyperpolarized [1-13C]alanine, Magn. Reson. Imaging 29 (2011b) 1035−1040, https://doi.org/10.1016/j.mri.2011.07.001.

[63] J.M. Park, C. Khemtong, S.-C. Liu, R.E. Hurd, D.M. Spielman, In vivo assessment of intracellular redox-state in rat liver using hyperpolarized [1-13C]alanine, Magn. Reson. Med. 77 (2017a) 1741−1748, https://doi.org/10.1002/mrm.26662.

[64] P. Viswanath, G. Batsios, J. Mukherjee, A.M. Gillespie, P.E.Z. Larson, H.A. Luchman, J.J. Phillips, J.F. Costello, R.O. Pieper, S.M. Ronen, Non-invasive assessment of telomere maintenance mechanisms in brain tumors, Nat. Commun. 12 (2021b) 92, https://doi.org/10.1038/s41467-020-20312-y.

[65] A. Radaelli, R. Gruetter, H.A.I. Yoshihara, In vivo detection of d-amino acid oxidase with hyperpolarized d-[1-13C]alanine, NMR Biomed. 33 (2020) e4303, https://doi.org/10.1002/nbm.4303.

[66] S. Meier, M. Karlsson, P.R. Jensen, M.H. Lerche, J.Ø. Duus, Metabolic pathway visualization in living yeast by DNP-NMR, Mol. Biosyst. 7 (2011) 2834−2836, https://doi.org/10.1039/C1MB05202K.

[67] P. Viswanath, G. Batsios, V. Ayyappan, C. Taglang, A.M. Gillespie, P.E.Z. Larson, H.A. Luchman, J.F. Costello, R.O. Pieper, S.M. Ronen, Metabolic imaging detects elevated glucose flux through the pentose phosphate pathway associated with TERT expression in low-grade gliomas, Neuro Oncol. 23 (2021a) 1509−1522, https://doi.org/10.1093/neuonc/noab093.

[68] M. Mishkovsky, O. Gusyatiner, B. Lanz, C. Cudalbu, I. Vassallo, M.-F. Hamou, J. Bloch, A. Comment, R. Gruetter, M.E. Hegi, Hyperpolarized 13C-glucose magnetic resonance highlights reduced aerobic glycolysis in vivo in infiltrative glioblastoma, Sci. Rep. 11 (2021) 5771, https://doi.org/10.1038/s41598-021-85339-7.

[69] A. Capozzi, T. Cheng, G. Boero, C. Roussel, A. Comment, Thermal annihilation of photo-induced radicals following dynamic nuclear polarization to produce transportable frozen hyperpolarized 13C-substrates, Nat. Commun. 8 (2017) 15757, https://doi.org/10.1038/ncomms15757.

[70] A. Capozzi, S. Patel, C.P. Gunnarsson, I. Marco-Rius, A. Comment, M. Karlsson, M.H. Lerche, O. Ouari, J.H. Ardenkjær-Larsen, Efficient hyperpolarization of U-13C-glucose using narrow-line UV-generated labile free radicals, Angew. Chem. Int. Ed. 58 (2019) 1334−1339, https://doi.org/10.1002/anie.201810522.

[71] J.M. Park, M. Wu, K. Datta, S.-C. Liu, A. Castillo, H. Lough, D.M. Spielman, K.L. Billingsley, Hyperpolarized sodium [1-13C]-Glycerate as a probe for assessing glycolysis in vivo, J. Am. Chem. Soc. 139 (2017b) 6629−6634, https://doi.org/10.1021/jacs.7b00708.

[72] E.P. Hackett, B.R. Shah, B. Cheng, E. LaGue, V. Vemireddy, M. Mendoza, C. Bing, R.M. Bachoo, K.L. Billingsley, R. Chopra, J.M. Park, Probing cerebral metabolism with hyperpolarized 13C imaging after opening the blood−brain barrier with focused ultrasound, ACS Chem. Neurosci. 12 (2021) 2820−2828, https://doi.org/10.1021/acschemneuro.1c00197.

[73] A. Kirpich, M. Ragavan, J.A. Bankson, L.M. McIntyre, M.E. Merritt, Kinetic analysis of hepatic metabolism using hyperpolarized dihydroxyacetone, J. Chem. Inf. Model. 59 (2019) 605−614, https://doi.org/10.1021/acs.jcim.8b00745.

[74] I. Marco-Rius, C. von Morze, R. Sriram, P. Cao, G.-Y. Chang, E. Milshteyn, R.A. Bok, M.A. Ohliger, D. Pearce, J. Kurhanewicz, P.E.Z. Larson, D.B. Vigneron, M. Merritt, Monitoring acute metabolic changes in the liver and kidneys induced by fructose and glucose using hyperpolarized [2-13 C]dihydroxyacetone, Magn. Reson. Med. 77 (2017) 65−73, https://doi.org/10.1002/mrm.26525.

[75] K.X. Moreno, S. Satapati, R.J. DeBerardinis, S.C. Burgess, C.R. Malloy, M.E. Merritt, Real-time detection of hepatic gluconeogenic and glycogenolytic states using hyperpolarized [2-13C]dihydroxyacetone, J. Biol. Chem. 289 (2014) 35859−35867, https://doi.org/10.1074/jbc.M114.613265.

[76] K.X. Moreno, C.E. Harrison, M.E. Merritt, Z. Kovacs, C.R. Malloy, A.D. Sherry, Hyperpolarized δ-[1-13C]gluconolactone as a probe of the pentose phosphate pathway, NMR Biomed. 30 (2017) e3713, https://doi.org/10.1002/nbm.3713.

[77] G. Batsios, C. Taglang, P. Cao, A.M. Gillespie, C. Najac, E. Subramani, D.M. Wilson, R.R. Flavell, P.E.Z. Larson, S.M. Ronen, P. Viswanath, Imaging 6-phosphogluconolactonase activity in brain tumors in vivo using hyperpolarized δ-[1-13C]gluconolactone, Front. Oncol. (2021) 11, https://doi.org/10.3389/fonc.2021.589570, 589570.

[78] H.A. Krebs, The history of the tricarboxylic acid cycle, Perspect. Biol. Med. 14 (1970) 154−172, https://doi.org/10.1353/pbm.1970.0001.

[79] P.-G. Henry, G. Adriany, D. Deelchand, R. Gruetter, M. Marjanska, G. Öz, E.R. Seaquist, A. Shestov, K. Uğurbil, In vivo 13C NMR spectroscopy and metabolic modeling in the brain: a practical perspective, Magn. Reson. Imaging 24 (2006) 527−539, https://doi.org/10.1016/j.mri.2006.01.003.

[80] M.A. Schroeder, H.J. Atherton, D.R. Ball, M.A. Cole, L.C. Heather, J.L. Griffin, K. Clarke, G.K. Radda, D.J. Tyler, Real-time assessment of Krebs cycle metabolism using hyperpolarized C magnetic resonance spectroscopy, FASEB J. 23 (2009) 2529−2538, https://doi.org/10.1096/fj.09-129171.

[81] M. Marjańska, A.A. Shestov, D.K. Deelchand, E. Kittelson, P.-G. Henry, Brain metabolism under different anesthetic conditions using hyperpolarized [1-13C]pyruvate and [2-13C]pyruvate, NMR Biomed. 31 (2018) e4012, https://doi.org/10.1002/nbm.4012.

[82] B.T. Chung, H.-Y. Chen, J. Gordon, D. Mammoli, R. Sriram, A.W. Autry, L.M. Le Page, M. Chaumeil, P. Shin, J. Slater, C.T. Tan, C. Suszczynski, S. Chang, Y. Li, R.A. Bok, S.M. Ronen, P.E.Z. Larson, J. Kurhanewicz, D.B. Vigneron, First hyperpolarized [2-13C]pyruvate MR studies of human brain metabolism, J. Magn. Reson. 309 (2019) 106617, https://doi.org/10.1016/j.jmr.2019.106617.

[83] D.S. Wishart, Y.D. Feunang, A. Marcu, A.C. Guo, K. Liang, R. Vázquez-Fresno, T. Sajed, D. Johnson, C. Li, N. Karu, Z. Sayeeda, E. Lo, N. Assempour, M. Berjanskii, S. Singhal, D. Arndt, Y. Liang, H. Badran, J. Grant, A. Serra-Cayuela, Y. Liu, R. Mandal, V. Neveu, A. Pon, C. Knox, M. Wilson, C. Manach, A. Scalbert, HMDB 4.0: the human metabolome database for 2018, Nucleic Acids Res. 46 (2018) D608−D617, https://doi.org/10.1093/nar/gkx1089.

[84] P.R. Jensen, T. Peitersen, M. Karlsson, R. in 't Zandt, A. Gisselsson, G. Hansson, S. Meier, M.H. Lerche, Tissue-specific short chain fatty acid metabolism and slow metabolic recovery after ischemia from hyperpolarized NMR in vivo, J. Biol. Chem. 284 (2009) 36077−36082, https://doi.org/10.1074/jbc.M109.066407.

[85] U. Koellisch, C.V. Gringeri, G. Rancan, E.V. Farell, M.I. Menzel, A. Haase, M. Schwaiger, R.F. Schulte, Metabolic imaging of hyperpolarized [1-13C]acetate and [1-13C]acetylcarnitine—investigation of the influence of dobutamine induced stress, Magn. Reson. Med. 74 (2015a) 1011−1018, https://doi.org/10.1002/mrm.25485.

[86] U. Koellisch, C. Laustsen, T.S. Nørlinger, J.A. Østergaard, A. Flyvbjerg, C.V. Gringeri, M.I. Menzel, R.F. Schulte, A. Haase, H. Stødkilde-Jørgensen, Investigation of metabolic changes in STZ-induced diabetic rats with hyperpolarized [1-13C]acetate, Phys. Rep. 3 (2015b) e12474, https://doi.org/10.14814/phy2.12474.

[87] J.A.M. Bastiaansen, T. Cheng, M. Mishkovsky, J.M.N. Duarte, A. Comment, R. Gruetter, In vivo enzymatic activity of acetylCoA synthetase in skeletal muscle revealed by 13C turnover from hyperpolarized [1-13C]acetate to [1-13C]acetylcarnitine, Biochim. Biophys. Acta BBA—Gen. Subj. 1830 (2013) 4171−4178, https://doi.org/10.1016/j.bbagen.2013.03.023.

[88] J.A.M. Bastiaansen, T. Cheng, H. Lei, R. Gruetter, A. Comment, Direct noninvasive estimation of myocardial tricarboxylic acid cycle flux in vivo using hyperpolarized 13C magnetic resonance, J. Mol. Cell. Cardiol. 87 (2015) 129−137, https://doi.org/10.1016/j.yjmcc.2015.08.012.

[89] D.R. Ball, B. Rowlands, M.S. Dodd, L. Le Page, V. Ball, C.A. Carr, K. Clarke, D.J. Tyler, Hyperpolarized butyrate: a metabolic probe of short chain fatty acid metabolism in the heart, Magn. Reson. Med. 71 (2014) 1663−1669, https://doi.org/10.1002/mrm.24849.

[90] J.A.M. Bastiaansen, M.E. Merritt, A. Comment, Measuring changes in substrate utilization in the myocardium in response to fasting using hyperpolarized [1-13C]butyrate and [1-13C]pyruvate, Sci. Rep. 6 (2016) 25573, https://doi.org/10.1038/srep25573.

[91] H.A.I. Yoshihara, J.A.M. Bastiaansen, M. Karlsson, M.H. Lerche, A. Comment, J. Schwitter, Detection of myocardial medium-chain fatty acid oxidation and tricarboxylic acid cycle activity with hyperpolarized [1−13C]octanoate, NMR Biomed. 33 (2020) e4243, https://doi.org/10.1002/nbm.4243.

[92] S. Cerdan, B. Künnecke, J. Seelig, Cerebral metabolism of [1,2-13C2]acetate as detected by in vivo and in vitro 13C NMR, J. Biol. Chem. 265 (1990) 12916−12926, https://doi.org/10.1016/S0021-9258(19)38247-X.

[93] W.J. Nicklas, D.D. Clarke, S. Berl, Decarboxylation studies of glutamate, glutamine, and aspartate from brain labelled with [1-14C] acetate, l-[U-14C]-aspartate, and l-[U-14C]glutamate, J. Neurochem. 16 (1969) 549−558, https://doi.org/10.1111/j.1471-4159.1969.tb06854.x.

[94] B. Lanz, K. Uffmann, M. T Wyss, B. Weber, A. Buck, R. Gruetter, A two-compartment mathematical model of neuroglial metabolism using [1-11C] acetate, J. Cerebr. Blood Flow Metabol. 32 (2012) 548−559, https://doi.org/10.1038/jcbfm.2011.162.

[95] J. Yang, S.S. Li, J. Bacher, J. Shen, Quantification of cortical GABA-glutamine cycling rate using in vivo magnetic resonance signal of [2-13C] GABA derived from glia-specific substrate [2-13C]acetate, Neurochem. Int. 50 (2007) 371−378, https://doi.org/10.1016/j.neuint.2006.09.011.

[96] E. Vinckenbosch, M. Mishkovsky, A. Comment, R. Gruetter, Modulations of cerebral TCA cycle activity studied by hyperpolarized acetate 13C MRS, Proc. Intl. Soc. Mag. Reson. Med. (2015) 114.

[97] D. Abdurrachim, C.C. Woo, X.Q. Teo, W.X. Chan, G.K. Radda, P.T.H. Lee, A new hyperpolarized 13C ketone body probe reveals an increase in acetoacetate utilization in the diabetic rat heart, Sci. Rep. 9 (2019b) 5532, https://doi.org/10.1038/s41598-019-39378-w.

[98] J.J. Miller, D.R. Ball, A.Z. Lau, D.J. Tyler, Hyperpolarized ketone body metabolism in the rat heart, NMR Biomed. 31 (2018a) e3912, https://doi.org/10.1002/nbm.3912.

[99] C. von Morze, M.A. Ohliger, I. Marco-Rius, D.M. Wilson, R.R. Flavell, D. Pearce, D.B. Vigneron, J. Kurhanewicz, Z.J. Wang, Direct assessment of renal mitochondrial redox state using hyperpolarized 13C-acetoacetate, Magn. Reson. Med. 79 (2018) 1862−1869, https://doi.org/10.1002/mrm.27054.

[100] D. Abdurrachim, X.Q. Teo, C.C. Woo, S.Y. Ong, N.F. Salleh, J. Lalic, R.-S. Tan, P.T.H. Lee, Cardiac metabolic modulation upon low-carbohydrate low-protein ketogenic diet in diabetic rats studied in vivo using hyperpolarized 13C pyruvate, butyrate and acetoacetate probes, Diabetes Obes. Metabol. 21 (2019a) 949−960, https://doi.org/10.1111/dom.13608.

[101] W. Chen, G. Sharma, W. Jiang, N.R. Maptue, C.R. Malloy, A.D. Sherry, C. Khemtong, Metabolism of hyperpolarized 13C-acetoacetate to β-hydroxybutyrate detects real-time mitochondrial redox state and dysfunction in heart tissue, NMR Biomed. 32 (2019) e4091, https://doi.org/10.1002/nbm.4091.

[102] G. Sharma, X. Wen, N.R. Maptue, T. Hever, C.R. Malloy, A.D. Sherry, C. Khemtong, Co-polarized [1-13C]pyruvate and [1,3-13C2]Acetoacetate provide a simultaneous view of cytosolic and mitochondrial redox in a single experiment, ACS Sens. 6 (2021) 3967−3977, https://doi.org/10.1021/acssensors.1c01225.

[103] C. Najac, M. Radoul, L.M. Le Page, G. Batsios, E. Subramani, P. Viswanath, A.M. Gillespie, S.M. Ronen, In vivo investigation of hyperpolarized [1,3-13C2]acetoacetate as a metabolic probe in normal brain and in glioma, Sci. Rep. 9 (2019) 3402, https://doi.org/10.1038/s41598-019-39677-2.

[104] L. Mazuel, R.F. Schulte, A. Cladière, C. Spéziale, M. Lagrée, M. Leremboure, B. Jean, F. Durif, C. Chassain, Intracerebral synthesis of glutamine from hyperpolarized glutamate, Magn. Reson. Med. 78 (2017) 1296−1305, https://doi.org/10.1002/mrm.26522.

[105] R.J. DeBerardinis, A. Mancuso, E. Daikhin, I. Nissim, M. Yudkoff, S. Wehrli, C.B. Thompson, Beyond aerobic glycolysis: transformed cells can engage in glutamine metabolism that exceeds the requirement for protein and nucleotide synthesis, Proc. Natl. Acad. Sci. USA 104 (2007) 19345−19350, https://doi.org/10.1073/pnas.0709747104.

[106] H. Dafni, P.E.Z. Larson, S. Hu, H.A.I. Yoshihara, C.S. Ward, H.S. Venkatesh, C. Wang, X. Zhang, D.B. Vigneron, S.M. Ronen, Hyperpolarized 13C spectroscopic imaging informs on hypoxia-inducible factor-1 and Myc activity downstream of platelet-derived growth factor receptor, Cancer Res. 70 (2010) 7400−7410, https://doi.org/10.1158/0008-5472.CAN-10-0883.

[107] F.A. Gallagher, M.I. Kettunen, S.E. Day, D.-E. Hu, J.H. Ardenkjaer-Larsen, R. in t Zandt, P.R. Jensen, M. Karlsson, K. Golman, M.H. Lerche, K.M. Brindle, Magnetic resonance imaging of pH in vivo using hyperpolarized 13C-labelled bicarbonate, Nature 453 (2008) 940−943, https://doi.org/10.1038/nature07017.

[108] F.A. Gallagher, M.I. Kettunen, S.E. Day, M. Lerche, K.M. Brindle, 13C MR spectroscopy measurements of glutaminase activity in human hepatocellular carcinoma cells using hyperpolarized 13C-labeled glutamine, Magn. Reson. Med. 60 (2008) 253−257, https://doi.org/10.1002/mrm.21650.

[109] W. Qu, Z. Zha, B.P. Lieberman, A. Mancuso, M. Stetz, R. Rizzi, K. Ploessl, D. Wise, C. Thompson, H.F. Kung, Facile synthesis [5- 13C-4- 2H 2]-L-glutamine for hyperpolarized MRS imaging of cancer cell metabolism, Acad. Radiol. 18 (2011) 932−939, https://doi.org/10.1016/j.acra.2011.05.002.

[110] C. Cabella, M. Karlsson, C. Canapè, G. Catanzaro, S. Colombo Serra, L. Miragoli, L. Poggi, F. Uggeri, L. Venturi, P.R. Jensen, M.H. Lerche, F. Tedoldi, In vivo and in vitro liver cancer metabolism observed with hyperpolarized [5-13C]glutamine, J. Magn. Reson. 232 (2013) 45−52, https://doi.org/10.1016/j.jmr.2013.04.010.

[111] L. Salamanca-Cardona, H. Shah, A.J. Poot, F.M. Correa, V. Di Gialleonardo, H. Lui, V.Z. Miloushev, K.L. Granlund, S.S. Tee, J.R. Cross, C.B. Thompson, K.R. Keshari, In vivo imaging of glutamine metabolism to the oncometabolite 2-hydroxyglutarate in IDH1/2 mutant tumors, Cell Metabol. 26 (2017) 830−841.e3, https://doi.org/10.1016/j.cmet.2017.10.001.

[112] M.R. Reynolds, A.N. Lane, B. Robertson, S. Kemp, Y. Liu, B.G. Hill, D.C. Dean, B.F. Clem, Control of glutamine metabolism by the tumor suppressor Rb, Oncogene 33 (2014) 556−566, https://doi.org/10.1038/onc.2012.635.

[113] M.M. Chaumeil, P.E.Z. Larson, H.A.I. Yoshihara, O.M. Danforth, D.B. Vigneron, S.J. Nelson, R.O. Pieper, J.J. Phillips, S.M. Ronen, Non-invasive in vivo assessment of IDH1 mutational status in glioma, Nat. Commun. 4 (2013) 2429, https://doi.org/10.1038/ncomms3429.

[114] M.M. Chaumeil, P.E.Z. Larson, S.M. Woods, L. Cai, P. Eriksson, A.E. Robinson, J.M. Lupo, D.B. Vigneron, S.J. Nelson, R.O. Pieper, J.J. Phillips, S.M. Ronen, Hyperpolarized [1-13C] glutamate: a metabolic imaging biomarker of IDH1 mutational status in glioma, Cancer Res. 74 (16) (2014) 4247–4257, https://doi.org/10.1158/0008-5472.CAN-14-0680.

[115] S.A. Butt, L.V. Søgaard, P.O. Magnusson, M.H. Lauritzen, C. Laustsen, P. Åkeson, J.H. Ardenkjær-Larsen, Imaging cerebral 2-ketoisocaproate metabolism with hyperpolarized 13C magnetic resonance spectroscopic imaging, J. Cerebr. Blood Flow Metabol. 32 (2012) 1508–1514, https://doi.org/10.1038/jcbfm.2012.34.

[116] S. Josan, T. Xu, Y.-F. Yen, R. Hurd, J. Ferreira, C.-H. Chen, D. Mochly-Rosen, A. Pfefferbaum, D. Mayer, D. Spielman, In vivo measurement of aldehyde dehydrogenase-2 activity in rat liver ethanol model using dynamic MRSI of hyperpolarized [1-(13) C]pyruvate, NMR Biomed. 26 (2013) 607–612, https://doi.org/10.1002/nbm.2897.

[117] M. Karlsson, P.R. Jensen, R. in 't Zandt, A. Gisselsson, G. Hansson, J.Ø. Duus, S. Meier, M.H. Lerche, Imaging of branched chain amino acid metabolism in tumors with hyperpolarized 13C ketoisocaproate, Int. J. Cancer 127 (2010) 729–736, https://doi.org/10.1002/ijc.25072.

[118] E.H. Suh, E.P. Hackett, R.M. Wynn, D.T. Chuang, B. Zhang, W. Luo, A.D. Sherry, J.M. Park, In vivo assessment of increased oxidation of branched-chain amino acids in glioblastoma, Sci. Rep. 9 (2019) 340, https://doi.org/10.1038/s41598-018-37390-0.

[119] R.E. Ericksen, S.L. Lim, E. McDonnell, W.H. Shuen, M. Vadiveloo, P.J. White, Z. Ding, R. Kwok, P. Lee, G.K. Radda, H.C. Toh, M.D. Hirschey, W. Han, Loss of BCAA catabolism during carcinogenesis enhances mTORC1 activity and promotes tumor development and progression, Cell Metabol. 29 (2019) 1151–1165.e6, https://doi.org/10.1016/j.cmet.2018.12.020.

[120] T. Nishihara, H. Nonaka, T. Naganuma, K. Ichikawa, S. Sando, Mouse lactate dehydrogenase X: a promising magnetic resonance reporter protein using hyperpolarized pyruvic acid derivative, Y. Chem. Sci. 3 (2012) 800–806, https://doi.org/10.1039/C1SC00919B.

[121] C. von Morze, R.A. Bok, M.A. Ohliger, Z. Zhu, D.B. Vigneron, J. Kurhanewicz, Hyperpolarized [(13)C]ketobutyrate, a molecular analog of pyruvate with modified specificity for LDH isoforms, Magn. Reson. Med. 75 (2016) 1894–1900, https://doi.org/10.1002/mrm.25716.

[122] T. Nishihara, H.A.I. Yoshihara, H. Nonaka, Y. Takakusagi, F. Hyodo, K. Ichikawa, E. Can, J.A.M. Bastiaansen, Y. Takado, A. Comment, S. Sando, Direct monitoring of γ-glutamyl transpeptidase activity in vivo using a hyperpolarized 13C-labeled, Molecular Probe 55 (2016) 10626–10629, https://doi.org/10.1002/anie.201603731.

[123] G. Batsios, C. Najac, P. Cao, P. Viswanath, E. Subramani, Y. Saito, A.M. Gillespie, H.A.I. Yoshihara, P. Larson, S. Sando, S.M. Ronen, In vivo detection of γ-glutamyl-transferase up-regulation in glioma using hyperpolarized γ-glutamyl-[1-13C]glycine, Sci. Rep. 10 (2020) 6244, https://doi.org/10.1038/s41598-020-63160-y.

[124] R.E. Hurd, Y.-F. Yen, D. Mayer, A. Chen, D. Wilson, S. Kohler, R. Bok, D. Vigneron, J. Kurhanewicz, J. Tropp, D. Spielman, A. Pfefferbaum, Metabolic imaging in the anesthetized rat brain using hyperpolarized [1-13C] pyruvate and [1-13C] ethyl pyruvate, Magn. Reson. Med. 63 (2010) 1137–1143, https://doi.org/10.1002/mrm.22364.

[125] J.J. Miller, J.T. Grist, S. Serres, J.R. Larkin, A.Z. Lau, K. Ray, K.R. Fisher, E. Hansen, R.S. Tougaard, P.M. Nielsen, J. Lindhardt, C. Laustsen, F.A. Gallagher, D.J. Tyler, N. Sibson, ^{13}C pyruvate transport across the blood-brain barrier in preclinical hyperpolarised MRI, Sci. Rep. 8 (1) (2018) 15082, https://doi.org/10.1038/s41598-018-33363-5.

[126] J. Chen, E.P. Hackett, J. Singh, Z. Kovacs, J.M. Park, Simultaneous assessment of intracellular and extracellular pH using hyperpolarized [1-13C] alanine ethyl ester, Anal. Chem. 92 (2020) 11681–11686, https://doi.org/10.1021/acs.analchem.0c01568.

[127] J. Singh, E.H. Suh, G. Sharma, J. Chen, E.P. Hackett, X. Wen, A.D. Sherry, C. Khemtong, C.R. Malloy, J.M. Park, Z. Kovacs, 13C-Labeled diethyl ketoglutarate derivatives as hyperpolarized probes of 2-ketoglutarate dehydrogenase activity, Anal. Sens. 1 (2021) 156–160, https://doi.org/10.1002/anse.202100021.

[128] P.R. Jensen, S.C. Serra, L. Miragoli, M. Karlsson, C. Cabella, L. Poggi, L. Venturi, F. Tedoldi, M.H. Lerche, Hyperpolarized [1,3-(13) C2]ethyl acetoacetate is a novel diagnostic metabolic marker of liver cancer, Int. J. Cancer J. Int. Cancer 136 (2014) E117–E126, https://doi.org/10.1002/ijc.29162.

[129] A. Lodi, S.M. Woods, S.M. Ronen, Treatment with the MEK inhibitor U0126 induces decreased hyperpolarized pyruvate to lactate conversion in breast, but not prostate, cancer cells, NMR Biomed. 26 (2013) 299–306, https://doi.org/10.1002/nbm.2848.

[130] F.A. Gallagher, M.I. Kettunen, D.-E. Hu, P.R. Jensen, R. in 't Zandt, M. Karlsson, A. Gisselsson, S.K. Nelson, T.H. Witney, S.E. Bohndiek, G. Hansson, T. Peitersen, M.H. Lerche, K.M. Brindle, Production of hyperpolarized [1,4-13C2]malate from [1,4-13C2]fumarate is a marker of cell necrosis and treatment response in tumors, Proc. Natl. Acad. Sci. USA 106 (2009) 19801–19806, https://doi.org/10.1073/pnas.0911447106.

[131] S.E. Bohndiek, M.I. Kettunen, D. Hu, K.M. Brindle, Hyperpolarized 13C spectroscopy detects early changes in tumor vasculature and metabolism after VEGF neutralization, Cancer Res. 72 (2012) 854–864, https://doi.org/10.1158/0008-5472.CAN-11-2795.

[132] M.R. Clatworthy, M.I. Kettunen, D.-E. Hu, R.J. Mathews, T.H. Witney, B.W.C. Kennedy, S.E. Bohndiek, F.A. Gallagher, L.B. Jarvis, K.G.C. Smith, K.M. Brindle, Magnetic resonance imaging with hyperpolarized [1,4-13C2]fumarate allows detection of early renal acute tubular necrosis, Proc. Natl. Acad. Sci. USA 109 (2012) 13374–13379, https://doi.org/10.1073/pnas.1205539109.

[133] C. Laustsen, P.M. Nielsen, H. Qi, M.H. Løbner, J. Palmfeldt, L.B. Bertelsen, Hyperpolarized [1,4-13C]fumarate imaging detects microvascular complications and hypoxia mediated cell death in diabetic nephropathy, Sci. Rep. 10 (2020) 9650, https://doi.org/10.1038/s41598-020-66265-6.

[134] J.J. Miller, A.Z. Lau, P.M. Nielsen, G. McMullen-Klein, A.J. Lewis, N.R. Jespersen, V. Ball, F.A. Gallagher, C.A. Carr, C. Laustsen, H.E. Bøtker, D.J. Tyler, M.A. Schroeder, Hyperpolarized [1,4-13C2]Fumarate enables magnetic resonance-based imaging of myocardial necrosis, Jacc Cardiovasc. Imaging 11 (2018b) 1594–1606, https://doi.org/10.1016/j.jcmg.2017.09.020.

[135] M. Singh, S. Josan, M. Zhu, A. Jhajharia, D. Mayer, Dynamic metabolic imaging of copolarized [2-13 C]pyruvate and [1,4-13 C2]fumarate using 3D-spiral CSI with alternate spectral band excitation, Magn. Reson. Med. 81 (2019) 2869–2877, https://doi.org/10.1002/mrm.27639.

[136] D.E. Korenchan, R.R. Flavell, Spatiotemporal pH heterogeneity as a promoter of cancer progression and therapeutic resistance, Cancers 11 (2019), https://doi.org/10.3390/cancers11071026.

[137] C. Ward, J. Meehan, M.E. Gray, A.F. Murray, D.J. Argyle, I.H. Kunkler, S.P. Langdon, The impact of tumour pH on cancer progression: strategies for clinical intervention, Explor Target Antitumor Ther. 1 (2020) 71−100, https://doi.org/10.37349/etat.2020.00005.

[138] C. Mu, D.E. Korenchan, S. Wang, D.M. Wilson, R.R. Flavell, Tumor microenvironment biosensors for hyperpolarized carbon-13 magnetic resonance spectroscopy, Mol. Imag. Biol. 23 (2021) 323−334, https://doi.org/10.1007/s11307-020-01570-0.

[139] X. Zhang, Y. Lin, R.J. Gillies, Tumor pH and its measurement, J. Nucl. Med. 51 (2010) 1167−1170, https://doi.org/10.2967/jnumed.109.068981.

[140] R.B. Reeves, Temperature-induced changes in blood acid-base status: pH and PCO2 in a binary buffer, J. Appl. Physiol. 40 (1976) 752−761, https://doi.org/10.1152/jappl.1976.40.5.752.

[141] D.E. Korenchan, R. Bok, R. Sriram, K. Liu, R.D. Santos, H. Qin, I. Lobach, N. Korn, D.M. Wilson, J. Kurhanewicz, R.R. Flavell, Hyperpolarized in vivo pH imaging reveals grade-dependent acidification in prostate cancer, Oncotarget 10 (2019) 6096−6110, https://doi.org/10.18632/oncotarget.27225.

[142] D.E. Korenchan, R.R. Flavell, C. Baligand, R. Sriram, K. Neumann, S. Sukumar, H. VanBrocklin, D.B. Vigneron, D.M. Wilson, J. Kurhanewicz, Dynamic nuclear polarization of biocompatible 13C-enriched carbonates for in vivo pH imaging, Chem. Commun. 52 (2016) 3030−3033, https://doi.org/10.1039/C5CC09724J.

[143] R.K. Ghosh, S.J. Kadlecek, M. Pourfathi, R.R. Rizi, Efficient production of hyperpolarized bicarbonate by chemical reaction on a DNP precursor to measure pH, Magn. Reson. Med. 74 (2015) 1406−1413, https://doi.org/10.1002/mrm.25530.

[144] N. Maptue, W. Jiang, C. Harrison, A.M. Funk, G. Sharma, C.R. Malloy, D. Sherry, C. Khemtong, Esterase-catalyzed production of hyperpolarized 13C-enriched carbon dioxide in tissues for measuring pH, ACS Sens. 3 (2018) 2232−2236, https://doi.org/10.1021/acssensors.8b01097.

[145] M.A. Schroeder, P. Swietach, H.J. Atherton, F.A. Gallagher, P. Lee, G.K. Radda, K. Clarke, D.J. Tyler, Measuring intracellular pH in the heart using hyperpolarized carbon dioxide and bicarbonate: a 13C and 31P magnetic resonance spectroscopy study, Cardiovasc. Res. 86 (2010) 82−91, https://doi.org/10.1093/cvr/cvp396.

[146] C.-M. Moon, Y.-H. Kim, Y.-K. Ahn, M.-H. Jeong, G.-W. Jeong, Metabolic alterations in acute myocardial ischemia-reperfusion injury and necrosis using in vivo hyperpolarized [1-13C] pyruvate MR spectroscopy, Sci. Rep. 9 (2019) 18427, https://doi.org/10.1038/s41598-019-54965-7.

[147] S. Düwel, C. Hundshammer, M. Gersch, B. Feuerecker, K. Steiger, A. Buck, A. Walch, A. Haase, S.J. Glaser, M. Schwaiger, F. Schilling, Imaging of pH in vivo using hyperpolarized (13)C-labelled zymonic acid, Nat. Commun. 8 (2017) 15126, https://doi.org/10.1038/ncomms15126.

[148] C. Hundshammer, S. Düwel, S.S. Köcher, M. Gersch, B. Feuerecker, C. Scheurer, A. Haase, S.J. Glaser, M. Schwaiger, F. Schilling, Deuteration of hyperpolarized (13) C-labeled zymonic acid enables sensitivity-enhanced dynamic MRI of pH, Chemphyschem 18 (2017) 2422−2425, https://doi.org/10.1002/cphc.201700779.

[149] T. Nishihara, Y. Kameyama, H. Nonaka, Y. Takakusagi, F. Hyodo, K. Ichikawa, S. Sando, A strategy to design hyperpolarized (13) C magnetic resonance probes using [1-(13) C]α-Amino acid as a scaffold structure, Chem. Asian J. 12 (2017) 949−953, https://doi.org/10.1002/asia.201700098.

[150] C. Hundshammer, S. Düwel, D. Ruseckas, G. Topping, P. Dzien, C. Müller, B. Feuerecker, J.B. Hövener, A. Haase, M. Schwaiger, S.J. Glaser, F. Schilling, Hyperpolarized amino acid derivatives as multivalent magnetic resonance pH sensor molecules, Sensors 18 (2018) 600, https://doi.org/10.3390/s18020600.

[151] G.D. Reed, C. von Morze, R. Bok, B.L. Koelsch, M. Van Criekinge, K.J. Smith, H. Shang, P.E.Z. Larson, J. Kurhanewicz, D.B. Vigneron, High resolution $^{13\$}$C MRI with hyperpolarized urea: in vivo $_2$ mapping and $^{15\$}$N labeling effects, IEEE Trans. Med. Imag. 33 (2014) 362−371, https://doi.org/10.1109/TMI.2013.2285120.

[152] A. Radaelli, D. Ortiz, A. Michelotti, M. Roche, R. Hata, S. Sando, O. Bonny, R. Gruetter, H.A.I. Yoshihara, Hyperpolarized (1-13C)alaninamide is a multifunctional in vivo sensor of aminopeptidase N activity, pH, and CO2, ACS Sens. 7 (2022) 2987−2994, https://doi.org/10.1021/acssensors.2c01203.

[153] R.R. Flavell, C. von Morze, J.E. Blecha, D.E. Korenchan, M. Van Criekinge, R. Sriram, J.W. Gordon, H.-Y. Chen, S. Subramaniam, R.A. Bok, Z.J. Wang, D.B. Vigneron, P.E. Larson, J. Kurhanewicz, D.M. Wilson, Application of Good's buffers to pH imaging using hyperpolarized (13)C MRI, Chem. Commun. Camb. Engl. 51 (2015) 14119−14122, https://doi.org/10.1039/c5cc05348j.

[154] D.E. Korenchan, C. Taglang, C. von Morze, J.E. Blecha, J.W. Gordon, R. Sriram, P.E.Z. Larson, D.B. Vigneron, H.F. VanBrocklin, J. Kurhanewicz, D.M. Wilson, R.R. Flavell, Dicarboxylic acids as pH sensors for hyperpolarized (13)C magnetic resonance spectroscopic imaging, Analyst 142 (2017) 1429−1433, https://doi.org/10.1039/c7an00076f.

[155] K.R. Keshari, J. Kurhanewicz, R. Bok, P.E.Z. Larson, D.B. Vigneron, D.M. Wilson, Hyperpolarized 13C dehydroascorbate as an endogenous redox sensor for in vivo metabolic imaging, Proc. Natl. Acad. Sci. USA 108 (2011) 18606−18611, https://doi.org/10.1073/pnas.1106920108.

[156] S.E. Bohndiek, M.I. Kettunen, D.-E. Hu, B.W.C. Kennedy, J. Boren, F.A. Gallagher, K.M. Brindle, Hyperpolarized [1-13C]-ascorbic and dehydroascorbic acid: vitamin C as a probe for imaging redox status in vivo, J. Am. Chem. Soc. 133 (2011) 11795−11801, https://doi.org/10.1021/ja2045925.

[157] D.M. Wilson, V. Di Gialleonardo, Z.J. Wang, V. Carroll, C. von Morze, A. Taylor, V. Sai, M. Vancriekinge, R. Bok, M.A. Ohliger, K.R. Keshari, Hyperpolarized (13)C spectroscopic evaluation of oxidative stress in a rodent model of steatohepatitis, Sci. Rep. 7 (2017) 46014, https://doi.org/10.1038/srep46014.

[158] K. Yamamoto, A. Opina, D. Sail, B. Blackman, K. Saito, J.R. Brender, R.M. Malinowski, T. Seki, N. Oshima, D.R. Crooks, S. Kishimoto, Y. Saida, Y. Otowa, P.L. Choyke, J.H. Ardenkjaer-Larsen, J.B. Mitchell, W.M. Linehan, R.E. Swenson, M.C. Krishna, Real-Time insight into in vivo redox status utilizing hyperpolarized [1-$^{13\$}$C] N-acetyl cysteine, Sci. Rep. 11 (2021) 12155, https://doi.org/10.1038/s41598-021-90921-0.

[159] C.E. Christensen, M. Karlsson, J.R. Winther, P.R. Jensen, M.H. Lerche, Non-invasive in-cell determination of free cytosolic [NAD+]/[NADH] ratios using hyperpolarized glucose show large variations in metabolic phenotypes, J. Biol. Chem. 289 (2014) 2344–2352, https://doi.org/10.1074/jbc.M113.498626.

[160] H. Qi, P.M. Nielsen, M. Schroeder, L.B. Bertelsen, F. Palm, C. Laustsen, Acute renal metabolic effect of metformin assessed with hyperpolarised MRI in rats, Diabetologia 61 (2018) 445–454, https://doi.org/10.1007/s00125-017-4445-6.

[161] A.R. Lippert, K.R. Keshari, J. Kurhanewicz, C.J. Chang, A hydrogen peroxide-responsive hyperpolarized 13C MRI contrast agent, J. Am. Chem. Soc. 133 (2011) 3776–3779.

[162] M. Giorgio, M. Trinei, E. Migliaccio, P.G. Pelicci, Hydrogen peroxide: a metabolic by-product or a common mediator of ageing signals? Nat. Rev. Mol. Cell Biol. 8 (2007) 722–728, https://doi.org/10.1038/nrm2240.

[163] A. Wibowo, J.M. Park, S.-C. Liu, C. Khosla, D.M. Spielman, Real-time in vivo detection of H2O2 using hyperpolarized 13C-thiourea, ACS Chem. Biol. 12 (2017) 1737–1742, https://doi.org/10.1021/acschembio.7b00130.

[164] H. Nonaka, R. Hata, T. Doura, T. Nishihara, K. Kumagai, M. Akakabe, M. Tsuda, K. Ichikawa, S. Sando, A platform for designing hyperpolarized magnetic resonance chemical probes, Nat. Commun. 4 (2013) 2411, https://doi.org/10.1038/ncomms3411.

[165] R. Hata, H. Nonaka, Y. Takakusagi, K. Ichikawa, S. Sando, Design of a hyperpolarized 15N NMR probe that induces a large chemical-shift change upon binding of calcium ions, Chem. Commun. Camb. Engl. 51 (2015) 12290–12292, https://doi.org/10.1039/C5CC04597E.

[166] A. Mishra, G. Pariani, T. Oerther, M. Schwaiger, G.G. Westmeyer, Hyperpolarized multi-metal 13C-sensors for magnetic resonance imaging, Anal. Chem. 88 (2016) 10790–10794, https://doi.org/10.1021/acs.analchem.6b03546.

[167] E. Ho, Y. Song, Zinc and prostatic cancer, Curr. Opin. Clin. Nutr. Metab. Care 12 (2009) 640, https://doi.org/10.1097/MCO.0b013e32833106ee.

[168] S. Wang, D.E. Korenchan, P.M. Perez, C. Taglang, T.R. Hayes, R. Sriram, R. Bok, A.S. Hong, Y. Wu, H. Li, Z. Wang, J. Kurhanewicz, D.M. Wilson, R.R. Flavell, Amino acid-derived sensors for specific Zn2+ detection using hyperpolarized 13 C magnetic resonance spectroscopy, Chem. Weinh. Bergstr. Ger. 25 (2019) 11842–11846, https://doi.org/10.1002/chem.201902771.

[169] E.H. Suh, J.M. Park, L. Lumata, A.D. Sherry, Z. Kovacs, Hyperpolarized 15N-labeled, deuterated tris (2-pyridylmethyl)amine as an MRI sensor of freely available Zn2+, Commun. Chem. 3 (2020) 185, https://doi.org/10.1038/s42004-020-00426-6.

[170] J.M. Sands, Mammalian urea transporters, Annu. Rev. Physiol. 65 (2003) 543–566, https://doi.org/10.1146/annurev.physiol.65.092101.142638.

[171] K. Golman, J.H. Ardenkjaer-Larsen, J.S. Petersson, S. Månsson, I. Leunbach, Molecular imaging with endogenous substances, Proc. Natl. Acad. Sci. U. S. A 100 (2003) 10435–10439, https://doi.org/10.1073/pnas.1733836100.

[172] C. von Morze, R.A. Bok, J.M. Sands, J. Kurhanewicz, D.B. Vigneron, Monitoring urea transport in rat kidney in vivo using hyperpolarized 13C magnetic resonance imaging, Am. J. Physiol. Ren. Physiol. 302 (2012a) F1658–F1662, https://doi.org/10.1152/ajprenal.00640.2011.

[173] E.S.S. Hansen, N.J. Stewart, J.M. Wild, H. Stødkilde-Jørgensen, C. Laustsen, Hyperpolarized 13 C,15 N2 -Urea MRI for assessment of the urea gradient in the porcine kidney, Magn. Reson. Med. 76 (2016) 1895–1899, https://doi.org/10.1002/mrm.26483.

[174] L.B. Bertelsen, P.M. Nielsen, H. Qi, T.S. Nørlinger, X. Zhang, H. Stødkilde-Jørgensen, C. Laustsen, Diabetes induced renal urea transport alterations assessed with 3D hyperpolarized (13) C,(15) N-urea, Magn. Reson. Med. 77 (2016) 1650–1655, https://doi.org/10.1002/mrm.26256.

[175] P.M. Nielsen, E.S. Szocska Hansen, T.S. Nørlinger, R. Nørregaard, L. Bonde Bertelsen, H. Stødkilde-Jørgensen, C. Laustsen, Renal ischemia and reperfusion assessment with three-dimensional hyperpolarized13C,15N2-urea, Magn. Reson. Med. 76 (2016) 1524–1530, https://doi.org/10.1002/mrm.26377.

[176] C.Ø. Mariager, P.M. Nielsen, H. Qi, S. Ringgaard, C. Laustsen, Hyperpolarized 13C,15N2-urea T2 relaxation changes in acute kidney injury, Magn. Reson. Med. 80 (2018) 696–702, https://doi.org/10.1002/mrm.27050.

[177] C. Laustsen, T. Stokholm Nørlinger, D. Christoffer Hansen, H. Qi, P. Mose Nielsen, L. Bonde Bertelsen, J. Henrik Ardenkjaer-Larsen, H. Stødkilde-Jørgensen, Hyperpolarized 13C urea relaxation mechanism reveals renal changes in diabetic nephropathy, Magn. Reson. Med. 75 (2016) 515–518, https://doi.org/10.1002/mrm.26036.

[178] G.D. Reed, C. von Morze, A.S. Verkman, B.L. Koelsch, M.M. Chaumeil, M. Lustig, S.M. Ronen, R.A. Bok, J.M. Sands, P.E.Z. Larson, Z.J. Wang, J.H.A. Larsen, J. Kurhanewicz, D.B. Vigneron, Imaging renal urea handling in rats at millimeter resolution using hyperpolarized magnetic resonance relaxometry, Tomogr. J. Imaging Res. 2 (2016) 125–135, https://doi.org/10.18383/j.tom.2016.00127.

[179] J.Y.C. Lau, A.P. Chen, Y.-P. Gu, C.H. Cunningham, Voxel-by-voxel correlations of perfusion, substrate, and metabolite signals in dynamic hyperpolarized 13C imaging, NMR Biomed. 29 (2016) 1038–1047, https://doi.org/10.1002/nbm.3564.

[180] M. Fuetterer, J. Busch, S.M. Peereboom, C. von Deuster, L. Wissmann, M. Lipiski, T. Fleischmann, N. Cesarovic, C.T. Stoeck, S. Kozerke, Hyperpolarized (13)C urea myocardial first-pass perfusion imaging using velocity-selective excitation, J. Cardiovasc. Magn. Reson. 19 (2017) 46, https://doi.org/10.1186/s12968-017-0364-4.

[181] N. Bahrami, C.L. Swisher, C. von Morze, D.B. Vigneron, P.E.Z. Larson, Kinetic and perfusion modeling of hyperpolarized (13)C pyruvate and urea in cancer with arbitrary RF flip angles, Quant. Imag. Med. Surg. 4 (2014) 24–32, https://doi.org/10.3978/j.issn.2223-4292.2014.02.02.

[182] H.-Y. Chen, P.E.Z. Larson, R.A. Bok, C. von Morze, R. Sriram, R. Delos Santos, J. DeLos Santos, J.W. Gordon, N. Bahrami, M. Ferrone, J. Kurhanewicz, D.B. Vigneron, Assessing prostate cancer aggressiveness with hyperpolarized dual-agent 3D dynamic imaging of metabolism and perfusion, Cancer Res. 77 (2017) 3207–3216, https://doi.org/10.1158/0008-5472.CAN-16-2083.

[183] R. Bok, J. Lee, R. Sriram, K. Keshari, S. Sukumar, S. Daneshmandi, D.E. Korenchan, R.R. Flavell, D.B. Vigneron, J. Kurhanewicz, P. Seth, The role of lactate metabolism in prostate cancer progression and metastases revealed by dual-agent hyperpolarized ^{13}C MRSI, Cancers 11 (2) (2019) 257, https://doi.org/10.3390/cancers11020257.

[184] H. Qin, V. Zhang, R.A. Bok, R.D. Santos, J.A. Cunha, I.-C. Hsu, J.D. Santos Bs, J.E. Lee, S. Sukumar, P.E.Z. Larson, D.B. Vigneron, D.M. Wilson, R. Sriram, J. Kurhanewicz, Simultaneous metabolic and perfusion imaging using hyperpolarized 13C MRI can evaluate early and dose-dependent response to radiation therapy in a prostate cancer mouse model, Int. J. Radiat. Oncol. Biol. Phys. 107 (2020) 887–896, https://doi.org/10.1016/j.ijrobp.2020.04.022.

[185] J.E. Lee, C.J. Diederich, R. Bok, R. Sriram, R.D. Santos, S.M. Noworolski, V.A. Salgaonkar, M.S. Adams, D.B. Vigneron, J. Kurhanewicz, Assessing high-intensity focused ultrasound treatment of prostate cancer with hyperpolarized 13C dual-agent imaging of metabolism and perfusion, NMR Biomed 317 (2018) e3962, https://doi.org/10.1002/nbm.3962.

[186] S. Düwel, M. Durst, C.V. Gringeri, Y. Kosanke, C. Gross, M.A. Janich, A. Haase, S.J. Glaser, M. Schwaiger, R.F. Schulte, R. Braren, M.I. Menzel, Multiparametric human hepatocellular carcinoma characterization and therapy response evaluation by hyperpolarized (13) C MRSI, NMR Biomed. 29 (2016) 952–960, https://doi.org/10.1002/nbm.3561.

[187] H. Qin, S. Tang, A.M. Riselli, R.A. Bok, R.D. Santos, M. Van Criekinge, J.W. Gordon, R. Aggarwal, R. Chen, G. Goddard, C.T. Zhang, A. Chen, G. Reed, D.M. Ruscitto, J. Slater, R. Sriram, P.E.Z. Larson, D.B. Vigneron, J. Kurhanewicz, Clinical translation of hyperpolarized 13C pyruvate and urea MRI for simultaneous metabolic and perfusion imaging, Magn. Reson. Med. 87 (2022) 138–149, https://doi.org/10.1002/mrm.28965.

[188] E. Milshteyn, G.D. Reed, J.W. Gordon, C. von Morze, P. Cao, S. Tang, A.P. Leynes, P.E.Z. Larson, D.B. Vigneron, Simultaneous T1 and T2 mapping of hyperpolarized 13C compounds using the bSSFP sequence, J. Magn. Reson. 312 (2020) 106691, https://doi.org/10.1016/j.jmr.2020.106691.

[189] J. Svensson, S. Månsson, E. Johansson, J.S. Petersson, L.E. Olsson, Hyperpolarized 13C MR angiography using trueFISP, Magn. Reson. Med. 50 (2003) 256–262, https://doi.org/10.1002/mrm.10530.

[190] C. von Morze, P.E.Z. Larson, S. Hu, H.A.I. Yoshihara, R.A. Bok, A. Goga, J.H. Ardenkjaer-Larsen, D.B. Vigneron, Investigating tumor perfusion and metabolism using multiple hyperpolarized (13)C compounds: HP001, pyruvate and urea, Magn. Reson. Imaging 30 (2012b) 305–311, https://doi.org/10.1016/j.mri.2011.09.026.

[191] A.K. Grant, E. Vinogradov, X. Wang, R.E. Lenkinski, D.C. Alsop, Perfusion imaging with a freely diffusible hyperpolarized contrast agent, Magn. Reson. Med. 66 (2011) 746–755, https://doi.org/10.1002/mrm.22860.

[192] C. von Morze, R.A. Bok, G.D. Reed, J.H. Ardenkjaer-Larsen, J. Kurhanewicz, D.B. Vigneron, Simultaneous multiagent hyperpolarized (13)C perfusion imaging, Magn. Reson. Med. 72 (2014) 1599–1609, https://doi.org/10.1002/mrm.25071.

[193] C. von Morze, J.A. Engelbach, G.D. Reed, A.P. Chen, J.D. Quirk, T. Blazey, R. Mahar, C.R. Malloy, J.R. Garbow, M.E. Merritt, 15 N-carnitine, a novel endogenous hyperpolarized MRI probe with long signal lifetime, Magn. Reson. Med. 85 (2021) 1814–1820, https://doi.org/10.1002/mrm.28578.

[194] C. Cudalbu, A. Comment, F. Kurdzesau, R.B. van Heeswijk, K. Uffmann, S. Jannin, V. Denisov, D. Kirik, R. Gruetter, Feasibility of in vivo 15N MRS detection of hyperpolarized 15N labeled choline in rats, Phys. Chem. Chem. Phys. PCCP 12 (2010) 5818–5823, https://doi.org/10.1039/c002309b.

[195] M. Durst, E. Chiavazza, A. Haase, S. Aime, M. Schwaiger, R.F. Schulte, α-trideuteromethyl[15N]glutamine: a long-lived hyperpolarized perfusion marker, Magn. Reson. Med. 76 (2016) 1900–1904, https://doi.org/10.1002/mrm.26104.

[196] H.U. Kauczor, R. Surkau, T. Roberts, MRI using hyperpolarized noble gases, Eur. Radiol. 8 (1998) 820–827, https://doi.org/10.1007/s003300050479.

[197] A. Capozzi, C. Roussel, A. Comment, Optimal glass-forming solvent brings sublimation dynamic nuclear polarization to 129Xe hyperpolarization biomedical imaging standards, J. Phys. Chem. C 119 (2015) 5020–5025, https://doi.org/10.1021/jp5124053.

[198] A. Comment, S. Jannin, J.-N. Hyacinthe, P. Miéville, R. Sarkar, P. Ahuja, P.R. Vasos, X. Montet, F. Lazeyras, J.-P. Vallée, P. Hautle, J.A. Konter, B. van den Brandt, J.-P. Ansermet, R. Gruetter, G. Bodenhausen, Hyperpolarizing gases via dynamic nuclear polarization and sublimation, Phys. Rev. Lett. 105 (2010) 018104, https://doi.org/10.1103/PhysRevLett.105.018104.

[199] M. Pourfathi, J. Clapp, S.J. Kadlecek, C.D. Keenan, R.K. Ghosh, N.N. Kuzma, R.R. Rizi, Low-temperature dynamic nuclear polarization of gases in Frozen mixtures, Magn. Reson. Med. 76 (2016) 1007–1014, https://doi.org/10.1002/mrm.26002.

[200] S. Fain, M.L. Schiebler, D.G. McCormack, G. Parraga, Imaging of lung function using hyperpolarized helium-3 magnetic resonance imaging: review of current and emerging translational methods and applications, J. Magn. Reson. Imaging JMRI 32 (2010) 1398–1408, https://doi.org/10.1002/jmri.22375.

[201] L. Ebner, J. Kammerman, B. Driehuys, M.L. Schiebler, R.V. Cadman, S.B. Fain, The role of hyperpolarized 129xenon in MR imaging of pulmonary function, Eur. J. Radiol. 86 (2017) 343–352, https://doi.org/10.1016/j.ejrad.2016.09.015.

[202] K. Sakai, A.M. Bilek, E. Oteiza, R.L. Walsworth, D. Balamore, F.A. Jolesz, M.S. Albert, Temporal dynamics of hyperpolarized 129Xe resonances in living rats, J. Magn. Reson. B 111 (1996) 300–304, https://doi.org/10.1006/jmrb.1996.0098.

[203] J.M. Wang, S.H. Robertson, Z. Wang, M. He, R.S. Virgincar, G.M. Schrank, R.M. Smigla, T.G. O'Riordan, J. Sundy, L. Ebner, C.R. Rackley, P. McAdams, B. Driehuys, Using hyperpolarized 129Xe MRI to quantify regional gas transfer in idiopathic pulmonary fibrosis, Thorax 73 (2018) 21–28, https://doi.org/10.1136/thoraxjnl-2017-210070.

[204] K. Qing, J.P. Mugler, T.A. Altes, Y. Jiang, J.F. Mata, G.W. Miller, I.C. Ruset, F.W. Hersman, K. Ruppert, Assessment of lung function in asthma and COPD using hyperpolarized 129Xe chemical shift saturation recovery spectroscopy and dissolved-phase MRI, NMR Biomed. 27 (2014) 1490–1501, https://doi.org/10.1002/nbm.3179.

[205] Z. Wang, E.A. Bier, A. Swaminathan, K. Parikh, J. Nouls, M. He, J.G. Mammarappallil, S. Luo, B. Driehuys, S. Rajagopal, Diverse cardiopulmonary diseases are associated with distinct xenon magnetic resonance imaging signatures, Eur. Respir. J. 54 (2019), https://doi.org/10.1183/13993003.00831-2019.

[206] J. Jayapaul, L. Schröder, Molecular sensing with host systems for hyperpolarized 129Xe, Molecules 25 (2020) 4627, https://doi.org/10.3390/molecules25204627.

[207] M. Rao, N.J. Stewart, G. Norquay, P.D. Griffiths, J.M. Wild, High resolution spectroscopy and chemical shift imaging of hyperpolarized (129) Xe dissolved in the human brain in vivo at 1.5 tesla, Magn. Reson. Med. 75 (2016) 2227–2234, https://doi.org/10.1002/mrm.26241.

[208] M.R. Rao, N.J. Stewart, P.D. Griffiths, G. Norquay, J.M. Wild, Imaging human brain perfusion with inhaled hyperpolarized 129Xe MR imaging, Radiology 286 (2018) 659–665, https://doi.org/10.1148/radiol.2017162881.

[209] Y. Shepelytskyi, F.T. Hane, V. Grynko, T. Li, A. Hassan, M.S. Albert, Hyperpolarized 129Xe time-of-flight MR imaging of perfusion and brain function, Diagnostics 10 (2020), https://doi.org/10.3390/diagnostics10090630.

[210] J. Chacon-Caldera, A. Maunder, M. Rao, G. Norquay, O.I. Rodgers, M. Clemence, C. Puddu, L.R. Schad, J.M. Wild, Dissolved hyperpolarized xenon-129 MRI in human kidneys, Magn. Reson. Med. 83 (2020) 262–270, https://doi.org/10.1002/mrm.27923.

[211] P.J. Niedbalski, C.S. Hall, M. Castro, R.L. Eddy, J.H. Rayment, S. Svenningsen, G. Parraga, B. Zanette, G.E. Santyr, R.P. Thomen, N.J. Stewart, G.J. Collier, H.-F. Chan, J.M. Wild, S.B. Fain, G.W. Miller, J.F. Mata, J.P. Mugler, B. Driehuys, M.M. Willmering, Z.I. Cleveland, J.C. Woods, Protocols for multi-site trials using hyperpolarized 129 Xe MRI for imaging of ventilation, alveolar-airspace size, and gas exchange: a position paper from the 129 Xe MRI clinical trials consortium, Magn. Reson. Med. 86 (2021) 2966–2986, https://doi.org/10.1002/mrm.28985.

Chapter 5

Technical considerations of MRI methods for validating DNP probes in small animals

Crystal E. Harrison[1], Jun Chen[1], Sung-Han Lin[1] and Jae Mo Park[1,2,3,*]

[1]Advanced Imaging Research Center, The University of Texas Southwestern Medical Center, Dallas, TX, United States; [2]Department of Biomedical Engineering, The University of Texas Southwestern Medical Center, Dallas, TX, United States; [3]Department of Radiology, The University of Texas Southwestern Medical Center, Dallas, TX, United States

**Corresponding author: E-mail: jaemo.park@utsouthwestern.edu*

List of abbreviations

ASCT Alanine-serine-cysteine transporter
BBB Blood-brain barrier
CSI Chemical shift imaging
DNP Dynamic nuclear polarization
EPI Echoplanar imaging
EPSI Echoplanar spectroscopic imaging
FID Free induction decay
FUS Focused ultrasound
HP Hyperpolarized
IDEAL Iterative decomposition of water and fat with echo asymmetry and least-squares estimation
LDH Lactate dehydrogenase
MCT Monocarboxylate transporter
MR Magnetic resonance
MRSI Magnetic resonance spectroscopic imaging
NAD(H) Nicotinamide adenine dinucleotide
NMR Nuclear magnetic resonance
PC Pyruvate carboxylase
PDH Pyruvate dehydrogenase
PDK Pyruvate dehydrogenase kinase
PEPCK Phosphoenolpyruvate carboxykinase
PET Positron emission tomography
RF Radiofrequency
SNR Signal-to-noise ratio
T1 Spin-lattice relaxation time
TCA Tricarboxylic acid

5.1 Introduction

Due to their essential roles in metabolic processes and ideal characteristics for tracing specific metabolic pathways, carbon and nitrogen have been targeted as nuclei for labeling with either stable isotopes or radioisotopes in developing probes that explore metabolic pathways. Stable isotopes such as carbon-13 (^{13}C) and nitrogen-15 (^{15}N) are of special interest in

The Chemistry of Hyperpolarized Magnetic Resonance Probes. https://doi.org/10.1016/B978-0-323-91842-8.00004-5

dissolution dynamic nuclear polarization (DNP) for their nuclear magnetic resonance (NMR) characteristics. The natural abundance of ^{13}C and ^{15}N is only 1.1% and 0.4%, respectively, and thus the administration of hyperpolarized (HP) agents that are labeled in certain carbon or nitrogen positions allows us to investigate the specific metabolic pathways without interference from background signals of intrinsic pools. Based on their in vivo utility, HP probes can be classified into metabolic substrates, biological sensors, and non-biological agents. As the unique strength of dissolution DNP is non-invasive visualization of metabolic processes, its applications are largely focused on assessing in vivo metabolism. Experimental design and data analysis are often not trivial for in vivo studies with HP probes. This chapter will summarize components to consider when designing in vivo experiments for HP probes and will discuss what to consider when analyzing and interpreting the in vivo data.

5.2 Considerations for MR acquisition

Metabolic assessment with HP substrates can be viewed as an in vivo enzyme assay (Fig. 5.1) [1]. For a proper assessment of the enzymatic reaction of interest, the HP agent needs to be delivered to the targeted cellular compartment at a sufficient concentration. Acquisition methods such as magnetic fields, pulse sequences, and radiofrequency (RF) coils should be carefully chosen to accurately measure the HP signals of interest. Establishing desired metabolic conditions by selecting an optimal animal model and metabolic modulation is another critical component for evaluating the performance of HP probes.

5.2.1 Administration of HP probes

Unlike other exogenous tracer-type probes such as positron emission tomography (PET) probes that are in the nM-μM, the concentration of HP solution is typically in the mM range when used for in vivo studies. This is primarily because the in vivo sensitivity of HP signals is limited even with hyperpolarization, and the metabolic capacity of most enzymatic reactions is large. Although such a high dosage is often super-physiological, it is assumed that in vivo metabolic profiles observed with HP probes reflect unperturbed metabolic status because the HP solution is administered as a bolus and the subsequent MR data acquisition takes only a couple of minutes for most in vivo experiments. HP solution is typically administered intravenously via a tail vein or jugular vein, or rarely intraperitoneally, but both intravenous (*i.v.*) and intraperitoneal (*i.p.*) are acceptable. The injection rate is typically 0.03–0.5 mL/s, but the relationship between the injection rate and the acquired data is underexplored [2–6].

Typically, the higher the concentration of the administered HP probe, the better the signal-to-noise ratio (SNR) of the resulting MR data is, unless the enzymatic reaction or the delivery pathway is saturated. In addition to toxicity, there are additional factors to consider when deciding the concentration of dissolved HP agents: the solubility of the sample, the plasma osmolarity, and the injection volume. First, the maximum concentration of dissolved HP probes is often restricted by the polarizer. The dissolution process in commercially available DNP equipment dilutes HP samples significantly, as the required amount of hot solvent to rapidly convert the frozen solid sample to a liquid state is substantial. Limited sample volumes and a minimum solvent-to-sample ratio allow the concentration of HP solutions to be no more than 200–300 mM, even for highly concentrated neat compounds such as pyruvic acid or lactic acid (>10 M). It is more challenging to achieve high concentrations for compounds with a large molecular weight, low solubility, or that require glassing matrices. Second, physiological factors such as osmolarity and blood volume should be considered. Total blood volume in rodents increases

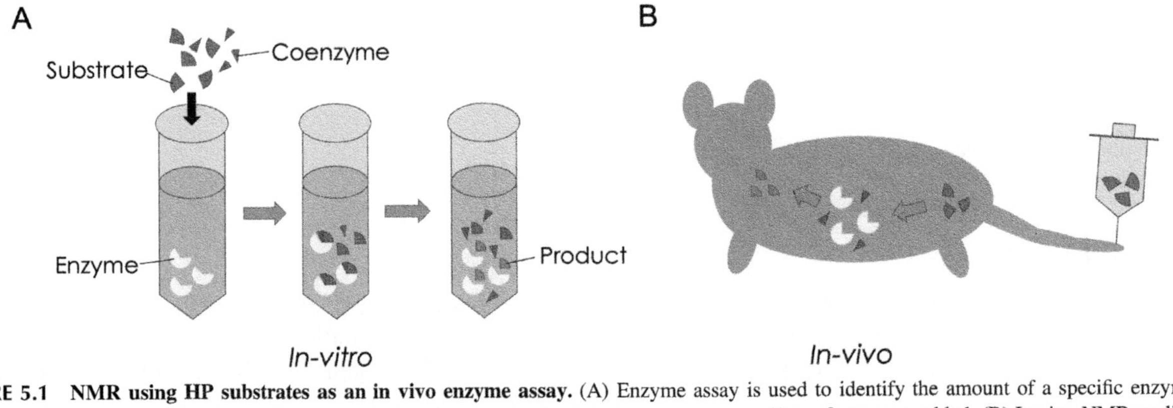

FIGURE 5.1 NMR using HP substrates as an in vivo enzyme assay. (A) Enzyme assay is used to identify the amount of a specific enzyme in a distinct specimen by measuring the product amount when the associated substrate, coenzymes, and/or cofactors are added. (B) In vivo NMR studies with the administration of HP substrates can be viewed as in vivo enzyme assays.

with body weight in early ages and slowly saturates in adults. The total blood volume of an adult rodent is 55–70 mL/kg for rats and 77–80 mL/kg for mice [7,8]. Osmolarity is 275–325 mOsm in plasma and ~200 mOsm in the cell. Third, the maximum volume of solution for a bolus injection is restricted to avoid haemodilution, increased venous pressure, abnormal acid-base equilibrium, and diuresis [9].

5.2.2 Delivery of HP probes to the targeted compartment

The observable time window of HP signals is shorter in vivo in comparison to in vitro because the apparent in vivo T_1 relaxation of HP signals is accelerated by blood circulation, biomembrane transport, and the associated metabolic conversions. Therefore, it is crucial to understand whether the HP probe is deliverable to the targeted compartment within the time window. One of the most fundamental requirements is tissue perfusion. Because the HP solution is delivered through blood vessels, vascular perfusion is essential for successful in vivo experiments. Well-perfused organs such as the heart and kidney are relatively easy targets to test with new probes.

From a cellular-level perspective, HP probes for assessing cellular metabolism in the cytosol or mitochondria should consider the transport mechanism over the biomembranes and the location of the targeted enzymes. For instance, monocarboxylate transporters (MCTs) rapidly deliver probes with a carboxylic group such as pyruvate and lactate to the intracellular space, allowing immediate observation of the products. Nonetheless, HP pyruvate can be rate-limited by MCT [10,11]. Alanine-serine-cysteine transporter (ASCT) is relatively slower than MCT [12], and therefore, HP pyruvate or lactate is a more effective probe than HP alanine for observing pyruvate oxidation via pyruvate dehydrogenase (PDH). The blood-brain barrier (BBB) is an additional membrane that needs to be considered in brain studies [13] and often plays a rate-limiting step for HP probe delivery [14]. Additional strategies are required to facilitate a hydrophilic or polar molecule to cross these membranes. Temporary disruption of the BBB using mannitol for HP glutamate was demonstrated by Mazuel et al. [15]. Increased cellular delivery of HP pyruvate using focused ultrasound (FUS) with microbubbles by transiently disrupting the BBB was also demonstrated by several groups [16–18]. For instance, metabolic products of HP glycerate, a hydrophilic molecule, were reportedly detected in the FUS-applied brain region, whereas no product was observed in the contralateral FUS-unapplied region (Fig. 5.2) [16]. Esterification is another strategy in designing HP probes

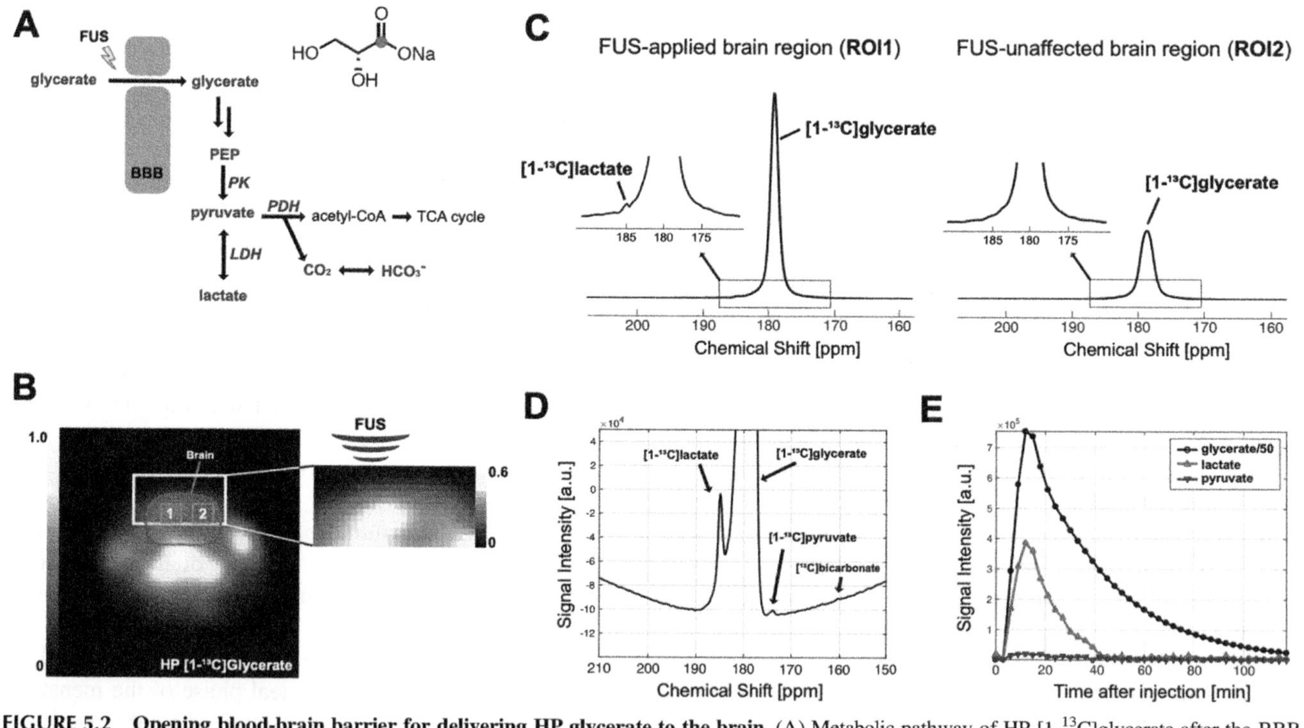

FIGURE 5.2 Opening blood-brain barrier for delivering HP glycerate to the brain. (A) Metabolic pathway of HP [1-^{13}C]glycerate after the BBB opening. (B) Axial ^{13}C image of rat brain, acquired immediately after a bolus injection of HP [1-^{13}C]glycerate. (C) A FUS-applied left brain hemisphere had a larger uptake of glycerate than the contralateral unaffected hemisphere. In addition to glycerate, the [1-^{13}C]lactate peak was detected in the FUS-applied brain region. (D) Time-averaged spectrum and (E) time curves of ^{13}C-metabolites were acquired using slice-selective dynamic ^{13}C MRS and demonstrate that an increased amount of glycerate was delivered to the FUS-applied brain region and was metabolized to lactate and pyruvate. *Adapted from Ref. [16].*

to facilitate cellular delivery across the biomembrane, which is composited with phospholipid bilayers. Multiple ethyl esters of ^{13}C-labeled probes, including 2-ketoglutarate [19], pyruvate [20], alanine [21], and succinate [22,23] were developed as HP probes to exploit the intracellular esterases.

5.2.3 Animal model, physiological condition, and metabolic modulation

Rodent models have been routinely used for in vivo evaluations of HP probes or MR pulse sequences. The majority of HP studies were performed under anesthesia for handling the animals and suppressing motion artifacts. However, the selection of anesthetic agents and the depth of anesthesia need to be done carefully and monitored throughout the study, as metabolic alterations associated with anesthesia may be introduced during the experiment [24–26]. Indeed, significant vasodilation and changes in product measurements were reported in HP studies of cerebral [27–30] and cardiac metabolism [31].

Metabolism is regulated by a network of physiological contributing factors such as nutritional condition and diet, respiration rate, cardiac rate, and gender, all of which contribute to metabolic flexibility [32]. For instance, the fractional contribution of carbohydrates to the tricarboxylic acid (TCA) cycle relative to fatty acids or other substrates in the myocardium or liver is highly sensitive to nutritional status, which can change as a daily cycle even without a controlled diet. Tissue perfusion and oxidative metabolism in skeletal muscle are extremely sensitive to exercise [33]. The amount of physical activity or even the duration of anesthesia prior to HP injection need to be considered as variables. Although controlling these variables to be consistent is crucial for reliable assessment of in vivo metabolism, metabolic sensitivity to these conditions can be conversely exploited to test the performance of related HP probes. For example, to demonstrate the feasibility of probing gluconeogenic pathways such as pyruvate carboxylase (PC) or phosphoenolpyruvate carboxykinase (PEPCK), HP substrates are often applied to fasted animals [34–37].

Metabolic modulation can be achieved by treating animals with well-determined drugs. Acute metabolic modulation is particularly powerful because comparing the baseline acquisition and post-modulation of the same subject via sequential injections of HP solution within a single MR session reduces multiple experimental variables such as body weight and coil positioning. Drugs that were utilized in HP studies for their acute metabolic effects (e.g., <1 h from the initial treatment) include ethanol for perturbing the cellular redox state by increasing the NADH-to-NAD^+ ratio [38–40], dichloroacetate for inhibiting PDH kinase (PDK) [41–44], dobutamine for accelerating the TCA cycle [41,45–47], 3-mercaptopicolinic acid for PEPCK inhibition [48,49], and hydrogen peroxide for oxidative stress [50].

Genetic modifications or surgical procedures can achieve enhanced and diverse metabolic modulation. Genetically modified rodent models are available to block genes that are associated with specific pathways and have been used for several HP studies to provide tailored pathway analysis [34,51–53]. Tumor implantation is a well-established surgical procedure for preclinical in vivo cancer studies. Although chemically induced models [54] (e.g., ethylnitrosourea [55,56] or diethylnitrosamine [57–59]) grow tumors spontaneously, the stochastic nature of tumor development over time is challenging for experiment management and data analysis. Because of rapid tumor growth and localization, orthotopic xenograft models are more commonly used to test HP substrates [3,60–68]. Subcutaneous tumor models were also used in HP studies [69–73] but often developed limited vascularization around the tumors, restricting the timely delivery of HP probes. As shown in Fig. 5.3 [74], discrepant HP delivery was observed in mice with subcutaneously injected H1299 cells. Contrast-enhanced ^1H MRI can verify the perfusion to the tumor but needs to be acquired after HP scans because the residual contrast agent shortens injected HP signals otherwise. Surgical procedures that ligate arterial blood flow were performed in small animals for myocardial [75–77], renal [78], and hepatic [79] ischemia-reperfusion studies with HP substrates.

For cardiac studies, larger animals (e.g., pigs) than rodents are desired for the slower cardiac rate and the larger size, which allow cardiac-gated data acquisition and structural analysis, respectively. Data acquired with cardiac- or respiratory-gating reduces variability induced by coronary flow and ventricular volume between injections. The feasibility of dual-gated cardiac studies in rodents was demonstrated previously [80]. Indeed, a recent dual-phase cardiac imaging study in humans reported a significant difference in observed HP signals when acquired at end-systole and end-diastole [81].

Sex needs to be considered as a biological variable [82–84]. Moreover, studies have reported that carbohydrate metabolism is altered in the menstrual cycle [85] and glucose metabolism is impaired in the luteal phase of the menstrual cycle [86]. Often, study subjects with a single gender are used to avoid the metabolic variation induced by the menstrual cycle or gender-derived metabolic features. However, metabolic sex differences were rarely investigated using HP substrates except pyruvate metabolism in kidneys [87].

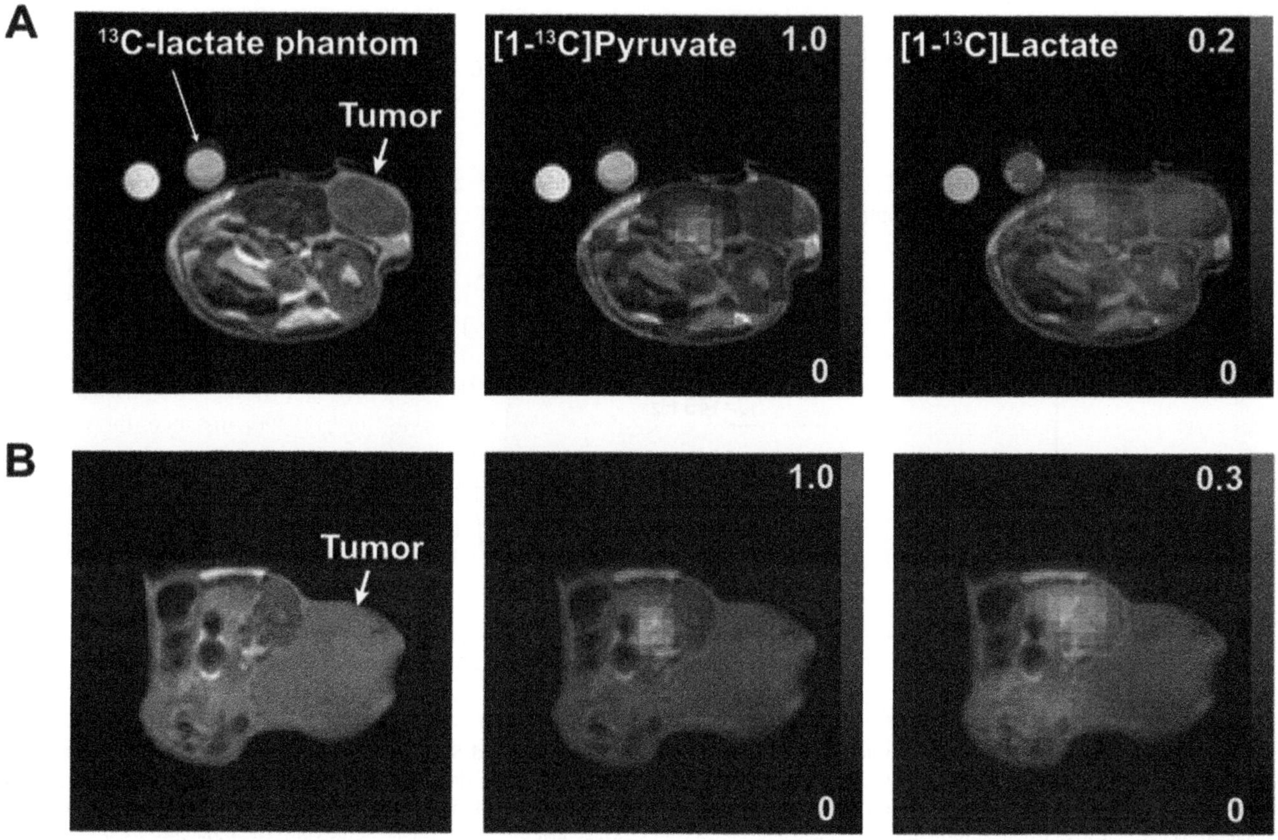

FIGURE 5.3 Perfusion of HP pyruvate in subcutaneous tumors. Rapid delivery of HP probes to targeted tissue is critical for HP imaging. In vivo assessment of tumor metabolism using HP pyruvate is successful for the mouse in (A) but not for the mouse in (B) due to perfusion discrepancy despite the successful xenograft establishment of subcutaneously injected H1299 in nu/nu mice. *Images in (a) are adapted from Ref. [74].*

5.2.4 Magnetic field strength and homogeneity

In conventional MRI, signal intensity is proportional to magnetic field strength (B_0). The signal sensitivity of the HP signal is, however, independent of the magnetic field strength of the scanner and is determined by the polarizer and other experimental parameters. Therefore, it makes sense to consider low-field systems for HP studies, but other potentially important components such as spin-lattice relaxation time (T_1) and spectral chemical shift dispersion can be significantly affected by the acquiring field strength. The T_1 is generally longer in lower fields, particularly when the relaxation mechanism is dominated by chemical shift anisotropy, whose relaxation rate is inversely proportional to the B_0^2, rather than dipole-dipole interaction. The T_1 relaxation mechanism is complex and varies with molecular structure or other parameters such as temperature and pH. The spectral distribution of HP substrates and products is also determined by B_0 based on the spin-spin coupling (J-coupling) constants and field-dependent chemical shift dispersion. If spectrally crowded peaks need to be resolved, a high field may be useful to exploit large chemical shift dispersion. For instance, products from [1-^{13}C] pyruvate along PC and PEPCK pathways such as C1 and C4 aspartate or malate could be readily resolved from large peaks such as C1 alanine and C1 pyruvate hydrate at 7T or higher fields [88]. On the contrary, J-couplings are more clearly observed at lower fields (<1T), and ultra-low fields can be strategically used to emphasize these effects [72,89–91] (Fig. 5.4).

5.2.5 Key acquisition parameters—MR spectroscopy

Non-imaging is the simplest approach yet a powerful acquisition method for capturing HP signals. By inspecting the acquired time-resolved spectra, the in vivo temporal behavior of each HP signal, which is also useful for developing optimal sampling and imaging strategies, can be identified with minimal RF perturbations. In addition, T_2* can be estimated by measuring the full-width half maximum of each peak, which is proportional to $1/\pi T_2^*$. One of the most basic

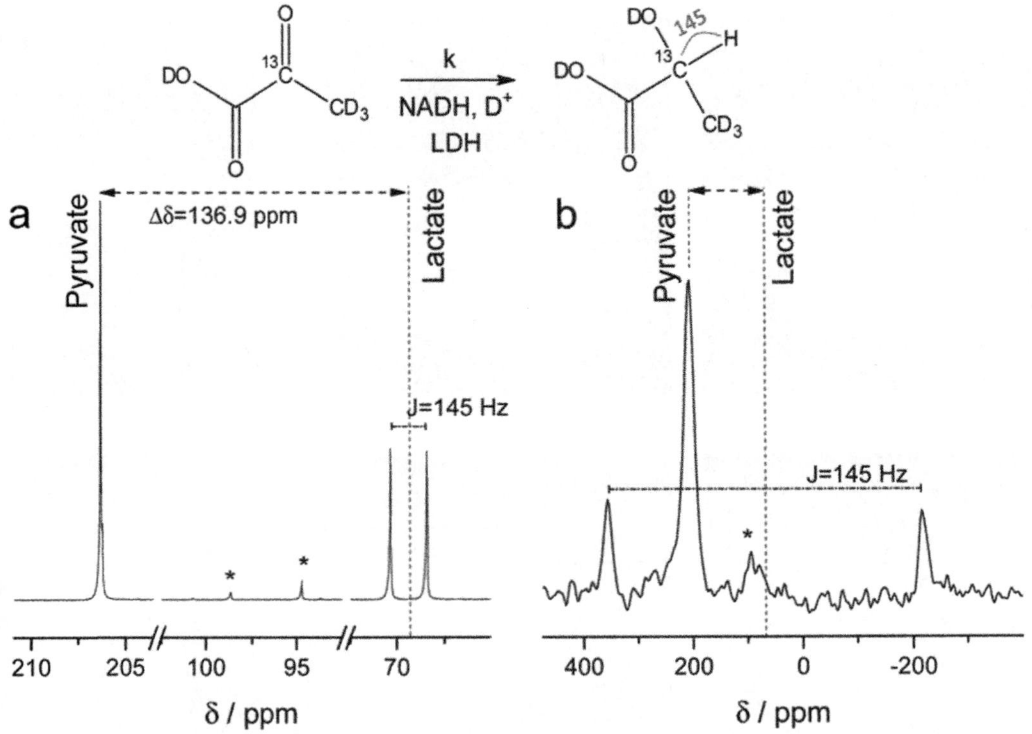

FIGURE 5.4 **^{13}C Spectra of HP [2-^{13}C]pyruvate-d_3 and [2-^{13}C]lactate-d_3 at high and low field.** Spectra acquired with a single scan during the enzymatic pyruvate-to-lactate conversion with lactate dehydrogenase at (A) $B_0 = 7$ T and (B) $B_0 = 24$ mT. The pyruvate is converted into lactate leading to a change in chemical shift of the ^{13}C spins. Additionally, a large J-coupling of 145 Hz becomes observable as one proton is introduced into the otherwise deuterated lactate. *denotes hydrated form of pyruvate and side products of cleavage that do not participate in the conversion. *Adapted from Ref. [91].*

non-imaging methods is pulse-and-acquire using a series of small tip-angle RF excitations. The use of small flip angles is primarily for conservative utilization of HP signals during time-resolved observation by preserving hyperpolarization of the injecting probe because HP signals excited by an RF pulse are disposable and non-recoverable. A constant flip angle is commonly used for dynamic MR spectroscopy. However, other flip angle schemes are also developed for optimal data acquisition and analysis, considering acquisition parameters such as temporal resolution, acquisition time window, and SNR of HP products [92–94]. RF pulses that are designed to excite with increased flip angles on products with no or minimal excitation on the injected HP material can be exploited to improve the SNR of the products without sacrificing the HP signal from the injected probe [95]. Non-selective (hard) pulses are typically shorter than spatially selective (soft) pulses, resulting in a shorter echo time and therefore a milder rolling baseline in the acquired absorption-mode spectra. Localization can be alternatively achieved by positioning a small RF coil over the targeted organ. A slice-selective RF pulse requires caution when prescribing the center frequency of excitation, as peaks at off-resonance may suffer chemical shift displacement artifacts, which are further discussed in Section 5.3.5.

5.2.6 Considerations for imaging approach

Imaging is often a more attractive option than spectroscopy for capturing metabolism, as it provides a spatial distribution of labeled metabolites. However, it requires careful selection of the method and parameters for appropriate and reliable imaging. Magnetic resonance spectroscopic imaging (MRSI) is a general term that refers to MR methods that simultaneously acquire spectral and spatial information. The most fundamental and straightforward MRSI option is chemical shift imaging (CSI). CSI measures the free-induction decay (FID) of individual phase-encoded voxels and is considered an unambiguous ground truth as it records voxel-by-voxel FID with a wide spectral bandwidth. However, CSI requires an excessive number of RF excitations, paired with FID readouts (#excitations = #FID = #voxels). This inefficient HP signal utilization and long data acquisition time result in poor spatial resolution, temporal resolution, and SNR. Due to the large amount of RF excitations, a single-slice and single-timepoint 2D CSI is typically used. Improved image quality and shortened acquisition times were demonstrated for CSI by utilizing variable flip angle schemes [96] and sparse k-space

sampling [97]. Other MRSI methods utilize multiecho imaging schemes with incremental echo times for improved data acquisition speed at the expense of spectral bandwidth. With a sufficient number of echoes, the Fourier transform of the multiecho images along the echo time can generate voxel-by-voxel spectra. Echoplanar spectroscopic imaging [98] and spiral CSI [99] use this type of acquisition scheme.

Imaging-focused methods can be exploited but require additional strategies to differentiate metabolites with distinct resonances. The first approach is using RF pulses that are both spectrally and spatially selective. Echoplanar imaging (EPI) [100,101] and spiral readouts [102] have been popularly utilized for superior k-space sampling speed and spatial resolution for limited acquisition time. This strategy accompanies metabolite-interleaved acquisition, as it images a single metabolite map at a time. Simultaneous imaging of multiple metabolites was also demonstrated by exploiting off-resonance artifacts along the phase-encoding direction of flyback EPI [103,104]. An alternative approach to distinguishing metabolites with imaging readouts is to acquire a series of images with several pre-calculated echo times, based on a priori knowledge of the targeted chemical shifts to measure, in combination with an iterative least-square reconstruction to distinguish metabolites [105−107].

Additional acceleration strategies are available via sparse sampling and parallel imaging. Sparse pseudorandom sampling in combination with compressed sensing in HP targets takes advantage of signal sparsity along the frequency domain but requires sufficient SNR, as this method often fails when the noise dominates the signal in the sparse domain. This type of acceleration has been demonstrated in MRSI and MRI methods [97,108−110]. Acceleration via parallel imaging is also available. To overcome the absence of background ^{13}C signals, parallel imaging techniques using fiducial markers [111], calibration-integrated [112−114] or calibrationless [114−116] approaches were demonstrated. Finally, CSI-MRSI hybrid data acquisition that utilizes spatiospectral correlation was proposed for HP imaging [117].

MR characteristics such as transverse relaxation time (T_2) and B_0 field inhomogeneity are deterministic factors for k-space sampling strategies. The dominant pulse sequence scheme is gradient echo, which is governed by T_2* and is therefore sensitive to B_0 field inhomogeneity. Heterogeneous regions such as cancerous or hemorrhagic tissues require careful evaluation of HP signal quantification as they experience accelerated T_2* decay due to heterogeneous B_0. Pulse sequences that utilize long T_2 of HP molecules were also proposed for increased SNR and improved immunity to heterogeneous B_0 [118−121].

Although area under the curves (AUCs) are favorably compared with kinetic modeling in terms of robustness, convenience, and independence of the arterial input function, kinetic parameters such as apparent conversion rate constants between HP metabolites may represent the dynamic behavior of HP signals with improved accuracy and sensitivity. For metabolic studies that focus on these kinetic observations, it is important to minimize the signal decay due to the RF pulsing of the metabolites involved, and for sufficiently long temporal observations, reliable HP signals are required. Analysis of dynamic aspects of HP signals is challenging as it requires compensating for the in vivo T_1 decay and the HP signal loss due to excitation. The more accurately the pulse decay and T_1 are understood or predetermined, the more accurate the kinetic measurement will be. Sequence components that significantly affect the kinetic parameters have been investigated for precisely estimating the true enzymatic conversion rate. For instance, modulating echo time and crusher gradients can improve the kinetic assessment by suppressing signals from flowing spins in the vasculature [122], Fig. 5.5. Analytical kinetic models are discussed in Section 5.3.2.

5.2.7 Repeatability, reproducibility, and longitudinal study

For most probes, multiple injections of the HP solution can be performed. To confirm repeatability, to enhance SNR by averaging, to assess acute metabolic modulation, or to compare acquisition schemes, the same HP probe can be injected into an animal within an MR session. To evaluate improved performance or to acquire multiple distinct metabolic assessments, two or more HP probes are administered separately. To avoid fluid overload (hypervolemia) or metabolic transition due to the first HP injection, a sufficient time interval to recover should be given to the animal between adjacent injections, and comparable physiological conditions should be maintained.

5.2.8 Co-polarization

Simultaneous polarization of multiple agents allows co-injection of the HP agents. The co-polarization approach can be considered when they do not interfere with each other in terms of chemical reactions and the spectral distribution of the peaks. As each HP sample has a distinct optimal preparation process, such as a mixture with glassing agent and free radical, these samples are often prepared separately within a sample container by either physical separation when the volumes are small or by placing each sample as a frozen layer or droplets (e.g., using liquid nitrogen). The co-polarization strategy has

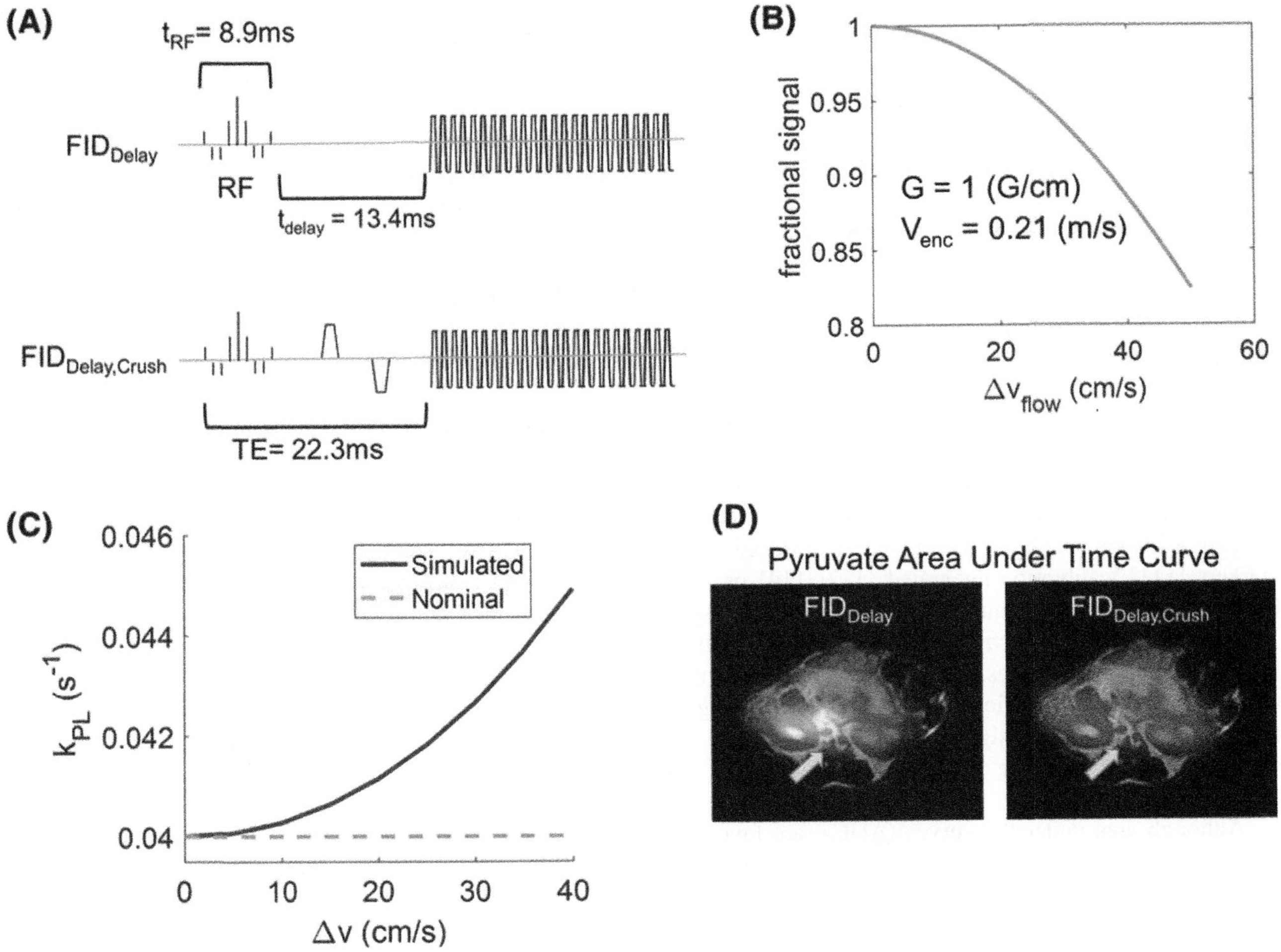

FIGURE 5.5 The Impact of crusher gradients in apparent conversion rate constants. (A) FID sequences with (FID$_{Delay, Crush}$) and without (FID$_{Delay}$) crusher gradients that were tested. (B) Fraction of signal remaining with the given crusher gradient strength in A. (C) Simulated k_{PL} estimates increased with higher flow velocity, where more flowing spins were suppressed. (D) Map of pyruvate AUC. Spatially dependent flow-suppression can be observed around the major vasculature. *Adapted from Ref. [122].*

been attempted for simultaneously assessing multiple metabolic pathways or biological conditions [123–127], bidirectional reactions [128], or simultaneous assessment of metabolism and perfusion [73,129]. However, the study should be carefully designed so as not to disturb labeling patterns (Fig. 5.6) [125].

5.3 Considerations for data analysis and interpretation

Typically, HP data is multidimensional (spectral, temporal, and spatial), multicompartmental, fragmentary in metabolism, and vulnerable to experimental parameters and artifacts. Due to the complexity and susceptibility of HP data, its analysis and interpretation are tricky and require careful validation and an in-depth understanding of MR physics, biochemistry, and physiology. If the contributing factors are not sufficiently taken into account, it is easy to overinterpret the results. This section will summarize key components of scrutinizing in vivo HP data and validation methods for reaching reasonable conclusions.

5.3.1 Quantification of HP signals and normalization

Determining how to quantify HP signals is critical in data analysis. Although not physically meaningful and described as an arbitrary unit, the most straightforward method to describe an HP signal is the raw MR signal (S_{raw}). To compensate for the variability of polarization level (Pol), measured by the polarizer or a separate spectrometer, and dissolution-to-injection time (t_{di}), a normalization factor can be applied to the raw signal using Eq. (5.1), where T_1 is the in vitro T_1 relaxation time

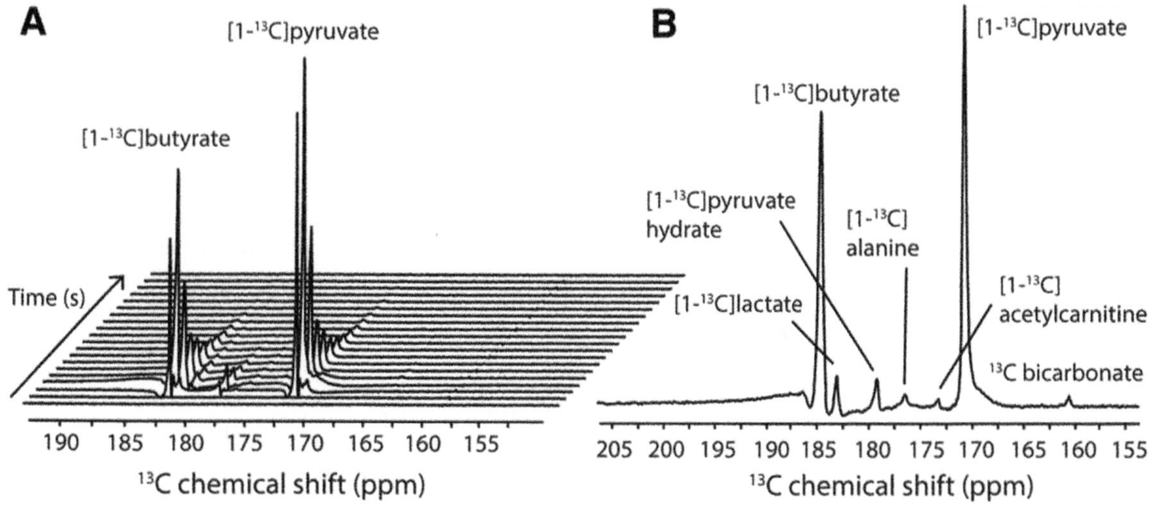

FIGURE 5.6 Assessing cardiac metabolism by co-injection of HP [1-^{13}C]pyruvate and HP [1-^{13}C]butyrate. In-vivo ^{13}C MRS measured in the healthy rat myocardium after the injection of an HP [1-^{13}C]pyruvate/[1-^{13}C]butyrate mixture. (A) ^{13}C spectral time course following the injection of the substrate mixture. (B) Time-averaged ^{13}C spectra. The myocardial metabolism of both pyruvate and butyrate could be observed through the formation of alanine, lactate, bicarbonate, and acetylcarnitine. *Adapted from Ref. [125].*

of the HP probe [33,130]. However, this type of metric is dependent on the polarizer and other environmental factors that may affect T_1, such as the variation of fringe magnetic field along the transport route of HP solution, and is therefore most appropriate for a single-site study with controlled experimental environments.

$$S_{\text{corrected}} = \frac{S_{\text{raw}}}{\text{Pol}} e^{-\frac{t_{\text{di}}}{T_1}} \quad \text{(Eq. 5.1)}$$

If HP data is acquired with a spatially inhomogeneous B_1 profile, additional intensity correction can improve the uniformity of HP signals spatially or between subjects [131].

An alternative to measuring raw HP signals, and perhaps the most commonly used metric is metabolite ratios. The ratio metric is unitless and independent of polarization level and experimental parameters. Another powerful aspect of the ratio metric is that it compensates for the B_1 field inhomogeneity. However, because the injected HP signal is typically strong in the vasculature, this approach can be significantly affected by the vascular density in the voxels of interest, particularly for non-imaging methods. For this reason, for HP substrates that generate multiple products, ratios between the products (e.g., bicarbonate-to-lactate ratio from HP [1-^{13}C]pyruvate) can be utilized to avoid the perfusion bias. Because all these ratio metrics can be extremely sensitive to the variation of the denominator, it is recommended to use the most reliable metabolite in the denominator, to create a reliability mask by setting a minimum threshold from the denominator, or to use a denominator that includes the signal in the numerator, such as total HP signals (*t*HP), Eq. (5.2), or sum of products (*tP*), Eq. (5.3). Both metrics are stable and can be considered as *normalized* product amounts.

$$\frac{\text{Product}_n}{t\text{HP}} = \frac{S_{\text{product}_n}}{S_{\text{substrate}} + \sum_{i=1}^{N} S_{\text{product}_i}} \quad \text{(Eq. 5.2)}$$

$$\frac{\text{Product}_n}{tP} = \frac{S_{\text{product}_n}}{\sum_{i=1}^{N} S_{\text{product}_i}} \quad \text{(Eq. 5.3)}$$

5.3.2 Kinetic analysis

In vivo HP signals are time-varying while experiencing T_1 decay, metabolic conversion, blood flow, membrane transport, and RF excitation. Compared to single-time-point measurements, dynamic measurements reflect the HP signals more precisely. Multiple methods were proposed to describe the dynamic behavior. The simplest and probably the most

straightforward measure is the AUC of each metabolite's time course, as shown in Eq. (5.4), where S_i indicates the measured signal of the metabolite at the ith time point. The AUC can also be normalized to tHP or tP, as shown in Eq. (5.2).

$$\mathrm{AUC} = \sum_{i=1}^{N} S_i \cdot \Delta t_i \qquad \text{(Eq. 5.4)}$$

However, the AUC does not describe the shape of the time course and omits the details of the dynamic characteristics. An alternative way to describe the kinetic behavior is by metabolite conversion rate constants. Given the complexities of in vivo conditions, *apparent* rate constants are typically targeted rather than the true constants. First-order two-site or multisite exchange models were initially developed to depict the apparent conversion of HP [1-^{13}C]pyruvate to [1-^{13}C] lactate [132], as shown in Eq. (5.5).

$$\frac{d}{dt}\begin{bmatrix} S_{\mathrm{pyr}} \\ S_{\mathrm{lac}} \end{bmatrix} = \begin{bmatrix} -\left(\frac{1}{T_{1,\mathrm{pyr}}} + k_{\mathrm{PL}}\right) & k_{\mathrm{LP}} \\ k_{\mathrm{PL}} & -\left(\frac{1}{T_{1,\mathrm{lac}}} + k_{\mathrm{LP}}\right) \end{bmatrix}\begin{bmatrix} S_{\mathrm{pyr}} \\ S_{\mathrm{lac}} \end{bmatrix} + \begin{bmatrix} u(t) \\ 0 \end{bmatrix} \qquad \text{(Eq. 5.5)}$$

In this simplified model, the apparent lactate-to-pyruvate conversion rate constant, k_{LP}, was assumed to be zero, and the arterial input function, $u(t)$, was derived from the pyruvate bolus [133]. Extensions to this highly simplified model have also been proposed in order to better accommodate the multiple underlying transports, compartments, and metabolic processes [133–136]. Other kinetic analyses include analytical models that focus on the initial appearance of HP products. For instance, the rate of signal production can be described as the slope of the peak during the time when the signal increases linearly following its appearance [33,137–140]. Moreover, it has been shown that using the AUC for a well-designed time-evolved dataset can be used to determine the apparent conversion rate [141]. More thorough reviews on kinetic models are also available [142–144].

5.3.3 Biological interpretation of HP metrics

Analyzing the HP data using these metrics and correlating them to biological characteristics requires careful inspection. First, as HP substrate is in high concentration and dosage, the dose-effect should be considered. The dose effects were proposed previously [134,145] as a Michaelis–Menten-like model with maximum reaction velocity ($V_{\max}$) and Michaelis constant (K_{m}) according to Eq. (5.6). The importance of including Michaelis–Menten corrections depends on the dose, and significant saturation effects have been clearly demonstrated [49,134,135,145,146].

$$k_{\mathrm{pyr}\to x} = \frac{V_{\max}}{K_{\mathrm{m}} + \mathrm{Dose}_{\mathrm{pyr}}} \qquad \text{(Eq. 5.6)}$$

Second, acquisition parameters and methods should be examined. The underlying assumption of these approaches is that the measured signals represent the amount of HP signals in the targeted region. This is often acceptable when a small RF flip-angle is used for all the HP metabolites, as the RF sampling does not affect the overall time course. However, the assumption might not be acceptable for HP data acquired with large flip-angle excitations, variable flip-angles, or heterogeneous flip-angles between metabolites. In addition, for HP data acquired in a metabolite-interleaved manner, the discordant timepoints between acquired data should be taken into account (Fig. 5.7).

Third, the regional and compartmental origins of HP signals should be identified. Data acquired using non-localized or slice-selective MRS can easily lead to erroneous conclusions or over-interpretations. For instance, one might conclude that HP alanine conversion from HP pyruvate in the brain MRS that was acquired with slice-selective RF excitation is a biomarker of alanine cycling that occurs subsequent to glutamate-glutamine neurotransmitter cycling. This is highly likely an incorrect insight, as the majority of the alanine signal is produced from peripheral muscle tissue rather than the brain [130,148–150] (Fig. 5.8). Another example is the quantification of bicarbonate production in the myocardium. While bicarbonate and alanine are predominantly myocardial, large portions of pyruvate and lactate are also from ventricles and blood pools [151]. Although somewhat effective in describing the oxidative pathway of pyruvate, the bicarbonate-to-lactate ratio from non-selective MRS may not represent the balance between LDH and PDH pathways in the myocardium. It has also been shown that lactate derived in the liver can increase the HP of lactate detected in the heart, leading to an overestimation of LDH [152]. Inspecting the spatial distribution of HP signals is essential to avoid a potentially fallacious assessment.

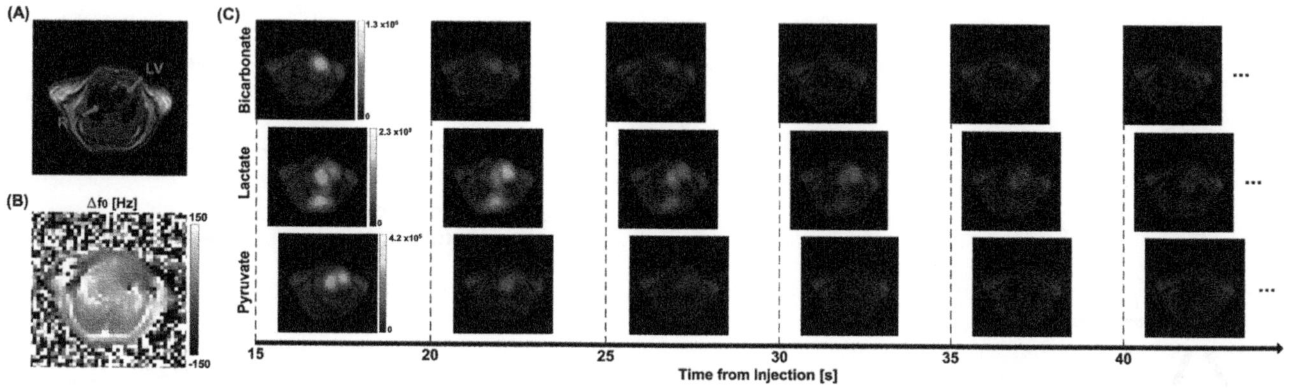

FIGURE 5.7 Temporal mismatch of dynamic metabolite-interleaved imaging. [^{13}C]Bicarbonate, [1-^{13}C]lactate and [1-^{13}C]pyruvate are sequentially excited using a spectral-spatial RF pulse and imaged with single-shot spiral readouts. (A) T_2-weighted image and (B) Δf_0 map were acquired in ^{1}H. (C) HP ^{13}C mages of [1-^{13}C]pyruvate, [1-^{13}C]lactate and [^{13}C]bicarbonate acquired every 5 s after an injection of HP [1-^{13}C]pyruvate. *Adapted from Ref. [147].*

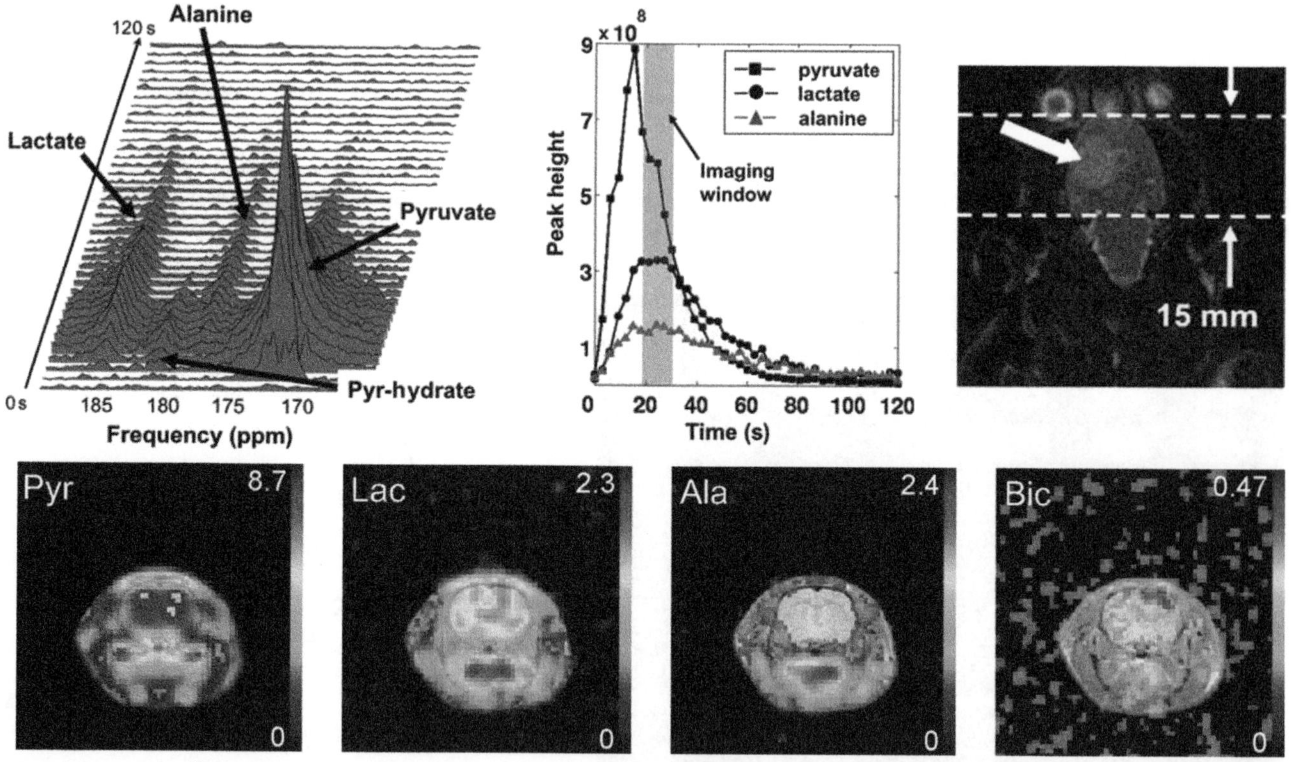

FIGURE 5.8 HP [1-^{13}C]alanine generated from HP [1-^{13}C]pyruvate in a rat head. (*Top row*) alanine detected in rat head using dynamic ^{13}C MRS (non-imaging) are from peripheral non-brain tissues as indicated by the absence of cerebral alanine signal in the high-resolution metabolite maps (*bottom row*). *Adapted from Refs. [130,150].*

5.3.4 Isotopic exchange and pool size

Metabolic conversion between HP signals over a unidirectional pathway (e.g., [1-^{13}C]pyruvate to [^{13}C]bicarbonate via PDH) reflects metabolic flux via the pathway. However, the interpretation of products via bidirectional pathways such as lactate from pyruvate via LDH can be ambiguous as the appearance of HP products from HP substrates is contributed by both flux and isotopic exchange. As illustrated in Fig. 5.9, an increased HP signal can be observed in the presence of a larger intrinsic pool of the product even without upregulated enzyme activity. The effect of intrinsic pool size on HP products was demonstrated by co-injecting an unlabeled exchange-linked dissolution agent [153] (Fig. 5.10). Decomposing metabolic flux and isotropic exchange have been attempted by several investigators [154–157] but distinguishing one from the other still remains under investigation.

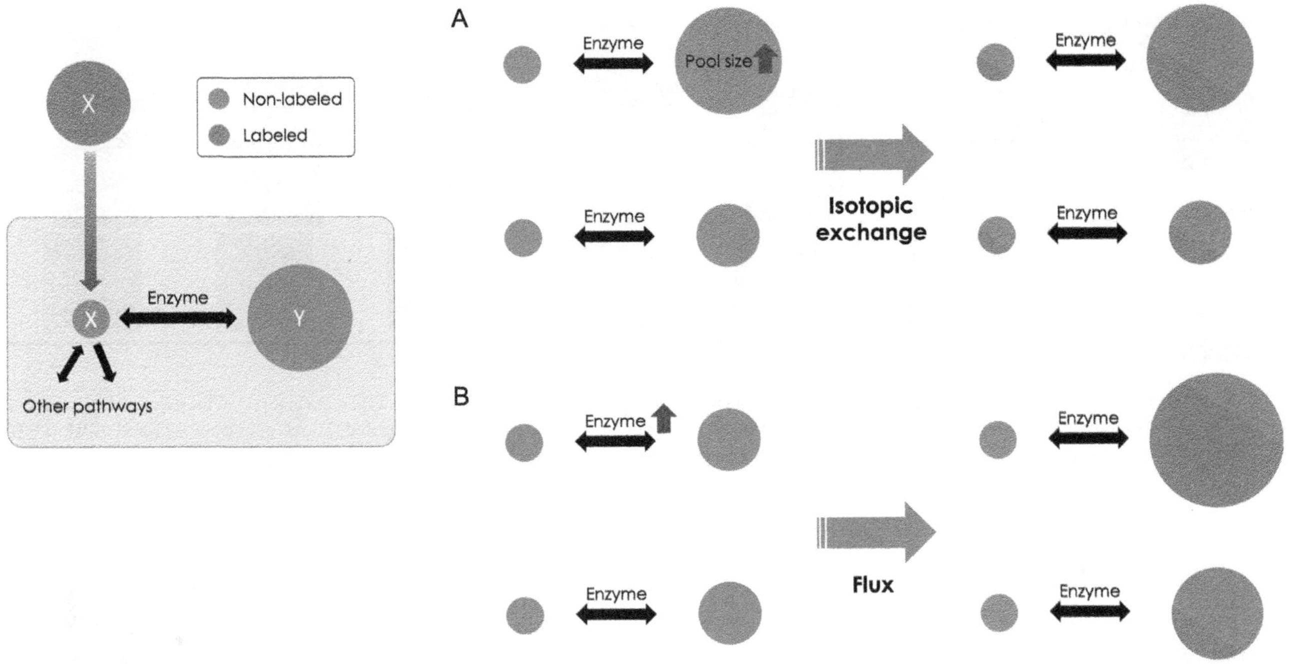

FIGURE 5.9 Isotope Exchange and metabolic net flux. For a bidirectional pathway, both isotope exchange and metabolic flux contribute to the signal intensity of HP products. Increased HP product signal can be observed in presence of a larger intrinsic pool of the product even without having upregulated enzyme activity.

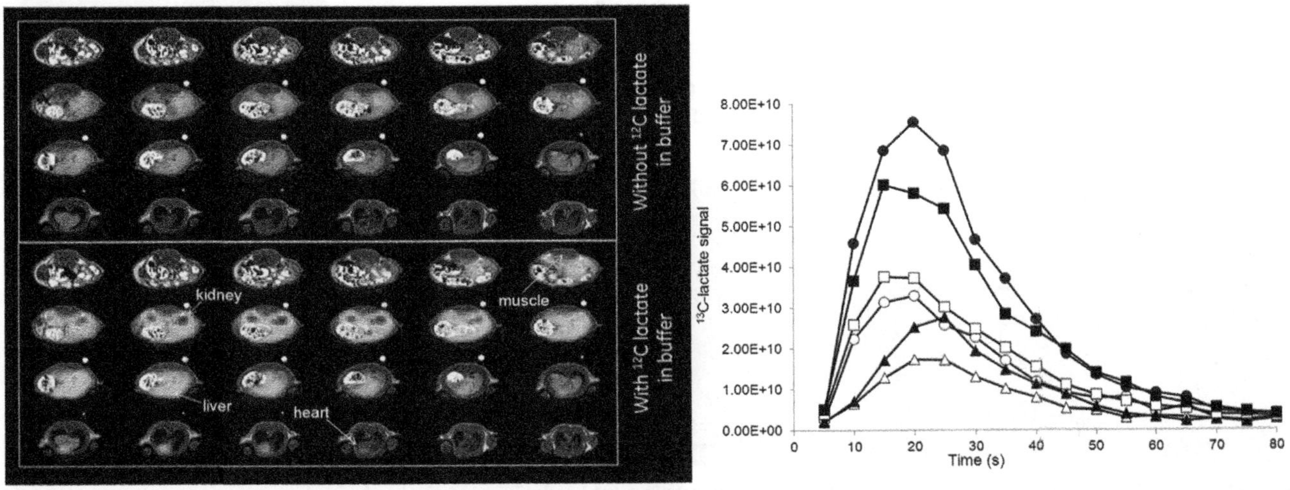

FIGURE 5.10 Impact of lactate pool size in HP lactate imaging. (*Left*) HP [1-^{13}C]lactate images of normal rat collected following an injection of 80 mM [1-^{13}C]pyruvate with and without 40 mM unlabeled sodium lactate in the dissolution buffer. (*Right*) Comparison of time-resolved [1-^{13}C]lactate signals from liver (*triangles*), kidney (*circles*) and vascular (*squares*) regions of interest with (*black filled*) and without (*open*) 40 mM sodium lactate added to the dissolution buffer. *Adapted from Ref. [153].*

5.3.5 Artifacts and undesirable signals

Data acquired with an injection of HP compounds often suffers from artifacts or turns out to be contaminated. Most of the artifacts and unwanted signals are predictable or can be identified. A simple example is the intrinsic lipid signal from the natural abundance of large subcutaneous fat, which can be non-negligible in overweighted subjects [33,158]. Off-resonance is probably the most common but most impactful source of artifacts. A mis-prescribed excitation frequency for a slice-selective RF pulse results in a shifted slice excitation from the prescribed slice. Even if the frequency was correctly prescribed, peaks that are far from the center frequency experience through-slice chemical shift displacement

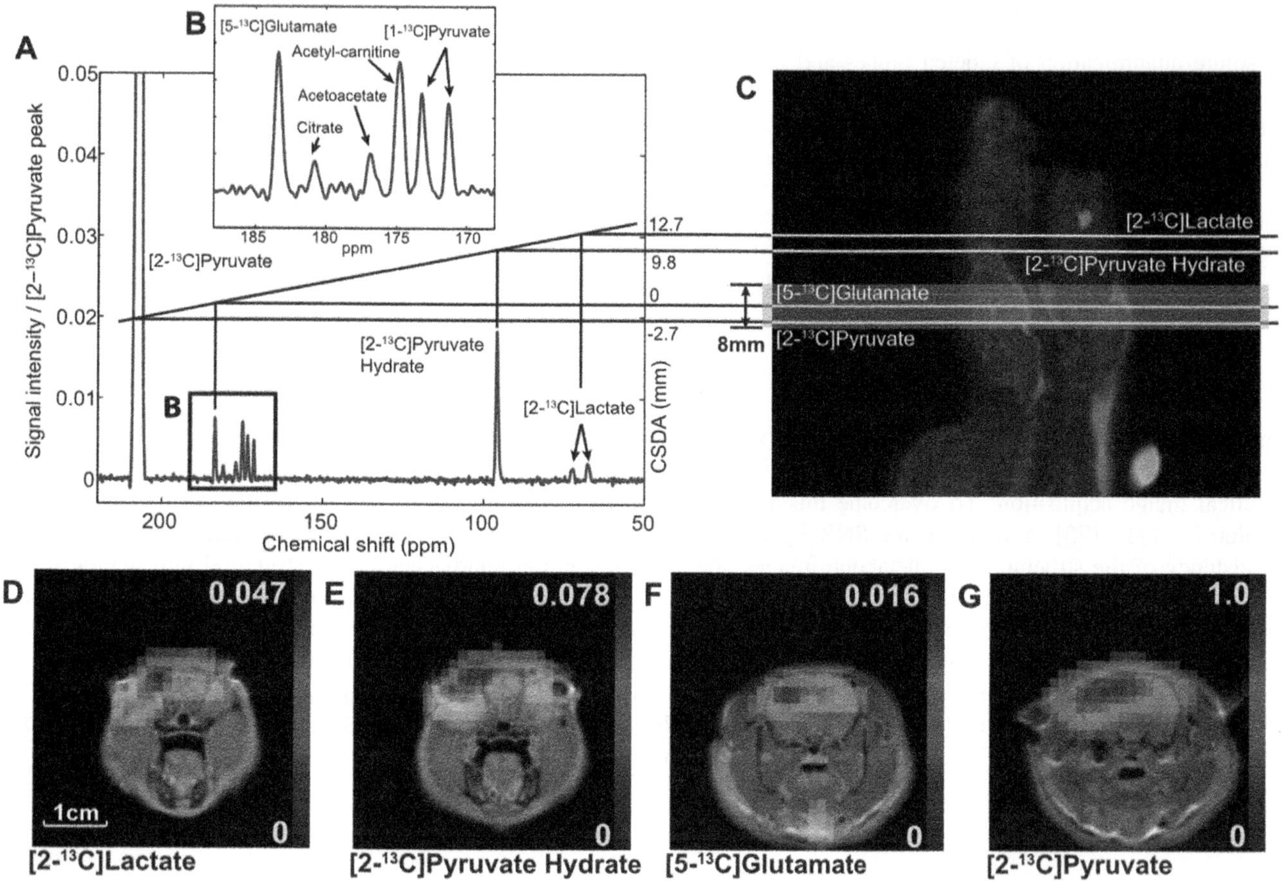

FIGURE 5.11 Chemical shift displacement artifact. (A) An averaged spectrum measured from a rat brain in vivo using a 2D slice-selective CSI after a bolus injection of 125-mM HP [2-^{13}C]pyruvate (post-dichloroacetate) at 3T. (B) Zoomed region containing most of mitochondrial metabolite peaks. (C) Shifted slice excitation profiles, shown on sagittal ^{1}H image of the rat brain, for [2-^{13}C]lactate, [2-^{13}C]pyruvate hydrate, [5-^{13}C]glutamate, and [2-^{13}C] pyruvate due to the wide spectral distribution of the metabolites when an 8-mm slice (shaded region) is prescribed at the resonance frequency of [5-^{13}C] glutamate. Reconstructed metabolite maps of (D) [2-^{13}C]lactate, (E) [2-^{13}C]pyruvate hydrate, (F) [5-^{13}C]glutamate, and (G) [2-^{13}C]pyruvate, overlaid on ^{1}H MR images at the corresponding slice locations. *Adapted from Ref. [43].*

artifacts as described in Eq. (5.7), where d is the resulting slice displacement of a peak with frequency offset, Δf_{CS}, the prescribed slice thickness, Δz, and RF pulse bandwidth, Δf [159,160] (Fig. 5.11). While chemical shift displacement artifacts are mostly unwanted, some acquisition methods have been developed to exploit these features [161,162] or avoid them [43,163].

$$d = \Delta z \cdot \frac{\Delta f_{CS}}{\Delta f} \quad \text{(Eq. 5.7)}$$

B_0 inhomogeneity is another critical component that can generate off-resonance artifacts. Since inhomogeneous B_0 is primarily caused by magnetic susceptibility differences, this is most prominent where two or more tissues have large magnetic susceptibility differences. For instance, cardiac MRI often shows signal losses due to the magnetic susceptibility variations from the heart-lung interface, such as the apex of the right ventricle or posterior wall of the myocardium [164,165]. Large spatial variations in magnetic susceptibility due to haemorrhages and calcification in cancer are also potential sources of artifacts. Correcting the B_0 inhomogeneity is relatively straightforward for CSI or other MR spectroscopic imaging methods or embedded for methods such as iterative decomposition of water and fat with echo asymmetry and least-squares estimation (IDEAL) reconstruction. However, imaging approaches that utilizes spectrally selective RF pulses may need additional reconstruction steps to correct the B_0 inhomogeneity [166,167]. If the passband or stopband of an RF pulse is narrow relative to variations of the peak resonances, the targeted peaks may not be excited or unintended peaks may be excited.

5.3.6 Ex vivo validation

Absolute quantification of labeled probes and products from their HP signals is extremely difficult, as it requires deciphering in vivo relaxation mechanisms, blood circulation, cellular transports, and complex metabolic dynamics. Moreover, it is often obscured that the increased product is due to its large intrinsic pool size, the upregulated activities of associated enzymes, or other cofactors. While biological assays of regulating enzyme activity or mRNA expression level can be examined, more integrated metabolic information with fractional enrichments of labeled isotopes in multiple metabolites can be evaluated via isotopomer analysis using NMR and mass spectrometry [59,168–170] (Fig. 5.12). Due to the limited signal sensitivities, this complementary analysis is performed ex vivo using strategically labeled (e.g., [U-^{13}C]pyruvate) probes that are designed to maximize metabolomics by distinguishing spectral or mass patterns of products along different pathways.

5.3.7 Postprocessing techniques

The non-recoverable, rapidly time-varying HP signals limit the achievable spatial resolution of in vivo HP images, resulting in an order of magnitude discrepancy in the spatial resolution between the ^{1}H structural and the HP images in practical image acquisition. To overcome this intrinsic limitation, post-processing techniques that augment the spatial resolution [171–175] or improve the SNR by denoising have been developed [176–178]. Exploiting the anatomical consistency of the structural and metabolic images, the super-resolution approaches utilize structural information that is available from the high-resolution ^{1}H MRI for enhancing spatial resolution of HP metabolic images [172–174]. For instance, the patch-based super-resolution algorithm reconstructs a high-resolution image with the guidance of a weighting matrix, which assumes the similarity of the adjacent voxels and is calculated from the moving patch. This approach has been demonstrated to enhance the spatial resolution of ^{13}C images while preserving quantification accuracy [175]. Denoising approaches for HP [177,178] and non-HP [176] data were also demonstrated using singular value decomposition or its higher-dimensional analog, tensor decomposition, by approximating the data rank.

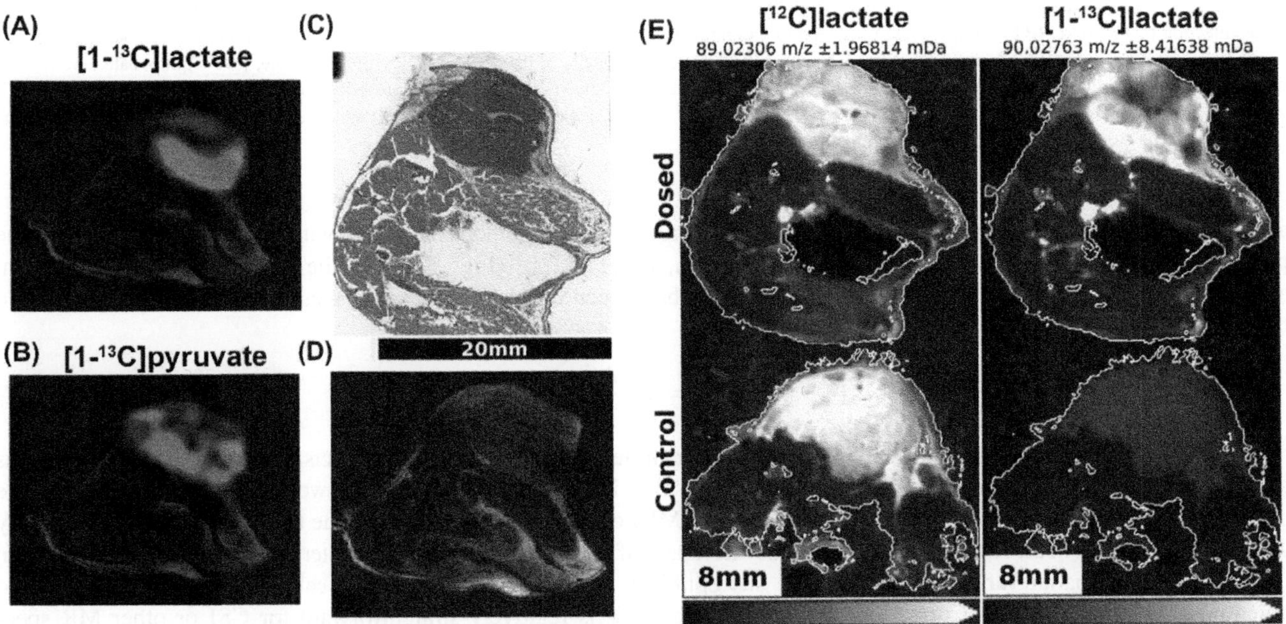

FIGURE 5.12 Comparison of In Vivo HP [1–^{13}C]lactate and ex vivo Mass spectrometry imaging. (A) HP [1-^{13}C]lactate and (B) HP [1-^{13}C] pyruvate images following injection of HP [1-^{13}C]pyruvate. The corresponding (C) histology and (D) anatomical T_2-weighted ^{1}H images. (E) Maps of [^{12}C] and [1-^{13}C]lactate generated by mass spectrometry imaging of a frozen section taken from the same slice from which the ^{13}C HP images are acquired. *Adapted from Ref. [168].*

5.4 Conclusion

The development of probes for in vivo metabolic analysis with DNP requires consideration of the probe's properties as well as the acquisition strategy to produce the desired result. The decay mechanism, pulsing strategy, and reaction kinetics must all be considered, as they will confound the signal dynamics. Interpretation of the results must take these into account in addition to metabolic influences due to the animal's condition, including nutritional state and metabolite pool sizes. The discussion above is not a comprehensive list of challenges but a guide to understanding how to identify the difficulties and considerations needed for developing new probes. A carefully designed experiment with a properly chosen probe has the ability to assess real-time metabolic states and changes to ascertain details of the enzymatic processes and ultimately be used for translation into the clinic for assessing disease state and treatment response.

References

[1] S. Meier, P.R. Jensen, M. Karlsson, M.H. Lerche, Hyperpolarized NMR probes for biological assays, Sensors 14 (2014) 1576–1597.

[2] A.P. Chen, et al., Hyperpolarized C-13 spectroscopic imaging of the TRAMP mouse at 3T - initial experience, Magn. Reson. Med. 58 (2007) 1099–1106.

[3] J.M. Park, et al., Metabolite kinetics in C6 rat glioma model using magnetic resonance spectroscopic imaging of hyperpolarized [1-13C]pyruvate, Magn. Reson. Med. 68 (2012) 1886–1893.

[4] T.H. Witney, et al., Detecting treatment response in a model of human breast adenocarcinoma using hyperpolarised 1- 13 Cpyruvate and 1,4- 13 C 2 fumarate, Br. J. Cancer 103 (2010) 1400–1406.

[5] M.A. Schroeder, et al., In vivo assessment of pyruvate dehydrogenase flux in the heart using hyperpolarized carbon-13 magnetic resonance, Proc. Natl. Acad. Sci. U. S. A. 105 (2008) 12051–12056.

[6] K. Golman, J.H. Ardenkjær-Larsen, J.S. Petersson, S. Månsson, I. Leunbach, Molecular imaging with endogenous substances, Proc. Natl. Acad. Sci. U. S. A. 100 (2003) 10435–10439.

[7] B.M. Mitruka, H.M. Rawnsley, Clinical Biochemical and Hematological Reference Values in Normal Experimental Animals, Masson Pub, USA, 1981, https://doi.org/10.1042/bst0070267a.

[8] J.E. Harkness, J.E. Wagner, Biology and husbandry, in: J.E. Harkness, P.V. Turner, S. VandeWoude, C.L. Wheler (Eds.), Harkness and Wagner's Biology and Medicine of Rabbits and Rodents, Wiley-Blackwell, 2010, pp. 23–106.

[9] D.B. Morton, et al., Refining procedures for the administration of substances. Report of the BVAAWF/FRAME/RSPCA/UFAW Joint Working Group on Refinement, Lab. Anim. 35 (2001) 1–41.

[10] T. Harris, G. Eliyahu, L. Frydman, H. Degani, Kinetics of hyperpolarized 13C1-pyruvate transport and metabolism in living human breast cancer cells, Proc. Natl. Acad. Sci. U. S. A. 106 (2009) 18131–18136.

[11] Y. Rao, et al., Hyperpolarized [1-13C]pyruvate-to-[1-13C]lactate conversion is rate-limited by monocarboxylate transporter-1 in the plasma membrane, Proc. Natl. Acad. Sci. U. S. A. 117 (2020) 22378–22389.

[12] S.K. Joseph, N.M. Bradford, J.D. McGivan, Characteristics of the transport of alanine, serine and glutamine across the plasma membrane of isolated rat liver cells, Biochem. J. 176 (1978) 827–836.

[13] E.A. Neuwelt, et al., Engaging neuroscience to advance translational research in brain barrier biology, Nat. Rev. Neurosci. 12 (2011) 169–182.

[14] J.J. Miller, et al., 13C pyruvate transport across the blood-brain barrier in preclinical hyperpolarised MRI, Sci. Rep. 8 (2018) 1–15.

[15] L. Mazuel, et al., Intracerebral synthesis of glutamine from hyperpolarized glutamate, Magn. Reson. Med. 78 (2017) 1296–1305.

[16] E.P. Hackett, et al., Probing cerebral metabolism with hyperpolarized 13C imaging after opening the blood-brain barrier with focused ultrasound, ACS Chem. Neurosci. 12 (2021) 2820–2828.

[17] Y. Takado, et al., Hyperpolarized 13C magnetic resonance spectroscopy reveals the rate-limiting role of the blood-brain barrier in the cerebral uptake and metabolism of l -lactate in vivo, ACS Chem. Neurosci. 9 (2018) 2554–2562.

[18] T.H. Peeters, et al., Imaging hyperpolarized pyruvate and lactate after blood-brain barrier disruption with focused ultrasound, ACS Chem. Neurosci. 10 (2019) 2591–2601.

[19] J. Singh, et al., 13C-Labeled diethyl ketoglutarate derivatives as hyperpolarized probes of 2-ketoglutarate dehydrogenase activity, Anal. Sens. 1 (2021) 156–160.

[20] R.E. Hurd, et al., Metabolic imaging in the anesthetized rat brain using hyperpolarized [1-13C] pyruvate and [1-13C] ethyl pyruvate, Magn. Reson. Med. 63 (2010) 1137–1143.

[21] J. Chen, E.P. Hackett, J. Singh, Z. Kovács, J.M. Park, Simultaneous assessment of intracellular and extracellular pH using hyperpolarized [1-13C] alanine ethyl ester, Anal. Chem. 92 (2020) 11681–11686.

[22] N.M. Zacharias, H.R. Chan, N. Sailasuta, B.D. Ross, P. Bhattacharya, Real-time molecular imaging of tricarboxylic acid cycle metabolism in vivo by hyperpolarized 1- 13C diethyl succinate, J. Am. Chem. Soc. 134 (2012) 934–943.

[23] K.L. Billingsley, et al., Hyperpolarized [1,4-13C]-diethylsuccinate: a potential DNP substrate for in vivo metabolic imaging, NMR Biomed. 27 (2014) 356–362.

[24] T. Maekawa, C. Tommasino, H.M. Shapiro, J. Keifer-Goodman, R.W. Kohlenberger, Local cerebral blood flow and glucose utilization during isoflurane anesthesia in the rat, Anesthesiology 65 (1986) 1440151.

[25] P.F. Conzen, et al., Systemic and regional hemodynamics of isoflurane and sevoflurane in rats, Anesth. Analg. 74 (1992) 79–88.

[26] O.Z. Chi, et al., Effects of isoflurane on transport across the blood-brain barrier, Anesthesiology 76 (1992) 426–431.

[27] S. Josan, et al., Effects of isoflurane anesthesia on hyperpolarized 13C metabolic measurements in rat brain, Magn. Reson. Med. 70 (2013) 1117–1124.

[28] M. Marjańska, A.A. Shestov, D.K. Deelchand, E. Kittelson, P.G. Henry, Brain metabolism under different anesthetic conditions using hyperpolarized [1-13C]pyruvate and [2-13C]pyruvate, NMR Biomed. 31 (2018) 1–8.

[29] E. Flatt, et al., Measuring glycolytic activity with hyperpolarized [h7, u-13 c6]2 d-glucose in the naive mouse brain under different anesthetic conditions, Metabolites 11 (2021).

[30] V. Hyppönen, et al., Metabolism of hyperpolarised [1−13C]pyruvate in awake and anaesthetised rat brains, NMR Biomed. 35 (2022) 1–10.

[31] J. Steinhauser, P. Wespi, G. Kwiatkowski, S. Kozerke, Assessing the influence of isoflurane anesthesia on cardiac metabolism using hyperpolarized [1-13C]pyruvate, NMR Biomed. 31 (2018) 2–10.

[32] S. Zhang, M.W. Hulver, R.P. McMillan, M.A. Cline, E.R. Gilbert, The pivotal role of pyruvate dehydrogenase kinases in metabolic flexibility, Nutr. Metab. 11 (2014) 1–9.

[33] J.M. Park, et al., Hyperpolarized 13C MR spectroscopy depicts in vivo effect of exercise on pyruvate metabolism in human skeletal muscle, Radiology 300 (2021) 626–631.

[34] M.E. Merritt, C. Harrison, A.D. Sherry, C.R. Malloy, S.C. Burgess, Flux through hepatic pyruvate carboxylase and phosphoenolpyruvate carboxykinase detected by hyperpolarized 13C magnetic resonance, Proc. Natl. Acad. Sci. U. S. A. 108 (2011) 19084–19089.

[35] K.X. Moreno, et al., Real-time detection of hepatic gluconeogenic and glycogenolytic states using hyperpolarized [2-13c]dihydroxyacetone, J. Biol. Chem. 289 (2014) 35859–35867.

[36] J.M. Park, et al., Hyperpolarized sodium [1-13C]-glycerate as a probe for assessing glycolysis in vivo, J. Am. Chem. Soc. 139 (2017) 6629–6634.

[37] H.A.I. Yoshihara, A. Comment, J. Schwitter, Assessment of aspartate and bicarbonate produced from hyperpolarized [1-13C]pyruvate as markers of renal gluconeogenesis, Front. Physiol. 12 (2021) 1–8.

[38] O.A. Forsander, P.H. Mäenpää, M.P. Salaspuro, Influence of ethanol on the lactate/pyruvate and beta-hydroxybutiate/acetoacetate ratios in rat liver experiments, Acta Chem. Scand. 19 (1965) 1770–1771.

[39] D.M. Spielman, et al., In vivo measurement of ethanol metabolism in the rat liver using magnetic resonance spectroscopy of hyperpolarized [1-13C]pyruvate, Magn. Reson. Med. 62 (2009) 307–313.

[40] J.M. Park, C. Khemtong, S.C. Liu, R.E. Hurd, D.M. Spielman, In vivo assessment of intracellular redox state in rat liver using hyperpolarized [1-13C]Alanine, Magn. Reson. Med. 77 (2017) 1741–1748.

[41] M.A. Schroeder, et al., The cycling of acetyl-coenzyme A through acetylcarnitine buffers cardiac substrate supply: a hyperpolarized 13C magnetic resonance study, Circ. Cardiovasc. Imaging 5 (2012) 201–209.

[42] D. Mayer, et al., Application of hyperpolarized [1-13C]lactate for the in vivo investigation of cardiac metabolism, NMR Biomed. 25 (2012) 1119–1124.

[43] J.M. Park, et al., Volumetric spiral chemical shift imaging of hyperpolarized [2-13c]pyruvate in a rat c6 glioma model, Magn. Reson. Med. 75 (2016) 973–984.

[44] A.Z. Lau, J.J. Miller, M.D. Robson, D.J. Tyler, Simultaneous assessment of cardiac metabolism and perfusion using copolarized [1-13C]pyruvate and 13C-urea, Magn. Reson. Med. 77 (2017) 151–158.

[45] S. Josan, et al., In vivo investigation of cardiac metabolism in the rat using MRS of hyperpolarized [1-13C] and [2-13C]pyruvate, NMR Biomed. 26 (2013) 1680–1687.

[46] U. Koellisch, et al., Metabolic imaging of hyperpolarized [1-13C]acetate and [1-13C]acetylcarnitine - investigation of the influence of dobutamine induced stress, Magn. Reson. Med. 74 (2015) 1011–1018.

[47] L. Menichetti, et al., Assessment of real-time myocardial uptake and enzymatic conversion of hyperpolarized [1-13C]pyruvate in pigs using slice selective magnetic resonance spectroscopy, Contrast Media Mol. Imaging 7 (2012) 85–94.

[48] N.W. DiTullio, et al., 3-Mercaptopicolinic acid, an inhibitor of gluconeogenesis, Biochem. J. 138 (1974) 387–394.

[49] E. Can, et al., [13C]bicarbonate labelled from hyperpolarized [1-13C]pyruvate is an in vivo marker of hepatic gluconeogenesis in fasted state, Commun. Biol. 5 (2022) 1–9.

[50] A. Wibowo, J.M. Park, S.C. Liu, C. Khosla, D.M. Spielman, Real-time in vivo detection of H2O2 using hyperpolarized 13C-thiourea, ACS Chem. Biol. 12 (2017) 1737–1742.

[51] R. Bok, et al., The role of lactate metabolism in prostate cancer progression and metastases revealed by dual-agent hyperpolarized 13 C MRSI, Cancers 11 (2019).

[52] A.J. Bakermans, et al., Myocardial energy shortage and unmet anaplerotic needs in the fasted long-chain acyl-Co A dehydrogenase knockout mouse, Cardiovasc. Res. 100 (2013) 441–449.

[53] G. Sharma, et al., Real-time hyperpolarized 13C magnetic resonance detects increased pyruvate oxidation in pyruvate dehydrogenase kinase 2/4−double knockout mouse livers, Sci. Rep. 9 (2019) 1–11.

[54] A. Onaciu, et al., Spontaneous and induced animal models for cancer research, Diagnostics 10 (2020) 1–36.

[55] H. Druckrey, S. Ivanković, R. Preussmann, Teratogenic and carcinogenic effects in the offspring after single injection of ethylnitrosourea to pregnant rats [37], Nature 210 (1966) 1378–1379.

[56] J.M. Park, et al., Initial experiences of simultaneous in vivo metabolic imaging using MRI, PET, and hyperpolarized 13C MRSI from rat glioma models, in: International Society of Magnetic Resonance in Medicine, 2016, p. 674.

[57] Y. fang Ding, Z. hui Wu, Y. jie Wei, L. Shu, Y. ru Peng, Hepatic inflammation-fibrosis-cancer axis in the rat hepatocellular carcinoma induced by diethylnitrosamine, J. Cancer Res. Clin. Oncol. 143 (2017) 821–834.
[58] E. Bliemsrieder, et al., Hyperpolarized 13C pyruvate magnetic resonance spectroscopy for in vivo metabolic phenotyping of rat HCC, Sci. Rep. 11 (2021) 1–10.
[59] J. Chen, et al., Profiling carbohydrate metabolism in liver and hepatocellular carcinoma with [13C]-Glycerate probes, Anal. Sens. 1 (2021) 196–202.
[60] I. Park, et al., Detection of early response to temozolomide treatment in brain tumors using hyperpolarized 13C MR metabolic imaging, J. Magn. Reson. Imag. 33 (2011) 1284–1290.
[61] M.M. Chaumeil, et al., Hyperpolarized 13C MR spectroscopic imaging can be used to monitor Everolimus treatment in vivo in an orthotopic rodent model of glioblastoma, Neuroimage 59 (2012) 193–201.
[62] C.L. Mallett, et al., Longitudinal anatomical and metabolic MRI characterization of orthotopic xenograft prostate tumors in nude mice, J. Magn. Reson. Imag. 40 (2014) 848–856.
[63] S. Düwel, et al., Multiparametric human hepatocellular carcinoma characterization and therapy response evaluation by hyperpolarized 13C MRSI, NMR Biomed. 29 (2016) 952–960.
[64] V.C. Sandulache, et al., Evaluation of hyperpolarized [1-13C]-pyruvate by magnetic resonance to detect ionizing radiation effects in real time, PLoS One 9 (2014).
[65] R. Sriram, et al., Non-invasive assessment of lactate production and compartmentalization in renal cell carcinomas using hyperpolarized 13C pyruvate MRI, Cancers 10 (2018) 1–14.
[66] E.H. Suh, et al., In vivo assessment of increased oxidation of branched-chain amino acids in glioblastoma, Sci. Rep. 9 (2019) 1–9.
[67] R. Mair, et al., Metabolic imaging detects low levels of glycolytic activity that vary with levels of C-MyC expression in patient-derived xenograft models of glioblastoma, Cancer Res. 78 (2018) 5408–5418.
[68] F. Ahamed, et al., Modeling hyperpolarized lactate signal dynamics in cells, patient-derived tissue slice cultures and murine models, NMR Biomed. 34 (2021) 1–12.
[69] S. Düwel, et al., Imaging of pH in vivo using hyperpolarized 13C-labelled zymonic acid, Nat. Commun. 8 (2017).
[70] C. Hundshammer, et al., Simultaneous characterization of tumor cellularity and the Warburg effect with PET, MRI and hyperpolarized 13C-MRSI, Theranostics 8 (2018) 4765–4780.
[71] R.F. Schulte, et al., Saturation-recovery metabolic-exchange rate imaging with hyperpolarized [1-13C] pyruvate using spectral-spatial excitation, Magn. Reson. Med. 69 (2013) 1209–1216.
[72] S.S. Tee, et al., Sampling hyperpolarized molecules utilizing a 1 Tesla permanent magnetic field, Sci. Rep. 6 (2016) 1–6.
[73] J.Y.C. Lau, A.P. Chen, Y.P. Gu, C.H. Cunningham, Voxel-by-voxel correlations of perfusion, substrate, and metabolite signals in dynamic hyperpolarized 13C imaging, NMR Biomed. 29 (2016) 1038–1047.
[74] S.-S. Tee, et al., PKM2 activation sensitizes cancer cells to growth inhibition by 2-deoxy-D-glucose, Oncotarget 8 (2017) 90959–90968.
[75] M.S. Dodd, et al., Impaired in vivo mitochondrial krebs cycle activity after myocardial infarction assessed using hyperpolarized magnetic resonance spectroscopy, Circ. Cardiovasc. Imaging 7 (2014) 895–904.
[76] H.A.I. Yoshihara, J.A.M. Bastiaansen, C. Berthonneche, A. Comment, J. Schwitter, An intact small animal model of myocardial ischemia-reperfusion: characterization of metabolic changes by hyperpolarized13C MR spectroscopy, Am. J. Physiol. Heart Circ. Physiol. 309 (2015) H2058–H2066.
[77] D. Oh-lci, et al., Hyperpolarized metabolic MR imaging of acute myocardial changes and recovery after ischemia-reperfusion in a small-animal model, Radiology 278 (2016) 742–751.
[78] C. Baligand, et al., Hyperpolarized 13C magnetic resonance evaluation of renal ischemia reperfusion injury in a murine model, NMR Biomed. 30 (2017) 1–12.
[79] C.M. Moon, et al., Metabolic alterations in a rat model of hepatic ischaemia reperfusion injury: in vivo hyperpolarized 13 C MRS and metabolic imaging, Liver Int. 38 (2018) 1117–1127.
[80] M.H. Lauritzen, et al., Imaging regional metabolic changes in the ischemic rat heart in vivo using hyperpolarized [1-13C]pyruvate, Tomogr. (Ann Arbor, Mich.) 3 (2017) 123–130.
[81] J. Ma, et al., Dual-phase imaging of cardiac metabolism using hyperpolarized pyruvate, Magn. Reson. Med. 87 (2022) 302–311.
[82] J.B. Rubin, The spectrum of sex differences in cancer, Trends Cancer 8 (2022) 303–315.
[83] B. Tramunt, et al., Sex differences in metabolic regulation and diabetes susceptibility, Diabetologia 63 (2020) 453–461.
[84] E. Murphy, G. Amanakis, N. Fillmore, R.J. Parks, J. Sun, Sex differences inmetabolic cardiomyopathy, Cardiovasc. Res. 113 (2017) 370–377.
[85] W.N. Spellacy, K.L. Carlson, S.L. Schade, Menstrual cycle carbohydrate metabolism: studies on plasma insulin and blood glucose levels during an intravenous glucose tolerance test, Am. J. Obstet. Gynecol. 99 (1967) 382–386.
[86] M.P. Diamond, D.C. Simonson, R.A. DeFronzo, Menstrual cyclicity has a profound effect on glucose homeostasis, Fertil. Steril. 52 (1989) 204–208.
[87] Y. Wen, et al., Sex differences in kidney function and metabolism assessed using hyperpolarized [1-13c]pyruvate interleaved spectroscopy and nonspecific imaging, Tomography 6 (2020) 5–13.
[88] P. Lee, et al., In Vivo hyperpolarized carbon-13 magnetic resonance spectroscopy reveals increased pyruvate carboxylase flux in an insulin-resistant mouse model, Hepatology 57 (2013) 515–524.
[89] D.A. Barskiy, et al., Zero-field nuclear magnetic resonance of chemically exchanging systems, Nat. Commun. 10 (2019) 1–9.

[90] A.M. Coffey, et al., High-resolution low-field molecular magnetic resonance imaging of hyperpolarized liquids, Anal. Chem. 86 (2014) 9042–9049.

[91] S. Korchak, A.P. Jagtap, S. Glöggler, Signal-enhanced real-time magnetic resonance of enzymatic reactions at millitesla fields, Chem. Sci. 12 (2020) 314–319.

[92] K. Nagashima, Optimum pulse flip angles for multi-scan acquisition of hyperpolarized NMR and MRI, J. Magn. Reson. 190 (2008) 183–188.

[93] Y. Xing, G.D. Reed, J.M. Pauly, A.B. Kerr, P.E.Z. Larson, Optimal variable flip angle schemes for dynamic acquisition of exchanging hyperpolarized substrates, J. Magn. Reson. 234 (2013) 75–81.

[94] C.M. Walker, et al., Effects of excitation angle strategy on quantitative analysis of hyperpolarized pyruvate, Magn. Reson. Med. 81 (2019) 3754–3762.

[95] M. Mishkovsky, et al., Measuring glucose cerebral metabolism in the healthy mouse using hyperpolarized 13C magnetic resonance, Sci. Rep. 7 (2017) 4–11.

[96] Y.F. Yen, et al., Imaging considerations for in vivo 13C metabolic mapping using hyperpolarized 13C-pyruvate, Magn. Reson. Med. 62 (2009) 1–10.

[97] J.X. Wang, M.E. Merritt, A.D. Sherry, C.R. Malloy, Accelerated chemical shift imaging of hyperpolarized 13C metabolites, Magn. Reson. Med. 76 (2016) 1033–1038.

[98] C.H. Cunningham, et al., Design of flyback echo-planar readout gradients for magnetic resonance spectroscopic imaging, Magn. Reson. Med. 54 (2005) 1286–1289.

[99] D. Mayer, Y.S. Levin, R.E. Hurd, G.H. Glover, D.M. Spielman, Fast metabolic imaging of systems with sparse spectra: application for hyperpolarized 13C imaging, Magn. Reson. Med. 56 (2006) 932–937.

[100] C.H. Cunningham, W. Dominguez Viqueira, R.E. Hurd, A.P. Chen, Frequency correction method for improved spatial correlation of hyperpolarized 13C metabolites and anatomy, NMR Biomed. 27 (2014) 212–218.

[101] J.W. Gordon, et al., Translation of Carbon-13 EPI for hyperpolarized MR molecular imaging of prostate and brain cancer patients, Magn. Reson. Med. 81 (2019) 2702–2709.

[102] A.Z. Lau, et al., Rapid multislice imaging of hyperpolarized 13C pyruvate and bicarbonate in the heart, Magn. Reson. Med. 64 (2010) 1323–1331.

[103] G.D. Reed, et al., A method for simultaneous echo planar imaging of hyperpolarized 13C pyruvate and 13C lactate, J. Magn. Reson. 217 (2012) 41–47.

[104] J. Ma, E.P. Hackett, R.F. Schulte, J.M. Park, Simultaneous assessment of hyperpolarized [1-13C] lactate and [13C] bicarbonate with fly-back echo planar imaging, in: International Society of Magnetic Resonance in Medicine, 2019, p. 6733.

[105] S.B. Reeder, J.H. Brittain, T.M. Grist, Y.F. Yen, Least-squares chemical shift separation for 13C metabolic imaging, J. Magn. Reson. Imag. 26 (2007) 1145–1152.

[106] F. Wiesinger, et al., IDEAL spiral CSI for dynamic metabolic MR imaging of hyperpolarized [1-13C]pyruvate, Magn. Reson. Med. 68 (2012) 8–16.

[107] K. Deh, et al., Dynamic volumetric hyperpolarized 13C imaging with multi-echo EPI, Magn. Reson. Med. 85 (2021) 978–986.

[108] S. Hu, et al., Compressed sensing for resolution enhancement of hyperpolarized 13C flyback 3D-MRSI, J. Magn. Reson. 192 (2008) 258–264.

[109] S. Hu, et al., 3D compressed sensing for highly accelerated hyperpolarized 13C MRSI with in vivo applications to transgenic mouse models of cancer, Magn. Reson. Med. 63 (2010) 312–321.

[110] P.E.Z. Larson, et al., Fast dynamic 3D MR spectroscopic imaging with compressed sensing and multiband excitation pulses for hyperpolarized 13C studies, Magn. Reson. Med. 65 (2011) 610–619.

[111] A.Z. Lau, J.Y.C. Lau, A.P. Chen, C.H. Cunningham, Simultaneous multislice acquisition without trajectory modification for hyperpolarized 13C experiments, Magn. Reson. Med. 80 (2018) 1588–1594.

[112] A. Arunachalam, et al., Accelerated spectroscopic imaging of hyperpolarized C-13 pyruvate using SENSE parallel imaging, NMR Biomed. 22 (2009) 867–873.

[113] M.A. Ohliger, et al., Combined parallel and partial fourier MR reconstruction for accelerated 8-channel hyperpolarized carbon-13 in vivo magnetic resonance Spectroscopic imaging (MRSI), J. Magn. Reson. Imag. 38 (2013) 701–713.

[114] Y. Feng, et al., Development and testing of hyperpolarized 13C MR calibrationless parallel imaging, J. Magn. Reson. 262 (2016) 1–7.

[115] P.J. Shin, et al., Calibrationless parallel imaging reconstruction based on structured low-rank matrix completion, in: Magnetic Resonance in Medicine, vol 72, 2014, pp. 959–970.

[116] J.W. Gordon, et al., 3D hyperpolarized C-13 EPI with calibrationless parallel imaging, J. Magn. Reson. 289 (2018) 92–99.

[117] H. Lee, et al., High resolution hyperpolarized 13C MRSI using SPICE at 9.4T, Magn. Reson. Med. 80 (2018) 703–710.

[118] C.H. Cunningham, et al., Double spin-echo sequence for rapid spectroscopic imaging of hyperpolarized 13C, J. Magn. Reson. 187 (2007) 357–362.

[119] S. Josan, et al., Application of double spin echo spiral chemical shift imaging to rapid metabolic mapping of hyperpolarized [1 - 13C]-pyruvate, J. Magn. Reson. 209 (2011) 332–336.

[120] C. von Morze, et al., Frequency-specific SSFP for hyperpolarized 13C metabolic imaging at 14.1 T, Magn. Reson. Imaging 31 (2013) 163–170.

[121] Y.-F. Yen, et al., Non-CPMG echo-train sequence for T2 mapping and large SNR gain in hyperpolarized 13C imaging, in: Proceedings of ISMRM 20th Annual Meeting, 2012, p. 4295.

[122] H.Y. Chen, et al., Pulse sequence considerations for quantification of pyruvate-to-lactate conversion kPL in hyperpolarized 13C imaging, NMR Biomed. 32 (2019) 1–14.

[123] A. Eldirdiri, A. Clemmensen, S. Bowen, A. Kjær, J.H. Ardenkjær-Larsen, Simultaneous imaging of hyperpolarized [1,4-13C2]fumarate, [1-13C] pyruvate and 18F−FDG in a rat model of necrosis in a clinical PET/MR scanner, NMR Biomed. 30 (2017) 1−9.

[124] M. Singh, S. Josan, M. Zhu, A. Jhajharia, D. Mayer, Dynamic metabolic imaging of copolarized [2-13C]pyruvate and [1,4-13C2]fumarate using 3D-spiral CSI with alternate spectral band excitation, Magn. Reson. Med. 81 (2019) 2869−2877.

[125] J.A.M. Bastiaansen, et al., Probing cardiac metabolism by hyperpolarized 13C MR using an exclusively endogenous substrate mixture and photo-induced nonpersistent radicals, Magn. Reson. Med. 79 (2018) 2451−2459.

[126] D.M. Wilson, et al., Multi-compound polarization by DNP allows simultaneous assessment of multiple enzymatic activities in vivo, J. Magn. Reson. 205 (2010) 141−147.

[127] G. Sharma, et al., Co-polarized [1-13C]pyruvate and [1,3-13C2]acetoacetate provide a simultaneous view of cytosolic and mitochondrial redox in a single experiment, ACS Sens. 6 (2021) 3967−3977.

[128] J.M. Park, et al., Differentiation of pyruvate flux versus exchange in rat liver in vivo using a three-site exchange model, in: International Society of Magnetic Resonance in Medicine, 2013, p. 655.

[129] H. Qin, et al., Clinical translation of hyperpolarized 13C pyruvate and urea MRI for simultaneous metabolic and perfusion imaging, Magn. Reson. Med. 87 (2022) 138−149.

[130] D. Mayer, et al., Dynamic and high-resolution metabolic imaging of hyperpolarized [1- 13C]-pyruvate in the rat brain using a high-performance gradient insert, Magn. Reson. Med. 65 (2011) 1228−1233.

[131] W. Dominguez-Viqueira, et al., Intensity correction for multichannel hyperpolarized 13C imaging of the heart, Magn. Reson. Med. 75 (2016) 859−865.

[132] S.E. Day, et al., Detecting tumor response to treatment using hyperpolarized 13C magnetic resonance imaging and spectroscopy, Nat. Med. 13 (2007) 1382−1387.

[133] J.A. Bankson, et al., Kinetic modeling and constrained reconstruction of hyperpolarized [1-13C]-pyruvate offers improved metabolic imaging of tumors, Cancer Res. 75 (2015) 4708−4717.

[134] A. Comment, M.E. Merritt, Hyperpolarized magnetic resonance as a sensitive detector of metabolic function, Biochemistry 53 (2014) 7333−7357.

[135] J. Chen, E.P. Hackett, Z. Kovacs, C.R. Malloy, J.M. Park, Assessment of hepatic pyruvate carboxylase activity using hyperpolarized [1-13C]-l-lactate, Magn. Reson. Med. 85 (2021) 1175−1182.

[136] P.A. Gómez Damián, et al., Multisite kinetic modeling of 13 C metabolic MR using [1- 13 C]pyruvate, Radiol. Res. Pract. 2014 (2014) 1−10.

[137] M.A. Schroeder, et al., Extramitochondrial domain rich in carbonic anhydrase activity improves myocardial energetics, Proc. Natl. Acad. Sci. U. S. A. 110 (2013).

[138] M.A. Schroeder, et al., Hyperpolarized 13C magnetic resonance reveals early- and late-onset changes to in vivo pyruvate metabolism in the failing heart, Eur. J. Heart Fail. 15 (2013) 130−140.

[139] J.M. Park, et al., Hyperpolarized 13C NMR observation of lactate kinetics in skeletal muscle, J. Exp. Biol. 218 (2015) 3308−3318.

[140] J. Ma, et al., Dynamic 13C MR spectroscopy as an alternative to imaging for assessing cerebral metabolism using hyperpolarized pyruvate in humans, Magn. Reson. Med. 87 (2022) 1136−1149.

[141] D.K. Hill, et al., Model free approach to kinetic analysis of real-time hyperpolarized 13C magnetic resonance spectroscopy data, PLoS One 8 (2013) 1−9.

[142] C. Harrison, et al., Comparison of kinetic models for analysis of pyruvate-to-lactate exchange by hyperpolarized 13C NMR, NMR Biomed. 25 (2012) 1286−1294.

[143] C.J. Daniels, et al., A comparison of quantitative methods for clinical imaging with hyperpolarized 13C-pyruvate, NMR Biomed. 29 (2016) 387−399.

[144] D.M. Spielman, J.M. Park, Kinetic modeling of enzymatic reactions in analyzing hyperpolarized NMR data, Dyn. Hyperpolarized Nucl. Magn. Reson. (2021) 103−121, https://doi.org/10.1007/978-3-030-55043-1_5.

[145] M.L. Zierhut, et al., Kinetic modeling of hyperpolarized 13C1-pyruvate metabolism in normal rats and TRAMP mice, J. Magn. Reson. 202 (2010) 85−92.

[146] T. Xu, et al., Quantification of in vivo metabolic kinetics of hyperpolarized pyruvate in rat kidneys using dynamic 13C MRSI, NMR Biomed. 24 (2011) 997−1005.

[147] J. Ma, et al., Cardiac T2* measurement of hyperpolarized 13C metabolites using metabolite-selective multi-echo spiral imaging, Magn. Reson. Med. 86 (2021) 1494−1504.

[148] V. Eraković, G. Župan, J. Varljen, J. Laginja, A. Simonić, Altered activities of rat brain metabolic enzymes caused by pentylenetetrazol kindling and pentylenetetrazol - induced seizures, Epilepsy Res. 43 (2001) 165−173.

[149] J.M. Park, et al., Hyperpolarized 13C-lactate to 13C-bicarbonate ratio as a biomarker for monitoring the acute response of anti-vascular endothelial growth factor (anti-VEGF) treatment, NMR Biomed. 29 (2016) 650−659.

[150] I. Park, et al., Hyperpolarized 13C magnetic resonance metabolic imaging: application to brain tumors, Neuro Oncol. 12 (2010) 133−144.

[151] K. Golman, et al., Cardiac metabolism measured noninvasively by hyperpolarized 13C MRI, Magn. Reson. Med. 59 (2008) 1005−1013.

[152] P. Wespi, J. Steinhauser, G. Kwiatkowski, S. Kozerke, Overestimation of cardiac lactate production caused by liver metabolism of hyperpolarized [1-13C]pyruvate, Magn. Reson. Med. 80 (2018) 1882−1890.

[153] R.E. Hurd, et al., Exchange-linked dissolution agents in dissolution-DNP 13C metabolic imaging, Magn. Reson. Med. 70 (2013) 936−942.

[154] B.W.C. Kennedy, M.I. Kettunen, D.E. Hu, K.M. Brindle, Probing lactate dehydrogenase activity in tumors by measuring hydrogen/deuterium exchange in hyperpolarized L -[1- 13C,U- 2H]lactate, J. Am. Chem. Soc. 134 (2012) 4969−4977.

[155] T.H. Witney, M.I. Kettunen, K.M. Brindle, Kinetic modeling of hyperpolarized 13C label exchange between pyruvate and lactate in tumor cells, J. Biol. Chem. 286 (2011) 24572–24580.

[156] M.I. Kettunen, et al., Magnetization transfer Measurements of exchange between hyperpolarized [1-13C]pyruvate and [1-13C]lactate in a murine lymphoma, Magn. Reson. Med. 63 (2010) 872–880.

[157] J.M. Park, et al., Differentiation of flux and isotopic exchange using co-administration of hyperpolarized [2-13C]pyr and [1-13C]lac, in: Proceedings of the International Society of Magnetic Resonance in Medicine, 2012, p. 4298.

[158] S. Hu, et al., In vivo measurement of normal rat intracellular pyruvate and lactate levels after injection of hyperpolarized [1- 13C]alanine, Magn. Reson. Imaging 29 (2011) 1035–1040.

[159] J.M. Park, et al., Measuring mitochondrial metabolism in rat brain in vivo using MR spectroscopy of hyperpolarized [2-13C]pyruvate, NMR Biomed. 26 (2013) 1197–1203.

[160] C. Von Deuster, et al., Controlling through-slice chemical-shift artifacts for improved non-fat-suppressed musculoskeletal turbo-spin-echo magnetic resonance imaging at 7 T, Invest. Radiol. 56 (2021) 545–552.

[161] S. Yang, et al., Metabolite-selective hyperpolarized 13C imaging using extended chemical shift displacement at 9.4 T, Magn. Reson. Imaging 34 (2016) 535–540.

[162] G.J. Topping, et al., Hyperpolarized 13C spectroscopy with simple slice-and-frequency-selective excitation, Biomedicines 9 (2021) 1–13.

[163] S. Josan, et al., Dynamic metabolic imaging of hyperpolarized [2-13C]pyruvate using spiral chemical shift imaging with alternating spectral band excitation, Magn. Reson. Med. 71 (2014) 2051–2058.

[164] P.F. Ferreira, P.D. Gatehouse, R.H. Mohiaddin, D.N. Firmin, Cardiovascular magnetic resonance artefacts, J. Cardiovasc. Magn. Reson. 15 (2013) 1–39.

[165] M.K. Atalay, B.P. Poncelet, H.L. Kantor, T.J. Brady, R.M. Weisskoff, Cardiac susceptibility artifacts arising from the heart-lung interface, Magn. Reson. Med. 45 (2001) 341–345.

[166] J. Traechtler, V. Vishnevskiy, M. Fuetterer, S. Kozerke, Joint image and field map estimation for multi-echo hyperpolarized 13C metabolic imaging of the heart, Magn. Reson. Med. 86 (2021) 258–276.

[167] G.D. Reed, et al., Characterization and compensation of f0 inhomogeneity artifact in spiral hyperpolarized 13C imaging of the human heart, Magn. Reson. Med. 86 (2021) 157–166.

[168] M. Fala, et al., Comparison of 13C MRI of hyperpolarized [1-13C]pyruvate and lactate with the corresponding mass spectrometry images in a murine lymphoma model, Magn. Reson. Med. 85 (2021) 3027–3035.

[169] C.Y. Lee, et al., Correlation of hyperpolarized 13C-MRI data with tissue extract measurements, NMR Biomed. 33 (2020) 1–11.

[170] E.S. Jin, et al., Metabolism of hyperpolarized [1-13C]pyruvate through alternate pathways in rat liver, NMR Biomed. 29 (2016) 466–474.

[171] N. Dwork, et al., Di-chromatic interpolation of magnetic resonance metabolic images, Magn. Reson. Mater. Physics, Biol. Med. 34 (2021) 57–72.

[172] J. Kasten, A. Klauser, F. Lazeyras, D. Van De Ville, Magnetic resonance spectroscopic imaging at superresolution: overview and perspectives, J. Magn. Reson. 263 (2016) 193–208.

[173] G. Farkash, S. Markovic, M. Novakovic, L. Frydman, Enhanced hyperpolarized chemical shift imaging based on a priori segmented information, Magn. Reson. Med. 81 (2019) 3080–3093.

[174] M.J. Ehrhardt, F.A. Gallagher, M.A. McLean, C.B. Schönlieb, Enhancing the spatial resolution of hyperpolarized carbon-13 MRI of human brain metabolism using structure guidance, Magn. Reson. Med. 87 (2022) 1301–1312.

[175] J. Ma, J.M. Park, Super-resolution hyperpolarized 13C imaging of human brain using patch-based algorithm, Tomography 6 (2020) 343–355.

[176] J.R. Brender, et al., Dynamic imaging of glucose and lactate metabolism by 13 C-MRS without hyperpolarization, Sci. Rep. 9 (2019) 1–14.

[177] H.Y. Chen, et al., Tensor image enhancement and optimal multichannel receiver combination analyses for human hyperpolarized 13C MRSI, Magn. Reson. Med. 84 (2020) 3351–3365.

[178] Y. Kim, et al., Denoising of hyperpolarized 13C MR images of the human brain using patch-based higher-order singular value decomposition, Magn. Reson. Med. 86 (2021) 2497–2511.

Index

'*Note:* Page numbers followed by "f" indicate figures and "t" indicate tables.'

A

B

C

D

E

N

O

P

Q

R

S

CPI Antony Rowe
Eastbourne, UK
June 05, 2024